MAMMOGRAPHY AND BREAST IMAGING
PREP

SECOND EDITION

Olive Peart, MS, RT(R)(M)
Program Chair
Radiologic Technology Program
Fortis College
Landover, Maryland

McGraw Hill Education

New York   Chicago   San Francisco   Athens   London   Madrid
Mexico City   Milan   New Delhi   Singapore   Sydney   Toronto

Mammography and Breast Imaging: Program Review and Exam Preparation, Second Edition

4 5 6 7 8 9 LWI 22 21

ISBN: 978-1-25-985945-8
MHID: 1-25-985945-2

Notice

Medicine is an ever-changing science. As new research and clinical experience broaden our knowledge, changes in treatment and drug therapy are required. The author and the publisher of this work have checked with sources believed to be reliable in their efforts to provide information that is complete and generally in accord with the standards accepted at the time of publication. However, in view of the possibility of human error or changes in medical sciences, neither the author nor the publisher nor any other party who has been involved in the preparation or publication of this work warrants that the information contained herein is in every respect accurate or complete, and they disclaim all responsibility for any errors or omissions or for the results obtained from use of the information contained in this work. Readers are encouraged to confirm the information contained herein with other sources. For example and in particular, readers are advised to check the product information sheet included in the package of each drug they plan to administer to be certain that the information contained in this work is accurate and that changes have not been made in the recommended dose or in the contraindications for administration. This recommendation is of particular importance in connection with new or infrequently used drugs.

This book was set in Minion Pro by Graphic World, Inc.
The editors were Susan Barnes and Christina M. Thomas.
The production supervisor was Jeffrey Herzich.
Project management was provided by Graphic World, Inc.

Library of Congress Cataloging-in-Publication Data

Names: Peart, O. J. (Olive J.), author.
Title: Mammography and breast imaging PREP : program review and exam prep /
   Olive Peart.
Other titles: Mammography and breast imaging program review and examination
   preparation
Description: Second edition. | New York : McGraw Hill Education, [2018] |
   Includes bibliographical references and index.
Identifiers: LCCN 2017016147| ISBN 9781259859458 (pbk.) | ISBN 1259859452
   (pbk.)
Subjects: | MESH: Mammography—methods | Ultrasonography, Mammary—methods |
   Breast Neoplasms—diagnosis | Breast—diagnostic imaging | Radionuclide
   Imaging
Classification: LCC RG493.5.R33 | NLM WP 815 | DDC 618.1/907572—dc23 LC
record available at https://lccn.loc.gov/2017016147

McGraw-Hill Education books are available at special quantity discounts to use as premiums and sales promotions, or for use in corporate training programs. To contact a representative please visit the Contact Us pages at www.mhprofessional.com

Reviewers

Robin Smith, MS, RT(R)(CT)(M), CPN
Assistant Professor
Clinical Coordinator
Radiology Program
St. Vincent's College
Bridgeport, Connecticut

Kalam Chowdhury, MD RDMS (AB), RDMS (OB-GYN) RDMS (BR), RDCS (AE), RVT (VT)
Program Director
Diagnostic Medical Sonography (DMS)
Medical Career Institute
County College of Morris
Randolph, New Jersey
Adjunct Faculty
Harlem Hospital Center School of Radiologic Technology
New York, New York

Yanely Peguero, RT(R)(M)(CT)(MR)
Mammography Technologist
Valley Hospital Breast Center
MRI/CT Technologist, Affinity Radiology
Ridgewood, New Jersey

Caren Greenstein, MD
Director of Breast Imaging Services
Rye Brook, New Rochelle, and Armonk, New York
White Plains Hospital
White Plains, New York

Alicia Giaimo, MHS, RT(R)(M), ROT
Clinical Assistant Professor
Quinnipiac University
Hamden, Connecticut

Kei Doi, MD, PhD
Vice President
Northeast Radiology
Mount Kisco, New York

Cynthia Silkowski, MAEd, RDMS, RVT
Chairperson
Associate Professor
Department of Medical Imaging Sciences
Director
Diagnostic Medical Sonography
Rutgers, The State University of New Jersey
School of Health Professions
Scotch Plains, New Jersey

Dunstan Abraham, MPH, RDMS, RPA-C
Physician Assistant/Sonographer
Lincoln Hospital
Bronx, New York

Contents

Color Insert can be found after page 398

Reviewers..*iii*
Preface ....*xv*
Acknowledgments...*xvii*

1. The History of Breast Cancer and Breast Imaging...........................1
  Keywords and Phrases 2
  Cancer 3
  Early Incidence of Breast Cancer 4
  Breast Cancer Treatment Throughout the Ages.............................4
  The Discovery of X-Rays 7
  Mammography in the United States—Quality Versus Quantity.................11
  American College of Radiography Accreditation—
      Standardized Mammography Equipment and Processing 12
  The Mammography Quality Standards Act 13
  Breast Cancer Screening Update...13
  Summary...*18*
  Review Questions..*19*
  Chapter Quiz..*20*

2. Patient Education and Assessment.....................................23
  Keywords and Phrases 24
  Patient Care and Communication 27
  The Environment..29
  Medical History and Documentation 31
      Subjective Versus Objective Data....................................32
      Examples of Nonverbal Communication.................................32
      History-Taking Techniques...33
      Documentation...33
  Breast Examinations..34
      Breast Self-Examination...36
      Clinical Breast Examination 39
  Breast Examination Guidelines 39
  Benefits Versus Risks of Breast Imaging 41
      Mortality Reduction...41
      Risk from Radiation Exposure in Mammography Screening...............41
  Breast Cancer Screening 44
  Advantages and Disadvantages of Mammography Screening...................44
  Risk Factors for Breast Cancer 45
      Types of Risk Assessments 45
      Major Risk Factors..47
      Minor Risk Factors..51
  Facts Versus Myths and Misconceptions...................................52
  Summary...*59*
  Review Questions..*60*
  Chapter Quiz..*61*

3. Anatomy, Physiology, and Pathology of the Breast 63

    Keywords and Phrases 64

    The Breast 66

    Location of the Breast 66

    Surface Anatomy of the Breast 66

    Localization Terminology 67

    Deep Anatomy of the Breast 68

    Vascular Circulation 70

        Arterial Circulation 70

        Venous Drainage 70

    The Lymphatic Drainage of the Breast 71

    Nervous Supply 73

    Breast Tissue Composition 73

    Factors Affecting Tissue Composition 74

        Hormone Fluctuation 75

        Menopause 75

        Hormone Therapy 77

        Lactation 78

        Parity 78

    The Male Breast 78

        General Description 79

        Male Breast Cancer—Signs and Symptoms ...79

        Risk Factors for Breast Cancer in Males 79

        Diagnosis, Treatment, and Prognosis 80

    Summary 80

    Review Questions 82

    Chapter Quiz 83

4. Benign and Malignant Diseases of the Breast 85

    Keywords and Phrases 86

    Changes in the Breast 91

        Breast Augmentation 93

        Reduction Mammoplasty 94

    Characteristics of Benign and Malignant Breast Lesions 94

        Circular or Oval Lesion 95

        Spiculated or Stellate Lesion 97

        Asymmetric Breast Tissue 100

        Calcifications 101

        Thickened Skin Syndrome 108

    Types of Breast Cancer 108

        Ductal Carcinoma 109

        Lobular Carcinoma 109

        Other Breast Cancers 110

    Summary 112

    Review Questions 113

    Chapter Quiz 114

5. Mammography Equipment 117

    Keywords and Phrases 118

    X-Ray Imaging System 120

        The Cathode 121

        The Anode 122

        Heat Dissipation 122

        X-Ray Tube 123

    Design Characteristics of Mammography Units 123

Electron Interaction With Matter—X-Ray Production 124
    Characteristic or Photoelectric Radiation 124
    Bremsstrahlung Radiation 125
X-Ray Interaction With Matter 125
Mammography Imaging 128
Mammography Tube 129
    General Radiography Versus Soft Tissue Imaging 129
    Filtration 129
    The Anode and Target 132
    Line-Focus Principle 134
    Focal Spot Size 136
    Anode Heel Effect 136
    Kilovoltage Range 137
    Exposure Time, Milliamperes, and Milliampere-Seconds 137
    Compression Devices 138
    Compression Plate 138
    Automatic Exposure Control 140
    Density Compensation and Backup Timer 140
    Grids 141
    Beam-Restricting Devices 142
    System Geometry (Source-to-Image Receptor
      Distance, Object-to-Image Receptor Distance) 143
Magnification 143
Monitors 144
    Acquisition Workstation 146
    Review Workstation 147
Print and Storage Options 148
Picture-Archiving and Communication System 149
    Advantages and Disadvantages 149
DICOM, HIS, RIS, EMR, and HL7 150
Teleradiography 151
Lossy and Lossless Compression 152
Computer-Aided Detection 152
Image Quality: Digital and Computed Mammography 154
Summary *155*
Review Questions *158*
Chapter Quiz *159*

6. Breast Imaging Mammography 161
Keywords and Phrases 162
Breast Imaging Mammography 163
    Breast Compression 164
    Applying Compression 165
Standard Projections 166
    Positioning Guide 168
    MLO—The Mediolateral Oblique Projection 169
    CC—The Craniocaudal Projection 172
Supplementary Projections 173
    ML—Mediolateral or 90 Degrees Lateral 175
    LM—Lateromedial 175
    LMO—Lateromedial Oblique 177
    TAN—Tangential Projection 178
    XCCL or XCCM—Exaggerated Lateral or Medial
      Craniocaudal Projection 178
    FB—From Below 179

CV—Cleavage or Valley View 181

AT—Axillary Tail 182

SIO—Superior–Inferior Oblique 183

Rolled Positions 184

Magnification 185

Spot Compression 187

Imaging the Nonconforming Patient...188

Imaging Small Breasts 188

Male Breast Imaging..189

Sectional Imaging of the Extremely Large Breast or
   Mosaic Breast Imaging 189

Imaging Patients With Obese Upper Arms 190

Patient With an Inframammary Crease That Is Not Horizontal.......190

Patient With Protruding Abdomen 190

Patient With Thick Axilla 190

The Kyphotic Patient..191

Patient With Adhesive Capsulitis (Frozen Shoulder) 191

Elderly or Unstable Patients..191

Mentally Challenged Patient 192

Patients With Pectus Excavatum (Sternum Depressed) or
   Pectus Carinatum (Pigeon Chest)...192

Barrel Chest 192

Delicate Inframammary Fold..193

Imaging Implants 193

Imaging the Postsurgical Site—Postlumpectomy Imaging 196

Postmastectomy Imaging 196

Mammograms of the Irradiated Breast...196

Imaging the Stretcher Patient 197

Imaging the Wheelchair Patient 198

Specific Imaging Problems...199

Nipple Not in Profile 199

Skin Folds or Wrinkling of the Breast...199

Patients With Uneven Breast Thickness...201

Moles and Scars...201

Pacemakers and Implantable Venous Access Systems 201

Triangulation Techniques 202

Before Signing Off 204

Summary...205

Review Questions..209

Chapter Quiz...210

7. Breast Imaging—Digital Mammography, Ultrasound,
   and Magnetic Resonance Imaging 213

Keywords and Phrases 214

Ultrasound Terminology...217

MRI Terminology..220

Analog Mammography 222

Digital Mammography 222

Direct Flat-Panel Detector Technology...231

Indirect Flat-Panel Detector Technology...232

Photostimulable Storage Phosphor Technology 233

Photon-Counting Image Capture Technology 235

Advantages of Digital Mammography...235

Disadvantages of Digital Mammography...236

Ultrasound Imaging of the Breast 237

Brief History of Ultrasound...237

The Principle of Ultrasound 238
The Transducer (Probe) 239
Common Ultrasound Terminologies....................................241
Uses of Ultrasound 246
Management of Nonpalpable Masses 246
Ultrasound Evaluation of Palpable Masses.........................247
Doppler Effect 247
Color Doppler 248
Ultrasound Characteristics of Benign Lesions 248
Ultrasound Characteristics of Malignant Masses................248
Ultrasound Characteristics of Intermediate Lesions 249
Typical Appearance of Lesions on Ultrasound 249
Imaging Implants...252
Ultrasound Artifacts 252
Limitations of Breast Ultrasound Imaging 253
Imaging Problems and Solutions..254
Risks of Ultrasound...254
Advantages of Breast Ultrasound 254
Magnetic Resonance Imaging of the Breast 255
Brief History of MRI...255
Electromagnetic Spectrum 255
Magnetism...256
The Principle of MRI 257
Equipment 259
Operating Console 261
Computer Parameters..261
MRI Parameters 262
Use of Contrast Agents 263
Pulse Sequences and Techniques.....................................264
Postprocessing Technique...265
Breast Magnetic Resonance Imaging 266
MRI Artifacts 267
Advantages and Uses of Breast MRI 272
MRI of the Augmented Breast...273
Limits of Breast MRI 273
Risks and Potential Complications 276
MRI Quality Assurance 276
Summary..277
Review Questions...280
Chapter Quiz..281

8. Breast Imaging—Adjunctive Modalities 283
Keywords and Phrases 284
Diagnostic Options...286
Digital Mammography Technologies 286
Digital Breast Tomosynthesis or Three-Dimensional Mammography 286
Digital Subtraction Mammogram 295
Dual-Energy Mammography..295
Ultrasound Technologies...295
Automated Ultrasound and Three-Dimensional
    Breast Ultrasound..296
Full-Field Breast Ultrasound 297
Elastography 298
Ultrasound Imaging Using Capacitive Microfabricated
    Ultrasonic Transducers...298
Contrast-Enhanced Ultrasound 299

Magnetic Resonance Imaging Technologies 299
  Magnetic Resonance Spectroscopy 299
Molecular Imaging Technologies...299
  Nuclear Medicine 300
  Fluorodeoxyglucose–Positron Emission
    Tomography 301
  Positron Emission Mammography...303
  Radionuclide Imaging With Single-Photon Emission
    Computed Tomography..305
  Breast Scintigraphy, Scintimammography, or Breast
    Molecular Imaging 305
  Breast-Specific Gamma Imaging 306
  Lymphoscintigraphy—Sentinel Node Mapping 308
Radiation Dose Related to Nuclear Imaging......................................310
Other Imaging Modalities..311
  Optical Imaging 311
  Cone-Beam Computed Tomography...312
  Computed Tomographic Laser Mammography...........................312
  Transscan or T Scan 313
  Microwave Imaging Spectroscopy...313
  Neutron-Simulated Emission Computed
    Tomography 314
  Summary...*314*
  Review Questions..*317*
  Chapter Quiz..*318*

9. Interventional Procedures 321
Keywords and Phrases..322
The Need for Interventional Techniques 323
Cytological Analysis...324
Histological Analysis..324
General Patient Preparation 324
Cyst Aspiration...325
Stereotactic Localization and Biopsy..326
  Add-On Units 327
  Dedicated Prone Units 328
  Blended Technology..328
  Stereotactic Procedure..328
  Limitations 329
Preoperative Needle Localization...329
  Localization Procedure Using Mammography 330
Breast Biopsy...331
  Fine Needle Aspiration Biopsy..332
  Core Biopsy Methods 334
Other Biopsy Methods 335
  Radiofrequency Biopsy 335
  The Mammotome Biopsy System..336
  The Automatic Tissue Excision and Collection 337
  Large Core Breast Biopsy..337
  Open Surgical Biopsy..338
  Modality-Based Core Biopsy..338
  Comparison of the Various Biopsy Methods 341
Specimen Radiography...343
  Specimen Radiography Units...343
Ductography or Galactography 344
Nipple Aspiration..344

Ductal Lavage...344
Pathology Review..345
Mammoplasty...346
    Cosmetic Intervention..346
    Reduction Mammoplasty...346
    Mastopexy..347
    Augmentation Mammoplasty.......................................347
    Summary..*349*
    Review Questions..*353*
    Chapter Quiz...*354*

10. Treatment Options 357
Keywords and Phrases...358
Breast Cancer Treatment..361
    Staging Breast Cancer..363
    Mastectomy as an Option..364
    Preventative Surgical Options.....................................365
    Lumpectomy as an Option...366
    Lymph Node Biopsy...367
    Comparisons of Surgical Treatment Options...................368
    Other Options to Consider...368
Radiation Therapy...368
    Indications for Radiation Therapy................................370
    Classifications of Radiation Therapy............................370
    Side Effects of Radiation Therapy...............................371
    Intensity-Modulated Radiation Therapy.........................371
    Image-Guided Radiation Therapy................................372
    Internal Radiation or Partial Radiation..........................373
    Intraoperative Electron Radiation Therapy.....................375
Chemotherapy..375
    The Cell Cycle...376
    Uses of Chemotherapy...376
    Drugs Used in Chemotherapy.....................................376
    Chemotherapy Regimens..377
    Method of Giving Chemotherapy.................................377
    Side Effects of Chemotherapy....................................377
Targeted Treatment...379
    Molecular Treatment..379
    Hormone Therapy...381
Other Treatment Options...383
    Ablative Hormone Therapy..383
    Antiangiogenesis..383
    Circulating Breast Cancer Cells..................................384
Gene Therapy..384
    Methods of Gene Therapy...385
Hematopoietic Stem Cell Transplantation...............................385
Immunotherapy or Biotherapy..386
Photodynamic Therapy...386
Thermal Ablation..387
Cryotherapy..388
Prophylactic Surgery..388
Surgical Reconstruction Using Implants.................................389
    Nipple and Areola Reconstruction...............................389
    Implant Procedures...389
    Preparation for Surgery..389
    Immediate Reconstruction...390

Delayed Reconstruction 390
Other Reconstruction Options...390
Autologous Tissue Reconstruction...392
Pain Medication and Pain Management 393
Conclusion..394
Summary..*394*
Review Questions..*397*
Chapter Quiz..*398*

11. Quality Assurance and Quality Control 401
Keywords and Phrases 402
Quality Assurance 403
Benefits of Quality Assurance 404
Agencies Responsible for Safety and Quality of Imaging
    Practices Within the United States 404
Responsibilities of the Various Quality Control Personnel 405
The Radiologist 405
Lead Interpreting Physician (Radiologist) 406
Interpreting Physician (Radiologist) 406
The Technologist 406
The Medical Physicist 406
Quality Control Testing..407
General QC Tests 408
Physicist Tests 422
Summary..*434*
Review Questions..*438*
Chapter Quiz..*439*

12. The US Mammography Quality Standards Act (MQSA) 441
Keywords and Phrases 442
Purpose of the Mammography Quality Standards 444
Interim Regulation 444
Final Regulations..445
Accreditation and Certification 445
Lawful Mammography Services in the
    United States 446
Quality Assurance Regulations of MQSA 446
Clinical Images 446
Levels of Findings or Noncompliance 449
Withdrawal of Accreditation..449
Suspension or Revocation 451
Adverse Reporting 452
Digital Mammography...452
Quality Control..455
Printers and Monitors 455
Monitors 455
Mobile Units...456
Infection Control 457
Xeroradiography 457
Certificate Placement..458
Consumer Complaints Mechanism..458
Self-Referrals 459
Record Keeping..459
Content of Records and Reports 459
Assessment Categories..460

MQSA Assessment Categories 460
Communication of Results to the Patient 461
Medical Audit 462
MQSA and Health Insurance Portability and Accountability Act 462
Final Regulation Documentation 463
Requirements for the Radiologic Technologists 464
Requirements for the Interpreting Physician 466
Requirements for the Medical Physicist 467
Breast Cancer Screening Today 474
Enhancing Quality Using the Inspection Program (EQUIPT Initiative) 475
Details on EQUIP Questions 476
Summary 477
Review Questions 480
Chapter Quiz 481

Appendix A: Quality Control Tests—Digital Breast Tomosynthesis 483
Digital Breast Tomosynthesis 483
Accreditation and Certification Options for Facilities Using a 3D System
    With Either 2D FFDM Images or 2D Images Generated from the
    3D Image Set (i.e., 2D Synthesized Images) 484
Reimbursement for DBT Imaging 484
Mammography Quality Standards Act Inspection Questions Under
    the Final Regulations 485

Appendix B: US MQSA Facility Certification Extension
Requirements for Digital Breast Tomosynthesis System 493
Qualified Personnel 495
Interpreting Physicians 495
Personnel Qualifications: Interpreting Physicians Who are Qualified
    to Interpret DBT Mammograms 495
Radiology Technologists 495
Medical Physicists 496
Lead Interpreting Physician Attestation to Staff Personnel Qualifications 496

Appendix C: Monitor and Laser Printer Test Procedures 497
Printer and Monitor Quality Control—if No Manual is Provided 497

Appendix D: Quality Control Forms 499
Contents 499

Appendix E: Analog Imaging, Quality Assurance and Quality Control 515
Keywords and Phrases 516
Analog Imaging 517
Double Versus Single Emulsion 517
Mammography Films 518
Mammography Screens 519
The Characteristic Curve 519
The Cassette 520
Image Quality: The Screen/Film System 521
Kilovoltage Range 524
Automatic Exposure Control 524
Quality Assurance 525
Benefits of Quality Assurance 526
Quality Control Testing 526
Technologists 526
General Technologist Quality Control Tests 527

Testing Details—General 542
    Phantom Images—Analog Imaging 542
    Visual Checklist—Analog 544
    Repeat/Reject Analysis—Analog 544
    Viewbox and Viewing Conditions—Analog....................................545
    Compression—Analog 547
    Light Field and X-Ray Field Congruence—Analog 548
    Darkroom Cleanliness...548
    Processor Quality Control 549
    Screen Cleaning 554
    Analysis of Fixer Retention in Film..554
    Darkroom Fog and Safelight Testing...555
    Screen/Film Contact and Identification 556
Summary..558
Review Questions...560
Chapter Quiz...561

Answers to Review Questions 565
Chapter 1: The History of Breast Cancer and Breast Imaging............565
    Chapter 1 Quiz 566
Chapter 2: Patient Education and Assessment 566
    Chapter 2 Quiz 568
Chapter 3: Anatomy, Physiology, and Pathology of the Breast 569
    Chapter 3 Quiz 570
Chapter 4: Benign and Malignant Diseases of the Breast.................570
    Chapter 4 Quiz 571
Chapter 5: Mammography Equipment 572
    Chapter 5 Quiz 574
Chapter 6: Breast Imaging Mammography...................................575
    Chapter 6 Quiz 576
Chapter 7: Breast Imaging—Digital Mammography, Ultrasound, and
    Magnetic Resonance Imaging 576
    Chapter 7 Quiz 577
Chapter 8: Breast Imaging—Adjunctive modalities........................578
    Chapter 8 Quiz 579
Chapter 9: Interventional Procedures...580
    Chapter 9 Quiz 580
Chapter 10: Treatment Options 581
    Chapter 10 Quiz 581
Chapter 11: Quality Assurance and Quality Control.......................582
    Chapter 11 Quiz 583
Chapter 12: The US Mammography Quality StandardS Act (MQSA).....584
    Chapter 12 Quiz 585
Index 587

Preface

Welcome to the second edition of *Mammography and Breast Imaging Prep*. This edition has undergone a thorough update and revision. The text is based on the latest American Registry of Radiologic Technologists (ARRT) content specifications and can serve as a valuable reference for students, practicing technologists, educators, and managers.

Beginning in 2017, claims submitted for x-rays performed on analog equipment in the United States will receive a 20% reduction in payment. Beginning in 2018, claims for x-rays performed using computer radiography (CR) technology will be reduced by 7%. The cuts will increase to 10% beginning in 2023. As a result of these new regulations, these old technologies are slowly being phased out in x-ray departments, with a corresponding removal of most analog information from the ARRT examination. Although breast imaging was the last modality to go digital, analog breast imaging is definitely a modality of the past. In keeping with this trend, this text focuses on digital imaging and the new technologies. Some sections from the previous edition have been moved or rearranged to provide a smoother transition, whereas other sections, including all analog content, have been deleted from the chapters and moved to the appendix. Overall, information on digital imaging has been updated and revised to match the curriculum guide published by the ARRT.

Despite the remarkable benefits of digital imaging systems, the need to protect our patients and the ability to acquire the theoretical knowledge necessary to provide good patient care remain vital. The standard projections or positions needed to image the breast have not changed significantly, which means that the basic element of breast imaging remains constant. Therefore, even with the tremendous advantages of digital imaging, the positioning skill of the technologist, including the application of compression, remains critical. This text serves to integrate these elements with radiation protection, theoretical skills, and the practical application of these skills.

The text is also a study and reference resource, providing comprehensive information on breast imaging, including breast anatomy, physiology, and pathology; methods of imaging; interventional techniques; and treatment options. It has been critically reviewed by technologists and radiologists and will be of value to those interested in up-to-date information on the Mammography Quality Standards Act (MQSA) and the new Enhancing Quality Using the Inspection Program (EQUIP) initiative. EQUIP is designed to identify image quality concerns early so that they can be quickly corrected, which is beneficial not only for patients but also for facilities.

In methods of imaging, the text covers the adjunctive breast-imaging modalities such as ultrasound, magnetic resonance imaging (MRI), digital breast tomosynthesis (DBT), cone-beam breast computed tomography, and molecular imaging modalities such as positron emission mammography (PEM) and breast-specific gamma imaging (BSGI).

Advanced treatment options provided by internal-beam radiation, intensity-modulated radiation therapy (IMRT), and image-guided radiation therapy (IGRT) are presented using a concise and practical approach.

Each chapter begins with an outline followed by the objectives and key words. The chapter content ends with a summary and review questions. The review questions include both long-answer questions and multiple-choice questions, giving students the opportunity to test their knowledge of the core concepts presented in the chapter.

Acknowledgments

I especially thank my husband, George Peart, for his steady support and his willingness to accommodate my writing hours. Thanks also for the encouragement given by my children Dione, Nadine, and George, who even served as models when I needed them.

Thanks to past and present students and instructors who have used this text and have sent comments or questions. I truly appreciate the candid reviews and constructive criticism from my reviewers and the comments and suggestions from my readers. This feedback is critical. A simple question can spur the revamping of an entire chapter. As one reader said, because the word "density" can have a totally different meaning in the digital world, interpretation of concepts in which this word is used can become challenging.

I also extend appreciation and thanks to those who gave permission to reproduce photos, images, and illustrations.

And last but not least: no text is possible without the assistance of a variety of individuals working behind the scenes. I would like to express genuine appreciation and gratitude to the editors and project managers associated with McGraw-Hill. These include Jennifer Pollock, Christina Thomas, Cassie Carey, and Sunil Kumar, to mention just a few.

I always love to hear from readers and can be contacted through the editors, from my website at http://www.opeart.com, by email at olive@opeart.com, or by writing to P.O. Box 13, Shrub Oak, NY, 10588.

The History of Breast Cancer and Breast Imaging | 1

Keywords and Phrases
Cancer
Early Incidence of Breast Cancer
Breast Cancer Treatment Throughout the Ages
The Discovery of X-Rays
Mammography in the United States—Quality Versus Quantity
American College of Radiography Accreditation—Standardized
    Mammography Equipment and Processing
The Mammography Quality Standards Act
Breast Cancer Screening Update
Summary
Review Questions
Chapter Quiz

Objectives

On completing this chapter, the reader will be able to:

1. Describe the oldest known recorded incidence of breast cancer
2. Discuss how breast cancer treatment has changed since the Middle Ages
3. Discuss why the early physicians developed erroneous theories on the human anatomy
4. Identify the physician who corrected the errors in human anatomy
5. Identify the surgeon who recognized that cancer could spread to the regional axillary nodes
6. Identify the pathologist who named many of our common tumors
7. Identify the scientist credited with the discovery of x-rays
8. Discuss the changes that have occurred in breast cancer treatment since the discovery of x-rays
9. Explain why screening mammography was recommended for asymptomatic women
10. Discuss the need for standardized mammography screening
11. Explain the drawbacks of the American College of Radiography (ACR) program

12. Explain the benefits of the Mammography Quality Standards Act (MQSA)
13. Explain the controversy surrounding the new mammography screening recommendations in the United States from the US Preventive Services Task Force (USPSTF)

KEYWORDS AND PHRASES

- **Anesthetic** is a medication that produces anesthesia, the partial or complete loss of sensation with or without loss of consciousness.
- **Baseline mammogram** refers to a woman's very first mammogram. Future mammograms can be compared against this baseline.
- **Cautery** is a device or agent used to coagulate blood or tissue by heat, electricity, freezing, or other caustic substances.
- **Chemotherapy** is the use of drugs to treat cancer that has spread through the body.
- **Chromosomes** are normally found in pairs, and every human typically has 23 pairs of chromosomes in every cell. Within the chromosomes are genes that provide genetic information of the individual.
- **DNA** (deoxyribonucleic acid) is a double helix, or two strands, of genetic material spiraled around each other to form chromosomes.
- **Glandular dose** in mammography refers to the mean or average dose to the glandular tissue of the breast.
- **HER2 protein** is the result of certain patterns of DNA. A certain segment of DNA called an oncogene codes for HER2/neu. The HER2/neu oncogene then produces the HER2/neu protein. If this segment of DNA becomes damaged as the cells reproduce, it can cause cancer cells that overexpress HER2/neu, resulting in the overproduction of HER2/protein. Tumors are tested to determine their HER2/neu status (the oncogene, the protein, or both) because HER2 indicates a more aggressive cancer. However, tumors that are HER2/neu can respond to Trastuzumab (Herceptin). Herceptin is a monoclonal antibody that targets the HER2/neu protein while leaving normal cells alone. Other names for this gene are erb-b2 receptor tyrosine kinase, CD340, proto-oncogene NEU, and ERBB2.
- **Humor** refers to any body fluid or semifluid substance in the body, such as blood or lymph.
- **Hyperplasia** is excess proliferation or increase in the amount of normal cells in normal tissues or organs.
- *In situ* describes a cancer that has not metastasized or invaded surrounding tissues.
- **Ischemia** is the temporary deficiency of blood flow to an organ or tissue.
- **Lymph nodes** are kidney-shaped organs that are part of the lymphatic system. They lie at intervals along the lymphatic vessels.
- **Malignant** is the descriptive term for a cancerous tumor.
- **Mastectomy** is the surgical removal of the breast. There are two types: modified radical mastectomy, which is the removal of the entire breast and some of the underarm lymph nodes; and radical mastectomy, which is the removal of the entire breast, lymph nodes, and

chest wall muscles under the breast. The radical mastectomy is not generally practiced today because the modified radical mastectomy is less debilitating and deforming, yet just as effective.

- **Metaplasia** is a change in the makeup of tissue from normal to abnormal.
- **MUC1** is a cell-associated mucin glycoprotein that is found in more than 90% of mammary gland tumors. The overexpression of MUC1 proteins has been associated with increased invasive behavior of cancer cells.
- **Nitrous oxide (N_2O)** is also called "laughing gas." It is a colorless, almost odorless gas, which was first discovered in 1793 by the English scientist and clergyman Joseph Priestley. Initially the gas was used for recreational enjoyment and public shows, but in the 1840s the anesthetic and analgesic properties of the gas were discovered. Since then, nitrous oxide is often used as a dental anesthesia.
- **Optical density** describes the overall blackness of the radiograph image.
- **Paget's disease of the nipple** is a rare malignant cancer of the nipple and areola. It is not associated with Paget's disease of the bone.
- **Quality assurance** describes the total overall management of actions taken to consistently provide high image quality in the radiology department, the primary objective being to enhance patient care. In general, quality assurance must be a planned and systematic series of actions applied to monitor key components of operation for consistent performance.
- **Quality control** deals specifically with the management and maintenance of any technical components of quality assurance, such as equipment or personnel.
- **Radiation therapy** is the use of radium or its radioactive emissions to treat diseases.
- **Radium (Ra)** is a radioactive metallic element with an atomic number of 88. There are four radium isotopes that occur naturally. One is formed by the disintegration of uranium 238. It has a half-life of 1620 years, and one of its first decay products is radon. The final stable decay product is lead.
- **Sensitivity** measures the ability to respond to or register small changes or difference.
- **Specificity** is the quality of being precise rather than general.
- **Staging** refers to the extent of the cancer spread based on factors such as location and size of tumors as well as lymph node involvement.
- **Tumors/lesions** refer to any abnormal swelling, enlargement, or growth. They can be benign or malignant.
- **Uranium (U)** is a radioactive metallic element with an atomic number of 92. Uranium is the heaviest naturally occurring element.

CANCER

Cancer is a disease caused by an uncontrolled growth of a single cell. In normal growth the division of one cell into two is carefully regulated and arrested by specific signals. In cancer, the cells have a potential for limitless cell divisions and survival. Cancer cells also have the capacity

> Edwin Smith Papyrus, dated 1600 BC, documents 48 surgical cases, describing detailed examinations, diagnoses, treatments, and prognoses.
>
> A breast tumor is listed as case 45, indicating that the ancient Egyptians were fully aware of this devastating disease.

to invade, metastasize, and mutate. Unfortunately, because malignant growth and normal growth are so intertwined, separating and differentiating between the two may be the only way to "cure" cancer.

EARLY INCIDENCE OF BREAST CANCER

Breast cancer is not a new disease. In 1862, the American Egyptologist Edwin Smith bought an ancient manuscript that later proved to contain the first recordings of a breast tumor. The document, called the "Edwin Smith Papyrus," dates back to the 17th century BC (1600 BC). It is thought to contain the collected teachings of Imhotep, a great Egyptian physician who lived around 2625 BC. It is one of the oldest known medical papyri and documents 48 surgical cases, describing in detail the examination, diagnosis, treatment, and prognosis in each case. A breast tumor is listed as case 45, described as a "bulging mass on the breast, large hard and cool to the touch," indicating that the ancient Egyptians were fully aware of this devastating disease. Unfortunately, in earlier times there was no documented cure for breast cancer, and the treatment then involved cauterization of the diseased tissues.

"Histories," written by the Greek historian Herodotus, describes an illness that struck Atossa, the queen of Persia. Atossa was also the daughter of Cyrus and wife of Darius. She noticed a bleeding lump on her breast. The tumor was excised by a Greek slave named Democedes. The narrative indicates that she survived the surgery, but we do not know if the tumor recurred or how or when she died.

A documented early death from breast cancer was Theodora, a prostitute known throughout Byzantium, the ancient Greek city that was later renamed Constantinople. She later became the wife of the Justinian 1, emperor of Byzantium. Theodora discovered a lump in her breast at about age 48 and died shortly after in June 548 BC.

Other evidence of early breast cancer and the devastating nature of the disease is the painting *Bathsheba at Her Bath* (1654) (Fig. 1–1) by Rembrandt. Rembrandt Harmenszoon van Rijn (1606–1669) was a Dutch painter. In 1967, an Italian surgeon, T. C. Greco, was touring the Rijksmuseum while vacationing in Amsterdam. He noticed that the woman in the painting had an asymmetric left breast with swelling, discoloration, pitted skin, and fullness near her armpit. Greco felt that all were clear signs of advanced breast cancer. He later investigated and found that the woman in the painting was Rembrandt's mistress, Hendrickje Stoffels, and she later died after a long illness. Greco wrote that he is confident that her death was due to metastatic breast cancer.

BREAST CANCER TREATMENT THROUGHOUT THE AGES

One early surgeon to advocate the surgical removal of a breast tumor was Aetios of Amida. His procedure was to remove the diseased parts of the breast and cauterize the incision. Aetios of Amida was the sixth-century court

Figure 1–1. Bathsheba at Her Bath (1654)—painting by Rembrandt. (Reproduced from the National Library of Medicine.)

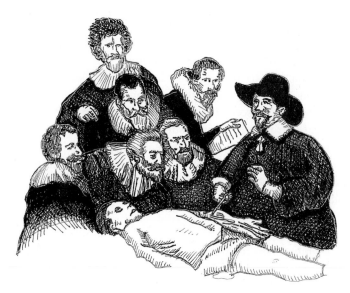

Figure 1–2. Surgical techniques used in the past.

physician to Justinian I and Theodora, emperor and empress of Byzantium. Justinian became emperor in 527 BC, 2 years after the death of his wife. Within days, he married his 25-year-old mistress Theodora. Twenty-three years later, she discovered a lump in her breast. Unfortunately, her cancer may have already metastasized. Aetios of Amida wrote that he tried his surgical procedure often, yet by June 548 BC, Theodora was dead. In those times surgery was a rare and radical treatment because there was no anesthesia or antisepsis available (Fig. 1–2).

Hippocrates, considered the father of Western medicine, supplied the earliest descriptive details of breast cancer. He named cancer "*karkinos*." *Karkinos,* or *Carcinus*, is a transliteration of the Greek word for "crab," so named because of the tendency of the cancer to reach out like the legs of a crab, grasping normal tissue. *Karkinos* eventually evolved into the term "carcinoma." It was Hippocrates who first outlined the "humoral" theory of the disease. The four body humors were blood, phlegm, yellow bile, and black bile, and any imbalance of these humors would result in illness. Hippocrates argued that cancer was from an excess of black bile. The body's tendency to produce vomit, diarrhea, blood, mucus, or pus during an illness only reinforced the supposed link between excess fluid and poor health and led to the continued belief in the humoral theory well into the seventh century.

As times advanced, the writings of Aristotle (384–322 BC) (Fig. 1–3) and the works of another Greek physician Claudius Galen (130–217 AD) had a profound influence on medical science. Aristotle was a noted philosopher and scientist, and his writings covered a wide range of subjects, from human and animal anatomy to metaphysics, physics, chemistry, botany, psychology, political theory, and ethics. Galen considered anatomy the foundation of medical knowledge, but because of the taboos, both religious and social, against human dissection, he was forced to experiment with the corpses of animals. He used what later turned out to be erroneous information as a basis for many of his theories on human anatomy. Galen was a prolific writer who also accepted the humoral theory and argued that human health required equilibrium between the four main bodily fluids or humors (blood, phlegm,

Figure 1–3. Aristotle (384–322 BC). (Reproduced from the National Library of Medicine.)

Figure 1–4. Andreas Vesalius (1514–1564). (Reproduced from the National Library of Medicine.)

Figure 1–5. Rudolf Virchow (1821–1902). (Reproduced from the National Library of Medicine.)

yellow bile, and black bile). Imbalances of these humors could be located in a specific organ or the body as a whole. This theory actually aided what is considered precise diagnosis for the times by allowing specific remedies to be prescribed. For example, Galen considered melancholia the chief contributing factor in the development of breast cancer, which he described as hard, painless knots that would eventually ulcerate through the skin, issuing black bile, and causing death. He recommended special diets in the treatment of the disease and lancing (bloodletting), whereby the vein was punctured to drain blood and excess fluids in order to restore humoral stability. Other treatments included expectorants to induce vomiting and laxatives to clear the bowels, all with the aim of relieving vascular pressure and simulating humoral stability.

Other physicians included exorcism and the use of topical ointment in their treatment plans. Exorcism was considered because serious diseases were thought to be the result of a spell cast on the victim by some enemy or the work of an offended god who had projected something into or extracted something out of the victim. Exorcism then was used to cast out the evil intruder or lure the soul back to its proper place within the body.

Galen's influence on medicine, or "humoral pathology," dominated for the next 1400 years. It took the works of Flemish physician, Andreas Vesalius (1514–1564), to correct the errors in Galen's human anatomy.

The European Renaissance, a period that roughly spanned the 14th century to the 17th century, was a period of great discovery, and with the fall of many social and religious taboos, Vesalius pioneered a new style in teaching anatomy (Fig. 1–4). He invited students to watch him dissecting cadavers or to view his illustrations as he lectured. Vesalius became known as the father of human anatomy, and he soon began questioning the medical doctrines of Aristotle and Galen. His famous work, *De Humani Corporis Fabrica Libri Septem (The Seven Books on the Structure of the Human Body),* was published in 1542. It contained accurate, beautifully illustrated, and detailed drawings of the human body and forever changed the study of anatomy and medicine. Vesalius's major contribution to breast cancer treatment was his suggestion that mastectomy and ligatures (sutures) were preferable to the use of cautery in controlling bleeding.

Further advances in the understanding of breast cancer came after influential French surgical writer Henri François Le Dran (1685–1770) recognized that the cancer could spread to the regional axillary nodes under the arm. With this discovery, the radical mastectomy became the preferred method of treatment for breast cancer. Although the study of anatomy improved in the 18th century, the prognosis for women with breast cancer remained poor because of infections and lack of good anesthesia. The discovery of nitrous oxide (N_2O) as an anesthetic in 1846 and antiseptic techniques in 1867 would, therefore, prove momentous.

Then, in 1859, German cellular pathologist Rudolf Virchow (1821–1902) wrote his book *Cellular Pathology* that became the foundation for all microscopic study of disease (Fig. 1–5). Virchow believed that diseases started not in organs or tissues of the body but in their individual cells. His beliefs were actually aided by the works of Italian anatomist Giovanni Battista Morgagni who, in 1761, showed that diseases were caused by lesions in organs and not humors of the body. Virchow developed the theory that malignant tumors were a result of conversion (metaplasia) of connective tissue, and although this later proved to be false, Virchow also coined the words "hyperplasia" and "ischemia" and named many of our common tumors. Based on his works, physicians began to speculate that a complete cure for breast cancer could only be achieved by the local removal of tumor along with adjacent healthy tissue. At about the same time, another noted surgeon and pathologist, British born Sir James Paget (1814–1899), was also detailing breast cancer. In 1874, Paget's research led to the description of the breast cancer known today as Paget's disease (an inflammatory disease of the nipple).

Many surgeons began keeping detailed records of their treatment of breast cancer. The super radical mastectomy was proposed as the cure by the surgeon William Halsted at Johns Hopkins Hospital in the 1890s. The common treatment included the removal of the breast, pectoral muscle and the axillary lymph nodes, chest wall and occasionally ribs, parts of the sternum, the clavicle, and lymph nodes inside the chest. However, because these surgeries were carried out only on advanced cancer cases, the survival rate remained low. Statistics kept at the Johns Hopkins Hospital reveal that the 10-year survival rate was only 12%, with a 30% local recurrence rate. Despite these poor survival rates, some surgeons went even further and began evacuating ribs and amputating a shoulder or clavicle to remove the chain of lymph nodes. These mastectomies permanently disfigured the bodies of patients. The removal of the pectoral major allowed the shoulder to cave inward and made it impossible to move the arm forward or sideways. With the removal of the lymph nodes, the disruption of lymph flow resulted in lymphedema of the entire arm, or "surgical elephantiasis."

Stephen Paget, an English surgeon (1855–1926), concluded that cancer cells spread by way of the bloodstream to all organs of the body. Paget's hypothesis was a key element in understanding the limitations of breast cancer surgery and led to the development of other systemic treatment for cancers.

THE DISCOVERY OF X-RAYS

The discovery of x-rays in 1895 by Wilhelm Conrad Roentgen (1845–1923), the discovery of radioactivity by Henri Becquerel (1852–1908), and the works of Pierre (1859–1906) and Marie Curie (1867–1934) in isolating radium from uranium allowed new advances in the treatment of breast cancer (Figs. 1–6 to 1–8).

Within a few years of these discoveries, radiation therapy was being used to improve the therapeutic response to breast cancer. These therapy treatment machines used radium or relatively low-voltage x-ray machines designed to destroy malignant tumors. Scientists now had another advantage in the battle with breast cancer. Between 1924 and 1928, Geoffrey Keynes, a young London doctor, pioneered a technique that used minimal surgery to remove the lump and then buried 50 mg of radium in the tumor site. Keynes and his colleagues found that their

Figure 1–6. Wilhelm Conrad Roentgen (1845–1923). (Reproduced from the National Library of Medicine.)

Figure 1–7. Henri Becquerel (1852–1908). (Reproduced from the National Library of Medicine.)

Mammography in the early 1960s:

- Conventional x-ray tubes with filters removed
- Unpredictable examination
- High dose—with reports of skin erythema after repeated imaging

cancer recurrence rate was at least comparable to rates obtained in New York and Baltimore, where radical surgeries without radiation treatment were still being performed. This technique was later called lumpectomy. Another technique, called simple mastectomy without the use of radiation, was pioneered by American surgeon George Crile.

The radical mastectomy would reign supreme until results of a randomized trial by the National Surgical Adjuvant Breast and Bowel Project (NSABP) were published in 1981. The trial followed three treatment groups: patients received radical mastectomy, simple mastectomy, or surgery followed by radiation. The results showed that the breast cancer recurrence, relapse, death, and distal metastasis rates for the three groups were statistically identical and that groups treated with the radical mastectomy accrued no benefits in survival, recurrence, or mortality.

Between 1930 and 1950, there was a noticeable improvement in the treatment of breast cancer. During World War I, it was discovered that agents used to combat the effects of mustard gas were also effective against a certain type of cancer of the lymph nodes. This led to similar drugs—chemotherapy treatment—being used to kill cancer cells by damaging their DNA. A staging system was developed: stages I and II represented operable or curable cases, stage III indicated advanced disease for which surgery was not a viable option, and stage IV described patients with distant metastases. The 10-year survival rate improved from 10% in 1920 to about 50% in 1950. But, although x-rays were now being used to detect the tumors, it was still thought that the low incidence of early detection was a factor in the high recurrence rate.

Even with the discovery of x-rays, mammography examination of the breast was unpredictable. Typically, the breast was imaged using conventional x-ray tubes, often with the filter removed to allow low-energy x-ray to exit. Unfortunately, the dose to the patient was also extremely high because the imaging was done on direct exposure films without a grid. In fact, in the early 1960s, there were even reports of skin erythema after repeat radiographs of the breast.

Figure 1–8. Pierre (1859–1906) and Marie Curie (1867–1934) with their daughter Irene. (Reproduced from the National Library of Medicine.)

In the 1950s, the Houston radiologist Robert L. Egan began imaging the breast using a high-milliamperage and low-kilovoltage technique. By using a fine-grain intensifying screen and industrial film, he was able to generate mammographic images that were clearer and easier to interpret. However, surgeons remained skeptical of mammography and often refused to operate if they could not palpate a lesion, even if the lesion was demonstrated on x-ray.

In the 1960s, mammography finally became a subspecialty within radiology, and a growing body of literature demonstrating the detection of small breast cancers using a mammogram finally convinced surgeons that they could not ignore the new technology.

Another tool in the war against breast cancer came with the use of xeroradiography machines in the 1960s. Xeroradiography was first developed in the late 1930s by Chester F. Carlson. But, although the technology existed, it took a request from the American College of Radiology (ACR) to prod the Xerox Corporation into producing the first Xero machine. In addition to the regular tungsten target x-ray tube, xerography used a dry process system, which consisted of two units: the conditioner and the processer. The conditioner prepared an aluminum plate by coating it with selenium and giving it an electrical charge. This plate, now a photoconductor, was placed in a cassette-like holder for exposure. After exposure, the holder was placed in the processer. The processer removed the plate from the cassette and exposed the plate to a blue powder of charged particles, which made the latent image visible. The powder image was transferred to a sheet of a plastic-coated paper to produce a mirror image. The paper was then heated to embed the toner particles into the plastic.

Xeromammography printed blue-and-white images on paper and needed no view boxes to view the images. The technique used low contrast, which allowed for even densities. It had a larger recording latitude than screen-film mammography and demonstrated both breast tissue and the skin line. Because the breast was not vigorously compressed under xeroradiography, the ribs could also be imaged. However, as with mammograms of this era, the main disadvantage of the Xero system was the radiation dose to the patient, which could be as high as 4R per projection (Fig. 1–9).

Scientists were, therefore, still facing a major problem. The only method they had for detecting breast cancer involved ionizing radiation, which was strongly associated with producing cancer. Also, breast cancer can be totally asymptomatic. It could take up to 5 years for a lesion to grow to a palpable mass, but by then the lesion would be approximately 10 mm in size. With lesions this large, the cancer could have already spread to the lymph nodes or could involve one or two lobes of the breast. The key, therefore, was to convince the public that the need to find cancers far outweighed any risk for developing a cancer from the radiation involved in its detection. In 1967, the Center for Genomics Research (CGR) in France introduced the first dedicated mammography machine. This unit featured a molybdenum anode, 0.7-mm focal spot, and a beryllium window port. In addition, the unit had a built-in device to compress the breast. As a result of these technical improvements, the contrast and resolution in the breast image improved, although the exposure was still high because of the continued use of direct exposure to x-ray films (Table 1–1).

Xeromammography had a larger recording latitude than screen-film mammography.

TABLE 1–1. Declining Mean Glandular Dose Trends by Years

Year	Dose
1974	13.8
1976	4.2
1980	2.5
1985	2.2
1988	1.8
1992	1.5
1995	1.5
1996	1.6
1997	1.6
1998	1.6
1999	1.7
2000	1.7
2001	1.8
2002	1.8

Source: US Food and Drug Administration. Division of Mammography Quality and Radiation Programs.

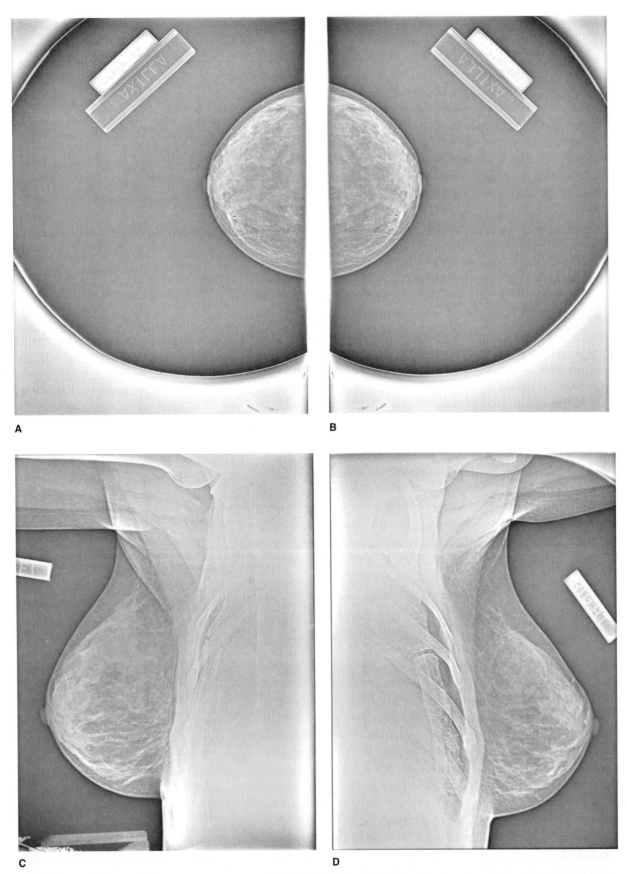

Figure 1–9. Xeroradiograph series showing a wider latitude image (when compared with a mammogram) and a clear image of the patient's ribs. **(A)** L CC; **(B)** R CC; **(C)** Left lateral; and **(D)** Right lateral.

MAMMOGRAPHY IN THE UNITED STATES—QUALITY VERSUS QUANTITY

With the new advances in mammography imaging in the mid-1970s, considerable evidence emerged that screening mammograms could significantly reduce the mortality rate of breast cancer in women. In 1963 the Health Insurance Plan (HIP) of greater New York launched a massive screening trial. In the course of 8 years, they screened more than 62,000 women. In 1971 positive results of the clinical trial showed that the possible carcinogenic effects of mammography were outweighed by the benefits, namely a 31% reduction in the mortality rate from breast cancer and a 25% reduction in the proportion of stage II or higher, advanced tumors detected in women older than 50 years. This was proved conclusively despite the limits of equipment and technique of that era (Fig. 1–10). Lesions discovered on a mammogram were smaller, usually involving only a single duct, and some were considered *in situ*, which presented a very good prognosis. Lesions were also discovered that never formed a palpable mass but showed up on a mammogram as calcifications (Table 1–2).

Within a few years the National Cancer Institute and the American Cancer Society (ACS) Breast Cancer Detection and Demonstration Project (BCDDP) pushed to take mammography screening nationwide. With the results from the HIP studies in mind, the first dedicated screen/film mammography machine was developed in the early 1970s. This system used a single-emulsion film and a single-emulsion calcium tungstate screen that was placed in a light-safe polyethylene bag and vacuum sealed. With the use of screens, radiation dose during the mammography examination was now at acceptable levels. The screen unfortunately reduced the resolution compared with direct x-ray exposure and xero-mammography; however, further research on the screen, the film, and the cassette system soon dramatically improved image quality. Because the breast tissue has little inherent contrast, it is of extreme importance to maximize that contrast. This is done through a combination of low-kilovoltage, vigorous compression of the breast tissue, and proper maintenance of the processing equipment. The introduction of grids in the late 1970s introduced a new generation of dedicated mammography machines. These new x-ray tubes have molybdenum or rhodium anodes and target. Such systems, combined with the routine use of grids and automated exposure control, are capable of producing high-quality mammograms at much lower kilovolt values than was previously thought possible. Chapter 5 will discuss the mammography unit in detail.

With new low-dose equipment available by the early 1980s, the ACS, the National Cancer Institute (NCI), and other organizations began promoting what soon was to become an accepted and widespread practice: routine mammographic screening of asymptomatic women. The schedule recommended was a baseline mammogram

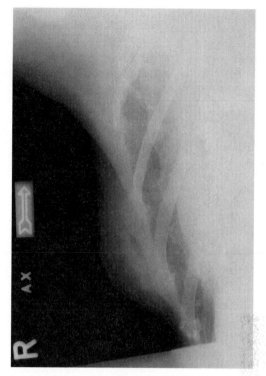

Figure 1–10. In the original method of imaging the MLO, the patient was imaged on a tabletop.

ACR Accreditation Requirements for Radiologists, Medical Physicists, and Radiologic Technologists:

- Boards, licensure, or certification
- Initial equipment training
- Continuing education
- Continuing experience

TABLE 1–2. HIP Study, Women Aged 40–64 (with 18-Year Follow-Up)

	Study	Control
Women assigned to	31,000	31,000
Breast cancer deaths	126	163

between 35 and 40 years of age and periodic follow-up examinations annually or biennially (depending on the recommendations of the patient's physician) until age 50 years, then annually thereafter.

Such was the response to the ACS recommendations that within 3 years (between 1985 and 1988) the number of mammographic facilities in the United States had more than doubled to a record 6400; however, it soon became evident that the widespread promotion of mammographic screening had encouraged a proliferation of unregulated facilities run by unqualified doctors, thus undermining the public's confidence not only in the procedure but also in mammography in general.

AMERICAN COLLEGE OF RADIOGRAPHY ACCREDITATION—STANDARDIZED MAMMOGRAPHY EQUIPMENT AND PROCESSING

In an effort to counteract the abuse, the ACR adopted a voluntary accreditation program in 1987. It seemed pointless to promote nationwide mammography screening based on the statistics of research done, unless the effectiveness of such mammographic screening ensured an accuracy and quality equal to that in the research studies. Shocking statistics showed that the detection rate for breast cancer was as low as 10% at some facilities.

Under the ACR accreditation process, the radiographing and processing of breast images were carefully monitored. All facilities performing mammography used only state-of-the-art equipment (described in detail in Chapter 5) (Fig. 1–11). In addition, personal requirements were established for radiologists, medical physicists, and radiologic technologists that included initial training, continuing education, and continuing experience. The technologist was required to obtain special education and training in breast anatomy, physiology, positioning, compression, quality assurance, and imaging techniques. Radiologists were required to be board certified and have completed a minimum of 40 hours of training in mammography, with an additional requirement of 15 hours of continuing education in the modality every 3 years.

With the need to keep glandular dose at a minimum while maintaining the highest possible contrast on the radiograph, the ACR adopted tests on these dedicated mammography units and processors to be done daily, weekly, monthly, and yearly. Facilities were also required to undergo periodic inspection of their units by a medical physicist. By checking the processor every day and plotting the data on an x-ray processing control chart, significant processing differences, as opposed to the normal daily fluctuations associated with a properly operating automatic processor, could be identified and corrected before processing clinical mammograms.

The ACR accreditation placed great emphasis on the duties of the technologist. Apart from maintaining quality assurance, the technologist is responsible for developing critical skills in patient positioning, applying adequate compression to the breast, ensuring proper technical factors, and examining the resultant radiographs. These duties cannot be overemphasized in mammography because a cancer cannot be detected if it is not present on the radiograph produced. Also, considering that a

Figure 1–11. General Electric's first mammography machine. (Used with permission from GE Medical Systems.)

woman will be undergoing routine yearly mammograms, all efforts should be made to keep the glandular dose to the minimum.

The ACR also recognized the importance of radiologist training. Unlike other x-rays in which a body part has definite normal or abnormal appearance, the breast tissue is unique to each patient and can vary considerably from patient to patient. Also, nonpalpable lesions in an asymptomatic patient will present no clinical appearance. Perception is, therefore, extremely important in mammography interpretation.

In 1990, the ACR published the first mammography quality control manual, which outlined quality control testing detail for the technologist, the interpreting physician, and the medical physicist. The 1992 and 1994 quality control manuals were adopted and referenced under the US Food and Drug Administration (FDA) Interim Rules for the Mammography Quality Standards Act (MQSA).

> **Responsibilities of the mammography technologist:**
> - Maintaining quality assurance
> - Developing critical skills in patient positioning
> - Applying adequate compression to the breast
> - Ensuring proper technical factors
> - Examining the resultant radiographs

THE MAMMOGRAPHY QUALITY STANDARDS ACT

The main drawbacks of the ACR program were its voluntary nature, the lack of enforcement or on-site evaluation of the facility, and possible conflicts of interest. To be effective, quality control in mammography must be performed consistently by a designated quality control technologist, who should periodically review all data with the radiologist and a medical physicist. This ensures that the frequency of the various tests can be adapted or modified to the facility's needs.

On October 27, 1992, the US Congress passed the MQSA in response to the growing public awareness and concern about the devastating effects of breast cancer. The MQSA was passed to ensure national quality standards for mammography services in the United States and to maximize the efficiency of mammography in the early detection of breast cancer. The regulations, which incorporated many of the ACR voluntary accreditation guidelines, were stated as a set of statutory requirements of quality standards, accreditation, certification, annual inspections, and sanctions. The FDA was given the authority to implement the act in June 1993. The act required that all mammography facilities be certified by October 1, 1994. Without certification, a facility could not lawfully perform mammography in the United States. The MQSA standards are discussed in Chapter 12.

> **Drawbacks of the ACR program:**
> - Voluntary nature
> - Lack of enforcement
> - Lack of on-site evaluation
> - Conflict of interest

BREAST CANCER SCREENING UPDATE

Recent published findings have thrown a lot of confusion on recommendations for breast cancer screening. The US Preventive Services Task Force (USPSTF), a group of independent health experts, is now suggesting that routine screening of the average-risk women could begin at age 50 years, instead of age 40 years. The recommendation is based on a number of studies, including a clinical trial conducted in Sweden, the Malmö Mammography Study, which was reported in 1988. The study showed a reduction in breast cancer deaths by 20% for women older than 55 years. However, screening with mammography showed no deaths averted for younger women. These findings also suggested that screening could end at age 74 years, that women could

USPSTF determined that the harms of BSE included:

- High anxiety
- Radiation exposure
- Likelihood of false-positive testing
- Uncalled-for biopsies and treatments
- Overestimation of benefits and the impact in overall breast cancer mortality

be screened every 2 years instead of every year, and that breast self-examination (BSE) has little value. The USPSTF followed with recommendations against the teaching of BSE, although they did not find sufficient evidence to recommend for or against performing clinical breast examinations (CBEs).

The USPSTF panels of experts are convened by the Department of Health and Human Services. They conduct scientific evidence reviews of a broad range of clinical preventive health care services and develop recommendations for primary care clinicians and health systems.

In many instances, the USPSTF concluded that the current evidence was insufficient to assess the benefits versus harm of the various mammography services. The panel also found evidence lacking or conflicting with no net benefit. For example, researchers concluded that under ideal circumstances, the benefits of mammography beyond age 79 years would be too low relative to cost to justify continued screening. The task force determined some of the harms of BSE and unnecessary mammography screening included high anxiety, radiation exposure, the likelihood of false-positive testing leading to uncalled-for biopsies, and treatments that may not affect overall breast cancer mortality.

The USPSTF did, however, conclude that the decision to start biennial screening before age 50 years should be a personal one or decided with clinician consult and should take into account an individual's genetics and family history.

The USPSTF recommendations on breast cancer screening have not changed the position of the ACS, which continues to recommend annual mammography screening for all healthy women starting at age 45 years, with the option to start at age 40 years as determined by clinician consult and the person's genetics and family history.

According to the ACR, current evidence still supports the substantial benefits for mammography screening even for women in their 40s. The ACR also recognized the limitation of mammography screening; however, it has concluded that the potential benefits of lives saved far outweigh the probable harm.

With older women, the ACR position is that age alone should not be the reason to stop regular mammograms. Other factors, such as the woman's health and any other serious illness, including congestive heart failure, end-stage renal disease, chronic obstructive pulmonary disease, and moderate-to-severe dementia, should be factored in any decision to stop screening.

- The ACR has concluded that the potential benefits of CBE and mammography screening far outweigh the probable harm.
- Breast cancer is more easily treated and cured when found before the cancer has spread.

The ACR, therefore, has not moved on its position. The ACR recommends that women in their 20s and 30s should have a CBE as part of a periodic (regular) health examination by a health professional, at least every 3 years. After age 40 years, women should have a breast examination by a health professional every year. The recommendation is that the CBE should take place shortly before the mammogram and that women should also be instructed on how to perform a BSE and should be advised on the limits of the CBE and the BSE. Research has shown that most breast cancers are found by chance instead of during a BSE, and whereas some women feel very comfortable doing BSE regularly, others become overanxious and stressed over the normal breast anatomy.

There have been significant accomplishment in breast cancer research (Box 1–1), and all researchers will agree the evidence clearly

Box 1–1. U.S. Department of Health and Human Services Women's Health Timeline & Timeline of FDA Accomplishments in Women's Health

1985–1991

1985   Reported lumpectomy combined with radiation therapy is as effective a treatment as mastectomy for many breast cancers. **(NIH)**

1987   Began collecting state-specific data on mammogram use and frequency through the *Behavioral Risk Factor Surveillance System.* **(CDC)**

1988   Required manufacturers of breast implants to submit safety and effectiveness data or have their product removed from the market. **(FDA)**

1991   Established the *National Breast and Cervical Cancer Early Detection Program* to provide free or low-cost mammograms, Pap tests, and follow-up services to low-income women for the early detection and control of breast and cervical cancers. **(CDC)**

1991   Launched the *Women's Health Initiative,* which was proposed by Dr. Bernadine Healy. A set of clinical trials and an observational study was conducted on the effects of postmenopausal hormone therapy, diet modification, and calcium and vitamin D supplements on heart disease, fractures, and breast and colorectal cancer. More than 160,000 postmenopausal women participated in this landmark study. A May 2014 report entitled *Economic Return From the Women's Health Initiative Estrogen Plus Progestin Clinical Trial: A Modeling Study* supports a substantial return on investment of public funds for this large study. **(NIH)**

1993–1999

1993   Established the *National Action Plan on Breast Cancer,* a public-private partnership to establish a comprehensive national plan to address breast cancer. **(OWH)**

1994   Established the CDC Office of Women's Health (OWH) to protect and advance the health of women through policy, science, and outreach, and to advocate for the participation of women in clinical trials.

1994   Implemented Mammography Quality Standards Act (MQSA) Program to ensure U.S. mammography facilities meet national standards for equipment, personnel, and quality control.

1994   Approved the first diagnostic test to detect Her2/Neu gene as an indicator of breast cancer recurrence.

1994   Discovered the *BRCA1 gene,* the first breast cancer gene, through the efforts of NIH researchers. Diagnostic tests can now identify women who have inherited defective copies of the gene and are more likely to develop breast cancer. **(NIH)**

1995   Established the OWH research program to support regulatory decision-making by providing a greater understanding of the impact of sex differences on medical product safety and efficacy.

1996   Approved a liquid-based cytology preparation instrument for use in Pap smears that revolutionizes detection of cervical cancer.

1998   Launched the Take Time to Care Program to disseminate FDA health and safety information to women and their families. Take Time to Care has since reached over 50 million women throughout the U.S. and Puerto Rico.

1998   Approved the first computer software that assists radiologists in detecting mammogram abnormalities.

1998   Approved the first product for chemoprevention of breast cancer in premenopausal and postmenopausal women who are breast cancer-free but at high risk for developing the disease. **(FDA)**

1998   Conducted the *Breast Cancer Prevention Trial (BCPT)* and the *Study of Tamoxifen and Raloxifene (STAR).* They demonstrated that half of breast cancers can be prevented with a medical intervention and provided a beginning from which a new paradigm for breast cancer prevention is evolving. Women at increased risk of this common disease now have preventive options where none had existed. **(NIH)**

1998   Approved the first genetically engineered cancer drug for treating *HER2*-positive metastatic breast cancer in women. **(FDA)**

1998   Approved an *in vitro* diagnostic and drug combination for identifying and targeting the *HER2* gene, which is expressed by about 25 percent of all breast cancer patients. **(FDA)**

1998   Launched the *National Women's Health Information Center,* the first combined website and toll-free phone number to provide reliable, accurate, commercial-free information on the health of women. **(OWH)**

2000–2005

2000   Approved digital mammography systems to detect subtle differences in breast tissue for the screening and diagnosis of breast cancer.

2002   Halted the estrogen-plus-progestin element of the *Women's Health Initiative* (WHI) trial because researchers found the risks of long-term estrogen-plus-progestin therapy outweighed its protective benefits. Study participants taking estrogen plus progestin were

(continued)

Box 1–1. *(Continued)*

at increased risk of heart attacks, stroke, invasive breast cancer, and blood clots as compared to women taking placebo pills. **(NIH)**

2003 Added a boxed warning to labeling of estrogen and estrogen-plus-progestin products to treat menopausal symptoms. Warning alerts women to increased risk for heart attack, stroke, invasive breast cancer, and blood clots.

2003 Launched the *Diethylstilbestrol (DES) Campaign* to raise awareness of possible health problems associated with DES exposure in women who were prescribed DES between 1938 and 1971 while they were pregnant. The campaign also focused on the women's daughters and sons. **(CDC)**

2003 Published results of *Women's Contraceptive and Reproductive Experiences (CARE)* study that found no increased risks of invasive breast cancer among women who used oral contraceptives and hormone replacement therapy (HRT). **(CDC)**

2004–2005 Created a partnership between OWH, other government agencies and women's organizations to educate 15 million women about benefits and risks associated with use of hormone therapy for menopause.

2006–2016

2006 Approved the first gene expression diagnostic test to assess risk of breast cancer recurrence. **(FDA)**

2007 Completed a national tour of *The Changing Face of Women's Health* exhibit. It explored menopause, society and body image, puberty, osteoporosis, breast health, heart disease,

sexually transmitted diseases, smoking, research and gender issues, real women's stories, and women's health history. This first national exhibit dedicated to women's health began its tour to different cities around the United States in 1999. **(CDC)**

2010 Established the *Advisory Committee on Breast Cancer in Young Women,* a federal advisory committee established by the Education and Awareness Requires Learning Young (EARLY) Act. **(CDC)**

2011 Revoked approval of Avastin's (bevacizumab) indication to treat breast cancer after two confirmatory trials submitted to FDA concluded the drug had not been shown to be safe and effective for that use. **(FDA)**

2011 Released a safety communication for women with breast implants describing a small but increased risk of developing anaplastic large cell lymphoma (a cancer) near the implant. **(FDA)**

2011 Approved the first 3D mammography imaging system (digital breast tomosynthesis) for use with 2D mammography.

2011 Co-sponsored the Dialogues on Diversifying Clinical Trials conference to promote best practices for the inclusion of women and minorities in clinical research.

2012 Approved the first automated breast ultrasound device for breast cancer screening in women who have dense breasts, a negative mammogram, and no symptoms. **(FDA)**

2013 Co-sponsored the FDA CDRH Health of Women conference to discuss recruitment & retention, analysis and communication,

and research roadmap for medical devices.

2013 Approved first non-hormonal treatment for moderate to severe hot flashes.

2013 Approved the first cohesive silicone gel-filled breast implant, providing additional options for breast reconstruction and breast augmentation in women at least 22 years of age. **(FDA)**

2015 Approved the first targeted treatment for advanced breast cancer specifically for use in postmenopausal women.

2016 FDA permitted marketing of the first cooling cap to reduce hair loss (alopecia) in female breast cancer patients undergoing chemotherapy.

2016 FDA allowed marketing of a new tissue expander system for soft tissue expansion in two-stage breast reconstruction following mastectomy and in the treatment of underdeveloped breasts and soft tissue deformities.

2016 OWH released the Women's Health Research Roadmap to provide a strategic plan to address seven priority areas where new or enhanced research is important to FDA regulatory decision making on products impacting women's health.

2016 OWH in collaboration with the NIH Office on Research in Women's Health launched the Diverse Women in Clinical Trial Initiative to raise awareness about the importance of participation of diverse groups of women in clinical research, and to share best practices about clinical research design, recruitment, and subpopulation analyses. www.fda.gov/womeninclinicaltrials.

1. Timeline of FDA Accomplishments in Women's Health. Available at: https://www.fda.gov/ForConsumers/ByAudience/ForWomen/FreePublications/ucm414927. htm#1993-1999. Accessed August 28, 2017

2. U.S. Department of Health and Human Services Women's Health Timeline. Available at: https://www.google.com/url?sa=t&rct=j&q=&esrc=s&source=web&cd=1 &ved=0ahUKEwjm2dTgqfvVAhVI0FQKHU5gDxkQFggoMAA&url=https%3A%2F%2Fwww.womenshealth.gov%2Ffiles%2Fassets%2Fdocs%2Fabout-us%2Fhhs-womens-health-timeline.pdf&usg=AFQjCNFHbCxSLirTbPxgg9aOeZN32s4HSg. Accessed August 2017

shows that early detection saves lives. Breast cancer is more easily treated and cured when found before the cancer has spread (Table 1–3; Fig. 1–12).

The position of the USPSTF on breast cancer screening is not new. In the later 1977s, the NCI commissioned a series of internal reports to test the validity of mammography screening. The general consensus was that the HIP study did not support the use of screening mammography for women aged 40 to 50 years. This position was supported by the ACS until 1983, when it decided to advocate a baseline mammogram for women aged 35 to 39 years and regular screening mammograms every 1 to 2 years for women aged 40 to 49 years. In an attempt to create a consensus, the NCI finally joined the ACS in 1988 and agreed to screening mammograms for younger women. However, an analysis of a series of eight randomized trials of mammography throughout the 1990s reversed that decision. The conclusion was that mammography had few demonstrable benefits for younger women. The idea of withholding mammography from women in their 40s did not sit well with the public, and by 1997 the NCI was again recommending screening at least once every other year for women aged 40 to 50 years. The USPSTF continues to question regular screening, and to a lesser extent the NCI continues to document the harms of screening. These include the dangers of false-negative results, false-positive results, overdiagnosis, and radiation risks. Overall the NCI considers unnecessary additional interventions to be a major result of false-positive results. Added interventions can include additional mammographic imaging, ultrasound, fine-needle aspiration, core biopsy, or excisional biopsy. The other danger is related to the sensitivity and accuracy of the mammogram as a diagnostic tool. Because the mammogram is not 100% accurate, a normal mammogram could potentially result in a misdiagnosis. And for the patient, a false-negative report could erroneously lead to a false sense of security, especially if the woman ignores a symptom based on her belief in the negative mammogram. Radiation exposure is considered a risk because younger women have breast tissue that will be more radiosensitive. Another consideration is that the high incidence of false-positive results can cause psychological distress and anxiety, which often lead to additional tests. The final area of concern is overdiagnosis. This has been suggested by some clinical studies because breast cancer incidence rates generally increased at the initiation of screening, but there is no compensatory drop in later years. This has led to speculation that some of the tumors discovered on mammography screening would never progress to cancer during the woman's lifetime. Studies have shown that carcinoma *in situ* does not always progress. However, the problem for clinicians is that cancers that will progress cannot be distinguished with certainty from those that will not; therefore, *in situ* diagnoses are often treated with surgery, radiation, chemotherapy, or hormonal therapy, or a combination.

TABLE 1–3. Current Breast Cancer Survival by Stage

Stage		5-Year Relative Survival Rate (%)
0	Carcinoma *in situ*	100
I	Tumor ≤2 cm. Axillary node negative	98
IIA	Tumor 2–5 cm with or without positive nodes	88
IIB	Tumor >5 cm without positive nodes	76
IIIA	Tumor >5 cm with positive nodes	56
IIIB	Any size of tumor if spread to breast skin, chest wall, or internal breast lymph nodes	49
IV	Any size if there is distant metastasis, e.g., to one, lungs	16

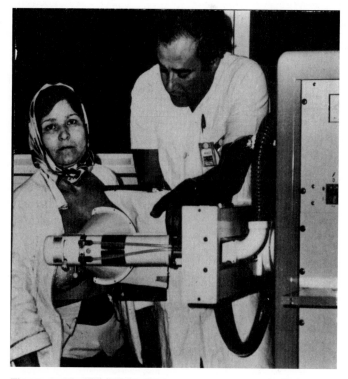

Figure 1–12. Woman having a mammogram at the Salah Azaïz Cancer Institute, World Health Organization (WHO)/ Ministry of Health, Tunisia. (Used with permission from the World Health Organization.)

Summary

- Breast cancer is not a new disease. There has been documented evidence of the disease dating back to 1600 BC.

- The detection and treatment of breast cancer have changed dramatically throughout the ages. The most significant changes took place during the European Renaissance period.

- The discovery of x-rays brought a major advance in both the diagnosis and treatment of breast cancer.

- The Mammography Quality Standard Act (MQSA) introduced the concept of an accreditation process whereby standardized mammography imaging and equipment would always be used and the technologist, physicist, and the radiologist would undergo specialized training.

- The US Preventive Services Task Force (USPSTF) continues to debate regular screening, and to a lesser extent the National Cancer Institute (NCI) continues to document the harms of screening.

- The diagnostic and treatment options available in the fight against breast cancer continue to improve (Box 1–1). Currently, the 5-year survival rate is still dependent on the stage at which the cancer is found. Survival rates, however, are just a relative number; these rates exclude patients dying of other diseases and do not take into account each woman's individual or unique genetic characteristics (Table 1–3).

REVIEW QUESTIONS

1. What factor or factors led to the early erroneous theories on the human anatomy?

2. Who was most responsible for correcting the errors in human anatomy?

3. In the early 18th century, the prognosis of women with breast cancer remained poor. What was a major contributing factor?

4. Name the pathologist who coined the terms "hyperplasia" and "ischemia."

5. How did the discovery of x-rays change the treatment options for breast cancer?

6. What conclusive finding led to the routine screening of asymptomatic women for breast cancer?

7. The American College of Radiography (ACR) used a three-pronged approach to monitoring breast facilities. This approach carefully assessed which three areas of breast imaging?

8. Give three disadvantages of the ACR accreditation program.

9. What is the mammography screening guideline recommended by the American Cancer Society (ACS)?

10. The devastating effects of breast cancer and the public's loss of confidence in the diagnostic options led to the passage of which act by the US Congress?

11. What process is now required before a facility can lawfully perform mammograms in the United States?

12. What are the current screening guidelines recommended by the US Preventive Services Task Force (USPSTF)?

CHAPTER QUIZ

1. One of the earliest cases of breast cancer was discovered by the American Egyptologist Edwin Smith in:
   (A) 1600 BC
   (B) 1862
   (C) 1895
   (D) 1995

2. The social and religious taboo of the time forced Claudius Galen to experiment and dissect _____ instead of humans.
   (A) Animals
   (B) Birds
   (C) Mammals
   (D) Reptiles

3. The father of human anatomy, recognized for his famous work, *The Seven Books on the Structure of Human Body*, is:
   (A) Claudius Galen
   (B) Henri Le Dran
   (C) Rudolf Virchow
   (D) Andreas Vesalius

4. Which of the following scientists was known for his or her work in the discovery of radioactivity and in isolating radium from uranium?
   (A) Marie Curie
   (B) Pierre Curie
   (C) Wilhelm Roentgen
   (D) Henri Becquerel

5. The terms "hyperplasia" and "ischemia" were coined by:
   (A) Henri Le Dran
   (B) Andreas Vesalius
   (C) Chester Carlson
   (D) Rudolf Virchow

6. The breast imaging equipment that used a two-step process, conditioner and processer, and printed blue and white images on paper was called:
   (A) Radiography
   (B) Radiation therapy
   (C) Xeroradiography
   (D) Mammography

7. One of the first studies to demonstrate that the benefits of mammography far outweighed the risks was conducted by:
   (A) MQSA
   (B) HIP
   (C) ACS
   (D) NCI

8. What technique, used in the early 1970s, proved effective in bringing the radiation dose during the mammography examination to acceptable levels?
   (A) Use of grids
   (B) Use of a molybdenum or rhodium anode
   (C) A single-emulsion film and screen system
   (D) Low-kilovoltage systems

9. The voluntary accreditation program was recommended by the:
   (A) MQSA
   (B) FDA
   (C) NCI
   (D) ACR

10. The mandatory mammography quality assurance act passed by the US Congress in 1992 is:
    (A) MQSA
    (B) ACS
    (C) ACR
    (D) FDA

BIBLIOGRAPHY

Adler A, Carlton R. *Introduction to Radiologic Science and Patient Care.* 6th ed. St. Louis, MO: Elsevier Saunders; 2016.

American College of Radiology. Mammography accreditation. http://www.acr.org/quality-safety/accreditation/mammography. Accessed August 2016.

American College of Radiology. *Mammography Quality Control Manual.* Reston, VA: Author; 1999.

Berns EA, Baker JA, Barke LD, et al. *Digital Mammography Quality Control Manual.* Reston, VA: American College of Radiology; 2016.

Bushong SC. *Radiologic Science for Technologists—Physics, Biology and Protection.* 10th ed. St. Louis, MO: Mosby; 2012.

Calonge N, Petitti DB, DeWitt TG, et al. Screening for breast cancer: U.S. Preventive Services Task Force recommendation statement. *Ann Intern Med.* 2009;151:716–726.

Encyclopædia Britannica. Andreas Vesalius. https://www.britannica.com/biography/Andreas-Vesalius. Accessed August 2016.

Encyclopædia Britannica. Aristotle. https://www.britannica.com/biography/Aristotle. Accessed August 2016.

Encyclopædia Britannica. Edwin Smith papyrus. https://www.britannica.com/topic/Edwin-Smith-papyrus. Accessed August 2016.

Encyclopædia Britannica. Galen of Pergamum. https://www.britannica.com/biography/Galen-of-Pergamum. Accessed August 2016.

Encyclopædia Britannica. Sir George Paget Thomson. https://www.britannica.com/biography/George-Paget-Thomson. Accessed August 2016.

Food and Drug Administration. Accreditation and certification overview. https://www.fda.gov/radiation-emittingproducts/mammographyquality-standardsactandprogram/documentarchives/ucm114207.htm. Accessed April 2017.

GE Healthcare. Mammography at 50: past, present, and future. http://news-room.gehealthcare.com/mammography-at-50-past-present-and-future/. Accessed August 2016.

Kolata G, Moss M. X-ray vision in hindsight: science, politics and the mammogram. *New York Times.* February 11, 2002:A18.

Lerner BN. *The Breast Cancer Wars: Hope, Fear and the Pursuit of a Cure in Twentieth-Century America.* New York, NY: Oxford University Press; 2003:41–92.

Maandelblatt JS, Cronin KA, Bailey S, et al. Effects of mammography screening under different screening schedules: model estimates of potential benefits and harms. *Ann Intern Med.* 2009;151(10):738–747.

Mukherjee S. *The Emperor of All Maladies.* New York, NY: Simon & Schuster; 2010.

National Breast Cancer Foundation (NBCF). The NBCF story. http://www.nationalbreastcancer.org/about-nbcf/nbcf-story. Accessed August 2016.

Olson JS. *Bathsheba's Breast: Women Cancer, and History.* Baltimore, MD: The Johns Hopkins University Press; 2005:1–14.

Peart O. *Lange Q & A: Mammography Examination.* 2nd ed. New York, NY: McGraw-Hill; 2008.

Sadawsky NL, Semine A, Harris JR. Breast imaging: a critical aspect of breast conserving treatment. *Cancer.* 1990;65:2113–2118.

Venes D, Biderman A, Adler E. *Taber's Cyclopedic Medical Dictionary.* 22nd ed. Philadelphia, PA: F.A. Davis; 2013.

Patient Education and Assessment | 2

Keywords and Phrases
Patient Care and Communication
The Environment
Medical History and Documentation
   Subjective Versus Objective Data
   Examples of Nonverbal Communication
   History-Taking Techniques
   Documentation
Breast Examinations
   Breast Self-Examination
   Clinical Breast Examination
Breast Examination Guidelines
Benefits Versus Risks of Breast Imaging
   Mortality Reduction
   Risk from Radiation Exposure in Mammography Screening
Breast Cancer Screening
Advantages and Disadvantages of Mammography Screening
Risk Factors for Breast Cancer
   Types of Risk Assessments
   Major Risk Factors
   Minor Risk Factors
Facts Versus Myths and Misconceptions
Summary
Review Questions
Chapter Quiz

Objectives

On completion of the chapter, the reader will be able to:

1. Define the term breast self-examination and describe how the examination is done
2. Define the term clinical breast examination and describe how the examination is done
3. Describe the breast cancer screening guideline as outlined by the American Cancer Society (ACS)
4. List the benefits versus risks of mammography screening
5. Identify the dose received during a routine mammogram
6. Discuss the advantages and disadvantages of mammography imaging
7. Identify the major and minor risk factors for breast cancer
8. Describe myths and misconceptions associated with breast cancer imaging and screening

KEYWORDS AND PHRASES

- **Accessory breast** is any breast tissue present in the axillary region or inferior to the actual breast.
- **Background radiation** comes from various sources, including space and the earth itself. The earth is constantly bombarded by particles from space. Fortunately, most are stopped before reaching the earth's surface, but all soil and rock contain trace quantities of naturally occurring radioactive elements.
- *BRCA1, BRCA2* are the symbols used to indicate an abnormal breast cancer gene. Normally, the breast cancer gene produces a protein that protects against unwanted cell growth. If the gene is defective, the protein produced is unable to prevent the growth of abnormal cancer cells. Patients with these genes are at a higher risk for developing breast cancer as well as other cancers.
- **Breast augmentation** is also called *augmentation mammoplasty* or *augmentation*. It is a method used to increase breast size by inserting an implant. Common implants are made of saline, a saline solution contained in a silicon shell, or silicone, which is a silicone shell filled with silicone gel. Mammoplasty is the surgical procedure used to reduce, reconstruct, or reshape the breast.
- **Breast reduction** is the removal of excess breast fat, glandular tissue, and skin and the shifting or repositioning of the nipple and areola. It can be performed for cosmetic reasons, to achieve a breast size in proportion with body size, or for medical reasons, to alleviate the discomfort or pain associated with overly large breasts.
- **Breast self-examination** is a thorough examination of the breast by oneself.
- **Clinical Breast Examination (CBE)** is a thorough examination of the breast, including the lymph nodes in the axilla and clavicular area, by a qualified health professional.
- **Curie** (symbol: Ci) was named in honor of the French physicist Marie Curie. The Curie measures the amount of radioactivity emitted

from a radioactive source and is mainly used in nuclear medicine. The SI unit is the becquerel, symbol: Bq. Radioactivity (3.7×1010 Bq = 1 Ci).

- **Digital imaging** is an electronic and computerized detection system that is used to form digital images of the breast. It replaces the screen/film combination system in mammography. The resultant image can be transmitted, manipulated, and efficiently stored using a variety of methods.

- **Eczema** refers to a general term for a rash, which may be itchy or red and is due to allergy, chemicals, or drugs. Scratching or rubbing a skin rash can cause oozing or a thick scaly crust to develop.

- **Effective dose** takes into account the different types of radiation and their different biologic effects. Effective dose also considers that various tissues and organs have different radiosensitivities and are affected more or less by radiation doses.

- **Entrance skin exposure (ESE)** is often referred to as the *patient dose* and is easiest to measure. It is a measure of the dose to the individual's skin from the radiation source.

- **False negative** is a test result that falsely indicates that a condition is not present when in fact it is.

- **False positive** is a test result that falsely indicates the presence of a condition when in fact the condition is not present.

- **Glandular dose** is used in mammography to describe radiation dose to the glandular tissues of the breast.

- **Gray (Gy)** is the international unit (SI unit) measuring the amount of radiation energy absorbed in a medium (e.g., body tissue). The biologic effects of radiation will vary by the type and energy of the radiation and the organism and tissues involved. The older unit is the rad, Radiation Absorbed Dose, symbol: rad (1 Gy = 100; rad = 1 J/kg).

- **Hodgkin's disease** is named after Thomas Hodgkin, an English physician (1798–1866). It is a malignant disorder characterized by painless progressive enlargement of lymphoid tissue. Untreated, it will lead to death.

- **Hormone therapy** is the use of hormones as a therapeutic treatment.

- **Lymphedema** is the accumulation of lymph in the soft tissues causing swelling and inflammation. It can be caused by obstruction or by the removal of lymph channels.

- **Milk ridge or line** extends from the armpits in the axilla to the groin region of the body. Breast tissue can form anywhere along the milk ridge, also known as the mammary line.

- **MRI,** magnetic resonance imaging, can be used to show the interaction of body tissues with radio waves in a magnetic field. These interactions result in high- or low-intensity signals that appear as bright or dark areas on an image. MRI does not use x-rays or sound waves but rather uses very complex magnetic properties of elements.

- **Nephrogenic systemic fibrosis and nephrogenic fibrosing dermopathy** (NSF/NFD) are rare conditions associated with some gadolinium-based contrast used in MRI. They were first identified in 1997 and most often occur in patients with impaired renal function. The conditions can develop within 2 days to 18 months after receiving the contrast injection and are characterized by tight, rigid skin that will eventually make any movement at the joints difficult or impossible and may lead to multiorgan failure and death.

- **Non-Hodgkin's lymphoma** is the growth of malignant or benign lymphoid tumors. The symptoms and treatment of the disease are similar to those for Hodgkin's lymphoma. The disease can be differentiated only at the cellular level.
- **Parity** is the terminology used if a woman carries a pregnancy to a point of viability (20 weeks of gestation) regardless of the outcome.
- **Parturition** is the act of giving birth.
- **Polythelia** is the presence of more than one nipple on the breast.
- **Rad** (Radiation Absorbed Dose, symbol: rad) measures the amount of radiation energy absorbed in a medium (e.g., body tissue). The biologic effects of radiation will vary by the type and energy of the radiation and the organs and tissues involved. The rad is still used in the United States. The SI unit of radiation-absorbed dose is the gray, symbol: Gy (1 rad = 0.01 Gy; 1 rad = 1×10^{-2} Gy) (Box 2–1).
- **Rem** (symbol: rem) is the "roentgen equivalent (in) man." The rem attempts to give the biologic effects of radiation. It gives the dose equivalent or occupational exposure as opposed to the absorbed dose. The dose equivalent (DE) is a measure of the radiation dose to tissues, whereas the absorbed dose measures the physical effects of radiation. The SI unit is the sievert, symbol: Sv. (1 rem = 0.01 Sv).
- **Roentgen** (symbol: R) is the measure of the ionization produced in air by x-rays or gamma radiation, or is a measurement of exposure to radiation. This is the measure made by a survey meter. The unit is named after the German physicist Wilhelm Röntgen. There is no SI unit for this measurement because it can be expressed in units of coulomb/kilogram (C/kg) (2.58×10^{-4} C/kg = 1 R).
- **Self-referral** refers to a patient who comes for diagnostic testing without a referral from a health care provider.
- **Sievert** (Sv) is the SI unit that attempts to give the biologic effects of radiation. It gives the dose equivalent or occupational exposure as opposed to the absorbed dose. The dose equivalent is a measure of the radiation dose to tissues, whereas the absorbed dose measures the physical effects of radiation. It is named after Rolf Sievert, a Swedish

Box 2–1. Standard Scientific Prefixes

Prefix	Symbol	1000^m	10^n	Decimal
Mega	M	1000^2	10^6	1000000
Kilo	K	1000^1	10^3	1000
Hecto	H	$1000^{2/3}$	10^2	100
Deca	Da	$1000^{1/3}$	10^1	10
		1000^0	10^0	1
Deci	D	$1000^{-1/3}$	10^{-1}	0.1
Centi	C	$1000^{-2/3}$	10^{-2}	0.01
Milli	M	1000^{-1}	10^{-3}	0.001
Micro	M	1000^{-2}	10^{-6}	0.000001
Nano	N	1000^{-3}	10^{-9}	0.000000001

medical physicist famous for work on radiation dosage measurement and research into the biologic effects of radiation. The older unit is the "roentgen equivalent: man," symbol: rem (1 Sv = 1 J/kg = 100 rem).

- **Thermoluminescence dosimeter (TLD)** is the instrument often used to measure the entrance skin dose. The TLD consists of lithium fluoride (LiF) in crystalline form, either as a chip or as powder. When exposed to radiation, the TLD absorbs energy and stores it. This energy is released in the form of light when the TLD is heated. The visible light released is proportional to the radiation dose received by the crystal.
- **Ulcers** occur where the skin or mucous membrane is damaged because of inflammation, necrosis, or trauma.
- **Ultrasound** uses sound waves to outline the shapes of various organs in the body in real time. Ultrasound gives rapid sequence of multiple images to duplicate motion.

PATIENT CARE AND COMMUNICATION

Breast cancer is the second leading cause of death among women in the United States, claiming the lives of thousands annually, and to date although there is no cure, breast cancer in its earliest stages can be effectively treated. The first line of defense against breast cancer is the screening mammogram. It is still the number one option.

Breast cancer screening has had a significant impact on the mortality rates due to breast cancer. With mammography screening, the rate of early breast cancer detection has increased significantly, and the mortality rate due to breast cancer has declined because of the emphasis on early detection. Studies have consistently shown that the death rate for breast cancer in women has decreased since the 1980s. Although the overall rate of decrease from 1998 through 2006 was 1.9% annually, the early years of screening showed significant decrease at the rate of 3.3% annually between 1995 and 1998. One clear fact is that early detection is associated with smaller lesions and higher survival rates, as well as an overall decrease in mortality rate. Each year, thousands of women have their first mammogram. They come with preconceived notions about the mammogram—stories that they have heard from friends, relatives, or coworkers. They seek compassion, reassurance, professionalism, education, and for some even counseling (Box 2–2).

The technologist will get just one chance to make a good impression on these patients. Remember, every step of the way, from the time the patient walks into the department to the time that the patient heads for home, both the technologist and the department are under assessment. The mammography examination in fact presents a unique opportunity for technologists to educate their patients.

Women are more likely to return for a routine mammogram and comply with a follow-up request after a pleasant experience. If the patient's first experience with mammography is painful and the technologist is unsympathetic, there is a greater chance that the patient will not return for future mammograms or additional studies. The patient will definitely not want to come back to the same facility. Patients whose first experience is unpleasant will also be reluctant to recommend a

With mammography screening, the rate of early breast cancer detection has increased significantly, and the mortality rate due to breast cancer has declined because of the emphasis on early detection.

Box 2–2. Factors Influencing a Woman's Commitment to Breast Cancer Screening

- Apprehension or anxiety concerning the results
- Cost of the mammogram
- Fear of compression
- Fear of radiation
- Myths and misconceptions
- Unpleasant, unsympathetic, or painful experience

Communication:

- Before the exam
- During the exam
- After the exam

Types of communication:

- Listening
- Encouraging comments
- Face-to-face questions
- Informal chats

Benefits of communication:

- Relaxes the patient
- Reduces the chance of suboptimal imaging of the pectoral muscles
- Identifies sensitive breast and the reason for the sensitivity
- Allows for possible rescheduling of the mammogram if needed
- Educates the patient
- Reveals fears and misconceptions

mammogram to their friends or family or will perpetuate the myths of mammography being a painful study.

The best way to combat the misconception about mammography is communication. Technologists should communicate with their patients before, during, and after the mammogram. Communication should not just be questions and answers from the technologist. Technologists should always invite questions from the patient, then listen and encourage further comments. Only by communicating with the patient will technologists be able to identify concerns that could be addressed and answer questions before the patient leaves. All patient communication should be face to face and as informal as possible. When going through routine mammography questioning, the technologist should also take the opportunity to explain the procedure to the patient and prepare the patient for the examination.

Technologists need to have a thorough understanding of the breast in order to understand the importance of medical history and documentation and to help patients with breast examinations. In the course of a casual conversation, the patient can reveal fears or misconceptions that are often a hindrance to routine breast cancer screening. A well-educated and informed technologist may well be the deciding factor, determining whether the patient will return for further screening or give up on diagnostic testing as a waste of time.

With millions of Americans having no health care coverage, cost can be a critical factor when making the decision to have a mammogram. The cost of mammograms can vary depending on the region of the country and the extent and nature of the examination. Costs in major cities are higher than in rural areas, and the cost of a routine screening mammogram (the four-projection series with no additional projections) will be less than the cost of a diagnostic mammogram, which can include additional projections (such as magnification, spot compression, or other supplementary projections of the breast). Thorough communication will identify these patient concerns, and the technologist can then help by informing the patient of the many low-cost screening mammograms available at many facilities during Breast Cancer Awareness Month or Mobile Mammography screening throughout the year. Most insurance policies will cover screening mammography for all women following the American College of Radiology (ACR) guidelines. The technologist should stick to the facts and should refrain from speculations or discussions about specific insurance information.

Another positive aspect of communication is that it helps to relax the patient. An uncooperative patient or one who is not relaxed is a difficult-to-image patient. It is almost impossible to get a good image of the pectoral muscle on the mediolateral oblique projection when the patient is tense or stiff. Suboptimal images will then prolong the examination, frustrating both the patient and the technologist.

By establishing a rapport with patients, the technologist can identify those with unusually sensitive breasts (Fig. 2–1). If the sensitivity is extreme and hormone related, the mammogram can perhaps be rescheduled.

Apprehension about the results of the mammogram can be such that some women might actually delay having the examination or even fail

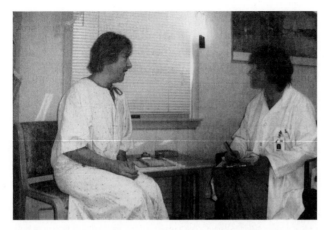

Figure 2–1. The technologist should sit and take a patient history before the mammogram.

to seek medical attention for a lump because they are afraid of the results. Technologists are not interpreting physicians and should not give the patient results unless authorized to do so by a physician. Patients, however, should always be given information on how and when they should expect to get their results. Mammography Quality Standards Act (MQSA) regulations state that all mammography patients—not just their physicians—must receive a copy of the mammogram results. This will relieve those patients who are self-referrals.

For patients in a panic about a possible lump, explain that not all lumps are cancerous. The lumps can be simple cysts—especially painful lumps. Although the diagnosis should be left to the radiologist, the technologist should be honest. The technologist must never lie to the patients by saying that there is nothing wrong or nothing to worry about and never leave the patient with misperceptions about the results or the procedure.

Before leaving the facility, the technologist should also alleviate patients' fears of additional projections, especially for screening mammograms. About one in ten patient may get a request to return to the facility for additional projections. A particularly nervous patient should be reassured about the possibility of a callback by informing the patient of the reasons for additional projections. This will avoid panic reactions and could alleviate some of the patient's anxiety (Box 2–3).

THE ENVIRONMENT

Remember that little things mean a lot. For perhaps 5 to 10 minutes, the technologist will have the patient as a captive audience. The attention of the technologist will not only make the patients feel good about having a mammogram but also make them feel comfortable and really cared for. Technologists will feel particularly appreciated when the patient comes back the next year and actually remembers not just a face but also a name.

The waiting room should provide comfortable chairs and reflect a calm and relaxing mood (Fig. 2–2). The lighting, wall coloring, and

Box 2–3. Reasons for Additional Projections

1. An area of the breast may need more compression.

2. The breast tissue in one area may be obscuring (covering or overlapping) the breast tissue of another area.

3. All breast tissue will undergo a certain amount of normal changes yearly. Sometimes, these changes are easy to track and explain, and sometimes they are not. This is when additional radiographs may be required to confirm that the changes are normal.

Ensuring patient compliance:

- Comfortable and relaxing wait area
- Pleasant experience
- Understanding technologist
- Informed technologist
- Compassionate technologist

Figure 2–2. The patients' waiting room should provide videos, newspapers, or magazines.

even pictures or the lack of pictures can all contribute to the mood of the patient waiting area. Many facilities now offer videos, information packets, or general magazines for the patient to view or read during the wait (Fig. 2–3). Often, tea or coffee can be provided, along with private, clean changing rooms with a mirror that will help the patient with reapplying makeup or for combing and fixing hair. Something as simple as a patient gown that does not fit or rips easily can have an impact on the patient's mammography experience (Box 2–4).

Figure 2–3. Patient information booklets should be readily available.

Box 2–4. Improving Patient Satisfaction During the Mammography Examination

To reduce breast discomfort during the mammogram:

- Selective scheduling
- Use a topical anesthetic gel before the examination
- Use oral pain medication

Some studies suggest that oral medication does not have a significant impact on breast discomfort, but women who received a topical application of 4% lidocaine gel reported significantly less breast discomfort during the mammogram. The Food and Drug Administration warns that topical anesthetic gels, such as lidocaine gel, can be absorbed into the bloodstream and, if used improperly, can cause life-threatening side effects such as irregular heartbeat, seizures, breathing difficulties, coma, or even death. When using topical anesthetic gels:

- Use the lowest strength
- Apply sparingly where pain exists or is expected to occur
- Never apply to broken or irritated skin
- Always consult a health care professional about possible side effects
- Avoid heating, wrapping, or covering the treated skin because this can increase the chances of serious side effects

Use of a cushion or pad on the mammography unit can help to:

- Hold very small breast in place
- Allow better positioning
- Allow a warm detector surface

MEDICAL HISTORY AND DOCUMENTATION

In the fight against breast cancer, technologists often play an important role. They are often the only link between the patient and the radiologist. The information that they convey can and often does have a profound effect on the diagnosis of the disease. The imaging technologist should take a few minutes to develop a rapport with the patient.

Breast cancer can be totally asymptomatic, and because a nonpalpable lesion in an asymptomatic woman will present no clinical findings, perception and analysis of the final image are of extreme importance. The interpreting physician, therefore, needs as much help as possible in making a final interpretation. Prior mammograms should be available for comparison and the technologist should document a compete clinical history.

Effective questioning techniques involve listening rather than asking questions. Listening will often reveal the patient's life experience and education background and help you to determine whether you can use medical or technical terms or overly simplified words for future questions. As long as the technologist is approachable, the patient will often tell her life experience, unasked.

Medical history documentation:

- Family or personal history of breast cancer
- History of breast surgery
- Biopsies, lumpectomy, augmentation, reduction
- History of breast trauma
- Physical symptoms
- Skin thickening
- Unusual lumps
- Dimpling or puckering
- Moles, eczema, or ulcers
- Nipple changes
- Abnormalities
- Accessory breast or nipple

Subjective Versus Objective Data

It is easy to assess objective data. This will include signs and symptoms that can be seen or felt and such things as a lump. However, the technologist cannot exclude subjective data, symptoms that are perceived by the affected individual only. If the patient feels a lump and the technologist does not feel the lump, the area should still be marked and the patient's clinical symptoms documented. The technologist should never disregard anything the patient says, even if it does not fit with the opinion you are forming about the patient's symptoms. The technologist can, however, ask for clarification of any information.

One of the most important aspects of history taking is proper communication skills. How the technologist interacts with the patient both verbally and nonverbally is often reflected in the feedback received from the patient in terms of the quality of information. Technologists should use suitable vocabulary. This includes the appropriate use of nontechnical language without "speaking down" to the patient. Sentences should be clear with concise language. Technologists should avoid sentence fragments, mumbling, and references to age, sex, and diseases. It is often best to avoid humor because it is easy to offend when the background, culture, or personality of a person is unknown.

Examples of Nonverbal Communication

Paralanguage refers to the pitch, tone, stress, pauses, speed, rate, volume, accent, or quality of our voice. Paralanguage instills interest in conversation and avoids monotony.

Body language can indicate interest in the patient's answers or an unwillingness to offer help. Studies have shown that the movement of your extremities, position of your body in relation to the patient's, and your facial expression can convey more meaning to a conversation than the actual words. For example, a puzzled look will not need words to convey confusion. When speaking to the patient, the technologist should maintain eye contact to ensure that instructions are understood and to monitor the patient's reactions.

Professional appearance and physical hygiene can help to project confidence and inspire the patient's confidence. Dirty nails, strong body odors, or untidy appearance will have a negative impact on effective communication. The technologist may also need to address specific patient needs of the visually impaired, speech and/or hearing impaired, or non–English-speaking patient. The visually impaired patients will need clear instructions before, during, and after the examination. The technologist can use gentle touch and provide a running commentary to keep the patient up to date throughout the examination. The speech- and/or hearing-impaired patients should not be shouted at. The technologist must determine the level of the patient's needs in order to utilize the most effective tool. This can include written instructions, pantomime, or sign language. In dealing with the non–English-speaking patients, facial expression and pantomime can be useful tools. Legally, the technologist should never utilize the services of friends or relatives as translators. It is especially important not to use a child. Most facilities now have legal telephone translation lines or specially trained employees who serve as translators.

Nonverbal communication:

- **Paralanguage:** the pitch, tone, stress, pauses, speed, rate, volume, accent, or quality of your voice.
- **Body language:** the movement or position of extremities, position of your body in relation to the patient's or facial expression.
- **Professional appearance:** clothing or dress, e.g., sloppy or untidy.
- **Physical hygiene:** dirty nails, strong perfumes, or strong body odors.

Communicating tips:

- **Visually impaired:** gentle touch and clear instructions before, during, and after the examination.
- **Speech and/or hearing impaired:** written instructions, pantomime, or sign language.
- **Non-English speaking:** official translators, facial expression and pantomime.

History-Taking Techniques

Open-ended questions (nondirected and nonleading; e.g., "Where is your pain?") are the best questions to use when starting any history taking. A leading question such as, "Is the pain in your left breast?" should be avoided. Whenever possible use facilitations, such as a nod, or say "yes," "okay," or "go on" to encourage elaboration. Silence can be used to give the patient time to remember, and it also facilitates accuracy. Probing questions must be used after the initial open-ended question to get more details (e.g., "How long have you had the pain?"). Repetition is basically a rewording of the questions and can be used to clarify information. Repeating the patient's response will also help to verify that the patient has not changed her mind. In repeating, the technologist should avoid jargon and should use precise and clear words. Before ending, the technologist should summarize the patient's information to verify the accuracy.

Documentation

The first step in evaluating asymptomatic or symptomatic patients is taking the complete medical history and performing a thorough visual inspection of the breast. Medical and family histories will provide information on symptoms or any abnormalities of the breast. Some developmental abnormalities are supernumerary nipples, accessory nipples, or polythelia. All of these refer to extra nipples formed along the mammary line or milk ridge (see Chapter 3). Patients may also develop accessory breast tissue in the axillary region or inferior to the breast. Occasionally, an extra nipple accompanies the accessory breast tissue. Because accessory breast tissue or nipples can also develop benign or malignant diseases, they should be well documented.

Other congenital abnormalities include asymmetry in breast size, shape, or position. Symptoms of note include inverted nipple or any discharge, such as bleeding from the nipple. The nipple can be normally inverted or flattened, but any recent change should be documented.

Medical history should also include any history of trauma to the breast or any breast surgery, including biopsies, breast augmentation, or reductions. Other breast symptoms are skin thickening, where the affected breast may feel heavier or larger than normal and the pores appear enlarged. Any unusual lumps in the breast or axilla or dimpling or puckering of the skin should be noted. Moles, eczemas, ulcers, and cysts can all be benign but must be documented on the clinical history sheet or marked with special radiopaque breast markers because they may or may not be visible on the final breast image and could affect the diagnosis (Fig. 2–4). Scar markers can be used to locate the site of past surgeries, eliminating the possibility of a misdiagnosis due to internal scaring. The patient's use of deodorant, talcum powder, and ointment should be checked because these sometimes have base metals, such as zinc or aluminum, that can cause artifacts on the radiograph that often mimic malignant breast calcifications. Documentation must also include any other health problems and risk factors for breast cancer and benign breast conditions (Fig. 2–5).

Many patients are not aware how important it is for the radiologist to compare the present mammography study with the old study. It is

---

Question types:

- **Open-ended questions (nondirected and nonleading):** used to start history taking, e.g., "Where is your pain?"
- **Facilitations:** used to encourage elaboration, e.g., a nod, 'yes', 'okay' or 'go on'.
- **Silence:** used to give the patient time to remember.
- **Leading questions:** should be avoided initially, e.g., "Is the pain in your left breast?"
- **Probing questions:** used to get details after the initial open-ended question, e.g., "How long have you had the pain?"
- **Repetition:** a rewording of the questions to clarify or verify information.
- **Summarize:** used to verify the accuracy.

The technologist should ensure that the patient understands the importance of prior studies.

The previous mammogram should be reviewed before a current study.

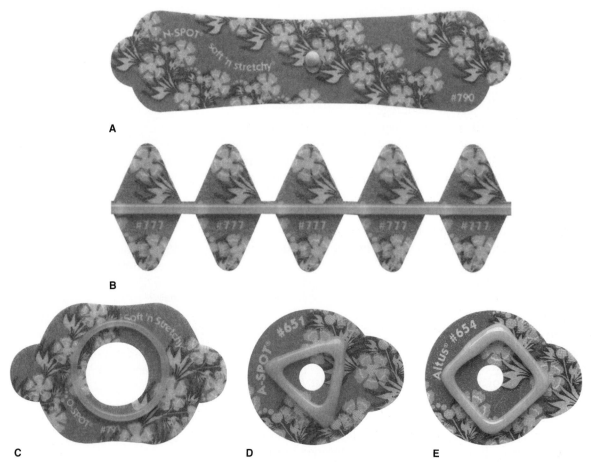

Figure 2–4. **(A)** Nipple marker (N-SPOT Designed for Digital Soft 'n Stretchy, 790). **(B)** Scar or surgical marker (S-SPOT Designed for Digital Soft 'n Stretchy, 777). **(C)** Mole or skin lesion marker (O-SPOT Designed for Digital Soft 'n Stretchy, 791). **(D)** Palpable lump marker (A-SPOT Light Image, 651). **(E)** Pain/nonpalpable area of concern marker (Altus Light Image, 654). (Used with permission from Beekley Medical.)

harder to miss subtle changes in the breast when there are old radiographs for comparison. The patient should always have the old radiographs available for comparison especially when they move from site to site to do their mammograms each year. In addition, the technologist should review the prior mammogram prior to performing the study.

BREAST EXAMINATIONS

There are two kinds of breast examination: the breast self-examination (BSE) and the clinical breast examination (CBE). Recommendations for CBE vary depending on the organization. Because of a lack of evidence supporting its benefits, many organizations have adopted a neutral or negative stance on the use of BSE or CBE.

The U.S. Preventive Services Task Force (USPSTF), the American Cancer Society (ACS), and the Canadian Taskforce on Preventive Health Care state that there is a lack of evidence to support BSE and CBE and no longer recommend the use of both for breast cancer screening. The USPSTF concludes that the current evidence is insufficient to assess the additional benefits and harms of the CBE and BSE. The American Academy of Family Physicians (AAFP) tends to follow the recommendations of the USPSTF.

Breast self-examination:

Recommended times:

   One week after the menstrual period ends

   If no period, the same day every month

Methods of assessment:

   Upright, to assess cancers in the upper half of the breast

   Supine, to assess cancers in the lower breast

Department of Radiology

Medical Record # _____   Date _____

Patient Name   _____   Age _____   Date of Birth _____

Telephone # (home) _____

        (work) _____

Referring Physician(s) name _____

Date of most recent breast examination by doctor _____   Regular period? Yes _____   No _____

Date of last period, or age at menopause _____   No. of pregnancies _____

Age at first pregnancy _____   Are you pregnant now? Yes _____   No_____

Date of last mammogram _____   Last mammogram done at _____

	YES	NO	Comments
Pain or discomfort in breast? Where? How long?	_____	_____	_____
Any lumps in your breast?	_____	_____	_____
If yes, is it a new lump, or how long has it been there?			
Are you taking hormone medication, or birth control pills? If yes, give name and state for how long			
	_____	_____	_____
Any weight gain or lost since last mammogram? If yes, how much?	_____	_____	_____
Any nipple discharge?	_____	_____	_____
Any breast surgeries? If yes, state type of surgery	_____	_____	_____
If yes, give date and indicate biopsy, implants, or reduction			
Circle any of the following that apply:			
Any history of breast cancer/chemotherapy/radiation treatment?	_____		
Any history of other cancers?	_____	_____	_____
Any family history of breast cancer?	_____	_____	_____

Patient signature _____

Technologist's initials _____   Comment _____

Radiologist's initials _____

Figure 2–5. Sample documentation sheet.

The American College of Obstetricians and Gynecologists (ACOG) and the American Medical Association (AMA) recommend starting CBE at age 40 years and continuing it annually thereafter.

The ACS does not currently recommend performing BSE, as it did in the past, and now recommends that women of all ages be told about the benefits and harms associated with BSE. The ACS and Susan G. Komen for the Cure Foundation still recommend that CBE be performed at least every 3 years starting between ages 20 and 39 years and annually starting at age 40 years.

The National Comprehensive Cancer Network (NCCN) uses the term *breast awareness* and recommends that women should be familiar with their breasts and promptly report any change to their health care provider. The NCCN suggests that periodic and consistent BSE may facilitate this breast awareness but does not advise a formal or specific education program.

Disadvantages of BSE include increased number of health care visits and increased number of benign biopsy results, leading to increased health care costs. When women detect changes in their breasts, they are more likely to seek professional help and more definitive testing to rule out cancer, which increases health care costs. In addition, they are likely to have increased feelings of anxiety and depression, often requiring counseling or treatment for what is often a benign condition.

The sensitivity of CBE ranges from 40% to 70%, and its specificity ranges from 86% to 99%. However, the sensitivity and specificity values of the BSE are difficult to determine, and most studies have not shown it to have any effect on breast cancer mortality. However, the advantages of BSE include allowing women to gain a sense of control over their health and to become comfortable with their own breasts. Additionally, BSE is a simple, noninvasive procedure that can be performed by nonmedically trained individuals. The ACS also states that BSE can also help women recognize normal versus abnormal breast tissue. Most societies still promote breast awareness and encourage women to know the look and feel of their breasts. Studies have suggested that women will find cancer while showering, bathing, or dressing rather than during a specific BSE. According to the National Breast Cancer Foundation, up to 70% of breast cancers are found by women performing their own BSE.

> The sensitivity for CBE ranges from 40% to 70%, and its specificity ranges from 86% to 99%.

Breast Self-Examination

To recognize and identify abnormalities, the breast must be examined in a systematic way at regular intervals. The best time to examine breasts is 1 week after the menstrual period ends when the breast are the least tender or swollen. For women not having regular menstruation, the BSE should be done on the same day every month.

Both breasts should be examined in the upright and the supine positions because a tumor in the upper half of the breast is easier felt when the woman is upright, whereas a tumor in the lower half of the breast is best felt when the woman is supine. It is essential that a woman learn the important points in any BSE (Fig. 2–6). The examination involves two main criteria:

- Looking for changes in the breast
- Feeling for changes in the breast

Looking for Changes

Breasts must be examined upright and supine. To examine the breasts erect, the woman should stand in front of a mirror or sit with both arms relaxed by the side. The mirror should be large enough to allow the woman to view both breasts clearly. The woman should check for indentations, retracted nipples, dimpling, or prolonged skin conditions, such as eczema. Other visual changes can include the development of

> **Looking for changes:**
> - Indentations
> - Retracted nipples
> - Dimpling
> - Prolonged skin conditions (e.g., eczema)
> - Unequal breasts
> - Changes in texture or color
> - Redness or scaling
> - Moles or scars

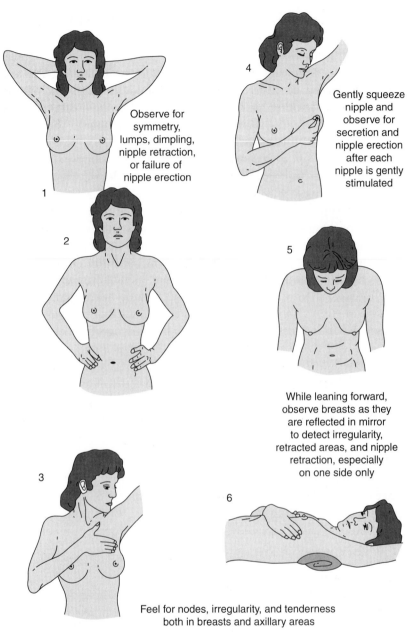

Figure 2–6. Conduction of a breast self-examination.

1. Observe for symmetry, lumps, dimpling, nipple retraction, or failure of nipple erection

4. Gently squeeze nipple and observe for secretion and nipple erection after each nipple is gently stimulated

5. While leaning forward, observe breasts as they are reflected in mirror to detect irregularity, retracted areas, and nipple retraction, especially on one side only

6. Feel for nodes, irregularity, and tenderness both in breasts and axillary areas

unequal breasts; changes in texture, color, or contour; redness; or scaling. Moles and scars should also be noted and recorded.

The entire procedure should then be repeated with both arms raised above the head. Next, the woman should place a hand on each hip, tense the chest muscles, and again look for changes or lumps. The woman can also lean forward, again checking for changes. Both nipples should be gently squeezed, as a check for discharge.

Feeling for Changes

The next step in any breast examination is feeling for changes. As with looking for changes in the breast, while feeling for changes, the woman should examine her breast in both the upright and the supine positions. While doing the upright examination, lotion or powder can be

Feeling for changes:

- Use of lotion or powder may be helpful
- Perform the examination in the shower if possible
- Examine one breast at a time
- Use the pads of the middle three fingers to apply different degrees of pressure
- Use more light to see lesions close to the skin and more pressure for deeper lesions
- For supine examination, use a pillow under one shoulder
- For upright examination, raise one arm

used to help the fingers glide across the breast. Some women also prefer to examine their breasts in the shower. Here, wet fingers will glide easily over the breast.

To begin the upright examination, start with one breast. The arm closest to the breast under examination should be raised. The pads of the fingers of the other hand are used to complete the breast examination, then the process is repeated for the other breast. The woman should use light, medium, and firm pressure with the pads of three fingers in either an up-and-down line, circular, or wedge pattern until the entire breast is checked. The woman has to decide which pattern she is most comfortable with.

For the supine examination, the woman should lie on her back with a pillow or other support under the shoulder closest to the breast under examination. The pads of the fingers of the other hand are then used to perform the examination. Regardless of the position employed, the entire breast and axilla should be examined using light, moderate, and firm pressure to locate lumps at different depths of breast tissue.

Up-and-Down Line Pattern

To examine the breast using the up-and-down line pattern, the woman should start in the axilla or most lateral aspect of the breast. Move the fingers downward little by little until the fingers have covered the entire breast. Start at the top again, a few centimeters medially, and repeat the downward movement of fingers. The entire process is repeated until the full breast is examined.

Circular Pattern

The circular examination method involves starting at the outermost edge and completing one full circle on the perimeter of the breast. Then go back to the starting point and move inward a few centimeters and complete the same process. Continue checking the entire breast by working toward the nipple.

Wedge Pattern

In an examination using the wedge pattern, the woman begins at the outer edge of the breast and moves inward to the nipple. Once at the nipple, the fingers are again placed at the outer edge of the breast a few centimeters from the starting point to begin another journey to the nipple. This process is repeated covering one wedge at a time until the entire breast is covered (Fig. 2–7).

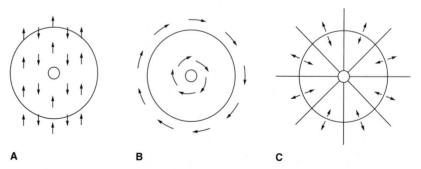

A **B** **C**

Figure 2–7. Breast examination patterns showing **(A)** vertical, **(B)** circular, and **(C)** wedge patterns of breast self-examination.

Clinical Breast Examination

The CBE can be performed to locate any lumps or suspicious areas and to examine the texture, size, and shape of the breast. This breast examination is a check of the breast by a qualified health professional. The CBE takes the same format as the BSE—looking and feeling for changes in the breast.

A thorough clinical examination can locate lumps or suspicious areas and changes in the nipples or skin of the breast. In a clinical breast examination, special attention is paid to lumps attached to the skin or deeper tissue. Because breast cancer leaves the breast through the lymph nodes, nodes in the axilla or above the clavicle must be checked for enlargement or firmness. Another point to note is that almost 75% of breast cancers are located in the upper quadrants of the breast, so the health professional will pay particular attention to this area.

After the medical and clinical examinations are completed, biopsies or breast imaging tests, such as mammography, can be performed as indicated (Fig. 2–8).

> **Clinical breast examination:**
> * A check of the breast by a qualified health professional
> * Involves looking and feeling for changes in the breast plus a check of the axilla and clavicle area

BREAST EXAMINATION GUIDELINES

Since the early 1970s, clinical studies have repeatedly shown that the periodic mammography screening of asymptomatic woman older than 50 years can reduce the mortality rate of breast cancer. Today, although not a perfect solution, the mammogram still remains the best detection tool in the treatment of breast cancer. However, in recent years the recommendations have become murky because various organization have proposed conflicting recommendations. The ACS suggests that other modalities are useful in imaging dense fibroglandular breast tissue, but it is also committed to mammography as offering the best screening solution for the general public. The ACS recommends that women who fall in high-risk categories—because of family history, genetic tendency, or other factors—be screened with MRI in addition to mammography.

Originally, the ACR recommended a 2-year screening schedule for women between 40 and 50 years; however, the earlier breast cancers are detected, the better are the treatment results. The ACS, therefore, suggested the following guidelines for routine mammography screening (Box 2–5).

In opposition to the ACS, the USPSTF, a group of independent health experts, is now suggesting that routine screening of the average-risk women should begin at age 50 years, instead of age 40 years. The USPSTF recommends screening every 2 years instead of every year. The USPSTF also gave a "C" grade to the recommendation of screening women before age 50 years. Some organizations fear that with this grade, screening may be selectively offered depending on individual circumstance and may no longer be covered by insurance. USPSTF suggests that screening should end at age 74 years. The USPSTF report suggested that for women older than 74 years, there is insufficient evidence regarding risks versus benefits, or that evidence of the benefits or harm of screening is lacking, of poor quality, or conflicting. The USPSTF does not recommend the BSE and CBE at any age and have concluded that it causes increased anxiety and stress in patients, and

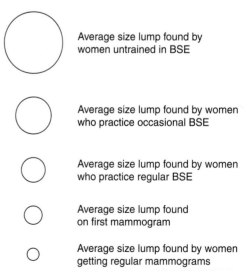

Average size lump found by women untrained in BSE

Average size lump found by women who practice occasional BSE

Average size lump found by women who practice regular BSE

Average size lump found on first mammogram

Average size lump found by women getting regular mammograms

Figure 2–8. Average size of lump found by women untrained in breast self-examination (BSE) compared with average size of lump found by women practicing occasional BSE.

Box 2–5. Breast Screening Guidelines

US Preventive Services Task Force

- Women aged 40–49 years
  - Individualized screening based on individual risk factors. The USPSTF recommends against routinely providing this service.
- Women aged 50–74 years
  - Screening begins
  - Mammogram every 2 years
- Women aged 75 years and older
  - No screening recommendations
- Breast self-examination (BSE) and clinical breast examination (CBE) are not recommended at any age

American College of Obstetricians and Gynecologists

- Screening every 1 or 2 years between 40 and 49 years of age, and every year after

American Cancer Society

- Women aged 40 to 44 years
  - Annual screening and CBE if desired
- Women aged 45 to 54 years
  - Mammograms recommended every year
  - CBE annually
- Women aged 55 years and older
  - Mammograms every 1 or 2 years
  - BSE and CBE not recommended.
- Screening should continue as long as a woman is in good health and is expected to live 10 more years or longer.
- Magnetic resonance imaging (MRI) or ultrasound screening only for women with high-risk factors or very dense breast tissue. For premenopausal women, the MRI should be performed within 7 to 15 days of the menstrual cycle. There are no timing recommendations for postmenopausal women.
- BSE is not recommended for women with an average risk, at any age.

American College of Radiology, the Society of Breast Imaging, and the National Comprehensive Cancer Network—an alliance of prominent cancer centers

- Yearly mammogram for all women starting at age 40 years
- Annual CBE starting at age 40 years

---

therefore its value has been downgraded. The USPSTF followed with recommendations against the teaching BSE, although they did not find sufficient evidence to recommend for or against performing clinical breast examination. The USPSTF and the National Cancer Institute (NCI) continue to review the benefits of regular screening. They consider the harmful effects of screening to include the dangers of false negatives, false positives, overdiagnosis, and radiation risks. The USPSTF recommends fewer screening mammograms because too many may lead to overdiagnosis and overtreatment. However, the ACS recommends more frequent screening based on the belief that more screenings are necessary to catch more tumors at an earlier stage.

BENEFITS VERSUS RISKS OF BREAST IMAGING

Mortality Reduction

Breast cancer in its early stages is asymptomatic. Since the advent of modern mammography in the late 1960s, studies have conclusively shown that regular screening mammograms will significantly reduce the mortality rate from breast cancer in women older than 50 years. Since 2002, some of these studies have been under attack, and questions have been raised about the efficacy and interpretive quality of mammography, especially for women younger than 50 years.

It is recognized that mammography cannot detect all breast cancers. Some cancers will present no imaging findings, and others are difficult if not impossible to detect when the woman has dense breasts. Mammography, therefore, should not be the only imaging solution when screening for breast cancer. Other modalities, such as ultrasound and MRI, are becoming increasingly useful in detecting cancers in dense fibroglandular breast tissue, especially in younger women who are at risk.

> **Entrance skin exposure versus glandular dose:**
> - Entrance skin dose: dose to the patient's skin
> - Glandular dose: dose to the glandular tissue of the breast

Risk from Radiation Exposure in Mammography Screening

Fear and misconception about the radiation dose in mammography are still a concern today, although the dose does not have the potential to produce any negative late effects. In general, patient dose from diagnostic imaging is measured as the entrance skin exposure (ESE) because it is easiest to measure. The ESE is also commonly referred to as the *patient dose*. A thermoluminescence dosimeter (TLD) can be taped to a patient's skin and used to take accurate readings during the exposure.

In mammography screening, however, because of the low-energy beam, the dose falls off very rapidly as the beam penetrates the breast. The biologic effect of a mammogram is, therefore, assumed to be more closely associated with the total energy absorbed by the glandular tissue of the breast. The *glandular dose* is, therefore, the dose of choice when calculating radiation doses associated with mammography.

The ESE for a typical single exposure during a mammogram with grid can be as high as 1000 mrad (10 mGy); however, the dose to the midline of the breast may only be 100 mrad (1.0 mGy). The ACR recommends that the average glandular dose on the mammogram should be no greater than 0.3 rad (300 mrad or 3 mGy) with a grid or 0.1 rad (100 mrad or 1 mGy) without a grid. With modern mammogram equipment, the patient will usually receive only approximately 0.1–0.2 rad (1–2 mGy) per projection. Despite these low glandular doses, negative misconceptions about radiation are still affecting the public's confidence in mammography screening.

> **Radiation dose in mammography:**
> ESE for a typical single exposure = 800–1000 mrad (8–10 mGy)
> - Glandular dose = 100 mrad (1.0 mGy)
> ACR recommends a glandular dose of:
> - 0.3 rad (300 mrad or 3 mGy) with a grid
> - 0.1 rad (100 mrad or 1 mGy) without a grid

Comparison of radiation dose:

Radiation is found in our atmospheres and in our soil. Radiation can also be inhaled or ingested through food or water

- The average person in North America receives about 3 mSv (300 mrem or 0.3 rem) per year of naturally occurring radiation
- Computed tomography of the abdomen gives about 10 mSv (1000 mrem) of radiation
- A lumbar spine x-ray gives about 3 mSv (300 mrem) of radiation
- We inhale 2 mSv (200 mrem or 0.2 rem) of radiation annually from radon in the atmosphere
- A single mammogram projection gives about 1 mSv (100 mrem) of radiation to the breast tissue
- We get about 0.1 mSv (10 mrem) from cooking with natural gas or from radon gas in the natural gas supply
- We get less than 0.1 mSv (10 mrem) per year from smoke detectors, which have radioactive elements
- A single chest x-ray projection gives about 0.08 mSv (8 mrem) of radiation
- During a mammogram, the radiation to the thyroid glands is about 0.04 mSv (4 mrem)
- A round-trip coast-to-coast airplane flight across the United States can give an average person 0.03 mSv (3 mrem) of radiation

One recent fear is the threat of thyroid cancer. The mediolateral-oblique projection gives the patient twice the amount of skin dose to the thyroid area as the craniocaudal projection; however, although the average skin dose to the thyroid is in the range of 0.022 to 0.039 rad (0.22 to 0.39 mGy) per projection, the dose falls off rapidly, leaving a tissue dose of approximately 0.004 rad (0.04 mGy). This is considered insignificant even compared with the average breast dose of 0.3 rad (3 mGy).

Because patients are unlikely to understand the scientific units for radiation, another way of looking at radiation is to take the effective dose from a single mammography projection and converting it into the amount of time it would take to accumulate the same effective dose from background radiation. The theory is that each would have the same potential for cancers and hereditary effects. Because the average background rate in the United States is approximately 300 mrem (3 mSv) per year, for mammography it would be approximately 4 months. The attraction of the effective dose is that a single value would then be used to compare one situation with another, or give an overall risk assessment. In reality, the various organs and tissues in the body receive different doses, and a complex calculation would be necessary to keep track of all the specific organs or tissue risk factors. In mammography, because only the breasts are irradiated, theoretically it could be possible to convert mammography doses to an effective dose using a female-based tissue-weighting factor. This method of explaining radiation is called *background equivalent radiation time* (BERT) and has been recommended by the National Council for Radiation Protection and Measurement (NCRP). This concept would help a patient who has difficulty understanding the radiation effects of mammography screening (see Box 2–6; JR Cameron [jrcameroo@facstaff.wisc.edu], Online Interview, 2003).

Box 2–6. Radiation Dose Made Easy

Using the background equivalent radiation time (BERT) to explain radiation to patients: The effective dose from mammography is approximately 1 mSv. The average background rate in the United States is approximately 3 mSv per year. Using the conversion, it would take approximately 4 months for the patient to accumulate the same effective dose of radiation as given in a single mammogram projection. This does not imply any risk; it simply emphasizes that radiation is a natural part of everyday life on the earth.

X-Ray Study	Effective Dose (mSv)	BERT (time to get dose from nature)
Single chest projection	0.08	10 days
Mammogram	1	4 months
Lumbar spine	3	1 year
UGI	3	1 year
Head CT	2	243 days
Abdominal CT	10	3.3 years

CT, computed tomography scan; UGI, upper gastrointestinal study.

Box 2–7. Radiation Doses from Typical Projections in Hologic Digital Mammography Unit

kVp	29	32
mAs	105	97.3
Entrance dose	14.8 mGy/1480 mrad	14.2 mGy/1420 mrad
Glandular dose	2.79 mGy/279 mrad	2.47 mGy/247 mrad
Compression	5 cm, 29 lb	6.9 cm, 31 lb
Filter	Molybdenum	Rhodium

Another aspect to consider is that there are no reported cases of breast cancer developing as a result of mammography even in the early days of breast imaging, with multiple examinations at doses many times higher than the current levels. Long-term studies have shown conclusively that regular mammography screenings result in a 30% reduction in breast cancer deaths in women older than 50 years. The mortality reduction benefits of mammography far outweigh any possible risks from screening. Also, patients should be made aware that the smaller the thickness of the area through which the radiation passes, the lower the dose of radiation the breast will receive. Simply by compressing the breast 1 cm more, the radiation dose to the patient can be reduced to almost half of its original value—yet another reason why the breast should be compressed.

Unfortunately, breasts of younger women—especially those with dense fibrous breasts—are more sensitive to radiation and usually receive a higher dose of radiation. Also, the skin dose is even higher during magnification mammography. At these high doses, there have been reports of skin erythema from repeated magnification imaging (Box 2–7). Technologists should always practice ALARA (As Low As Reasonably Achievable) and avoid unnecessary radiation to the patient by avoiding repeats. By carefully studying positioning and knowledge of all the routine and supplementary projection, the technologist will be able to ensure that the minimum number of images is taken during each patient's mammography examination (Fig. 2–9).

Skin dose in magnification mammography can often be much higher than 1000 mrad (higher than the skin dose during a lumbar spine x-ray).

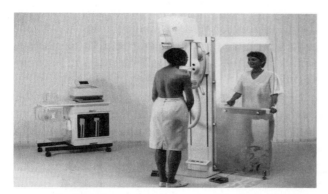

Figure 2–9. Positioning for a screening mammogram. (Used with permission from GE Medical Systems.)

BREAST CANCER SCREENING

The field of breast cancer detection is still evolving. The reality is there is no screening tool available that is 100% effective. There is still no known cure for breast cancer, so a woman's best strategy is early detection, reducing her known risks, and the judicious use of adjunctive modality imaging. The 5-year survival rate is lower for women whose cancer is diagnosed at a more advanced stage. Five-year survival rates are also lower for younger women because researchers suspect that younger women have tumors that are more aggressive and less responsive to therapies.

Depending on the interpretation of the mammogram, women can pursue many adjunctive imaging options. Women should also pursue adjunctive testing if they are at a greater risk for breast cancer or have dense breast tissue. In addition, women should seek educational resources. All women should be aware of the signs and symptoms of breast cancer. Breast cancer in its early stages is totally asymptomatic, but as the cancer grows there may be changes in the breast, such as thickening, swelling, skin irritation, or distortion, and nipple symptoms including spontaneous discharge, erosion, inversion, or tenderness.

The steep decline in the mortality rate from breast cancer over the past few years has been credited solely to mammography screening, but the benefits of emerging technologies cannot be overlooked. These technologies are offering improvement in detection and treatment options and, coupled with breast cancer research, may well serve the ultimate goal of eradicating breast cancer as a feared and dreaded disease.

ADVANTAGES AND DISADVANTAGES OF MAMMOGRAPHY SCREENING

The mammogram is the single most effective tool in early breast cancer detection. However, it is now acknowledged that the mammogram is not 100% effective. Cancer is generally visualized as a white area within the background density of the breast. If the background optical density is "black," as in fatty tissue, the cancer will be easily seen; but if the background optical density is "white," as in dense breast tissue, the cancer will be harder to detect. The sensitivity of the mammogram will be dependent on density, plus the age and hormone status of the patient.

Mammography also tends to understate the multifocality of a lesion. A perfect breast image is also critical in diagnosing breast cancer. Positioning is a very important part of imaging. If there is inadequate compression or if the breast tissue is not imaged because of poor positioning, a cancer can be missed.

Because of these stated disadvantages, mammography screening is generally acknowledged to have a 10% to 15% miss rate. As the patient ages, the optical density of the breast generally decreases; thus, the diagnostic sensitivity of the mammogram will increase. However other imaging modalities and computer-aided diagnosis (CAD) are often used as part of the workup to avoid unnecessary breast biopsies. These will be discussed in later chapters and include ultrasound, MRI, and molecular imaging, including positron emission mammography with

---

The mammogram is not 100% effective in detecting breast cancer.

Disadvantages of mammography imaging:

- Cancer is visualized as a white area within the background density of the breast
- Sensitivity will depend on breast density, patient age, and hormone status
- Mammography tends to understate the multifocality of a lesion
- Inadequate compression and poor positioning will affect interpretation

fluorodeoxyglucose (PEM–FDG), breast scintigraphy, and lymphoscintigraphy or sentinel node mapping.

RISK FACTORS FOR BREAST CANCER

Cancer is caused by DNA damage (i.e., mutations) in genes that regulate cell growth and division. When cells acquire mutations in specific
genes that control proliferation, such as protooncogenes or tumor
suppressor genes, these changes are copied with each new generation of
cells. Later, more mutations in these altered cells can lead to uncontrolled proliferation. This uncontrolled growth with limitless cell divisions and survival is termed *cancer*. Any factor that causes an increase
in the level of mutation or does not suppress the growth of abnormal
cell division will be considered a risk factor for cancer. Risk factors will
therefore increase a person's chance of getting the disease. There are
two general risk models that can be used to estimate risk factors. One
determines the likelihood that an individual is a carrier of a *BRCA1,
BRCA2*, or other gene mutations while the other calculates a woman's
absolute risk of developing breast cancer over a lifetime.

- Breast cancer risk estimates calculate the absolute breast cancer risk, which is the chance or probability of developing breast cancer in a defined age interval.
- Risk factors will therefore increase a person's chance of getting the disease. The average lifetime risk for breast cancer for women is about 12%.

Types of Risk Assessments

The average lifetime risk for breast cancer for women is about 12%.
Breast cancer risk estimates calculate the absolute breast cancer risk,
which is the chance or probability of developing breast cancer in a
defined age interval. An accurate risk assessment would correctly predict average risk in a group of women with the same risk factors and
age. A risk assessment tool used by the NCI is the Gail Model 2, which
estimates a woman's risk for developing breast cancer during the next
5-year period and up to age 90 years (lifetime risk) based on the woman's
age and the risk factor information provided. For comparison, the tool
will then calculate 5-year and lifetime risk estimates for a woman of the
same age who is at average risk for developing breast cancer. Lifetime
risk estimates are higher than 5-year age interval estimates because
breast cancer risk increases with years at risk.

The Breast Cancer Risk Assessment Tool is based on a statistical
model known as the *Gail Model*, which is named after Dr. Mitchell Gail,
Senior Investigator in the Biostatistics Branch of NCI's Division of
Cancer Epidemiology and Genetics. The model uses a woman's own
personal medical history, her own reproductive history, and the history
of breast cancer among her first-degree relatives (mother, sisters,
daughters) to estimate her risk for developing invasive breast cancer
over specific periods of time. The original Gail Model was developed in
1989 from data derived from the Breast Cancer Detection and Demonstration Project (BCDDP), which was a joint NCI and ACS breast cancer
screening study that involved 280,000 women aged 35 to 74 years, and
from NCI's Surveillance, Epidemiology, and End Results (SEER) program. It was used for screening purposes to estimate the probability of
developing breast cancer over a defined age interval.

The Gail Model 2 includes an assessment of the history of first-
degree family members and also predicts the risk for invasive breast
cancer. The Gail Model 2 is often used in breast cancer prevention

trials; however, it is most accurate for non-Hispanic white women and tends to overestimate risk in younger women. Estimates for African American women were based on data from the Women's Contraceptive and Reproductive Experiences (CARE) study and from SEER data. CARE participants included 1607 women with invasive breast cancer and 1637 without. However, the model underestimates risk in African American women, especially those with previous biopsies, and does not fully calculate the association between breast density and family history when calculating breast cancer risks. Estimates for Asian and Pacific Islander women in the United States were based on data from the Asian American Breast Cancer Study (AABCS) and SEER data. AABCS participants included 597 Asian and Pacific Islander women with invasive breast cancer and 966 women without breast cancer. Another major limitation of the Gail Model 2 is the lack of assessment of male relatives, and there is no accounting for the age of onset of breast cancer.

Overall, although the model has been "validated" for white women, the model needs further validation for Hispanic women and other minority subgroups. The tool should not be used by women who have a diagnosis of breast cancer or a medical history of breast cancer. This includes history of ductal carcinoma *in situ* (DCIS) or lobular carcinoma *in situ* (LCIS), which increases the risk for developing invasive breast cancer. More specific methods of estimating risk are appropriate for women known to have a strong family history of breast cancer related to mutations in the *BRCA1* or *BRCA2* gene. The tool is also not appropriate for women with other hereditary conditions, such as Li-Fraumeni syndrome, that increase a woman's risk for breast cancer.

This risk estimator is not foolproof. Some woman have a high-risk estimate and do not develop breast cancer. Also, some women with low-risk estimates will develop breast cancer. The Breast Cancer Risk Assessment Tool is updated periodically as new data and research become available.

Other risk assessment tools are more appropriate for women who have a history of certain medical conditions, including DCIS or LCIS. The International Breast Intervention Study (IBIS) Breast Cancer Risk Evaluation Tool is a breast cancer prediction model incorporating familial and personal risk factors that was developed by Jonathan Tyrer, Stephen W. Duffy, and Jack Cuzick to estimate the risk for invasive breast cancer or DCIS. The program assumes that there is a gene predisposing to breast cancer in addition to the *BRCA* genes. The woman's family history is used to calculate the likelihood of her carrying an adverse gene, which in turn affects her likelihood of developing breast cancer.

Women with a known mutation in either the *BRCA1* or *BRCA2* gene can use the Breast and Ovarian Analysis of Disease Incidence and Carrier Estimation Algorithm (BOADICEA) model to estimate their breast cancer risk. BOADICEA is a computer program that is used to calculate the risks for breast and ovarian cancer in women based on their family history. It is also used to calculate the probability that they are carriers of cancer-associated mutations in the *BRCA1* or *BRCA2* gene. The major drawback from this breast cancer risk assessment is that no other genetic factor is checked.

Major risk factors are relatively outside of a woman's control.

Minor risks can be hormonal related, related to the environment, or relatively within a woman's control.

Having a risk factor or even several risk factors does not mean that a person will get the disease.

Box 2–8. Breast Cancer Risk Assessment Tool

The Breast Cancer Risk Assessment Tool is an interactive tool designed by scientists at the National Cancer Institute (NCI) and the National Surgical Adjuvant Breast and Bowel Project (NSABP) to estimate a woman's risk for developing invasive breast cancer.

1. Does the woman have a medical history of any breast cancer or of ductal carcinoma *in situ* (DCIS) or lobular carcinoma *in situ* (LCIS), or has she received previous radiation therapy to the chest for treatment of Hodgkin's lymphoma?

2. Does the woman have a mutation in either the *BRCA1* or the *BRCA2* gene, or a diagnosis of a genetic syndrome that may be associated with an elevated risk for breast cancer?

3. What is the woman's age? (This tool only calculates risk for women 35 years or older.)

4. What was the woman's age at the time of her first menstrual period?

5. What was the woman's age at the time of her first live birth of a child?

6. How many of the woman's first-degree relatives (mother, sisters, daughters) have had breast cancer?

7. Has the woman ever had a breast biopsy?

   a. How many breast biopsies (positive or negative) has the woman had?

   b. Has the woman had at least one breast biopsy with atypical hyperplasia?

8. What is the woman's race/ethnicity?

   a. What is the sub-race/ethnicity?

Source: http://www.cancer.gov/bcrisktool/Default.aspx

The biggest risk factor for breast cancer is gender, that is, being female.

Risk factors in breast cancer can be categorized as major or minor. Major risk factors are those beyond a woman's control; these factors cannot be changed. Minor risks can be lifestyle choices, hormonal related, or cancer-causing factors in the environment. Major risk factors carry a significantly higher risk for breast cancer than minor risk factors. However, having a risk factor or even several risk factors does not mean that a person will get the disease (Box 2–8).

Major Risk Factors

The biggest risk factor for breast cancer is gender, that is, being female. Breast cancer is a disease that primarily occurs in females, although some men develop the disease.

Gender is therefore considered the biggest risk factor for breast cancer (Box 2–9). Other major factors include the following:

1. *Aging.* Age is an important risk factor because as a woman ages the incidence of cancer increases. Approximately 18% of breast cancer is diagnosed in women in their 40s, whereas 77% of women are older than 50 years when they are diagnosed. Women younger than 30 years account for only 0.3% of breast cancer cases, and women in

---

Box 2–9. Risk Factors for Breast Cancer

Risks Factors Estimated (Relative Risk)

- Major risk factors
  1. Female gender
  2. Aging (>4)
  3. Genetic risk factors, including positive *BRCA1/BRCA2* mutation (>4)
  4. Family history of breast cancer
     - Two or more relatives (mother, sister) (>5)
     - One first-degree relative (>2)
     - Family history of ovarian cancer in women younger than 50 years (>2)
  5. Dense breast tissue (3–4)
  6. Personal history of cancer or breast cancer (3–4)
     - Breast biopsy with atypical hyperplasia (4–5)
     - Breast biopsy with lobular or ductal carcinoma *in situ* (8–10)
- Minor risk factors
  1. Never having given birth (2)
  2. Having a first child when older than 30 years (2)
  3. No breastfeeding (2)
  4. Early menarche (<12 years of age) (2)
  5. Late menopause (>51 years of age) (1.5–2)
  6. Use of hormone replacement therapy—combined estrogen/progesterone (1.5–2)
  7. Current or recent use of oral contraceptives (1.25)
  8. Adult weight gain (1.5–2)
  9. Sedentary lifestyle—no exercise (1.3–1.5)
  10. Alcohol consumption (1.5)

The probability of developing breast cancer is:

- 0.06% for a woman at age 20 years
- 1.44% at age 40 years
- 2.39% at age 50 years
- 3% for a woman at age 70 years

---

their 30s account for approximately 3.5% of cases. It is estimated that the probability of developing breast cancer is only 0.06% for a woman at age 20 years, and that percentage jumps to 1.44% at age 40 years, 2.39% at age 50 years, and more than 3% for a woman at age 70 years. All women have a lifetime risk of approximately 12%, meaning that women face a 12% probability of developing breast cancer. It is from this calculation that the 1 in 8 statistical risk number was developed.

2. *Genetic risk factors.* Approximately 5% to 10% of breast cancer cases are hereditary. A woman with a hereditary risk for breast cancer may carry the *BRCA1*, *BRCA2*, or *PALB2* genes. *BRCA1* and *BRCA2* are tumor suppressor genes. They keep cancer tumors from forming. If the genes are mutated, cells will no longer die when they should, and a cancer will grow. Some mutations are inherited, whereas others are caused by exposure to radiation or mutation-inducing chemicals. Mutations can also occur spontaneously as a result of mistakes that are made when a cell duplicates its DNA molecules before cell division.

Women with a normal gene have a protective effect because the genes make proteins that keep cells from growing abnormally. Therefore, with the mutated gene, there is an increased susceptibility to breast cancer. Approximately 1% of the general population carries the mutated gene, and molecular tests are available to identify someone with the *BRCA* mutations responsible for inherited forms of breast cancer.

Women with an abnormal *PALB2* gene have a 33% to 58% lifetime risk for developing breast cancer. *PALB2* interacts with *BRCA2*, and biallelic mutations in *PALB2* (also known as *FANCN*), similar to biallelic *BRCA2* mutations, have been linked to Fanconi's anemia and breast cancer predisposition.

For women who have a *BRCA1* or *BRCA2* abnormality, the risk for developing breast cancer is about 3 to 7 times greater than that of a woman who does not have the mutation. The lifetime risk is between about 40% and 85%. These women also have significant elevation of the risk for ovarian cancer: 16% to 44%, compared with just under 2% for the general population. Men with *BRCA* abnormalities are considered to have a higher lifetime risk for male breast cancer, especially if the *BRCA2* gene is affected. One study found that men with a *BRCA2* mutation have a 7% lifetime risk for developing breast cancer. They are also at increased risk for developing prostate cancer.

Other genetic mutations are much more rare. These include *ATM*, *TP53*, *CHEK2*, *PTEN*, *CDH1*, and *STK11* mutations. The *ATM* gene functions to repair DNA damage. In some families, inheriting one mutated copy of this gene carried a risk for breast cancer. The *TP53* gene gives instructions for making a protein called p53 that helps stop the growth of abnormal cells. Inherited mutations of this gene cause Li-Fraumeni syndrome and are associated with an increased risk for breast cancer as well as some other cancers, such as leukemia, brain tumors, and sarcomas. The *CHEK2* gene is also a tumor suppressor that is activated in case of DNA damage. Mutation of this gene increases breast cancer risks. The *PTEN* gene normally helps in the regulation and growth of cells. Mutations result in Cowden's syndrome, which is rare and often results in increased benign and malignant tumors in the breast, digestive tract, thyroid, uterus, and ovaries. The *CDH1* gene and the protein associated with it work to ensure that cells within tissues are bound together. Mutations are linked to an increased risk for invasive lobular breast cancer and for gastric, colorectal, thyroid, and ovarian cancers. Defects in the *STK11* gene can lead to Peutz-Jeghers syndrome. The result is pigmented spots on the lips and mouths, polyps in the urinary and gastrointestinal tracts, and a higher risk for many types of cancer, including breast cancer.

3. *Family history of breast cancer.* Breast cancer risk is higher among women whose close first-degree relatives have this disease. A first-degree relative can be from either the mother's or the father's side of the family, but the closer the relative, the higher the risk factor. For example, a mother, sister, or daughter with breast cancer will double the risk. The more first-degree relatives with the disease, the greater are the risks. Having two first-degree relatives with the disease increases the risk by threefold. Women are also at a higher risk if the

> **As a woman ages, the incidence of breast cancer increases:**
> - <30 years old: 0.3% of breast cancer cases
> - 30s: 3.5% of breast cancer cases
> - 40s: 18% of breast cancer cases
> - >50 years old: 77% of breast cancer cases

> About 5% to 10% of breast cancer cases are hereditary.

More than 80% of women who develop breast cancer have no family history of the disease.

Digital mammography is a more effective tool for imaging dense breast tissue.

breast cancer occurs in a relative before 50 years of age. Women with a family history of breast cancer in the male members of the family are also at risk. Although having a family history will increase breast cancer risks, more than 85% of women who develop breast cancer have no family history of the disease.

4. *Personal history of cancer or breast cancer.* A woman with cancer in one breast has a greater risk for developing a new cancer in the other breast. Also, women who have had the chest area radiated because of a previous cancer treatment, such as for Hodgkin's disease (or non-Hodgkin's lymphoma), are at a slightly increased risk for breast cancer.

5. *Dense breast tissue.* The breast is made up of a varying mixture of fatty tissue and glandular or dense tissue. Dense breast is not a disease; however, it is genetic, and often a woman will be told of her breast density after the first mammogram. Although the variation of fat and glandular tissue is due to genetics, the amounts can also vary with age. As a woman ages, the breast slowly loses its glandular tissue in a process called *involution.* The glandular tissue is replaced with fat. Glandular tissue, therefore, predominates in younger women, whereas fatty tissue predominates in older patients. However, weight gain at any age will increase the fat content of the breast, leaving the breast less dense. Similarly, as a woman loses weight, the fat content of the breast will decrease resulting in an overall denser breast (Fig. 2–10).

If the breast is mainly fatty tissue with very few lobules or little glandular tissue to begin with, even with weight loss, it will not fit the definition of a dense breast. If the breast is dense to begin with, gaining weight will disperse the dense tissue, and the breast will seem less dense, but overall it will still be considered a dense breast.

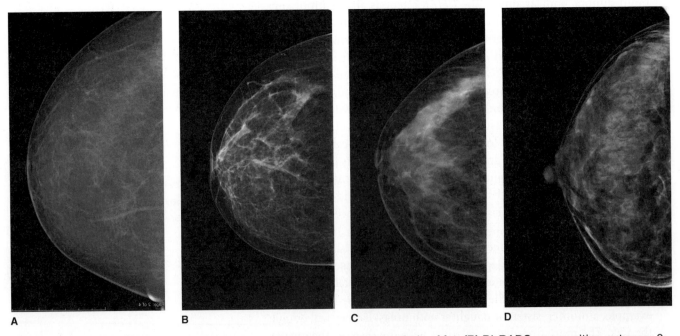

Figure 2–10. (A) BI-RADS composition category 1 composed almost entirely of fat. **(B)** BI-RADS composition category 2: scattered fibroglandular densities. **(C)** BI-RADS composition category 3: heterogeneously dense breast tissue. **(D)** BI-RADS composition category 4: extremely dense breast tissue.

Recent studies are leaning toward density being a factor in breast cancer risks in younger women. One reason is that breast cancers tend to develop in the glandular tissue of the breast. Increased glandular tissue often means denser breast tissue. Therefore, if a woman has dense breast tissue, she is at higher risk for developing breast cancer than a woman of the same age with fatty breast. Other studies also show that women with dense breast tissue face a higher risk for missed breast cancer if the mammogram is the only screening tool used. Analog mammography does not image dense breast tissue very well. Digital mammography imaging is a more effective tool for imaging dense breast tissue. Women with dense breasts also have other options. Although mammography is still the gold standard for breast imaging, women with dense breast tissue should utilize any of the adjunctive modalities, including breast ultrasound and breast MRI. Ultrasound and MRI are not effective screening tools for breast cancer; however, when used with digital mammography, they improve the odds and can help to prevent missed breast cancer. Breast density notification is now the law in about 27 states. These mandates vary by state but serve to inform women who have undergone mammography about the risks posed by breast density. The laws give women the necessary information to decide if they should take further action if they have dense breast tissue.

> Breast density is a factor in breast cancer risks in younger women.

Minor Risk Factors

Many minor breast cancer risk factors are associated with hormonal changes in the woman's body. Studies have suggested that although estrogen, one of the main female hormones, does not appear to directly cause the DNA mutations that trigger the development of human cancer, estrogen does stimulate cell proliferation, therefore promoting cancer growth. In other words, if breast cells already possess a DNA mutation that increases the risk for developing cancer, these cells will proliferate (along with normal breast cells) in response to estrogen stimulation. All factors that affect the reproductive hormones in a woman's body will also increase risks for breast cancer.

These risks include the following:

> **Minor risks:**
> - Not giving birth or having first child after 30 years of age
> - Not breastfeeding
> - Having early menarche or late menopause
> - Using hormone replacement therapy or birth control

1. Women who have not given birth and those having a first child after 30 years of age are at increased minor risk. Pregnancy is thought to have a protective effect against breast cancer because during the pregnancy the woman is not menstruating and there are no regular cycles of increasing estrogen levels. Therefore, it follows that the longer a woman goes without getting pregnant, the greater are her risks for breast cancer.
2. Some studies suggest that breastfeeding has a protective effect, especially if the feeding continues for 1½ to 2 years. A possible reason is that breastfeeding will reduce the total number lifetime menstrual cycles. Therefore, it follows that not breastfeeding will increase the risk for breast cancer.
3. Women who started menstruating at an early age (before age 12 years) or went through menopause at a late age (after age 55 years) are at an increased risk for breast cancer. As long as a woman is menstruating, there is a risk for mutating genetic material. The greater the amount of menstrual cycles in the woman's lifetime,

> The use of hormone therapy can have a significant impact on breast cancer risks. There are two main types of hormone therapy:
> - Combination hormone therapy (HT or HRT) with estrogen and progesterone, used by women who still have a uterus;
> - Estrogen only therapy (ET) or estrogen replacement therapy (ERT).

the greater would be the risk for breast cancer. Breast cancer risk rises by approximately 3% for each year of delayed menopause. Delayed generally means onset after age 51 years, which is the average age of menopause for women in the United States.

4. Another risk associated with hormone use includes the use of birth control pills. Studies have found that women using oral contraceptives (birth control pills) have a slightly higher risk for breast cancer than women who have never used them. After the pills are stopped, this risk seems to go back to normal over time. The exception would be for women carrying the *BRCA1* or *BRCA2* gene and women younger than 21 years. Younger women have an increased risk if the contraception pill is taken for more than 10 years. Older birth control pills were associated with a slightly increased risk for developing breast cancer. Depo-Provera is an injectable form of progesterone that is given once every 3 months as birth control. Women currently using birth control injections seem to have an increase in breast cancer risk, but it appears that there is no increased risk in women 5 years after they stop receiving the injections.

5. The use of hormone therapy (HT) has been proved to have a significant impact on breast cancer risks. There are two main types of hormone therapy: combination hormone therapy or hormone replacement therapy (HRT) with estrogen and progesterone, used by women who still have a uterus; and estrogen therapy (ET) or estrogen replacement therapy (ERT). It is thought that the combination of estrogen and progesterone in HRT may increase the risk for developing cancer. It may also increase the chances of dying from breast cancer. Combining estrogen and progesterone for 5 years appears to double breast cancer risks. Estrogen therapy alone lowers the risk for breast cancer but increases the risk for endometrial cancer. Most studies found that the risks associated with HRT return to normal within 5 years of stopping the treatment. ET taken for longer than 10 years has been found to increase the risk for ovarian and breast cancer in some studies. The administration of hormones generally increases the proliferation of glandular tissue, and because cancers are harder to detect in dense breast, the increase in glandular tissue caused by the HRT reduces the effectiveness of the mammogram. There have been no studies of bioidentical hormone therapy, which is the use of versions of estrogen and progesterone with the same chemical structure as those found naturally in the body. However, bioidentical hormones are thought to have the same health risks as other types of hormone therapy.

> On an average, mammography detects only about 90% of breast cancers.
>
> Mammography is more accurate in postmenopausal women compared with perimenopausal women.

FACTS VERSUS MYTHS AND MISCONCEPTIONS

Abortion. The National Cancer Institute (NCI) has concluded that having an abortion or miscarriage does not increase a woman's subsequent risk for developing breast cancer. The theory, however, is that terminating an unintended pregnancy sacrifices the protective effects of a term delivery. The effect would be the same as delayed childbearing; therefore, the net effect would be an increased risk.

Accuracy of the mammogram. Many women believe totally in the mammogram. On an average, the mammogram will detect only approximately 90% of breast cancer. Mammography is also more accurate in postmenopausal women compared with menopausal women. The accuracy or lack of accuracy of mammograms can actually have an impact on the doctor–patient relationship. The authority figure of the physician is well established, and many women will undergo a mammogram on the recommendation of their physician even if they themselves are not entirely convinced that the procedure is necessary. The most meaningful factor affecting many women's decision to have a mammogram is often the physician's recommendation. However, the stability of the doctor–patient relationship is one of trust. The patient will not respect the opinion or advice of a physician she cannot trust, and patients will question a physician's advice when they or their friends are repeatedly subjected to false-positive readings or unnecessary biopsies. Judicious use of the latest technology in reducing the level of false-positive reports should be the goal for all interpreting physicians. Also, because radiology is a dynamic field with technology that is constantly changing, proper initial training and regular in-service or continuing education courses are essential so that both the radiologists and technologists can keep the patient informed.

Alcohol. Some studies have suggested that moderate alcohol consumption can cause a modest increase in breast cancer risk. Heavy drinking, however, will double the risk. Heavy drinking is considered more than four drinks per day. Studies have also shown that drinking 3 to 4 glasses of wine per day will increase the risk for breast cancer by 37%. Women who drank the equivalent of a half glass of wine a day were 6% more likely to develop breast cancer. Women who drank a glass or two a day faced a 21% increased risk for breast cancer. The theory is that alcohol increases the circulation levels of estrogen, or changes the way the body metabolizes estrogen, which in turn causes cancer to grow. The risks were found to be greater in menopausal women. Menopausal women who drank a half glass of wine daily increased their risk for breast cancer by 18%.

In addition to the link between alcohol and estrogen, research has identified several genes that are involved with alcohol metabolism. One enzyme, alcohol dehydrogenase (ADH), is responsible for breaking down alcohol into acetaldehyde, which is carcinogenic in animals. Postmenopausal women with a variation in the gene that codes for ADH were at greater risk.

Aspirin. The *Journal of Breast Cancer Research* has shown that aspirin on a daily basis may lower risk factors. The study showed a reduction of 16% in ER-positive breast cancers. Approximately 75% of breast cancers are ER-positive. These cancers have receptor for the female hormone estrogen on their surface. Aspirin can block an enzyme called *cyclooxygenase* (COX), an activity that disrupts breast cancer development by reducing the amount of estrogen produced in the body.

Augmented breast. Having a silicone or saline implant can cause scarring of the breast tissue; however, neither type of implant is linked to any significant increase in breast cancer risks. In the past, breast cancers were missed in the augmented breast because of the limits of the technology; however, newer methods of mammogram imaging techniques (discussed in Chapters 7–9) and current

Approximately 90% of women who develop breast cancer do not have a first-degree relative with the disease.

The greatest risk for breast cancer is gender, that is, being female.

The next is increasing age, and this is the most important reason the patient should be returning for yearly mammograms despite past negative results.

imaging technology, including ultrasound and MRI, can be used to complement mammography screening and are effective in diagnosing cancer in the augmented breast. Recently, the U.S. Food and Drug Administration (FDA) announced a possible association between saline and silicone gel–filled breast implants and anaplastic large cell lymphoma (ALCL), a very rare type of cancer. ALCL can appear in different parts of the body, including the lymph nodes and skin. The data suggest that patients with breast implants may have a very small but significant risk for ALCL in the scar capsule adjacent to the implant. The risk is associated with textured silicone implants.

Biopsy. Women with a previous biopsy have a slightly increased risk for breast cancer. Some conditions have higher risks than others. Nonproliferative lesions linked to an overgrowth of breast tissue do not affect breast cancer risks. These include fibrosis and simple cysts, fibrocystic changes, mild hyperplasia, nonsclerosing adenosis, benign phyllodes tumor, papilloma, fat necrosis, duct ectasia, periductal fibrosis, squamous and apocrine metaplasia, epithelial-related calcifications, lipoma, hamartoma, hemangioma, neurofibroma, adenomyoepithelioma, and mastitis. Other proliferative lesions without atypia increase a woman's risk for breast cancer slightly. These include unusual or ADH, fibroadenoma, sclerosing adenosis, having several papillomas (papillomatosis), radial scars, atypical lobular hyperplasia (ALH), and LCIS.

Breast cancer. Breast cancer does occur primarily in women; however, men do have breast tissue and can get breast cancer. Breast tissue in women is constantly undergoing changes due to fluctuating hormones in the women's body. Women also have more breast tissue than men, and the male breast generally does not have lobules. These are some of the reasons that women are more at risk for breast cancer than men; however, approximately 1600 cases of male breast cancer are diagnosed every year in the United States.

Breast cancer risk. Many women believe that if they do not have a first-degree relative with breast cancer they need not worry about getting the disease. The fact is, however, that approximately 90% of women who develop breast cancer do not have a first-degree relative with the disease. The greatest risk for breast cancer is gender, that is, being female. The next is increasing age, and this is the most important reason the patient should be returning for yearly mammograms despite past negative results. Only 17% of breast cancers are diagnosed in women younger than 30 years, whereas 50% of all breast cancers are diagnosed in women 65 years and older.

Chemicals. Some studies suggest certain chemicals in the environment have estrogen-like properties and could affect breast cancer risks. Chemicals of this nature are known as endocrine disruptors. They interfere with the synthesis, secretion, transport, binding, action, or elimination of natural hormones in the body, including estrogen. The effects of exposure to endocrine disruptors are permanent, especially if exposure occurs during development. Chemicals known as endocrine disrupters are bisphenol A (BPA), often used to make plastic, including baby bottles; dichlorodiphenyldichloroethylene (DDE), which is a breakdown product of DDT, an organochlorine insecticide that is now banned in the United States; polychlorinated

Endocrine disruptors are chemicals in the environment that have estrogen-like properties and could affect breast cancer risks.

They can interfere with the synthesis, secretion, transport, binding, action, or elimination of natural hormones in the body, including estrogen.

biphenyls (PCBs), a class of organic compounds used in transformers and coolants that has also been banned; and diethylstilbestrol (DES), a nonsteroidal estrogen used in the past to prevent spontaneous abortions. It was discovered that children whose mothers used DES have an increased risk for breast and vaginal cancers.

Dense breast tissue. Dense breasts often mean that the breast has more glandular and fibrous tissue and less fatty tissue. Studies show that women with dense breasts have a risk of breast cancer that is 1.2 to 2 times that of women with average breast density. Breast density is affected by age, menopausal status, use of hormone therapy, pregnancy, and genetics.

Deodorants and antiperspirant. Some research studies have focused on the preservatives used in antiperspirants and deodorants. In laboratory tests, these preservatives, para-hydroxybenzoic acids or parabens, have been shown to act like estrogen in the body. Some scientists, therefore, fear that, like estrogen, they could increase breast cancer risks, especially because of the proximity of the underarm to the breast. Parabens are also found in many other cosmetic products, such as shampoo and makeup. They are also found in medication and food items. A second theory suggests that antiperspirants, which inhibit the sweat glands, will prevent toxins from leaving the body, and a third suggestion is the increased breast cancer risk for women using a blade (nonelectric) razor with an underarm antiperspirant or deodorant, or for women using an underarm antiperspirant or deodorant within 1 hour of shaving with a blade razor. The NCI is not aware of any conclusive evidence linking the use of underarm antiperspirants or deodorants and the subsequent development of breast cancer. The FDA, which regulates food, cosmetics, medicines, and medical devices, also does not have any evidence or research data that ingredients in underarm antiperspirants or deodorants cause cancer.

> The FDA, which regulates food, cosmetics, medicines, and medical devices, does not have any evidence or research data that ingredients in underarm antiperspirants or deodorants cause cancer.

In 2002, the results of a study looking for a relationship between breast cancer and underarm antiperspirants and deodorants did not show an increased risk for breast cancer in women who reported using an underarm antiperspirant or deodorant. The results also showed no increased breast cancer risk in women who reported using a blade (nonelectric) razor and an underarm antiperspirant or deodorant, or for women who reported using an underarm antiperspirant or deodorant within 1 hour of shaving with a blade razor. These conclusions were based on interviews with 813 women with breast cancer and 793 women with no history of breast cancer.

A different study examining the frequency of underarm shaving and antiperspirant or deodorant use among 437 breast cancer survivors was released in 2003. This study found that the age at development of breast cancer was significantly lower in women who used these products and shaved their underarms more frequently. Furthermore, women who began both of these underarm hygiene habits before 16 years of age were diagnosed with breast cancer at an earlier age than those who began these habits later. Although these results suggest that underarm shaving with the use of antiperspirants or deodorants may be related to breast cancer, it does not demonstrate a conclusive link, and a serious limitation of the study was the absence of a control

The ACS recommends 45 to 60 minutes of physical activity 5 days or more per week.

The mammogram should be scheduled 7 to 10 days after the start of the period when breasts are least sensitive.

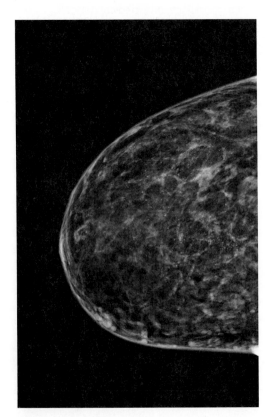

Figure 2–11. Fibrocystic changes can mimic malignant calcifications.

group without breast cancer. That means that there could be a simple explanation for the findings: younger women use antiperspirant and shave more often than older women.

Diet. The link between diet or weight and breast cancer can often give conflicting results. Being overweight is often linked with a higher risk for cancer, especially if the weight gain took place after menopause. Before menopause, the ovaries produce most of the body's estrogen, but after menopause, estrogen comes from fat. This means that having more fat will raise the estrogen levels in the body. Studies also suggest that weight gain as an adult carries a higher risk than being overweight since childhood. In addition, weight gain in the waist will affect the risk more than the same amount of fat in the hips and thighs. Higher body fat also leads to higher insulin levels, which is also linked to breast cancer during adulthood. However, the link is complex, and researchers are still not sure of the relationship. Because a high percentage of breast tumors are ER-positive, excess estrogen can potentially increase the growth rate of the cancer.

Exercise. Physical activity helps to reduces breast cancer risk. A study from the Women's Health Initiative showed that as little as 1¼ to 2½ hours per week of brisk walking reduced a woman's risk by 18%. Walking 10 hours a week reduced the risk a little more. The ACS recommends that adults get at least 150 minutes of moderate-intensity or 75 minutes of vigorous-intensity activity each week (or a combination of these), preferably spread throughout the week.

Fibrocystic breast. In the past, fibrocystic breast was thought to be a breast disease; however, studies have proved conclusively that fibrocystic changes are not cancerous and do not affect a woman's breast cancer risk.

Gender. Males are less likely to get breast cancer. Only about 1600 men get a diagnoses of breast cancer each year in the United States. However, the incidence of breast cancer in males has increased 25% in the past 20 years, and one study suggested that breast cancer kills 25% of men who develop it. Some reasons given were the lack of knowledge about the disease, the tendency to ignore breast lumps, and reporting for treatment when the cancers are invasive or have already metastasized (Fig. 2–11).

Implants. Silicone breast implants can cause scar tissue to form in the breast but are not linked to breast cancer. There have been a few reports of a rare disease, ALCL, which has been linked to breast implants. ALCL affects cells in the immune system and can be found around the breast implant or in the skin or lymph nodes. It is not a type of breast cancer. ALCL was first noted by the FDA in 2011. It is associated with textured surface implants. Symptoms are fluid buildup, hardening or a mass around their implants, swelling, and redness around the breast implants. ALCL is described as slow-growing cancer that is treatable when detected early.

Injury to the breast. Trauma or injury to the breast has not been proved to cause breast cancer. When breast tissue is injured, fat necrosis can occur, and as the body heals, a scar can develop. The scar tissue is sometimes mistaken for cancer on the mammogram. Also, some women discover a tumor because they are examining their breast after an injury. There is, however, no link between the two.

Mammograms are painful. Mammograms are uncomfortable, but the compression takes less than a minute. Mammograms are more uncomfortable for women with sensitive breasts. To reduce the sensitivity of the breast, the mammogram should be scheduled 7 to 10 days after the start of the period when the breasts are least sensitive. Although eliminating caffeine from a woman's diet for a month before the mammogram could reduce the sensitive of the breast, it is nearly impossible to remove caffeine completely because it is found in so many products, including tea, chocolate, soda, and coffee, just to mention a few.

Mastectomy. Approximately 8% to 10% of women will have a recurrence in the scar after a mastectomy. Also, before the mastectomy, the cancer could have already spread to the other areas of the body. Most research suggests that patients should have visual inspection of the site by an oncologist, or that imaging should be done yearly within the first 5 years of the mastectomy.

Obesity. Overweight or obesity, especially after menopause, increases breast cancer risk. Before menopause, most of the estrogen in the body comes from the ovaries, with very little linked to fat tissue. After the ovaries stop estrogen production at menopause, most of a woman's estrogen comes from fat tissue. Fat tissue after menopause can therefore raise estrogen levels and increase the chance of getting breast cancer. The risk appears to be increased for women who gained weight as an adult, but may not be increased among those who have been overweight since childhood. Also, excess fat in the waist area may affect risk more than the same amount of fat in the hips and thighs. Another factor to consider is that women who are overweight tend to have higher blood insulin levels, which has been linked to some cancers, including breast cancer.

Prophylactic mastectomy. A prophylactic mastectomy is the surgical removal of the breast for women with a very high risk for breast cancer. Some women opt to have their breasts removed to reduce their risk for developing the disease. Some prophylactic mastectomies might add years to a woman's life; however, the risk for breast cancer remains, and having the breast removed does not guarantee that the cancer will not develop in the small amount of breast tissue remaining after the surgery. Some studies show that having a prophylactic mastectomy can reduce the breast cancer risks by up to 90%. However, a woman is at risk as long as there is breast tissue present, and breast tissue that extends to the neck, under the arms, and into the chest wall is not removed during a mastectomy.

Race. Some studies confirm that white women have a slightly higher breast cancer risk than black women, but black women are more likely to die from this cancer. Studies show a more complicated picture, however, and in fact race actually plays a very minimal part in mortality rates. Factors such as the environment, lifestyle, staging, and treatment are often more critical in the prognosis of the disease. Further investigation often determines that the high death rate among blacks can be attributed to finding cancers at a later stage, limited treatment options, and poor treatment planning; moreover, socioeconomic factors and discrimination can and will have a profound influence in both the detection and treatment of breast cancer. Research also shows that although the majority of breast cancers in United States

> **Prophylactic mastectomy:**
> - Could reduce the breast cancer risks by up to 90%.
> - However, risk exists as long as there is breast tissue present (breast tissue extends to the neck, under the arms, and into the chest wall and is not removed during a mastectomy).

Factors affecting breast cancer diagnosis and treatment among minorities:

- Environment
- Discrimination
- Socioeconomic factors
- Finding cancers at a later stage
- Lifestyle
- Limited treatment or wrong treatment
- Poor staging and treatment planning

are ER-positive, this is often not the case in black women, who tend to develop the more aggressive HER2 subtype of cancer.

One study showed that blacks were more likely than whites to be diagnosed with large tumors and distant-stage disease. Another clinical study found that women were 1.4 to 3.6 times more likely to be diagnosed at a later and more likely fatal stage of the disease if they were of African American, American Indians, Hawaiian, Indian, Pakistani, Mexican, Puerto Rican, or South and Central American descent. According to the study, women of African American, Mexican, or Puerto Rican descent were 20% to 50% more likely to receive initial treatment that was inappropriate or inadequate, and women of African American, American Indian, Hawaiian, Vietnamese, Mexican, Puerto Rican, or South and Central American descent had a 20% to 200% greater risk for dying after a breast cancer diagnosis. In fact, for all minorities, the proportion of disease diagnosed at advanced stage and with larger tumor size is greater than for whites.

Risk factors. A risk factor gives only the probability of breast cancer, and even a strong factor is not a certainty. Also, studies have shown that 80% of women who get breast cancer have no identifiable risk factors.

Radiation. Past chest wall irradiation for cancer treatment, such a Hodgkin's disease or non-Hodgkin's lymphoma, can lead to the development of bilateral breast cancer in patients as young as 30 years. The risk varies with the patient's age when irradiated, with the highest risk associated with radiation during adolescence when the breasts are still developing. One study however, suggested that if chemotherapy was given, it stopped the production of ovarian hormones, which would lower the risks. Radiation treatment after 40 years of age does not seem to increase breast cancer risk.

Symptoms versus no symptoms. Despite the increased public awareness about breast cancer, there is still a general feeling among many women that any illness will manifest in a physical or visible form. Unfortunately, breast cancer in its early state is symptomless. As the cancer grows, some symptoms may appear. These can include lumps in the breast, nipple discharge, thickening of the breast skin, puckering or dimpling of the breast, and inverted nipples or a discharge from the nipple. Women who closely monitor their health status in terms of diet and routine exercise or who adopt other preventative health measures often feel more in control of their health. These women are likely to have routine mammograms because doing so is an extension of their health-conscious behavior and need to have control of their lives. Women least likely to obtain a mammogram are those who are less health conscious or who have uncertain attitudes about health care and preventive medicine, including mammography.

Tobacco smoke. Both mainstream and secondhand smoke contain chemicals that, in high concentrations, cause breast cancer in rodents. Recent studies have shown that heavy smoking over a long time is linked to a higher risk for breast cancer. In some studies, the risk was highest in women who started smoking before they had their first child. Studies have shown that chemicals in tobacco smoke reach breast tissue and are found in breast milk of rodents. The 2014 U.S. Surgeon General's report on smoking concluded that there is "suggestive but not sufficient" evidence that smoking increases the risk for breast cancer.

Summary

- Technologists should be able to provide compassionate and proficient care. Patients coming for a mammogram often need reassurance, counseling, and education in addition to professional service.

- In ensuring patient compliance, the technologist needs to be understanding, informed, and compassionate. Often, the technologist is the only link between the radiologist and the patient, and action or lack of action can have repercussions in conveying information or helping in diagnosis.

- Communication should take place before, during, and after the examination. It should be face to face and as informal as possible, and should include listening and facilitator skills to encourage comments and to relax and perhaps educate the patient. There is a lower chance of suboptimal imaging of the pectoral muscles when the patient is relaxed, and a relaxed patient is more likely to confide details of personal history that can affect the diagnosis, to identify sensitive breasts and the reason for the sensitivity, or to reveal myths and misconceptions that can be addressed.

- Patients should consult with their physician on the mammography screening guide that is best for them. The screening can depend on the type of breast tissue (fatty versus dense), the patients' family history or personal history, or the genetic makeup of the patient. In addition, imaging patients can request that their physician perform a CBE.

- Before any mammography examination, the technologist needs to take a detailed clinical history. The clinical history documentation includes any family or personal history of breast cancer; history of breast surgery, including biopsies, lumpectomies, augmentations, or reductions; and any history of breast trauma. Technologists will also need to document clinical symptoms, such as skin thickening, unusual lumps, dimpling or puckering, moles, eczema, ulcers, and nipple changes or nipple discharge. Other breast abnormalities, such as accessory breast or nipples, should be documented.

- The radiation dose from a one-projection mammogram is a measure of the glandular dose. It should be no more than 0.3 rad (3 mGy) with a grid and no more than 0.1 rad (1 mGy) without a grid. The entrance skin dose, however, is generally much higher, in the range of 8 to 1.4 rad (800–1400 mrad or 8–10 mGy).

- Breast imaging with mammography is still the gold standard. However, breast screening is becoming more individualized, and adjunctive imaging modalities are often used in the workup of a patient based on clinical findings or a suspicious lesion in the breast in order to avoid an unnecessary breast biopsy. Despite better imaging techniques, improved contrast, increased resolution, reduced radiation dose to the breast, and increased sensitivity of imaging techniques, myths and misconceptions about breast imaging abound. It is the technologist's responsibility to educate the patient and be an informed and responsible imager.

REVIEW QUESTIONS

1. List three reasons for taking and documenting the patient's medical history.

2. What is the best way to combat misconceptions about mammography?

3. State some of the benefits of communication.

4. Give examples of nonverbal communication.

5. Give an example of an open-ended question.

6. How does aging affect breast cancer risk?

7. What is the recommended guideline for routine mammography screening as suggested by the ACS?

8. What is the maximum average glandular dose for a single mammography projection as recommended by the ACR?

9. State two disadvantages of screening mammography.

10. What is the difference between a major and a minor risk factor?

11. What is the biggest risk factor for breast cancer?

12. Name four other major risk factors for breast cancer.

13. Name two minor risk factors for breast cancer.

14. How does exercise affect breast cancer risk?

15. How can trauma to the breast lead to the discovery of a cancer?

16. What is the concern, for children in particular, to avoid radiation to the chest and breast tissue?

17. What is the Gail Model 2?

18. Why is the mammogram so dependent on breast density?

CHAPTER QUIZ

1. The glandular dose is:
   - (A) The dose received on the skin of the breast
   - (B) Associated with dose to the radiosensitive cells of the breast
   - (C) The significant background dose recorded in ultrasound
   - (D) A record of the dose to the gonads

2. Factors that should be reported on the patient's medical history documentation because of their significant impact on diagnosis include:
   - (1) History of breast trauma
   - (2) Painful lumps or masses
   - (3) Sudden nipple retraction
   - (A) 1 only
   - (B) 1 and 2 only
   - (C) 2 and 3 only
   - (D) 1, 2, and 3

3. On an average, how accurate is the mammogram?
   - (A) 100%
   - (B) 90%
   - (C) 40%
   - (D) 20%

4. If the doctor writes, "Patient has a palpable lump in her left breast," this information is:
   - (A) Objective
   - (B) Informative
   - (C) Subjective
   - (D) Noninformative

5. Which of the following is an example of nonverbal communication?
   - (A) Speaking
   - (B) Nodding
   - (C) Talking
   - (D) Whispering

6. Which of the following are positive skills that the radiographer can use when obtaining a patient history?
   - (1) Leading questions
   - (2) Probing questions
   - (3) Repeating information
   - (A) 1 and 2 only
   - (B) 1 and 3 only
   - (C) 2 and 3 only
   - (D) 1, 2, and 3

7. Why is obesity considered a risk factor for breast cancer?
   - (A) A person who is overweight is more likely to have dense breast
   - (B) Overweight, especially after menopause, can result in increased estrogen in the body
   - (C) Overweight, especially before menopause, can result in increased estrogen in the body
   - (D) Overweight, especially after menopause, can result in decreased estrogen in the body

8. Minor risk factors of breast cancer are associated with:
   - (1) Use of HRT
   - (2) Family history of breast cancer
   - (3) Nulliparity
   - (A) 1 and 2 only
   - (B) 2 and 3 only
   - (C) 1 and 3 only
   - (D) 1, 2, and 3

9. Major risk factors for breast cancer include:
   - (1) Use of birth control pills
   - (2) Age
   - (3) Genetic risks
   - (A) 1 and 2 only
   - (B) 2 and 3 only
   - (C) 1 and 3 only
   - (D) 1, 2, and 3

10. A risk factor is any:
    - (A) Factor within a person's control
    - (B) Factor outside a person's control
    - (C) Significant factor that increases a person's chance of getting a disease
    - (D) Factor that decreases a person chance of getting a disease

BIBLIOGRAPHY

American Cancer Society (ACS). *Breast Cancer Facts and Figures 2009–2010.* Atlanta, GA: American Cancer Society; 2010.

American Cancer Society (ACS). Experimental breast imaging tests. http://www.cancer.org/treatment/understandingyourdiagnosis/examsandtest-descriptions/mammogramsandotherbreastimagingprocedures/mammograms-and-other-breast-imaging-procedures-newer-br-imaging-tests. Accessed September 2016.

American Cancer Society (ACS). Learn about cancer: breast cancer. http://www.cancer.org/cancer/breastcancer/index. Accessed September 2016.

American College of Radiology (ACR). ACR supports Senate bill to extend protection of women's access to annual mammography. http://www.acr.org/About-Us/Media-Center/Press-Releases/2016-Press-Releases/20160610-ACR-Supports-Senate-Bill-to-Extend-Protection-of-Womens-Access-to-Annual-Mammography. Accessed September 2016.

Bushong SC. *Radiologic Science for Technologists—Physics, Biology and Protection.* 10th ed. St. Louis, MO: Mosby; 2012.

Centers for Disease Control and Prevention. Breast cancer rates by race and ethnicity. http://www.cdc.gov/cancer/breast/statistics/race.htm. Accessed September 2016.

National Cancer Institute. Breast cancer risk assessment tool. http://www.cancer.gov/bcrisktool/about-tool.aspx. Accessed August 2016.

National Cancer Institute. Cancer topics. https://www.cancer.gov/types/breast/hp. Accessed September 2016.

National Center for Biotechnology Information. Screening for breast cancer: U.S. Preventive Services Task Force recommendation statement. http://www.ncbi.nlm.nih.gov/pubmed/19920272%20?tool=bestpractice.bmj.com. Accessed September 2016.

Peart O. *Lange Q&A: Mammography Examination.* 2nd ed. New York, NY: McGraw-Hill; 2008.

Rahman N, Rahman Seal S, Thompson D, et al. *PALB2*, which encodes a BRCA2-interacting protein, is a breast cancer susceptibility gene. *Nat Genet.* 2006;39:165–167. http://www.nature.com/ng/journal/v39/n2/abs/ng1959.html. Accessed August 2016.

Venes D, Biderman A, Adler E. *Taber's Cyclopedic Medical Dictionary.* 22nd ed. Philadelphia, PA: FA Davis; 2013.

Anatomy, Physiology, and Pathology of the Breast | 3

Keywords and Phrases
The Breast
Location of the Breast
Surface Anatomy of the Breast
Localization Terminology
Deep Anatomy of the Breast
Vascular Circulation
    Arterial Circulation
    Venous Drainage
The Lymphatic Drainage of the Breast
Nervous Supply
Breast Tissue Composition
Factors Affecting Tissue Composition
    Hormone Fluctuation
    Menopause
    Hormone Therapy
    Lactation
    Parity
The Male Breast
    General Description
    Male Breast Cancer—Signs and Symptoms
    Risk Factors for Breast Cancer in Males
    Diagnosis, Treatment, and Prognosis
Summary
Review Questions
Chapter Quiz

Objectives

On completion of the chapter, the reader will be able to:

1. Describe the exact location of the breast
2. Describe the skin of the breast
3. Describe the surface anatomy of the breast, including the position of the nipple and areola

4. Discuss the various location terminologies used when describing lesions in the breast
5. Discuss the makeup of the deep anatomy of the breast
6. Identify the arteries and veins supplying and draining the breast
7. Discuss the significance of lymph drainage from the breast
8. Describe the terminal ductal lobular unit (TDLU)
9. Discuss the factors that will affect breast tissue composition, including hormone fluctuation, menopause, and lactation
10. Describe the male breast
11. Identify the signs and symptoms of male breast cancer
12. List the risk factors for breast cancer in males
13. Describe the primary treatment for male breast cancer

KEYWORDS AND PHRASES

- **Acinus** (plural, acini) is the milk-producing element of the breast, located within the alveolar glands.
- **Chemotherapy** is an adjunctive therapy involving the use of drugs to treat cancer that may have spread throughout the body.
- **Cooper's ligaments** are also called *suspensory ligaments*. They are strands of connective tissue that run between the skin and deep fascia to support the lobes of the breast. They start at the most posterior portion (base) of the breast and extend outward to attach to the anterior fascia of the skin.
- **Fibrocystic breast** is a common condition in females. The breast develops thickening of the alveoli in addition to one or more cysts. Cysts are fluid-filled sacs. Fibrocystic breasts are often lumpy, swollen, and tender, especially a week before menstruation begins.
- **Gynecomastia** represents enlargement of the male breast due to growth of the ducts and supporting tissue.
- **Hormones** are substances that are released by a gland or organ in one part of the body, then travel through the bloodstream to regulate the activity or secretions of cells in another part of the body.
- **Inframammary fold or crease** is the most inferior aspect of the breast where it meets the anterior abdominal wall.
- **Involution** describes a process that begins at menopause as the breast loses its supportive tissue to fat, to produce a smaller breast or a larger, more pendulous one.
- **Klinefelter's syndrome** was first identified in 1942 by Dr. Harry Klinefelter. Patients with this syndrome are often referred to as *XXY males* because, instead of having 46 chromosomes, they have 47 chromosomes with two Xs and one Y. The doubling of the X chromosome results in symptoms such as enlarged breast, small testes, sparse facial and body hair, and inability to produce sperm. XXY males often experience deficits in specific cognitive functions, including language, concept formation, and problem solving, similar to that of dyslexic children. Many of the symptoms can be reversed by regular testosterone treatments starting at puberty and continuing for life.
- **Lactiferous ducts** are ducts within the breast conveying milk to the nipple.

- **Lactiferous sinus** is a widening or ampulla in the connecting duct immediately behind the nipple orifice. The ampulla is a pouch-like structure that holds milk (when it is being produced).
- **Latissimus dorsi muscle** is a broad triangular muscle located on the inferior part of the back. It arises from the lower half of the spine and the iliac crest and inserts at the front of the upper part of the humerus. It is the widest and most powerful muscle of the back.
- **Lobule** is also called the *terminal ductule lobular unit* (TDLU). It lies at the end of the branching tree of ducts. The TDLU holds the alveolar glands, which are the milk-producing elements of the breast.
- **Lymph node** is a small kidney-shaped organ, which is a part of the lymphoid tissue. Lymph nodes lie at intervals along the lymphatic vessel and are seen mammographically as ovoid structures with a lucent center or hilum, generally a few centimeters in length.
- **Montgomery's glands** are also called *glands of Montgomery* or *areolar glands*. These are large modified sebaceous glands on the areola of the breast. They secrete a fatty fluid that lubricates and protects the nipple during nursing and also have smooth muscles that cause the nipples to become erect when stimulated.
- **Morgagni's tubercles** are named for the Italian pathologist Giovanni Morgagni (1682–1771), and refer to protrusions on the nipple.
- **Palliative treatment** refers to treatment used to relieve or alleviate the pain or symptoms of a disease without actually curing the disease.
- **Pectoralis major muscle** is a large, thick, fan-shaped muscle located on the anterior portion of the upper parts of the chest, immediately posterior to the breast. The pectoralis consist of a group of at least six muscle fibers originating on the medial clavicle and lateral sternum and stretching upward and laterally to insert in the bicipital groove of the humerus.
- **Pectoralis minor muscle** is a thin, flat, triangular muscle lying below the pectoralis major. It arises from the upper margins and outer surfaces of the third, fourth, and fifth ribs and passes superior and laterally to insert in the medial border of the upper surface of the coracoid process of the scapula.
- **Prostate cancer** is a malignant tumor of the prostate gland. The prostate surrounds the neck of the bladder and urethra in the male.
- **Retromammary space** is an area of fatty tissue separating the breast from the pectoral muscle.
- **Sebaceous glands** are oil-secreting glands in the skin.
- **Serratus anterior** is a large, flat, fan-shaped muscle. It originates on the surface of the upper eight or nine ribs on the lateral aspect of the chest and inserts along the entire anterior length of the medial (vertebral) border of the scapula.
- **Sternum** is a narrow flat bone in the anterior and middle portion of the thoracic cavity.
- **Tail of Spence** is the upper outer quadrant of the breast, which extends toward the axilla. It is also the thickest portion of the breast.
- **TDLU or lobule** is the terminal ductal lobular unit and is the most distal end of the ductal system. It can be further divided into the extralobular terminal ductal (ETD) unit and the intralobular ductal (ITD) unit.

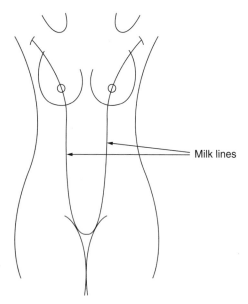

Figure 3–1. The milk line.

Location of the breast:
- Upper extent
  - Second or third rib
- Lower extent
  - Sixth or seventh rib
- Medial extent
  - Mid-sternum
- Lateral extent
  - Mid-axillary line (latissimus dorsi muscle)

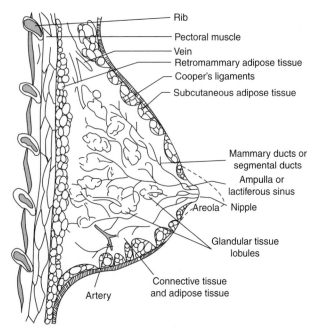

Figure 3–2. Anatomy of the breast.

THE BREAST

The breast is a mammary gland that is actually a modified sudoriferous (sweat) gland, or modified sebaceous gland that produces milk. The mammary glands of all mammals are the milk-producing organs used to provide nourishment for the newborn. In humans and in many cultures, the mammary glands are also regarded as sexual assets, the loss or scarring of which can have devastating psychological implications for women.

LOCATION OF THE BREAST

The breast extends vertically from the clavicle (the second or third rib) to meet the abdominal wall at the level of the sixth or seventh rib and horizontally from the mid-sternum to the mid-axillary line (the latissimus dorsi muscle); however, breast tissue can form anywhere along the milk ridge or line, also called the *mammary line* (Fig. 3–1). This line extends, one on each side of the body, from the armpits in the axilla to the groin region of the body.

Breast tissue lies on the superficial fascia, which covers the pectoralis major, serratus anterior, and external oblique muscles. Separating the breast from the pectoral muscle is a layer of adipose tissue and connective fascia referred to as the *retromammary space*. Deep to the pectoralis major is the pectoralis minor muscle, a thin, flat, triangular muscle. The retromammary space is filled with a layer of adipose or fatty tissue as opposed to the supporting and connective tissue (stroma), blood vessels, and various ductal structures that make up the glandular and fibrous tissues of the breast. The retromammary space is seen on the mediolateral oblique projection in mammography imaging, but only on about 20% of the craniocaudal projections.

SURFACE ANATOMY OF THE BREAST

In the female, breasts are accessory glands of the reproductive system with the function of secreting milk for nourishment of the newborn (Fig. 3–2). The female breast is spherical in shape, and its size varies with age, menstrual cycle, and lactation. Weight gain and loss can also affect breast size, and any rapid increase in size often results in the appearance of stretch marks. In females, hormonal stimulation generally results in the developed breast at puberty. The breast is loosely attached to the fascia covering the pectoralis major muscle (Fig. 3–3A and B). The breast is most secure at the superior and medial aspects. The lateral borders of the breast and the inferior aspect are the most mobile portions. The inframammary fold (or line or crease) represents the point of attachment of the most inferior portion of the breasts at the chest wall.

The *skin* of the breast, like the skin of the body, is filled with sweat glands, sebaceous glands (oil glands), and hair follicles that open to form the skin pores. The skin of the breast is thickest at the base, thinning as it approaches the nipple.

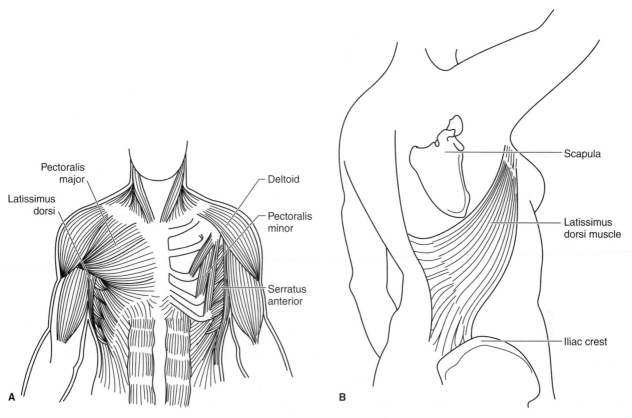

Figure 3–3. Muscles around the breast. **(A)** Viewed from the anterior; **(B)** viewed from the posterior.

The *nipple* lies at the center point of the breast. The nipple itself contains multiple crevices, within which are 15 to 20 orifices, or collecting ducts, that transfer milk from the lactiferous ducts. Normally, the nipple can be either flattened or inverted, both unilaterally and bilaterally; however, a nipple that suddenly becomes inverted or flattened can indicate malignancy.

The *areola* is the smooth, circular, darkened area surrounding the nipple and contains many small protrusions on its surface called *Morgagni's tubercles*. These tubercles are formed by openings of the ducts of Montgomery's glands, which are specialized sebaceous-type glands found on the areola. Montgomery's glands usually become more prominent during pregnancy and lactation. The glands secrete a fatty lubricant, which protects the nipple during nursing. The pigmentation of the areola is partly dependent on estrogen levels. In younger women, it is more prominent, but it tends to fade at menopause. Estrogen use at any age, however, will increase pigmentation of the areola.

Surface anatomy of the breast:
- Skin
  - Sebaceous glands
  - Hair follicles
- Nipple
  - Center point of the breast
  - 15–20 orifices
- Areola
  - Circular, darkened area around the nipple
  - Montgomery's glands

LOCALIZATION TERMINOLOGY

Each breast can be divided into four quadrants: the upper outer quadrant (UOQ), upper inner quadrant (UIQ), lower outer quadrant (LOQ), and lower inner quadrant (LIQ) (Fig. 3–4). The exact locations within the quadrants can be represented by viewing each breast separately as a clock face.

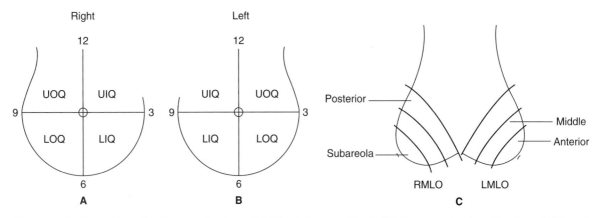

Figure 3–4. Breast localization terminology. **(A)** Clock-face method; **(B)** four-quadrant method; and **(C)** region method. The breast can be described as the face of a clock; using the four-quadrant method; or in terms of region, such as "subareola." Note that the 4-o'clock position on the right breast represents the LIQ but on the left breast would indicate the LOQ.

> **Deep anatomy:**
> - Lobules with alveolar glands
> - Mammary ducts
> - Lactiferous sinus (ampulla)
> - Lactiferous duct (collecting duct), exits to the nipple

The upper outer quadrant, which extends toward the axilla, is known as the axillary tail, tail of the breast, or tail of Spence. Similar to a clock, there are four clock positions (12, 3, 6, and 9 o'clock). The 4-o'clock position on the right breast (LIQ) will represent the 8-o'clock position on the left (LIQ).

The breast can also be divided into four regions. The posterior region is closest to the chest wall. The middle region refers to the middle of the breast, and the anterior region is located behind the nipple. There is also the subareolar region, which refers to the region immediately behind the areola. Lesions can be described in relation to the nipple (e.g., subareolar, or behind the nipple).

DEEP ANATOMY OF THE BREAST

The breast is composed of approximately 15 to 20 *lobes*. Each lobe consists of tree-like patterns of radiating ducts, exiting on the nipple. At the very beginning of each ductal structure is a widened area, the lobule, which holds the functional, milk-producing glands, called *alveolar glands*. Using the analogy of a tree, one lobe in the breast would be a single tree, with the lactiferous or collecting duct exiting at the nipple and represented by the root, the mammary ducts represented by the branches, and the lobules represented by the leaves. On the average female breast, each lobe has an independent opening on the nipple and can have 10 to 100 lobules held together by connective tissues, blood vessels, and branching ducts.

The *lobule* is also sometimes called a *ductule* or the *terminal ductal lobular unit* (TDLU) (Fig. 3–5). The TDLU is lined with a single layer of epithelial cells and a peripheral layer of myoepithelial cells. The TDLUs are further divided into the extralobular terminal duct (ETD), a small duct leading into the terminal ductules, and the intralobular terminal duct (ITD), located at the end of the terminal ductules. The ETD is surrounded by elastic tissue and lined by columnar cells. The ITD has no surrounding elastic tissue and contains cuboidal cells. The ITDs hold the alveolar glands. Within the alveolar glands is the *acinus*, which is the milk production element of the breast.

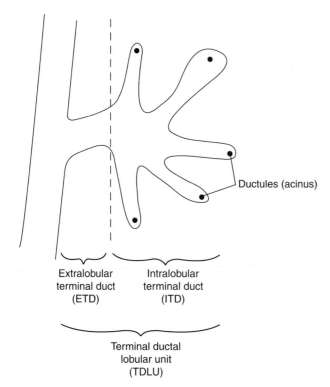

Ductules (acinus)

Extralobular terminal duct (ETD)

Intralobular terminal duct (ITD)

Terminal ductal lobular unit (TDLU)

Anatomy of the TDLU:
- Extralobular terminal duct (ETD)
  - Elastic tissue, lined by columnar cells
- Intralobular terminal duct (ITD)
  - Cuboidal cells
- Alveolar gland
  - Acinus: milk-producing elements

Figure 3–5. Diagram of the terminal ductal lobular unit (TDLU). The TDLU increases and decreases in size and number depending on menstrual cycles, pregnancy, lactation, and hormone use. The TDLUs are responsible for milk production, and it is here that most cancers originate. The ductal system ends at the TDLU. The unit is further divided into the extralobular terminal unit (ETD) and intralobular terminal unit (ITD).

Milk produced in the alveoli is propelled out by contraction of the myoepithelial cells lining the alveoli. The milk enters the ductal system, called the *segmental* or *mammary ducts*. The diameter of the ducts gradually increases in size. As the ducts approach the nipple, it widens to form a dilated structure called the *ampulla* or *lactiferous sinus*. In the past, it was assumed that the lactiferous sinus stored the milk before draining it into the last portion of the ductal structure—the *lactiferous duct*. This assumption is questioned by recent research. The lactiferous duct is a short segment of duct immediately behind the nipple (Fig. 3–6).

In the immature breast, the ducts and alveoli are lined by a two-layer epithelium of cells.

After puberty, the epithelium of the lobules proliferates, becoming multilayered and forming three alveolar cell types: superficial (luminal) A cells, basal B cells (chief cells), and myoepithelial cells. The innermost layer of the epithelium has the apical surface, which lines the lumen of the ducts. The other epithelial layer, the basal surface, adheres to the adjacent basement membrane. The basement membrane is a thin sheet of fiber that provides support and acts as a semi-permeable filter under the epithelium. The basal surface of the epithelium consists of myoepithelial cells, which are arranged in a branching, star-like fashion around the alveoli and excretory milk ducts. During lactation, it is the contraction of these cells that helps propel milk toward the nipples (Fig. 3–7). Beneath the epithelium and basement membrane is the connective tissue that helps to keep the epithelium in place.

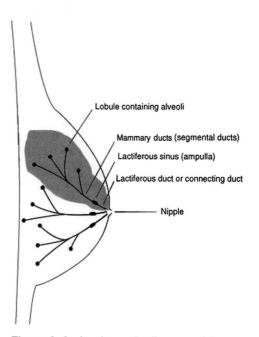

Lobule containing alveoli

Mammary ducts (segmental ducts)

Lactiferous sinus (ampulla)

Lactiferous duct or connecting duct

Nipple

Figure 3–6. A schematic diagram of the breast showing the branching distribution system of the collecting ducts. The nomenclature of the duct system varies. The gray area represents a single lobe.

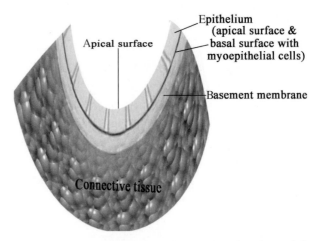

Figure 3–7. Cross-section of the ductal wall, showing the cellular makeup. The epithelium includes apical and basal surfaces. After puberty, the epithelium becomes multilayered.

Arterial supply:

Axillary artery

   Lateral thoracic artery

Acromiothoracic (thoracoacromial) branches

Internal thoracic artery

   Perforating branches

Intercostal arteries

   Perforating branches

VASCULAR CIRCULATION

The cardiovascular system consists of the heart, blood, and blood vessels. Blood will provide the cells in the body the needed oxygen and nutrients, and blood will also transport carbon dioxide and other waste. Blood circulates throughout the body by the pumping action of the heart. In general, arteries will carry the oxygen-rich blood from the heart to the cells of the body, while veins return oxygen-poor or deoxygenated blood to the heart.

An adequate blood supply is needed in the breast to provide nourishment for the newborn and to ensure that breast milk has a consistently high nutritional content.

Arterial Circulation

The breast gets most of its blood supply from the axillary artery through its lateral thoracic and acromiothoracic branches and from the internal thoracic artery (formally called the *internal mammary*) through its perforating branches. These branches travel through the first through fourth intercostal spaces and enter the breast along the medial margin. The largest of these are the first and second perforators. The smallest blood source for the breast is the intercostal arteries and their lateral perforating branches (Fig. 3–8).

A calcified artery will be outlined and clearly visible on a mammogram. Calcified arteries on a mammogram are now seen as a possible indication of an underlying cardiac problem.

Venous Drainage

The venous drainage from the breast corresponds with the arterial supply. In general, veins are larger than arteries. Although veins are usually peripherally located, there are deep and superficial networks draining the breast. The superficial veins run in longitudinal and transverse patterns just under the skin surface, eventually draining into the axillary veins. The deeper veins drain from behind the nipple. The largest of the

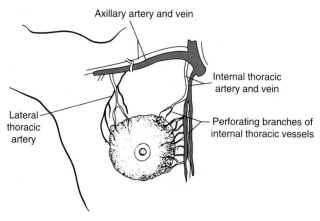

Figure 3–8. Schematic diagram showing arterial supply to the breast.

deep veins empty directly into the axillary veins, which continue as the subclavian vein. There is also drainage to the internal thoracic vein and the intercostal veins. Mammographically, veins appear as low optical density radiopaque vessels. Like arteries, they are not generally visualized, but calcified veins are visualized on the mammogram.

THE LYMPHATIC DRAINAGE OF THE BREAST

The lymphatic system is very closely aligned with the cardiovascular system. It consists of lymph fluid and lymphatic vessels, which transport the lymph. Lymph functions to drain waste and excess interstitial fluid from cells and tissue spaces and return it to the blood. The lymphatic system is also responsible for the transportation of hormones, glucose, and other nutrients, such as lipids and lipid-soluble vitamins A, D, E, and K to the cells in the body. The other function of the lymphatic system is to initiate specific responses against microbes or abnormal cells. This is an immune response whereby the lymph functions to either destroy intruders or protect the body by producing antibodies to fight the intruders. Although lymph begins as interstitial fluid or intercellular fluid, after the fluid enters the lymphatic system, it undergoes a name change and is called *lymph* (Fig. 3–9).

Lymphatic drainage:
- Lateral breast drains to pectoral group then to axillary nodes
- Medial breast drains to internal mammary lymph nodes
- Medial breast drains to mediastinal nodes, parasternal nodes, or the other breast

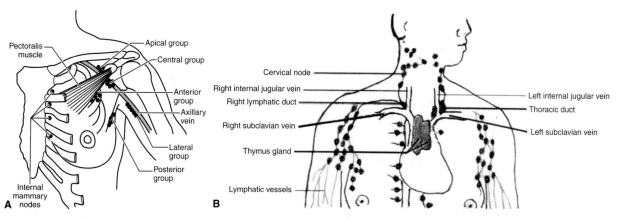

Figure 3–9. **(A)** Lymph drainage from the breast. **(B)** Lymph drainage to the subclavian lymph trunk.

The lymphatic vessels begin as lymphatic capillaries located in the spaces between cells. They are blind-ended (closed at one end) and unite to form larger lymphatic vessels. These vessels are thinner than veins and have more valves because there is no heart to pump the lymph. The valves therefore force a one-way flow. Lymph flow is also dependent on another pumping mechanism. The contraction of the skeletal muscles, along with pressure changes that occur during normal inhalation (breathing), acts to compress the lymphatic vessels and force the lymph toward the internal jugular and subclavian veins. The lymph then drains into the venous blood through the right lymphatic duct and the thoracic duct (left lymphatic duct) at the junction of the internal jugular and subclavian veins.

In the skin, the lymphatic vessels generally follow the same route as veins, whereas in the viscera, they follow arteries. At intervals along the lymphatic vessels are structures called *lymph nodes,* which act to trap or filter foreign particles. Most normal lymph nodes are less than 2 cm in size and have a kidney-shaped appearance.

Within the breast, there are both superficial and deep plexuses of lymphatic vessels. The superficial vessels drain from beneath the skin of the breast, except the nipple and areola, and carry lymph to the axillary nodes, through the pectoral node. The deeper lymphatic vessels drain the deeper glandular components. In general, lymph from the glandular tissues follows the path of the bloods vessels. The axillary lymph nodes receive about three fourths of the lymphatic drainage from the breast through the pectoral lymph nodes. There are generally 20 to 30 axillary lymph nodes arranged in five groups (Fig. 3–9A):

- *Anterior group*—lying deep below the pectoralis major muscle and along the lower border of the pectoralis minor and sometimes referred to as the *pectoral nodes*
- *Posterior group*—lying along the subscapular vessels and sometimes referred to as the *subscapular nodes*
- *Lateral group*—lying along the axillary vein
- *Central group*—lying in the axillary tissues; this is the most palpable group
- *Apical group*—lying above the pectoralis minor at the apex of the axilla and behind the clavicle

From the pectoral nodes, lymph can drain to any of the other axillary groups or to the interpectoral, deltopectoral, supraclavicular, or inferior deep cervical nodes. Lymph also travels along the tributaries of the internal thoracic vessels and the lateral perforating branches of the intercostal vessels to drain in to the internal mammary nodes.

The axillary nodes drain lymph into the infraclavicular (deltopectoral) and supraclavicular nodes, and from these into the subclavian lymphatic trunk. Lymph from the apical group can also drain directly into the subclavian lymph trunk. The subclavian trunk returns the lymph to the blood circulation through the right lymphatic duct or the thoracic duct (left lymphatic duct) (Fig. 3–9B).

The main direction of drainage from the lateral half of the breast tends to be into the pectoral nodes, whereas the medial breast tends to drain to the internal mammary nodes, parasternal nodes, or the opposite breast.

Five groups of axillary lymph nodes:

- **Anterior group:** deep below the pectoralis major muscle, along lower border of the pectoralis minor and sometimes referred to as the *pectoral nodes*
- **Posterior group:** along the subscapular vessels and sometimes referred to as the *subscapular nodes*
- **Lateral group:** along the axillary vein
- **Central group:** in the axillary tissues; this is the most palpable group
- **Apical group:** above the pectoralis minor at the apex of the axilla and behind the clavicle

The axillary lymph nodes drain not only to the breast but also to the pectoral region, upper abdominal wall, and upper limbs.

Lymph drainage from the breast is significant because it is by this route that a malignant disease may leave the breast. The axillary lymph nodes are the most common site for metastases from a breast cancer, but because of the abundant communication between the lymphatic vessels and the circulation system, cancer cells can be deposited in the opposite breast or the abdomen and therefore travel to the liver or pelvis.

NERVOUS SUPPLY

The nervous system has a network of specialized cells called *neurons*. There are three main functions of the nervous system.

- Sensory function responds to internal or external stimuli.
- Integrative function processes the sensory information by analyzing it, storing it, or deciding on an appropriate response.
- Motor function initiates a response to the stimuli. Motor responses can be muscle contraction or gland secretions.

The main divisions of the nervous system are the central nervous system (CNS) and the peripheral nervous system (PNS). The spinal cord and the brain make up the CNS. Its main job is to get the information from the body and send out instructions. The PNS transmits information to and from the body to the CNS. The peripheral nervous system can be further divided into the autonomic nervous system (ANS), responsible for involuntary actions, and the somatic nervous system (SNS), associated with voluntary control of body movements. The breast and the skin over it are supplied with sensory and sympathetic nerves by branches of the cervical plexus and by thoracic branches from the brachial plexus. The sympathetic and parasympathetic divisions of the ANS control most of the body's internal organs. The breast is also innervated by the anterior and lateral cutaneous branches of the second through sixth intercostal nerves.

BREAST TISSUE COMPOSITION

The breast is made up of a varying mixture of *adipose* or fatty tissues, *glandular* or secretory components, lymphatic vessels, and blood vessels. The fibrous and glandular tissues within the breast are generally described as fibroglandular densities. In general, the amount of fatty and glandular tissues varies with age. Most typically, glandular tissue predominates in younger women, whereas fatty tissue predominates in older women.

Adipose tissue is more radiolucent and, therefore, shows as higher optical density areas (black) on mammograms, whereas fibrous and glandular tissues are less radiolucent and show as lower optical density areas (white) on mammograms. The pattern and distribution of the glandular tissue are usually the same bilaterally, with most glandular breast tissues found centrally and extending laterally toward the axilla in the upper outer quadrant.

The *glandular tissue* is arranged within the lobules, whereas between the lobules are the fibrous and fatty tissues. The fibrous and connective

Nervous system response:

Sensory receptors detect internal and external stimuli

- Touch is an external stimulus
- Increase acidity in the blood is internal

Neurons carry that sensory information to the brain and spinal cord

Sympathetic division of the nervous system controls involuntary responses, such as ejection of milk in response to suckling action

tissues envelop the glandular components of the breast and act as a supporting framework for the entire breast. These fibrous bands or stands of connective tissue predominate in the upper parts of the breast, where they are called *suspensory ligaments* or *Cooper's ligaments*.

Cooper's ligaments run from the deep fascia and attach to the breast skin. They generally become looser with age and, like elastic bands, will also lose their elasticity over time, resulting in a drooping breast.

FACTORS AFFECTING TISSUE COMPOSITION

Although generally regarded as a single-function gland, the breast is actually very complex. Within the span of a year, changes in the breast due to hormones, weight gain or loss, or general aging can result in breast tissue patterns that are virtually unrecognizable from the prior years. Despite its complexity, however, a study of the anatomy and physiology of the breast reveals that regardless of its size, shape, or contour, the basic elements of the breast will remain virtually unchanged through time.

Increased or decreased glandular tissue of the breast is a part of the normal physiologic changes that take place within the breast and is generally mirrored on the opposite breast. It can be related to hormonal fluctuations (whether normal or synthetic), including menarche, pregnancy, lactation, or menopause. Increase in glandular tissue is also dependent on a woman's genetic predisposition. The young breast will have fibroglandular tissues with very little fat. Although this composition changes as a woman ages, it is still possible to find older women with extremely dense, glandular breasts. Weight gain and loss also increase or decrease the fat content of the breast tissue, thereby affecting the overall density of the breast (Fig. 3–10).

> **Factors affecting breast tissue composition:**
> - Hormones
>   - Menarche
>   - Hormonal fluctuation (normal or synthetic)
>   - Pregnancy & lactation
>   - Involution
>   - Perimenopause & menopause
> - Weight gain/loss

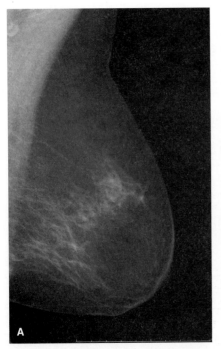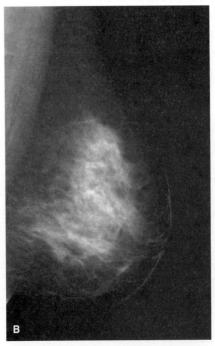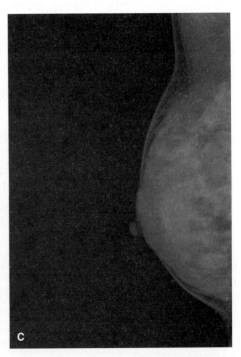

Figure 3–10. Mammogram showing **(A)** fatty breast tissue, **(B)** mixed-density breast tissue, **(C)** dense breast tissue.

The female breasts are inactive during childhood because there are no lobules. The breast will consist of small ducts within the fibrous tissue. As the female ages, the ovaries begin to produce estrogen and progesterone, and the process of puberty begins. Several hormones are responsible for the rapid changes that begin to take place. Prolactin, estrogen, and progesterone are produced by the female reproductive system; adrenal steroids by the adrenal glands; insulin by cells in the pancreas; growth hormone by the pituitary; and thyroid hormone by the thyroid glands (Box 3–1).

Estrogen produces the growth of the ducts, and prolactin/progesterone produces lobular development. Although the breast is fully developed by age 20 years, complete maturation of the breast tissue may not occur until the woman reaches 30 years of age.

Hormone Fluctuation

The system in the body that has the most impact in directing development and regulating body functions is the endocrine. The endocrine system controls body activities by releasing substances called *hormones*. Most hormones are released into the interstitial fluids, enter into the bloodstream, and will flow throughout the entire body. However, all hormones have a specific target cell, and only the target cell for a given hormone will have receptors that bind and recognize that hormone. When the hormones reach their target, they act by binding to protein or receptors in the target cells. The development of all female secondary sexual characteristics is the direct result of the action of hormones. The process starts at puberty.

The two most prominent hormones active in breast development are estrogen, which is responsible for ductal proliferation, and progesterone, which is responsible for lobular proliferation and growth. The gonadotropin-releasing hormones (GnRHs), secreted by the hypothalamus portion of the brain, stimulate the release of follicle-stimulating hormone (FSH) and luteinizing hormone (LH) from the anterior pituitary. FSH in turn initiates the secretion of estrogen by the growing ovarian follicles. LH also stimulates the corpus luteum within the female ovaries, which then produces and secretes estrogen, progesterone, and other hormones, such as relaxin and inhibin. After a woman starts producing estrogen, the changes in the breast can be intermittent, causing lumps or increased interstitial fluids (cysts), but will generally result in overall increased glandular tissue density (Fig. 3–11).

During each menstrual cycle, there is fluctuation in the hormone levels in the body, which causes structural changes in the breast tissue. The breast can increase in size, density, and nodularity, as well as sensitivity during the 3 to 4 days before the start of menstruation. Within 7 to 10 days after menstruation, this condition will diminish. The breast will, therefore, undergo repeated cycles of change during a woman's menstruating years. It is suggested that the greater the number of menstrual cycles, the greater is a woman's risk for breast cancer.

Menopause

Hormones secreted by the ovaries control a woman's menstrual cycle. At puberty, the secondary sex characteristics, including the breast, begin to develop, and a woman begins to menstruate. This first menses

---

Box 3–1. Hormones and Their Effect on the Breast

- Estrogen—growth of ducts
- Progesterone—lobular development
- Prolactin—lobular development during pregnancy
- Adrenal steroids—responsible for cell metabolism
- Insulin—necessary for glucose absorption. Even with a lot of food intake, the body is starved without insulin because many of our cells cannot easily access the calories contained in the glucose without the action of insulin.
- Growth hormone—controls the overall growth of all cells and tissues
- Thyroid hormone—responsible for cellular activity and metabolism

---

Although the breast is fully developed by age 20 years, complete maturation of the breast tissue may not occur until the woman reaches 30 years of age.

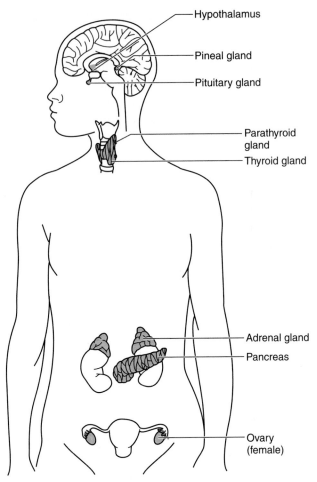

Figure 3–11. Location of some of the endocrine-producing glands in the body.

Perimenopause to menopause:

- Perimenopause
  - Irregular or heavy menstrual cycles
- Official start of menopause
  - No period for a year
  - Occurs between age 45 and 54 years
- Postmenopause
  - Involution and atrophy of breast tissue

is called *menarche*. Gradually, as the number of ovarian follicles becomes depleted, the ovaries become less responsive to hormonal stimulation. The percentages of ovulation cycles slowly decreases as a result of declining estrogen and progesterone levels, which can manifest as hot flashes, heavy sweating, headaches, hair loss, vaginal dryness, insomnia, depression, weight gain, or mood swings. It is also often a time of irregular or heavy menstrual cycles. This period for women is call perimenopause and can last several years. Eventually the menstrual period will stop, which marks the beginning of menopause.

However, menopause is not official until a women is period free, without spotting, being pregnant, breastfeeding, medications, or surgery for at least a year. Menopause generally occurs between 45 and 54 years of age, but it can begin as early as age 34 years. The average age in the United States is 51 years. Before menopause, women are termed *premenopausal*. After menopause, the term used is postmenopausal.

By the time a woman reaches menopause, the ovarian hormones are no longer stimulated, and the secretory cells and alveoli will degenerate. Generally, atrophy of mammary structures begins at menopause and ends 3 to 5 years later. With atrophy, the breast loses the glandular tissue and its supportive and connective tissue to fat, producing a smaller breast, or a larger, more pendulous one. This process is called *involution*. Involution will also occur normally with age and occurs faster and earlier in the multiparous woman than in the nulliparous woman. Atrophy of the glandular tissue generally begins first in the medial posterior aspect of the breast and then moves forward to the nipple; however, the tissue pattern of the breast will change with the introduction of hormone therapy.

Hormone Therapy

Normally, estrogen and progesterone, produced by the ovaries in women, are important hormones used to regulate menstrual cycle and pregnancy. With the beginning of menopause comes a drop in estrogen and progesterone, with resultant symptoms that can last for years and can vary tremendously from woman to woman. The most common symptoms are hot flashes or flushes, sweats, and sleep disturbances. Women can also experience mood swings, irritability, fatigue, vaginal dryness, memory loss, and insomnia. But the drop in estrogen also can contribute to other symptoms, such as changes in the vaginal and urinary tracts, which often cause painful intercourse, urinary infections, and the need to urinate more often.

Doctors in the past automatically prescribed hormone therapy involving the use of either estrogen alone or estrogen with another hormone called *progesterone* (or *progestin* in its synthetic form). The estrogen and progestin normally help to regulate a woman's menstrual cycle. Hormone Replacement Therapy (HRT), a combined estrogen and progestin therapy, was approved in 1945 to relieve perimenopausal symptoms. HRT is used by women who still have a uterus. Progestin is added to estrogen to prevent the overgrowth (or hyperplasia) of cells in the lining of the uterus. Because this overgrowth can lead to uterine cancer, women without a hysterectomy would be given estrogen plus progestin therapy. Women who have had their uterus removed cannot get endometrial cancer; therefore, these women can take estrogen alone. This is called *estrogen therapy* (ET) or *estrogen replacement therapy* (ERT). Hormones may be taken daily (continuous use) or only on certain days of the month (cyclic use) to treat menopausal symptoms.

HT will increase the amount of glandular tissue in the breast of menopausal women making the breast appear denser on mammograms. It also appears that HT inhibits the involution process. HT also causes increase in the size and the development of cysts and fibroadenomas in the breast.

In the past, it was thought that the benefits of HT would far outweigh the risks. Recent studies continue to find negative results with the long-term use of HT. The results of the Women's Health Initiative Memory Study (WHIMS) published in May 2003 in the *Journal of the American Medical Association* (JAMA) showed that women who use HT appear to be at increased risk for developing asthma. Although the risk for asthma is still low, the study also found that women older than 65 years were at an increased risk for dementia. HT is also linked with an increased risk for breast cancer, heart attacks, strokes, and blood clots.

Positive effects of HT:

- Relieves symptoms of menopause, such as:
  - Hot flushes
  - Sleep disturbance
  - Fatigue
  - Osteoporosis
  - Insomnia

Negative effects of HT can include:

- Breast or uterine cancer
- Asthma
- Dementia
- Heart attacks
- Strokes
- Blood clots

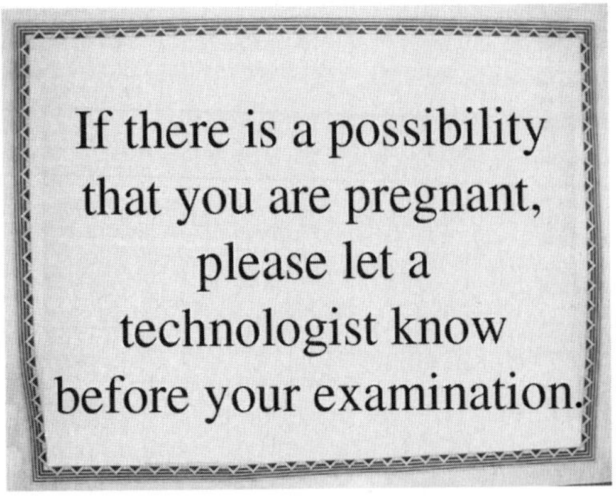

If there is a possibility that you are pregnant, please let a technologist know before your examination.

Figure 3–12. Example of a sign posted in the technologist waiting area.

- **Parity:** carried a pregnancy to a point of viability (20 weeks of gestation) regardless of the outcome.
- **Multiparity:** borne more than one child
- **Primaparity:** delivered a child of at least 500 g (or 29 weeks' gestation) regardless of viability.
- **Nulliparous:** never given birth to a viable offspring.

Lactation

Lactation refers to the normal production of milk, which is the primary function of the mammary glands. Within the first weeks of conception, the corpus luteum and the placenta both produce estrogen and progesterone and will influence the proliferation and development of the mammary tissues. Milk secretion is made possible by the production of the hormone prolactin and secretions from the adrenal cortex. The breast will begin to enlarge and become firmer, and the areola darkens as the pigmentation increases. The nipples become more enlarged and erect, and the lobules increase in size.

In the lactating breast, the alveoli are dilated, and the lactiferous ducts are distended with milk. As long as there is sucking action, the hormone prolactin will continue to stimulate milk production. Sucking action is also responsible for the production of other hormones, such as oxytocin, which allows contraction of the lobules and the ejection of milk.

Ideally, a mammogram should not be scheduled during lactation unless the patient is symptomatic, has a personal history of breast cancer, or is at very high risk. The milk in the lactating breast should be fully expressed before imaging. During lactation, the increased blood supply, milk production, and overall physiologic changes cause increased tissue density, which reduces the accuracy of the mammogram and makes it less exact as a diagnostic tool. Also, the increased density of the breast results in increased radiation exposure to the patient.

- The technologist should always confirm that the patient is not pregnant before the mammogram, and most facilities post signs as a reminder to patients and technologists (Fig. 3–12).
- Adjunct imaging modalities, such as ultrasound, are generally better suited for examining the lactating breast. If a mammogram must be done during lactation, the patient should nurse just before the mammogram to remove excess milk and to improve visualization of breast tissue.

Parity

Parity is the terminology used if a woman carries a pregnancy to a point of viability (20 weeks of gestation) regardless of the outcome. Multiparity is regarded as having borne more than one child, and primaparity as having delivered a child of at least 500 g (or 29 weeks' gestation) regardless of viability. A multiparous woman is generally considered to have a slightly lower risk for breast cancer than a nulliparous woman or one who has the first child after age 30 years. The nulliparous woman is described as one who has never given birth to a viable offspring. The nulliparous breast involutes more slowly than that of multiparous women, and multiparity is generally associated with breasts that involute earlier.

THE MALE BREAST

Male breast cancer is rare, and it is unusual for the technologist to encounter a male patient in need of a mammogram; however, because the incidence of male breast cancer is about 1% of all breast cancers, it will occasionally happen. Each year about 2000 new cases of cancer will be detected in men, and about 400 males will die of the disease. The male breast has less fatty tissue and can be harder to position, but imaging the male breast is very similar to imaging the small female breast.

General Description

The adult male breast is similar to a preadolescent girl. There may be a few branching ducts lined by flattened cells and surrounded by connective tissues, most often located behind the nipple and areola. In the female, these cells would eventually develop in response to hormonal secretions during puberty. However, in males, the male hormones prevent the growth of lobules keeping the growth of ducts to a minimum. Males, however, are also capable of responding to hormonal stimulation. Enlargement of the male breast due to growth of the ducts and supporting tissues is known as gynecomastia (Fig. 3–13). About 40% of adolescent boys will experience a temporary breast enlargement in response to hormones being secreted by the testes, but this response typically will disappear in 1 to 2 years.

In older men, gynecomastia can be triggered by drugs, disease, or the hormone estrogen, used to treat cancer. Drugs that can trigger gynecomastia include some prescribed for cardiovascular disorders or high blood pressure, migraines, or seizures. Gynecomastia can also be a result of diseases such as cancer of the testes or adrenal glands, cirrhosis of the liver, chronic renal dialysis, or Klinefelter's syndrome. Endocrine disorders can cause the oversecretion of estrogen, and diseases of the liver can disturb the hormone metabolism, therefore causing hormonal imbalance. Males who are overweight can also experience gynecomastia because of higher than normal estrogen levels. Klinefelter's syndrome is a rare disorder characterized by an abnormal chromosome pattern (XXY), poorly developed sex organs, and hormonal abnormalities. Men with this condition are 20 times more likely than the average man to develop breast cancer. Gynecomastia unrelated to Klinefelter's syndrome does not alter the male's risk for breast cancer.

Male Breast Cancer—Signs and Symptoms

Breast cancer in men, like breast cancer in women, will have similar signs and symptoms. Because the male breast is not very large, a painless lump in the breast is easily felt. Unfortunately, however, because of the smallness of the breast, even small cancers tend to rapidly metastasize from the breast. Another factor leading to the rapid spread of male breast cancer may be its location. Most male cancers are located around the areola, with its easy access to the mammary lymph pathway. Males also tend to be less aware of the possibility of breast cancer and do not often examine their breasts, leading to delayed diagnosis.

Risk Factors for Breast Cancer in Males

As with breast cancer in females, increasing age increases the risk for breast cancer in males. Other factors that affect male breast cancer include heredity of the breast cancer gene, radiation exposure,

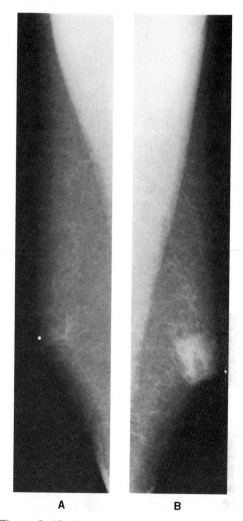

Figure 3–13. Radiographs of **(A)** a male breast without gynecomastia and **(B)** gynecomastia in the male breast.

Breast cancer in males is rare. Risk factors include:

- Age
- Heredity
- Breast cancer gene
- Radiation exposure
- Use of hormones
- Klinefelter's syndrome

hormones, and Klinefelter's syndrome. Male breast cancers tend to run in families, with reports suggesting that a man has an increased risk for breast cancer if his father or brother had breast cancer. The effect of hormones on male breast cancer has not been clearly established, although men who have taken estrogen to combat prostate cancer have been known to develop gynecomastia. Some researchers suggest that the breast tumors that develop in men who have been treated for prostate cancer may represent metastasis from prostate cancer rather than from a primary malignancy developing within the breast. However, a pathology workup should be able to differentiate prostate metastasis from primary breast cancer.

Diagnosis, Treatment, and Prognosis

All types of cancers seen in women can also occur in men, although lobular carcinomas are rare because lobules are normally absent from the male breast. Also, male breast cancer tends to occur more frequently in older men. The same procedures can be used to diagnose both male and female breast cancer. These would include a medical history, physical examination, mammography, and adjunctive testing.

Before any treatment, testing must be done to distinguish between a primary breast malignancy and a tumor that has metastasized to the breast. Treatment will then depend on the stage or extent of the cancer at the time of diagnosis and the status of the axillary nodes. The basic primary treatment, however, is a radical mastectomy. Radiation therapy is sometimes used for stage I cancer in males who are unable to tolerate anesthesia and surgery. In advanced cancers, radiotherapy alone is sometimes offered as a palliative treatment only. The use of chemotherapy is generally decided on an individual basis.

Distant metastasis in males occurs in patterns similar to that of females and will most often affect the bone, lungs, lymph nodes, liver, and brain.

The prognosis for men with breast cancer is similar to females with the same stage and type of cancer. However, because the male breast is typically much smaller than the female, the cancer very rapidly reaches the nipple, the skin, or muscles under the breast. Even small cancers in males are often found to have spread to the nearby tissues or lymph nodes. The extent of the spread can affect the prognosis of the disease. Another factor for males is the lack of awareness. Often, males will overlook common signs and symptoms, such as breast lumps, skin dimpling, or nipple retraction. This delay in detection can have a significant impact on the prognosis.

> The prognosis for men with breast cancer is similar to females with the same stage and type of cancer.
>
> However, because the male breast is typically much smaller than the female breast, the cancer very rapidly reaches the nipple, the skin, or muscles under the breast.

Summary

- The main purpose of the female breast is to provide nourishment for the newborn.
- The breast is located on the anterior chest wall, extending superior to inferior from the clavicle at the level of the second or third rib to the sixth or seventh rib and medially to laterally from mid-sternum to the latissimus dorsi muscle.

- The surface anatomy of the breast includes the skin, the nipple, and the areola.

- Location terminology used to describe sections of the breast includes the four-quadrant method, the four-region method, and the clock-face method.

- The deep anatomy describes the branching distribution system of collecting ducts. This includes the lobule or terminal ductal lobular unit, the segmental or mammary ducts, the lactiferous sinus, and the lactiferous duct. These structures are contained in a lobe. The breast can have 15 to 20 lobes and 10 to 100 lobules.

- The breast is extremely vascular with a rich blood supply and venous drainage. The main drainage is to the axillary veins.

- The main lymphatic drainage from the breast is to the axillary nodes.

- In the female, the breast receives its innervations from the lateral and anterior cutaneous branches of the second to the sixth intercostal nerves.

- The breast tissue is a mixture of adipose and glandular components. In addition to the vessels, there are fibrous bands called *suspensory* or *Cooper's ligaments* that support the breast.

- Factors affecting breast tissue include hormonal fluctuation, menarche, menopause, hormone replacement therapy, lactation, parity, and weight gain or loss.

- The structure of the male breast is very similar to the prepuberty female breast. Males can get breast cancer, and the signs and symptoms are very similar to the indications for female breast cancer. Male breast cancer can run in families. The risk factors include heredity, radiation exposure, hormones, and Klinefelter's syndrome. The diagnosis, treatment, and prognosis are the same for males and females.

REVIEW QUESTIONS

1. What is the upper extent of the breast?

2. The retromammary space is filled with a layer of what type of tissue?

3. On which two attachments is the breast most secure?

4. In which area is the skin of the breast thinnest?

5. Using clock notation, the 4-o'clock position (LIQ) on the right breast will represent which area on the left?

6. The fibrous bands that predominate in the upper breast and support the breast structure are called what?

7. Which vessels would be more superficial, arteries or veins?

8. What is the approximate size of a lymph node?

9. What is the name given to the milk-producing elements of the breast?

10. Where do most breast cancers originate?

11. What is the process by which the breast loses its supporting structure to fat?

12. What is the terminology used to describe a woman who has borne more than one child?

13. Males with what syndrome are more prone to develop breast cancer?

14. What effect will lactation have on the ducts and alveoli?

CHAPTER QUIZ

1. The shape and size of the breast could depend on:
   (1) Age
   (2) Weight gain or loss
   (3) Hormonal use
   (A) 1 and 2 only
   (B) 2 and 3 only
   (C) 1 and 3 only
   (D) 1, 2, and 3

2. The breast meets the abdominal wall at approximately the:
   (A) Mid-sternum
   (B) Mid-axillary line
   (C) Clavicle
   (D) Seventh rib

3. On average, how many lobes are found in an adult female breast:
   (A) 0–5
   (B) 5–15
   (C) 15–20
   (D) 25–30

4. Separating the breast from the pectoral muscle is a fatty area called the:
   (A) Inframammary fold
   (B) Retromammary space
   (C) Supporting and connective tissue
   (D) Milk line

5. The muscle immediately posterior to the breast is the:
   (A) Pectoralis minor muscle
   (B) Pectoralis major muscle
   (C) Serratus anterior muscle
   (D) Latissimus dorsi muscle

6. Parity refers to:
   (A) The condition of being pregnant or having delivered a child
   (B) Not having any children
   (C) Paring or removal of portions of the breast
   (D) Breast tissue that is not the same

7. What is the *main* purpose of hormone replacement therapy?
   (1) Relieve from the symptoms of menopause
   (2) Increase proliferation of glandular tissue
   (3) Control osteoporosis
   (A) 1 only
   (B) 1 and 2 only
   (C) 2 and 3 only
   (D) 1, 2, and 3

8. The specialized sweat glands found on the areola are called:
   (A) TDLU
   (B) Montgomery glands
   (C) Retromammary gland
   (D) Oil cyst

9. The duct that drains milk directly to the exterior is called the:
   (A) Lactiferous sinus
   (B) Ampule
   (C) Lactiferous duct
   (D) Segmental duct

10. Of the following, which mammogram would likely show very dense glandular breast?
    (A) Girl aged 10 years
    (B) Man aged 50 years
    (C) Woman aged 30 years
    (D) Woman aged 70 years

BIBLIOGRAPHY

American Cancer Society. Breast cancer in men. Available at http://www.cancer.org/cancer/breastcancerinmen/index. Accessed August 2016.

Andolina VF, Lille SL, Willison KM. *Mammographic Imaging: A Practical Guide.* 3rd ed. Philadelphia, PA: Lippincott Williams & Wilkins; 2010.

Ellis H. *Clinical Anatomy: Applied Anatomy for Students and Junior Doctors.* 11th ed. New York, NY: Wiley-Blackwell; 2006.

National Cancer Institute (NCI) Cancer Topics. Available at http://www.cancer.gov/types/breast/hp. Accessed August 2016.

Peart O. *Lange Q&A: Mammography Examination.* 2nd ed. New York, NY: McGraw-Hill; 2008.

Tabár L, Dean P. *Teaching Atlas of Mammography.* 3rd ed. New York, NY: Thieme Stuttgart; 2001.

Tortora GJ, Derrickson B. *Principles of Anatomy and Physiology.* 13th ed. New York, NY: John Wiley & Sons; 2011.

Tucker A, Ng Y, eds. *Textbook of Mammography.* 2nd ed. London, UK: Churchill Livingstone; 2001.

Venes D, Biderman A, Adler E. *Taber's Cyclopedic Medical Dictionary.* 22nd ed. Philadelphia, PA: F. A. Davis; 2013.

Benign and Malignant Diseases of the Breast | 4

Keywords and Phrases
Changes in the Breast
  Breast Augmentation
  Reduction Mammoplasty
Characteristics of Benign and Malignant Breast Lesions
  Circular or Oval Lesion
  Spiculated or Stellate Lesion
  Asymmetric Breast Tissue
  Calcifications
  Thickened Skin Syndrome
Types of Breast Cancer
  Ductal Carcinomas
  Lobular Carcinomas
  Other Breast Cancers
Summary
Review Questions
Chapter Quiz

Objectives

On completing this chapter, the reader will be able to:

1. Identify the categories of benign and malignant breast lesions
2. Describe the significance of a halo around a lesion
3. Differentiate between a radiolucent and a radiopaque oval lesion
4. Identify the characteristics of benign circular or oval lesions
5. Identify the characteristics of malignant circular or oval lesions
6. Identify the characteristics of benign spiculated or stellate lesions
7. Identify the characteristics of malignant spiculated or stellate lesions
8. Identify common characteristics of benign breast calcifications
9. Identify common characteristics of malignant breast calcifications
10. Identify common characteristics of thickened skin syndrome
11. Identify and describe the characteristics of the two common types of breast cancer
12. Identify and describe the characteristics of other cancers of the breast

KEYWORDS AND PHRASES

- **Abscess of the breast** can occur at any time but is more common in the lactating breast. It is generally located in the central or subareolar regions. A breast abscess can be a red, hot, and painful mass and is often associated with skin or nipple distortions. Ultrasound can be used to determine whether the mass contains any drainable fluid.

- **Adenosis, tumoral adenosis, or adenosis tumor** refers to the enlargement of breast lobules, which contain more glands than usual. If many enlarged lobules are concentrated in one area, the resultant mass is large enough to be felt. *Sclerosing adenosis* is a special type of adenosis in which the enlarged lobules are distorted by scar-like fibrous tissue.

- **Anasarca** is severe, generalized, and massive edema.

- **Asymmetric lesion** describes the presence of glandular tissue in one portion of the breast that is not present in a similar position in the contralateral breast. A focal asymmetric lesion would be limited to a small area, whereas global asymmetry could cover a larger portion of the breast.

- **Connective tissue disease** can refer to any disease of the connective tissue and often includes an autoimmune and/or inflammatory component. In autoimmunity, dysfunction of the immune system can cause the system to target connective tissue cells, resulting in inflammation. Connective tissue diseases can be genetic in origin or caused by gene abnormalities. Common connective tissue diseases include rheumatoid arthritis, systemic lupus erythematosus, and dermatomyositis.

- **Cysts** occur in the terminal ductal lobular units when the extralobular terminal duct becomes blocked. Fluid accumulates faster than it can be reabsorbed. Cysts vary in size and respond to hormonal fluctuations, but the development of a cyst also depends on a woman's genetic predisposition. Younger women, premenopausal women, and postmenopausal women taking estrogen are likely to have higher hormonal levels and, therefore, have an increased possibility of having cysts. The shape of the cyst depends on the amount of fluid within it. A tense cyst is round, whereas a lax cyst can be any shape depending on the compression applied.

- **Cytology** refers to the study of cells, including their formation, origin, structure, function, biochemical activity, and pathologic characteristics.

- **Duct ectasia** is commonly found in women in their 40s and 50s. When a duct becomes blocked, the area around the nipple becomes tender and red, causing a thick, green, black, or sticky discharge. Treatment usually involves removing the abnormal duct.

- **Ductal papillomas** are benign wart-like growths of glandular tissue and fibrovascular tissue, associated with milk ducts in the subareolar region. If seen mammographically, they are generally lobulated with fairly well-circumscribed borders. Papillomas are commonly solitary and are often centrally located but can be multiple and located along the peripheral area of the breast. Central papillomas often produce a bloody or serous nipple discharge and a palpable mass. If the duct containing a papilloma becomes cystically dilated, it is called an intracystic papilloma.

- **Eczema** is a superficial dermatitis of unknown cause. Lesions on the skin can become crusted, scaly, or weepy and can be exacerbated by psychological stress, illness, allergies, fibers, detergent, or perfumes.
- **Epidermoid cyst** is a pimple-like cyst that occurs in the oil glands of the skin. It may be radiolucent and well circumscribed with a smooth border, which can calcify. An infectious epidermoid cyst can mimic breast cancer. In the past it was called a epidermoid cyst, but the term was incorrect because these cysts do not always consist of sebaceous tissue.
- **Erythema** is the redness or inflammation of the skin or mucous membranes. The condition results from dilation or congestion of the superficial capillaries.
- **Fat necrosis** is death of fatty tissue in the breast that can occur spontaneously but is usually the result of biopsy, injury, or radiation therapy. At the site of the injury there may be hemorrhage; then as the tissue dies, liquefied fat and sometimes necrotic material and blood remain at the site. A capsule slowly forms around the injured area. Calcium slowly deposits on the inside rim of the capsule, which gives the eggshell-like appearance on the mammogram described as an *oil cyst*. The cavity can also fill with irregular-shaped calcifications, giving it a malignant appearance mammographically.
- **Fibroadenomas** are oval lesions that may contain calcifications. A fibroadenoma is a benign tumor common in women at any age. It is the most common benign lesion in women younger than 25 years. The fibroadenoma may be palpable, can be multiple, but often is identified by its characteristic mammographic appearance. The lesion can range in size from 1 cm and up. Occasionally, the fibroadenoma can calcify.
- **Fibroadenolipoma** (see hamartoma).
- **Fibrous nodules or focal fibrosis** can be asymmetric densities and round or oval lesions, and they most often occur in premenopausal women. They are prominent fibrous tissue, similar to ligament or scar tissue. Mammographically, the mass is benign appearing, and on ultrasound, it can be indistinguishable from a fibroadenoma.
- **Galactocele** is a benign milk-filled cyst with a high fat content. These lesions are generally associated with lactation. They are usually circular with sharply defined borders and have densities that are a combination of radiolucent and radiopaque. They are often left alone, but if painful they can be drained using a fine-needle puncture. They often contain a yellow fluid. Over time, the galactocele can calcify to a high optical density radiopaque lesion.
- **Granular cell tumors** are rare and almost always benign. In the breast they are felt as smooth, movable, and hard lumps about 1.3 to 2.5 cm (1/2 to 1 inch) in diameter. Having a granular cell tumor does not increase a woman's risks for breast cancer.
- **Gynecomastia** is a benign increase of tissue in the male breast. It can occur bilaterally or unilaterally. Gynecomastia does not increase the risk for breast cancer in male patients.

- **Halo** is seen mammographically as a complete or partial radiolucent ring around the circumference of the mass. The halo generally only surrounds a circular or oval lesion. It represents compression of fat adjacent to a mass. Circumscribed lesions with a halo are rarely malignant, although the presence of a halo does not absolutely rule out malignancy.

- **Hamartoma or fibroadenolipoma** is a benign tumor. It is the result of proliferation of fibrous, glandular, and fatty tissue and is generally surrounded by a thin capsule of connective tissue. Mammographically, it presents as a well-circumscribed, round, or oval mass that is both radiolucent and radiopaque combined. The mass can start as a very small growth; however, it is considered self-limited because the tumor consists of an overgrowth of normal tissue and the tumor cells do not reproduce.

- **Hemangioma** is a benign tumor containing a mass of blood vessels. It can slowly calcify, forming a large high optical density calcification in the breast.

- **Hematomas and microhematomas** are mixed optical density oval or circular lesions. A hematoma is formed by the pooling of blood as a result of trauma. Over time, a hematoma may calcify, resulting first in the formation of an oil cyst and later a high optical density radiopaque lesion. Hematomas sometimes have ill-defined borders and can mimic a malignant lesion. Clinical history is usually a significant factor in ruling out malignancy.

- **Hodgkin's disease** was first described by Thomas Hodgkin, a British physician (1798–1866). It is a malignant lymphoma affecting patients typically in their early 30s. Generally, the disease begins as enlarged lymph nodes in the neck, axilla, or groin but will metastasize gradually to the lymphatic tissue on both sides of the diaphragm or to tissues outside the lymph node.

- **Histology** refers to the microscopic analysis of organs and body tissues.

- **Hyperplasia or epithelial hyperplasia** can occur in the ducts (ductal hyperplasia) or the lobules (lobular hyperplasia). It is an overgrowth of cells lining the ducts or lobules. The increased cellular activity can sometimes result in microcalcifications. It is a proliferative breast disease and can be graded as mild, moderate, or florid (more extensive), depending on the cell appearance under the microscope. In atypical ductal hyperplasia (ADH) or atypical lobular hyperplasia (ALH), cells in the ducts or lobules are growing faster than normal, but the cells look normal. Hyperplasia usually develops naturally as the breast changes with age. It is not a form of breast cancer but is seen as a marker for women who may have a risk factor for developing breast cancer in the future. It can affect women of any age, but is more common in women over 35. The term ductal hyperplasia is sometimes used interchangeably with *papillomatosis*.

- **Intra** is a prefix meaning "occurring within."

- **Juvenile papillomatosis** refers to the entire spectrum of fibrocystic changes, including sclerosing adenosis, epithelial hyperplasia, cysts, duct stasis, and ductal hyperplasia. These lesions most often occur in women younger than 30 years. The hyperplasia may be significantly atypical but is not considered to be a premalignant condition. However, a patient with juvenile papillomatosis is considered at risk for developing breast cancer, especially if the disease is recurrent and/or bilateral or there is a family history of breast cancer.

- **Keloid scars** may form after surgery. Physically, they show as thick, darker pigmented irregular areas on the skin, and mammographically, they appear as irregular masses of varying densities. To avoid confusion with a malignancy, keloids should be identified and documented before a mammogram.

- **Keratoses** are skin lesions. They rarely calcify and mammographically appear as lobulated circular or oval masses with mixed densities.

- **Leptomeninges** is a term that comes from the Greek word *lepto*, meaning thin, fine, and slim. Leptomeninges are to the pia and arachnoid mater, two of the three layers covering the brain and spinal cord.

- **Lipoma** is a fatty tumor. It is radiolucent and may be huge, occupying the entire breast. It is easily seen mammographically and is not cancerous.

- **Lymphedema** refers to the accumulation of lymph in the soft tissues. It results in swelling of the area and can be a result of inflammation, obstruction of lymph canals due to malignant tumors, or removal of lymph nodes, for example, in a mastectomy.

- **Lymph nodes** are lesions of mixed optical density, and mammographically they generally have a radiolucent center corresponding to the hilus. Lymph nodes can be found in any quadrant of the breast and are not related to injury or malignancy. A lymph node is generally 2 cm in length. Enlarged lymph nodes, more than 3 cm, should be considered suspicious.

- **Mastitis** is an infection most often affecting women who are breast-feeding or have cracks in the breast skin. Bacteria entering the breast cause inflammation, redness, or swelling as the tissues increase blood flow to the infection. Mastitis can be painful and warm to touch and must be treated with antibiotics. If untreated, it can lead to breast abscesses.

- **Melanoma** is a malignant tumor of the melanocytes. Melanocytes are found in the basal cell layer of the epidermis and form melanin pigment. The tumor usually develops on the skin and has been linked to excessive exposure to ultraviolet light, especially sunlight.

- **Moles (nevi, singular nevus)** are growths on the skin that can be raised or flattened and are formed when the melanocytes cells in the skin grow in a cluster. Moles can be large and thick enough to be seen mammographically as a lesion similar to glandular tissue. Occasionally, with compression, air is trapped between the image receptor and the skin surrounding the skin mole forming a lucency around the mole. Moles should be labeled before the start of the mammogram to avoid errors in interpretation.

- **Morphology** refers to the science of the shape, form, and optical density of an individual structure. In mammography imaging, the term is often used when referring to an analysis of microcalcifications.

- **Oil cysts** are benign. Mammographically, they appear as high optical density tumors with lucent centers and eggshell-like calcifications.

They usually form as a result of fat necrosis or are calcified hematomas.

- **Papillomas** are a solid lump of wart-like growth in the breast ducts. Papillomas are composed mainly of mammary epithelium. Benign intraductal papillomas (IDPs) are papillomas without atypia. Occasionally, papillomas undergo atypical cellular proliferation, morphologically similar to ductal carcinoma *in situ* (DCIS) or atypical ductal hyperplasia (ADH). This is referred to as *papillomas with atypia.*

- **Parenchyma** refers to the main or functional tissue in an organ.

- **Peau d'orange** describes the skin of the breast when the breast skin thickens and develops prominent pores, resembling an orange skin. This condition occurs secondary to obstruction of the axillary lymphatics and may be a result of either benign or malignant conditions, such as inflammatory cancer.

- **Peri** is a prefix meaning "around."

- **Peritoneum** is the membrane lining the entire abdominal wall and covering the contents of the abdomen. It is divided into two layers, the parietal peritoneum and the visceral peritoneum, with the visceral layer against the abdominal wall and the parietal layer fanning out to support or separate the various abdominal organs. Between the two layers is the peritoneal cavity.

- **Phyllodes tumors** are rare fibroepithelial tumors. They are usually well circumscribed without a true capsule and mammographically will resemble a large lobulated fibroadenoma. The difference between the phyllodes tumor and the fibroadenoma is the overgrowth of fibroconnective tissue in the phyllodes tumor. Most phyllodes tumors are benign, but they can become malignant and metastasize.

- **Plasma cell mastitis, periductal mastitis, or ductal ectasia** is an inflammatory reaction characterized by the presence of plasma cells surrounding a dilated duct. Ductal ectasia may or may not cause nipple discharge or inversion. It is a benign condition with calcifications within the ducts (intraductal) or in the walls of the duct (periductal). Intraductal or periductal calcifications can be punctate, but some may be elongated, or sharply outlined, with smooth borders. These calcifications all have a high optical density.

- **Radial scar** is considered a risk for subsequent breast cancer. Generally, the radial scar is a lesion with no central tumor. Mammographically, a translucent, oval, or circular area is seen at the center of the lesion, and the lesion's appearance varies from one mammography projection to another. The radial scar can have a stellate configuration, with long spicules radiating from the center of the lesion; however, regardless of the size of the spicules, there is generally no associated skin thickening, dimpling, or discernible palpable mass.

- **Sarcomas** are any cancers arising from mesenchymal connective tissue, such as fat, nerves, muscles, cartilage, blood vessels, or even bone. They include *fibrosarcoma*—a sarcoma with a large amount of connective tissue; *liposarcoma*—a tumor derived from embryonal fat cells; *angiosarcoma*—a tumor arising from blood vessels; *osteosarcoma*—a tumor of the bone; and *fibrous histiocytoma*—a tumor containing histiocytes, which play a role in the body's immune system.

- **Sebaceous glands** are any of the glands that are found in the skin of mammals and secrete sebum into the hair follicles. Montgomery's glands on the areola are specialized sebaceous glands. The mammary glands, or breasts, are also modified sebaceous (sweat) glands designed to secrete milk. The sebaceous glands on the skin can calcify, forming ring-like calcifications with lucent centers.
- **Sclerosing adenosis (SA)** is a benign proliferative condition of the terminal duct lobular units (TDLU) characterized by an increased number of the acini and their glands. It can sometimes result in painful small lumps as a result of distorted lobules and stromal fibrosis. It is more common as the breast ages, most often in women in their 30s and 40s, but it can occur at any age.
- **Silhouette sign** is an indication of a benign process. Mammographically, lines through the breast will run through the middle of an abnormality. If the lines are seen into, through, and out of the lesion, they represent silhouetted structures in front of or behind the lesion.
- **Skin tags, acrochordon, or cutaneous papillomas** are benign skin growths consisting of fibers, nerve cells, fat cells, and an epidermis. The skin tag will project from the surface of the body yet remain attached by a narrow stalk. This gives the appearance of a soft, hanging piece of skin. Skin tags come in various sizes, and their origin is unknown.
- **Spiculated or stellate lesions** are usually malignant lesions with a solid distinct tumor center and sharp high optical density lines of variable length radiating in all directions away from the center. Usually, the spicules are not bunched together, and the larger the central tumor, the longer the spicules.
- **Stroma** is the supporting tissue of an organ, including the connective tissue, nerves, and blood vessels.

CHANGES IN THE BREAST

Breast changes can be either benign or malignant and refer to any pathologic change, or any change in the normal development and involution of the breast. Generally, breast changes will show mammographically as a mass with increased optical density. The optical density of a mass refers to the x-ray attenuation of the lesion relative to the expected attenuation of an equal volume of fibroglandular breast tissue. As a general rule, low optical density masses tend to indicate benign lesions, whereas high optical density masses indicate malignancy; however, it is often difficult to verify the optical density of a lesion when the breast is overall dense. On a radiograph of the breast, the less dense areas of fatty tissue will appear black or darker, and the denser glandular tissue will appear white or gray. Often, blood vessels can be seen, especially if they are calcified, and frequently lymph nodes are visualized within the breast as kidney-shaped oval densities with lucent centers.

Most breast cancers occur in the terminal duct lobular units (TDLUs); however, fibrous, connective tissue, and larger ducts can also be involved (Fig. 4–1). It is believed that epithelial cell carcinomas first

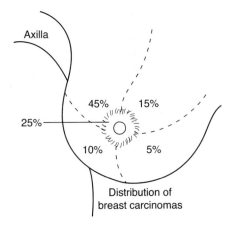

Figure 4–1. Most common breast cancer locations by quadrants.

develop as *epithelial hyperplasia,* in which the number of epithelial cells increases and changes. This stage is not considered malignant, and the process is reversible. The next stage is *atypical epithelial hyperplasia,* in which the cells may continue their rapid growth. There are conflicting studies on how carefully this stage should be monitored and whether it is indeed a precancerous lesion (Fig. 4–2). If the cell proliferation continues unabated, the next stage is *carcinoma in situ.* Although there is no invasion of cells outside of the duct or lobule walls, the cells are multiplying at such a fast rate that they often outstrip their blood supply. The result is cell necrosis at the very center of the growth. This necrosis or cell death may appear on the mammogram as microcalcifications. The changes due to this process are irreversible.

Carcinoma *in situ* is sometimes referred to as stage 0 carcinoma. The cell growth can slow or halt at this stage and can remain stable for years without further increase. If, however, the cells continue to multiply, they break out of the ductal or lobular walls and infiltrate or invade into the surrounding stroma. This is definitely a malignancy and is considered an infiltrating ductal or lobular carcinoma. At this point, the cancer can have access to lymph nodes and blood vessels.

Both benign and malignant lesions of the breast must be analyzed with either additional mammographic projections or magnification with or without spot compression, or by using adjunctive imaging

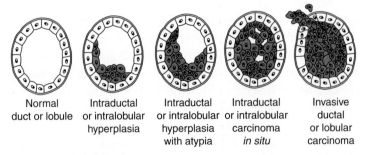

Figure 4–2. The progression of cell growth leading to invasive ductal and invasive lobular carcinoma.

modalities such as ultrasound, magnetic resonance imaging (MRI), or functional nuclear studies. It is also necessary to review all previous and present mammography images to identify visual changes occurring between exams, as well as to correlate imaging findings to the clinical assessment.

Changes can include a new lesion, an enlarging lesion, a change in the margins of the lesion, a developing spiculation, or developing microcalcifications. Malignant breast lesions are sometimes multicentric, and most are found in the upper outer quadrant of the breast. Because normal breast tissue is replaced by fat as it involutes, and the involution takes place from the posterior to the anterior, any new mass in the posterior areas of the breast—in a perimenopausal or menopausal woman—is a cause for concern.

In analyzing the mammogram, mammography interpreters will look for masses, focal or global asymmetric densities, architectural distortion, changes in the parenchymal contours, or calcifications. Although breast lesions have four distinct categories—circular or oval lesion, stellate or spiculated lesion, calcification, or thickened skin syndrome—breast cancer can involve any combination of two or more of these categories. Before any diagnostic finding, correlations must be made with the clinical findings, including comparison with previous mammograms if available, as well as a detailed comparison of the symmetry of left and right breast. It is equally important to make use of the available technology, such as mammography, ultrasound, or MRI, when ruling out a malignancy. Even so, whatever the diagnostic tool used to suggest a cancer—even if it is highly suspicious for cancer—only a microscopic diagnosis (cytologic and histology analysis) will confirm and reveal the exact type.

Breast Augmentation

Changes in the breast can also be associated with augmentation, a technique used to increase breast size. In the past, silicone was even injected directly into breast tissue (Fig. 4–3). Today, methods of augmentation include placing a silicone or saline prosthetic device in the front of (subglandular or retromammary) or behind (subpectoral or retropectoral) the pectoralis major muscle. These techniques are described further in Chapter 11. Implants placed behind the pectoralis major muscle in the retropectoral space are least likely to interfere with breast imaging. In 2006, the United States lifted a 14-year ban on implants containing silicone. The ban was originally put in place because of the controversy about the risk for connective tissue disorders associated with silicone. Mammographically, the denser silicone will obscure slightly more of the breast parenchyma than the more radiolucent saline; however, regardless of the type of implant, the presence of a calcified contracture capsule will interfere with breast compression, and, therefore, interpretation during a mammogram. Capsular contracture is the formation of a capsule lining around the implant. The process is the body's natural response to a foreign object. If it is going to occur, generally contracture takes place within the first few months of implant placement. The reasons are not fully understood. Often, the capsule, which is made up of fibrous tissue, will shrink and tighten,

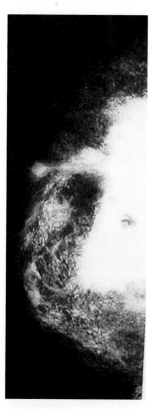

Figure 4–3. A mammogram of a 72-year-old with silicone injected directly into the breast.

squeezing on the implant. This produces a hard feel and can distort the appearance of the breast, giving it a ball-like appearance. In cases with calcified capsule, ultrasound and MRI have proved invaluable as adjunctive evaluation tools. Both ultrasonography and MRI can also be used to evaluate complications, such as rupture or leaking from the implants.

Reduction Mammoplasty

Reduction mammoplasty is a surgical technique used to reduce the total breast size for cosmetic or medical reasons, such as macromastia (excessive size of the breasts) or postsurgery after a lumpectomy to maintain equal breast size. The technology is described further in Chapter 9. Most techniques involve elevation of the nipple and removal of some of the glandular tissue. Reduction mammoplasty can lead to changes in the breast tissue. A baseline mammogram should be performed before any reduction mammoplasty, even though post-reduction changes can include asymmetric densities or calcifications, rendering the baseline useless as a comparison tool. The baseline, however, will establish that the breast is free of abnormal changes before the reduction mammoplasty. Palpable lumps after reduction mammoplasty can be caused by fat necrosis, hematomas, or fibrous scarring, but any palpable lesion should be evaluated to rule out malignancy.

CHARACTERISTICS OF BENIGN AND MALIGNANT BREAST LESIONS

The borders or margins of breast masses may be circumscribed, micro-lobulated, obscured, indistinct, or spiculated, and the shape can be round, oval, lobulated, or irregular (Figs. 4–4 and 4–5). A spiculated border is a strong indication of malignancy, whereas a circumscribed border is a strong indication of a benign abnormality; however, these are indicators only and will not necessarily determine the presence or absence of cancer. Some benign masses, such as postoperative scar, fat necrosis, hematoma, radial scars, or abscess, can mimic malignant lesions, and some malignant lesions, such as invasive ductal carcinoma (IDC), medullary cancer, mucinous cancer, fibrosarcoma, lymphoma, primary or secondary pseudolymphoma, and metastases, can mimic benign masses. Adjunctive testing, such as ultrasound or MRI, is there-fore important in the evaluation of any lesion in the breast; however, only histologic or cytologic analysis can definitively classify a suspicious breast lesion as malignant.

Benign diseases of the breast can manifest physically, such as in a painful cyst or nipple discharge. Some breast diseases, however, can be detected only by mammography or other imaging modalities. Depend-ing on the modality used, they will be seen as asymmetric densities, calcifications, circumscribed tumors, lesions, or skin thickening. Both benign and malignant breast lesions can be placed in five categories: circular or oval lesion, stellate or spiculated lesion, calcification, thickened skin syndrome, and any combination of the four.

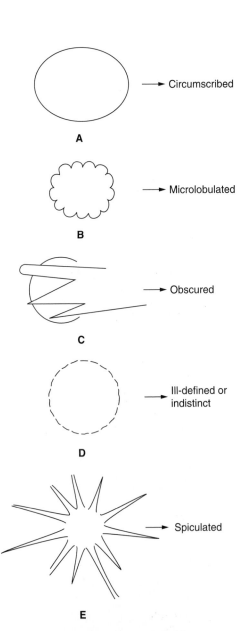

A → Circumscribed

B → Microlobulated

C → Obscured

D → Ill-defined or indistinct

E → Spiculated

Figure 4–4. Terminology used to describe the margins of breast masses: **(A)** circum-scribed; **(B)** microlobulated; **(C)** obscured; **(D)** indistinct; and **(E)** spiculated.

Circular or Oval Lesion

Circular or oval lesions can be poorly outlined, lobulated, solitary, or multiple. (Fig 4-6) These lesions can sometimes be symptomatic and are detected by the patient as a palpable mass. Circular or oval lesions can be detected mammographically, especially in fatty breast tissue, but are harder to detect in dense breasts. Ultrasound or MRI is much better at detecting masses in dense breast tissue. Mammographically, the circular or oval lesion is evaluated according to the optical density of the lesion in comparison to the surrounding parenchyma. Lesions can be radiolucent, combined radiolucent and radiopaque, low optical density radiopaque, or high optical density radiopaque. If the lesion is radiolucent, the parenchyma is visible through the lesion. Examples would be an oil cyst or a lipoma. High optical density radiopaque lesions will have a higher optical density than the surrounding parenchyma (e.g., cancer, abscess, hematoma, or epidermoid cyst). Examples of mixed optical density lesions include fibroadenolipoma, galactocele, inframammary lymph node, and hematoma. Fibroadenoma, keratosis, and cysts all tend to be low optical density radiopaque lesions. In general, all radiolucent lesions, all combined radiolucent and radiopaque lesions, and many low optical density radiopaque lesions are benign.

Because the borders of lesions are so important, magnification or spot compression can be used to enhance visualization of the lesion's borders as part of the mammographic workup. Circumscribed lesions are rarely malignant, and if the lesion has a "halo" sign, this can be indicative of a circular or oval lesion, which is often benign (e.g., a cyst) The absence of a halo does not necessarily rule out malignancy, and any circumscribed radiopaque tumor with unsharp borders and no demonstrable halo sign should lead to suspicion of malignancy, regardless of the optical density. Another characteristic of a benign lesion is the thin capsule—a curved radiopaque line—seen around lesions containing fat.

The most common circular oval lesions are cysts and fibroadenomas, and the breast ultrasound is the most valuable tool in assessing these and other circular or oval lesions. A cancer can develop within a cyst, such as an intracystic carcinoma; therefore, breast ultrasound should be

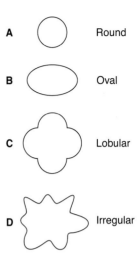

Figure 4–5. Common shapes of breast masses are **(A)** round; **(B)** oval; **(C)** lobular; and **(D)** irregular.

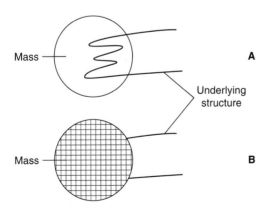

Figure 4–6. The silhouette sign occurs when **(A)** the underlying parenchyma is visible throughout a low optical density lesion and **(B)** in a high optical density mass, the underlying parenchyma is not visible through the abnormality.

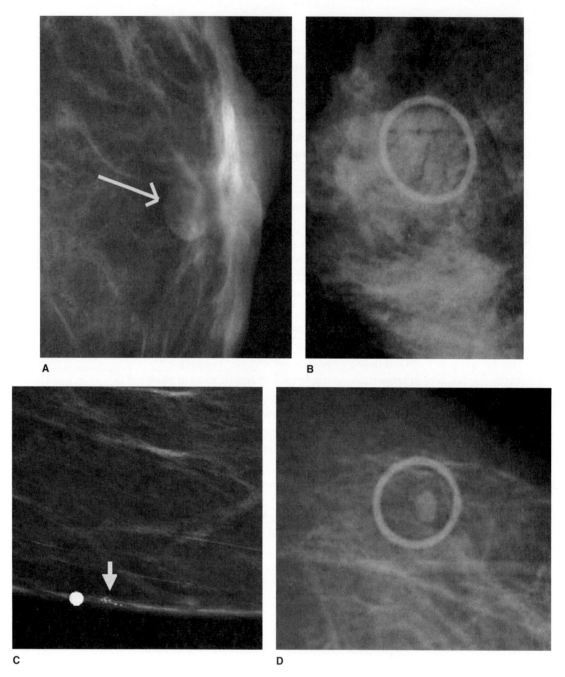

Figure 4–7. Skin lesions: **(A)** skin tag; **(B)** keratoses; **(C)** skin calcifications; and **(D)** skin mole.

used to differentiate solid versus cystic circular lesions. High-frequency ultrasound can also be used to assess the inside margins of the lesion.

Occasionally, a skin lesion can mimic a lesion in the breast on a mammogram because the mammographic images are two-dimensional. Skin lesions include skin tags, moles, keratoses, and epidermoid (sebaceous) cysts, and before a mammogram they should be noted and marked with a radiolucent ring to avoid confusion (Fig. 4–7). Mammographically, the tangential projection is often used to prove a lesion is in the skin.

Box 4–1. Mammographic Characteristics of Benign Circular or Oval Lesions

- Radiolucent (e.g., lipoma, oil cyst, galactocele)
- Radiolucent and radiopaque combined (e.g., lymph node, hamartoma, fibroadenolipoma, galactocele, hematoma)
- Low optical density-surrounding parenchymal structures can be seen through the lesion (e.g., fibroadenoma, cyst)
- Spherical or ovoid with smooth borders generally aligned in the direction of the nipple along the trabecular structure of the breast (e.g., cyst)
- Halo sign (e.g., cyst); malignant exceptions are intracystic cancer, papillary cancer, cancer within a fibroadenoma
- Capsule, a thin curved radiopaque line surrounding the lesion (e.g., fibroadenoma)
- Some exceptions are abscess, calcified galactocele, calcified hematoma, calcified epidermoid (sebaceous) cyst, and hemangioma

Generally, benign lesions will not change much in size or shape over time, whereas malignant lesions can grow significantly over a 1-year period; however, the use of estrogen alone is known to promote the enlargement of cysts and fibroadenomas, and a combination of estrogen and progesterone is often associated with diffuse increasing densities in the breast that can mimic cancer. If a benign-appearing mass develops new calcifications or changes significantly, it should be considered suspicious for malignancy, especially in postmenopausal women (Box 4–1 to 4–5; Fig. 4–8).

Spiculated or Stellate Lesion

Spiculated or stellate lesions have a solid central tumor with radiating structures and ill-defined borders (Fig. 4–9 & 4-10). Most breast cancers present mammographically as a spiculated lesion, although the very small spiculated lesions will be difficult to perceive. In a true spiculated lesion, the widest diameter of the radiating extensions occurs at the tumor margins and then tapers distally. The larger the central tumor, the longer the spicules. If the spicules reach the skin, or areola region, it

Box 4–2. Mammographic Characteristics of Malignant Circular or Oval Lesions

- High optical density-structures such as trabeculae of veins cannot be seen through the lesion (e.g., invasive ductal carcinoma, sarcoma)
- Smooth or lobulated and randomly orientated-not aligned along the trabecular structure of the breast (e.g., sarcomas)
- New lesion or new development of a solid mass

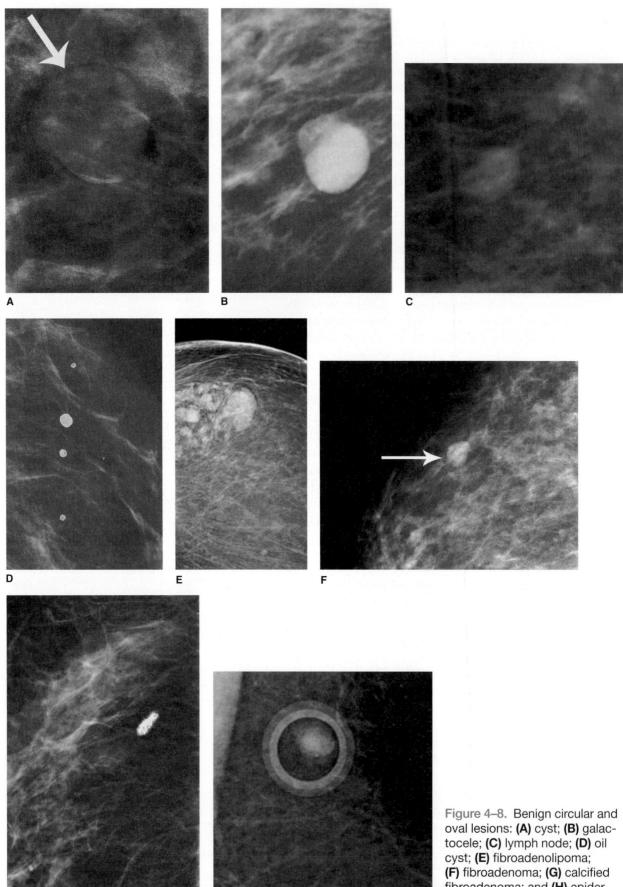

Figure 4–8. Benign circular and oval lesions: **(A)** cyst; **(B)** galactocele; **(C)** lymph node; **(D)** oil cyst; **(E)** fibroadenolipoma; **(F)** fibroadenoma; **(G)** calcified fibroadenoma; and **(H)** epidermoid cyst.

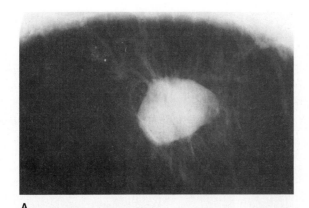

A

> **If the spicules reach the skin, or areola region, the result can be:**
> - dimpling
> - nipple retraction—if the lesion is behind the nipple
> - localized skin thickening—if the spicules disrupt the lymphatic drainage around the skin

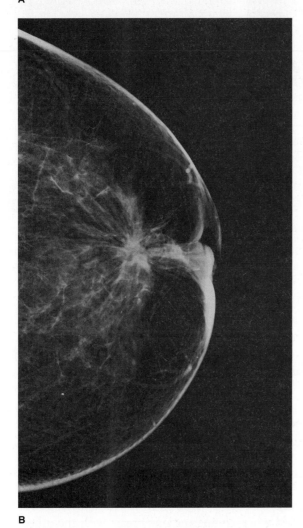

B

Figure 4–9. **(A)** Mammogram showing a spiculated lesion (stage II cancer). **(B)** Spiculated lesion resulting in nipple retraction.

can result in dimpling, or nipple retraction if it is behind the nipple. Localized skin thickening can result if the spicules disrupt the lymphatic drainage from areas around the skin. Spot compression or magnification is often helpful in identifying a true spiculated mass on a mammogram. If the spot compression identifies a central tumor, it is often an indication of malignancy. Benign lesions, such as postsurgical

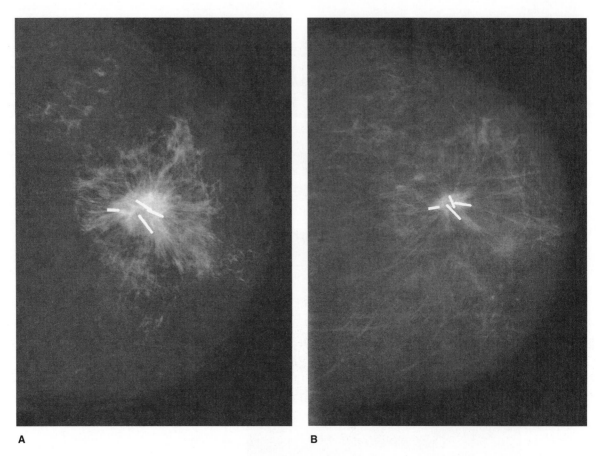

A   **B**

Figure 4–10. Mammogram showing a resolving postsurgical scar. **(A)** Postsurgical scar immediately after surgery. **(B)** Postsurgical scar slowly resolving over time. The minus density (optical density) artifacts on the radiograph represent surgical sutures.

fibrosis, fat necrosis, abscess, and hematomas, can mimic spiculated lesions. These, however, may sometimes be recognized as nonmalignant by the silhouette sign. If the lines are seen into, through, and out of the mass, then these lines represent silhouetted structures in front of or behind the mass. Invasive lobular or very small IDCs can present perception problems mammographically because these lesions do not form a central tumor and the spicules may be fine, lace-like linear densities that are difficult to detect.

Asymmetric Breast Tissue

Asymmetric breast tissue or focal architectural distortion (FAD) is usually identified when comparing one breast with the other. The breasts will present a mirror image, although 3% to 5% of normal breasts can show asymmetric lesions in the outer quadrant or axillary tail. If the optical density, even a symmetric lesion, corresponds to a palpable abnormality, it should be biopsied, especially if it appears in the posterior aspect of the postmenopausal breast because the process of involution occurs from the posterior to the anterior breast. A new lesion appearing in the posterior breast after involution begins would then be considered suspect

In some cases, FAD in the breast can be localized and present no definite mass; this can be a result of benign processes, such as surgery,

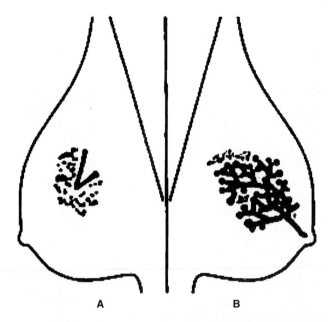

Figure 4–11. **(A)** In-tenting or in-drawing of glandular tissues often points to a malignancy that is sometimes hidden in the glandular tissue. **(B)** The area of architectural distortion on **B** represented by increased glandular tissue is not present on **A**.

radial scars, sclerosing lesions, or posttraumatic fat necrosis. The benign radial scar may have a radiolucent center, and the optical density will change appearance between different projections on the mammogram, whereas the spiculated mass will keep its dense center appearance on all projections. Radial scars, however, have been associated with significant increased risk for tubular carcinoma of the breast and malignant micro-calcifications.

Another form of asymmetric lesion is the in-drawing or tenting of the breast parenchyma. This condition is thought to reflect the parenchymal reaction to a tumor. Any tenting of the breast parenchyma should be investigated (Fig. 4–11).

Calcifications

Calcifications within the breast may or may not be associated with a tumor but should be evaluated separately from any associated tumor. The magnification mammography technique is extremely useful in evaluating calcifications. Calcifications must be analyzed based on their optical density, distribution, changes over time, number, morphology, and size.

In terms of their optical *density,* the calcifications can be low- or high-density radiopaque and any combination in between. In clusters of calcifications, the optical density can be consistently homogeneous or nonhomogeneous.

The *distribution* refers to the placement of the calcifications within the breast. Calcifications can be unilateral (in one breast), bilateral (in both breasts), clustered or grouped (three to five calcifications in a small-diameter area of less than 1 cm), diffuse or scattered (random

Box 4–3. Mammographic Characteristics of Benign Spiculated or Stellate Lesions

These lesions will often require a biopsy to rule out malignancy.

- No solid, dense, or distinct central mass
- May have translucent oval or circular area at the center (e.g., surgical scar) (Fig. 4–12)
- Very fine linear densities or lower optical density spicules (e.g., surgical scars or traumatic fat necrosis)
- Never associated with skin thickening or skin retraction (e.g., surgical scars; exception, traumatic fat necrosis)

Box 4–4. Mammographic Characteristics of Malignant Spiculated or Stellate Lesions

- Distinct central mass (e.g., invasive ductal carcinoma)
- Sharp, dense, fine lines of variable length radiating in all directions (the larger the central tumor mass, the longer the spicules; e.g., invasive ductal carcinoma)
- Spicules reaching the skin or muscle may cause localized skin thickening or skin dimpling (retraction)
- Commonly associated with malignant-type calcifications

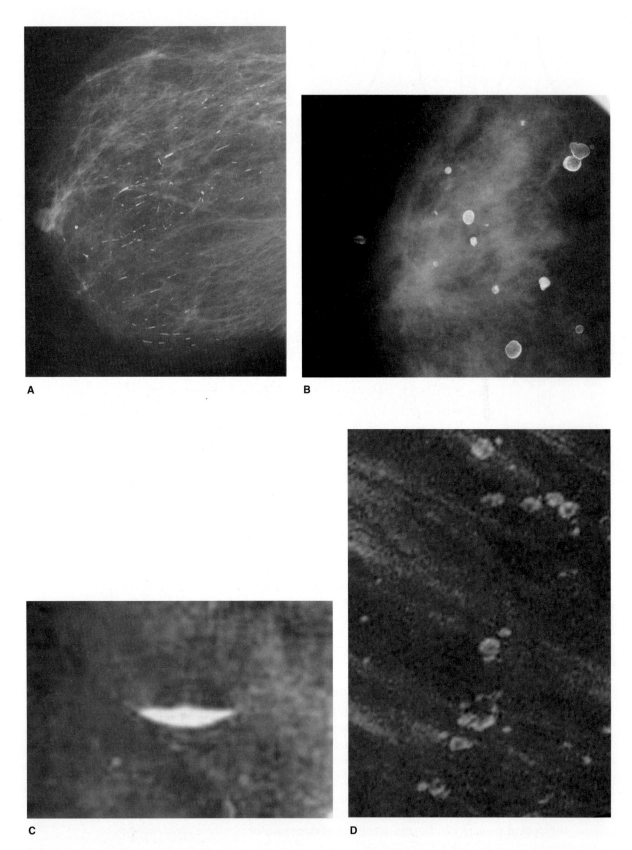

Figure 4–12. Benign breast calcifications: **(A)** plasma cell mastitis; **(B)** oil cyst; **(C)** milk of calcium; **(D)** sebaceous glands calcified;

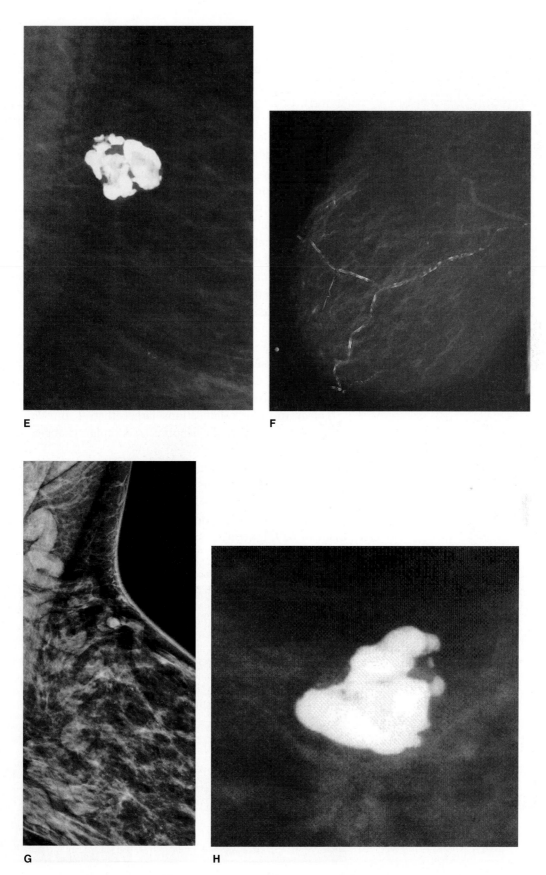

Figure 4–12. *(Continued)* **(E)** calcified hematoma; **(F)** vascular calcifications; **(G)** tortuous veins; and **(H)** hemangioma (unusual calcifications).

distribution throughout the breast), linear (calcifications arrayed in a line, suggesting deposits within a duct), segmental (calcifications within a duct and its branches suggesting an entire lobe), or regional (scattered within a large volume of breast tissue, without conforming to a lobe or duct).

Change over time refers to the variation in optical density, distribution, number, morphology, or size over time. In general, benign calcifications are relatively stable over time. The rapid metabolic activity of a malignancy can present radical change in malignant-type calcifications in a short period of time (Fig. 4–13).

The *number* of calcifications presented can vary. There can be a single calcification, or calcifications can appear in clusters of three or more either unilaterally or bilaterally.

Morphology deals with the study of the form and structure of any calcification, including its shape, pattern, or optical density. In mammography, calcifications can be described as coarse, linear, round, casting, amorphous, or pleomorphic.

Coarse calcifications are often larger than 0.5 mm in diameter. They are considered dystrophic calcifications if they occur in degenerated or necrotic tissue, such as after a trauma or irradiation. Popcorn-like coarse calcifications 2 to 3 mm in diameter can be produced by involuting fibroadenomas.

Linear rod-like calcifications larger than 1 mm in diameter are associated with ductal distribution and benign changes, such as ductal ectasia.

Round calcifications can vary in size. They are described as punctuate if they are smaller than 0.5 mm in diameter. Typically, round calcifications with diameters larger than 1 mm are benign.

Casting-type calcifications are fine, linear branching calcifications that can be fragmented with irregular contours. They can be irregular

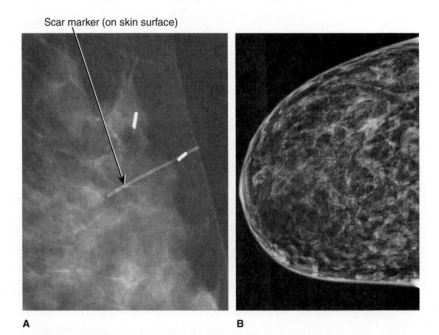

Scar marker (on skin surface)

A **B**

Figure 4–13. **(A)** Sutural calcifications. **(B)** Calcifications as a result of increased cellular activity.

in size and shape and smaller than 0.5 mm. The casts are produced when the calcifications form within the walls of the ducts—thus forming a "cast" around the ducts. The shape of the cast is determined by the uneven production of calcifications and the irregular necrosis of the cellular content. A calcification is seen as branching when it extends into adjacent ducts. Also, the width of the ducts determines the width of the castings. These calcifications are typically malignant. High-grade poorly differentiated (comedo) DCIS is frequently associated with casting-type microcalcifications (Fig. 4–14).

Amorphous or indistinct calcifications can form in multiple flake-like clusters. The individual calcifications can sometimes be less than 0.5 mm in diameter (and too small to categorize) or have an irregular appearance. They can also be greater than 0.5 mm in diameter, coarse, lacking a specific shape, and having variable optical density, shape, and pattern. Amorphous calcifications are of intermediate concern and can be malignant or benign.

Pleomorphic calcifications can be many different shapes and tend to be irregular in form, size, and optical density. They are sometimes referred to as granular calcifications because they are similar to granulated sugar or crushed stones. They can appear single or in multiple clusters. These are typically malignant calcifications (e.g., DCIS) (Fig. 4–15).

The *size* of calcifications in the breast can vary from microcalcifications in the millimeter range to macrocalcifications in the centimeter range. Malignant-type calcifications are often microcalcifications, sometimes smaller than 0.5 mm, but clusters of malignant calcifications can cover an area up to few centimeters in diameter.

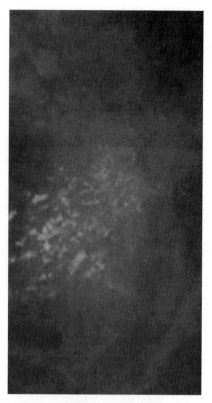

Figure 4–14. Casting-type calcifications (magnified).

Box 4–5. Mammographic Characteristics of Benign Calcifications

- Smooth contours, high uniform optical density (e.g., plasma cell mastitis, sutural calcifications)
- Evenly scattered homogenous (e.g., calcified arteries)
- Sharply outlined, spherical, or oval (e.g., oil cysts) (Fig. 4–15)
- Railroad tracks (e.g., calcified arteries)
- Pear-like densities—resemble teacups or pearl drops on the lateral projection (e.g., milk of calcium)
- Bilateral and evenly scattered following the course of the ducts throughout much of the parenchyma (e.g., plasma cell mastitis)
- Coarse, popcorn-like (e.g., involuting fibroadenomas)
- Linear, tubular, or rod-like (e.g., sutural calcifications)
- Ring-like, hollow with lucent centers (e.g., sebaceous gland calcifications)
- Eggshell-like (e.g., oil cyst, papilloma)
- Large, bizarre size (e.g., hemangiomas)

Calcifications analysis:

- Optical density—radiolucent or radiopaque
- Distribution
  - Unilateral—in one breast
  - Bilateral—in both breasts
  - Clustered or grouped—in a small area of less than 1 cm
  - Diffuse or scattered—random distribution throughout
  - Linear—arrayed in a line
  - Segmental—within a duct and its branches
  - Regional or scattered—within a large volume of tissue
- Change over time—rapid versus no change
- Number—single or clustered
- Morphology—study of the form and structure
- Size—millimeter to centimeter range

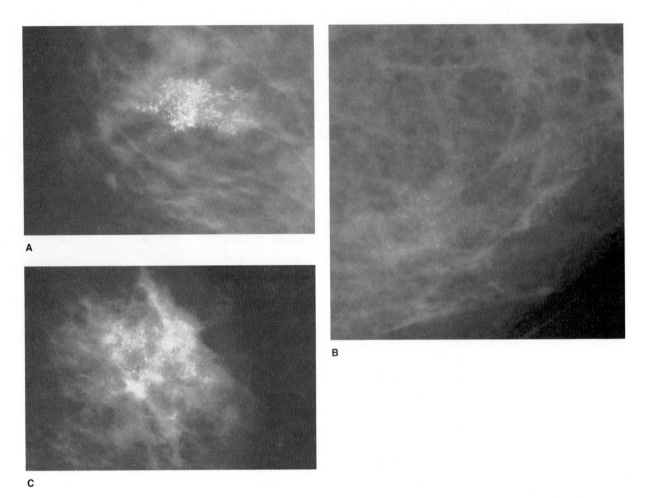

Figure 4–15. **(A–C)** The magnified appearance of granular and punctate-type malignant calcifications.

Causes of Breast Calcifications

Most breast calcifications are related to cystic changes. Benign conditions, such as *cystic hyperplasia, sclerosing adenosis,* and *ductal epithelia hyperplasia,* often cause increase cellular activity in the ducts or lobules, resulting in uniform, homogenous, and sharply outlined calcifications. These calcifications tend to be punctate or rounded but sometimes are malignant appearing (see Fig. 4–15). Another typical example of a benign calcification is the *milk of calcium*. Here, granules of calcified debris (milk of calcium) form in the lobules (see Fig. 4–15). Because these calcifications are mobile, they take the shape of the cavities where they are located. On the 90-degree lateral line, they will settle in the dependent portion of the lobules and are viewed as crescent shaped, resembling a teacup.

Calcifications associated with *ductal ectasia,* also *called plasma cell mastitis,* are also the result of cellular activity often following an infection with resultant inflammation. These are usually large, rod-shaped calcifications, sometimes 1 mm to 1 cm in length (see Fig. 4–15). They are the result of secretions within dilated benign ducts, whether periductal (calcifications within the walls of the ducts) or intraductal (calcifications within the duct itself). Periductal calcifications will have radiolucent centers representing the noncalcified centers of the duct. Intraductal calcifications are within the ducts and appear as linear and fragmental, forming along the long axis and pointing toward the nipple with only occasional branching. Often, these calcifications are bilateral and symmetric in distribution. *Comedo ductal carcinoma* also presents as ductal calcifications, but in this case the calcifications are variable in size, shape, and optical density, with many branching forms. The distribution here is often asymmetric and rarely bilateral.

Calcifications within the skin, or dermal calcifications, include calcified sebaceous glands or calcified apocrine cysts. The calcified sebaceous glands appear as very distinctive ring-like calcifications because of the calcium deposits within the inside walls or rims of the glands. Calcifications within the gland itself can appear on the mammogram as fine, punctuate calcifications.

Other calcifications can be related to breast surgery, and occasionally a fibroadenoma will undergo involution and calcify. A malignant calcification within a fibroadenoma is extremely rare, but there have been reported cases. Calcifications associated with surgery, such as dystrophic calcifications, can be difficult to analyze and may require biopsy, especially in the early stages. Sutural calcifications represent calcium deposited on sutures (stitches) and have a tubular appearance. They are occasionally seen in patients with a history of breast surgery or radiation therapy (see Fig. 4–16). Although not usual in United States, the parasitic nematode worms, common in tropical and subtropical countries, can result in filariasis, an infestation of the lymphatic system. These worm infestations—onechocerciasis, loiaisis, and trichinosis (an infection with the roundworm, *Trichinella spiralis*)—can sometimes migrate to the breast. Mammographically, they appear as low optical density tubular calcifications.

Oil cyst or eggshell-like calcifications can be the result of trauma or surgery. If a hematoma forms after the trauma or surgery, blood pools in the breast. Over time, the cells in the injured region die—possible resulting in fat necrosis. However, a capsule can slowly form around the injured area. Liquefied fat and necrotic material remain, and calcium gradually deposits on the inside wall of the capsule. The appearance on the mammogram is an irregular oval or eggshell-like calcification called an *oil cyst*.

Veins and, to a lesser extent, arteries can also become calcified. Vascular calcifications are very distinctive and are seen mammographically as two parallel lines or broken tubular patterns. If arterial calcifications are evident on a mammogram, studies suggest that the woman could be suffering from coronary artery disease (see Fig. 4–15). Women with significant arterial calcifications may therefore be referred for additional testing, to evaluate their cardiac risk factors.

Origin of Calcifications

Lobular calcifications can fill the acini, which are often dilated. These calcifications are often the result of cystic hyperplasia and are large and round, taking the shape of the lobule. However, with increased fibrosis, as in sclerosing adenosis, the calcifications can become less uniform and smaller, making them difficult to differentiate from malignant calcifications.

Morphology of calcifications:

- *Coarse dystrophic* calcifications are often larger than 0.5 mm in diameter.
- *Linear* rod-like calcifications are more than 1 mm in diameter.
- *Round* calcifications can vary in size.
- *Casting-type* calcifications are smaller than 0.5 mm; these are fine, linear branching, and fragmented with irregular contours.
- *Amorphous* or indistinct calcifications lack a specific shape. They are of intermediate concern and can be malignant or benign.
- *Pleomorphic* calcifications can be many different shapes. They are typically malignant.

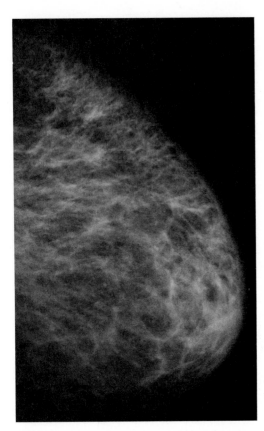

Figure 4–16. Mammographic imaging of thickened skin syndrome.

Causes of benign breast calcifications:

- Cystic changes
  - Cystic hyperplasia
  - Ductal epithelia hyperplasia
  - Ductal ectasia
  - Milk of calcium
  - Sclerosing adenosis
- Dermal calcifications
  - Calcified sebaceous glands
  - Calcified apocrine cysts
- Calcifications within the breast
  - Involuting (calcified) fibroadenomas
  - Sutural calcifications
  - Vascular calcifications

Intraductal calcifications are also the result of cellular activity. The cellular debris or secretions become calcified within the ducts. These calcifications can be variable in size, shape, and optical density, taking a fine, linear, fragmented, and branching form. Sometimes the calcifications form a "cast" around the ductal walls (casting calcifications), and these, like the intraductal calcifications, are often suspicious.

Thickened Skin Syndrome

Thickened skin may be present over the entire breast and may or may not be associated with an increased density (Fig. 4–16). The normal skin of the breast is 0.5 to 2 mm thick, except at the inframammary crease, near the cleavage, and in the periareolar region, where the skin is usually thicker. Skin thickening may be caused by either benign or malignant conditions and is generally a result of *edema*. The breast can present with bilateral, unilateral, or localized skin thickening. The skin will appear obviously thickened, most often in the lower dependent portion of the breast, and will take on the clinical appearance of an orange skin with prominent pores, hence the term—*peau d'orange*. The overall density of the breast is increased because of the high fluid content, which is seen as coarse, reticular patterns on the mammogram.

Localized or unilateral skin thickening can be a sign of a late malignancy either from direct invasion by the tumor or obstruction of the lymphatic or venous return. Lymphatic spread of breast cancer can block the intradermal and intramammary lymph nodes to the remaining breast, causing skin thickening. An example is inflammatory cancer, which will result in diffuse unilateral skin thickening because there is direct tumor invasion of the dermal lymphatics.

Most benign causes of skin thickening usually result in bilateral changes. Benign conditions that can produce bilateral skin thickening include cardiac failure and chronic renal failure that can lead to *anasarca*. Bilateral skin thickening can also be the result of bilateral axillary lymphatic obstruction. This condition can be secondary to breast cancer, metastases, lymphoma, advanced bronchial or esophageal cancer that leads to blockage of the mediastinal lymph drainage, and advanced gynecologic malignancies (e.g., ovarian or uterine). In any advanced gynecologic malignancy, the primary lymphatic drainage to the lesser pelvis is blocked, causing diversion to the thoracohypogastric collaterals, which in turn overloads the axillary and supraclavicular lymphatic drainage. Radiation treatment can also cause both unilateral and bilateral skin thickening with diffuse increased parenchymal densities, depending on whether one or both breasts are under treatment.

Causes of benign unilateral skin thickening include infection that can result in axillary lymphatic obstruction, which blocks the lymphatic drainage from the breast.

TYPES OF BREAST CANCER

Breast carcinomas arise from epithelial cells in the ducts or lobules, whereas cancers arising from the connective tissue are called *sarcomas*. Cancers can also develop in the skin, in fibrous tissue, and in other areas of the body and migrate to the breast. The two main classifications of breast carcinomas are ductal and lobular. Ductal carcinoma is most common and is diagnosed in about 90% of all breast cancers. Lobular carcinoma is diagnosed in 5% to 10% of cases. Often, these breast carcinomas will present mammographically as a stellate lesion or as microcalcifications. Other possible malignancies can present as asymmetric lesions within the breast tissue or in-drawing or tenting of the breast parenchyma.

Regardless of the cancer, perception is of vital importance in the initial detecting phase.

Ductal Carcinoma

DCIS describes a cancer that is confined to the duct and does not invade the duct walls (Fig. 4–17). Malignant epithelial cells proliferate within the ducts without invasion through the basement membrane. DCIS accounts for 22% to 23% of nonpalpable tumors detected mammographically. Studies have suggested that untreated DCIS will progress to invasive cancer, although the conclusions are controversial. Most cases of DCIS detected mammographically include microcalcifications. DCIS can be classified as poorly differentiated (high grade) or intermediate or well differentiated (low grade). High-grade, poorly differentiated (comedo) DCIS is more frequently associated with malignant casting-type microcalcifications. The presence of DCIS often increases the risk for an invasive cancer developing later, with a 30% chance of reoccurrence with 5 to 10 years.

IDC (not otherwise specified [NOS]) accounts for 80% of all breast cancers. In IDC, the cancer has spread from the ducts into the surrounding stromal tissue and may or may not extend into the pectoral fascia and muscle. This form of cancer is usually associated with a mass.

Lobular Carcinoma

LCIS is not generally seen mammographically. Some researchers suggest that LCIS is a precancerous lesion, called *lobular neoplasia*, because the condition does not indicate a cancer but rather only a collection of abnormal cells that have not penetrated through the lobule walls. It is 5 times more likely to occur in patients with DCIS, with or without invasion. Most often, LCIS is an incidental finding at biopsy for another mammographic or clinical abnormality. It is considered a risk factor for subsequent development of DCIS or IDC or invasive lobular carcinoma (ILC).

ILC is difficult to perceive on radiographs. It may have a spider-web appearance or cause skin retraction. Often, it is an asymmetric lesion not associated with microcalcifications. ILC accounts for only about 10% of all breast cancers in most studies, and because it is difficult to diagnose, this carcinoma accounts for a large percentage of potential malpractice suits for failure to detect breast cancer. It frequently metastasizes to the bone, *peritoneum,* gastrointestinal tract, ovaries, and *leptomeninges;* but if caught early, ILC responds well to hormone therapy. Adjunctive modalities, such as ultrasound and MRI, are much better at detecting ILC than mammography.

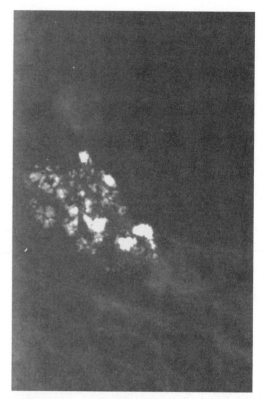

Figure 4–17. Magnified image of ductal carcinoma.

Other Breast Cancers

Although less common, there are a number of other breast cancers. Many have a better prognosis than IDC or ILC. In any breast cancer, however, the type of cancer and the stage of the cancer can have a significant impact on the long-term survival rate.

Bilateral breast cancer can occur simultaneously or may occur subsequent to the initial diagnosis. The incidence of simultaneous bilateral breast cancer is 4% to 5%. Bilateral breast cancer can be seen in patients who were irradiated at a young age for Hodgkin's disease. On average, the solid tumors will develop 15 years after treatment. Bilateral breast cancer can also be seen in patients who are younger than 40 years when diagnosed with their first breast cancer. This is often related to the *BRCA1* and *BRCA2* genes. Other studies have shown that the female breast is very susceptible to the cancer-causing effects of radiation when the exposure occurs before menopause. Girls treated as infants or children for nonmalignant conditions, such as enlarged thymus gland, or to monitor spinal curvature with scoliosis images or even to monitor tuberculosis treatment with multiple chest x-rays, are often at risk. It generally takes 5 to 10 years for radiation-induced cancer to develop, and the woman's age at the time of exposure is critical. Evidence suggests that the immature breast, especially at menarche, is most radiosensitive.

Inflammatory breast cancer (IBC) is a rare form of aggressive invasive cancer accounting for 1% to 5% of all breast cancers. IBC does not produce a distinct mass or lump that can be felt within the breast. The lack of a lump or mass also makes the cancer difficult to detect by mammograms. The disease is better detected by biopsy or clinical findings. Often, the disease will present as a benign inflammatory process with swelling, warmth, *erythema*, and skin thickening. The breast skin takes on the *peau d'orange* appearance. The symptoms result from vascular or lymphatic blockage as the cancer invades the skin. Treatment is often very aggressive and includes systemic hormonal and chemotherapy drugs and local radiation treatment with or without surgery. The 5-year survival rate is about 50%.

Lymphoma of the breast or primary lymphoma, non-Hodgkin's lymphoma, is extremely rare, representing 0.1% to 0.5% of all breast cancers. Secondary lymphoma is more common, representing 1.7% to 2.2% of all breast cancers, and may involve the axillary lymph nodes, with common primary sites being the gastrointestinal tract, the parotid gland, the mediastinum, or the lungs. Mammographically, there may be large single or multiple nodes that are incompletely circumscribed but not spiculated. There are no distinguishing sonographic features.

Medullary carcinomas account for 5% to 7% of all breast cancers and are often associated with a good prognosis even though they are quite large before they are physically detected. Although they often occur in younger women, these cancers rarely metastasize or tend to be confined within axillary lymph nodes. Mammographically, they are seen as round or lobulated noncalcified areas, which are fairly well circumscribed, with cells resembling the medulla (gray matter) of the brain. Sonographically, they are well circumscribed, frequently lobulated, and hypoechoic masses, often with posterior acoustic enhancement.

Other breast cancers:

- Bilateral breast cancer
- Inflammatory breast cancer
- Lymphoma
- Medullary carcinoma
- Metastasis to the breast
- Mucinous or colloid cancers
- Paget's disease
- Papillary carcinoma
- Phyllodes tumors
- Sarcomas
- Tubular carcinoma

Metastasis to the breast, usually lymphatic metastasis that crosses from one breast to another, is the most common form of metastatic involvement of the breast. Blood-borne metastasis is the next most common example and is from malignant *melanoma*, brachial carcinoma soft tissue sarcomas, and ovarian, gastrointestinal, and urogenital malignancies. Often, these metastases are bilateral rounded masses in patients with known primary tumors elsewhere. Although ovarian metastasis may demonstrate calcifications on a mammogram, other metastases rarely present calcifications mammographically.

Mucinous or *colloid cancers* account for less than 2% of all breast cancers. They may be associated with DCIS with or without microcalcifications and present mammographically as well-circumscribed masses but can be lobulated or ill-defined. The cancer cells often produce mucus and grow into a jelly-like tumor.

Paget's disease of the breast (first described by Jean Paget in 1874) is a special form of ductal carcinoma associated with *eczematous* changes of the nipple. It affects less than 5% of women in the United States. The early symptoms include scaly itchy nipple and a yellow or bloody nipple discharge with or without a rash. Late symptoms include inverted and crusty nipples leading to a total destruction of the nipple. Approximately 50% of these patients will develop a lump. This disease is difficult to detect because it is often mistaken for dermatitis and eczema.

Papillary carcinomas or, more often, *invasive papillary carcinomas* account for 0.9% of all breast cancers. They can be intracystic or intraductal. The term papilla describes the finger-like or thread-like shape of the microscopic cancer cell. They can present as circumscribed masses in older women. Often, they are indistinguishable from a cyst mammographically but on ultrasound will be seen as complex solid and cystic masses. If intraductal, nipple discharge can be an indication of malignancy in a single dilated duct. However, nipple discharge is often seen with benign papilloma. Fortunately, because papillary carcinoma stays within the ducts, it is a slow-growing tumor with very good prognosis.

Phyllodes tumors or cystosarcoma phyllodes are rare. Often, the tumors, which are similar in appearance to a large fibroadenoma, are benign. They present as a firm, mobile, well-circumscribed mass that will increase in size very rapidly within weeks. Approximately 16% to 30% of all phyllodes tumors are malignant. If malignant, the cancer will spread quickly to the lungs, followed by the long bones, heart, and liver. Treatment for the malignant disease is mastectomy. Often, the cancer does not respond well to chemotherapy or radiation therapy, and approximately 30% of patients with malignant phyllodes die from the disease.

Sarcomas of the breast are any cancers arising from mesenchymal connective tissue, such as fat, nerves, muscles, cartilage, blood vessels, or even bone. These cancers are extremely rare. They account for less than 0.1% of all malignant breast cancers. Primary sarcoma includes fibrosarcoma, angiosarcoma, pleomorphic sarcoma, leiomyosarcoma, myxofibrosarcoma, hemangiopericytoma, and osteosarcoma. Sarcomas spread very rapidly to the lungs, bones, liver, spleen, and skin. Survival rate is often dependent on the tumor size.

Tubular carcinoma is a well-differentiated form of IDC accounting for 1% to 2% of all breast cancers. The carcinoma is often nonpalpable, and mammographically it is indistinguishable from a benign radial scar. The tumor is usually small and made up of low-grade tube-shaped cells in the ducts. The neoplastic tubules mimic breast ductules. Pure tubular carcinoma is slow growing with good prognosis because it tends not to metastasize. The 10-year survival rate is 95%. Tubular carcinoma often starts in a radial scar; however, all radial scars do not develop into tubular carcinomas.

Summary

- The interpretation of breast images is a fine art that requires expertise and practice. In any mammographic evaluation, masking will improve perception of subtle alterations in the breast parenchyma by reducing extraneous light, therefore improving contrast. But only a thorough study of normal breast pattern, along with breast pathology, will accentuate how well the interpreter determines whether a lesion is normal, indeterminate, or suspicious.

- Changes in the breasts can include new lesions, enlarging lesions, or changes in the margin of the lesion. Changes can be developing calcifications or spiculations and can result from breast augmentation or reductions. Breast changes can also result in thickened skin syndrome and architectural distortion.

- Benign and malignant lesions have characteristic borders and shapes: benign lesions tend to be circumscribed and macrolobulated, and malignant lesions microlobulated, obscured, ill-defined, or spiculated. Oval or circular lesions tend to be benign and multiply lobulated, and irregular lesions tend to be malignant.

- Most breast calcifications are caused by cystic changes and the overproduction of fluids that are not reabsorbed. Calcifications can also form within a cyst, can form on the skin surface (dermal calcifications), or can be related to trauma, surgery, or calcification of blood vessels. Any calcification must be analyzed for optical density, distribution, changes over time, number, morphology, and size. Calcifications can be coarse and dystrophic, linear and rod-like, casting, amorphous, or pleomorphic. Calcifications can also be lobular or ductal in origin.

- The most common type of breast cancer is ductal carcinoma. Other, less common cancers include lobular carcinoma, bilateral breast cancer, inflammatory breast cancer, lymphoma, medullary carcinoma, metastasis to the breast, mucinous or colloid cancers, Paget's disease of the breast, papillary carcinoma, phyllodes tumors, sarcomas, and tubular carcinoma. Many of these cancers have a better prognosis than ductal carcinoma, with the exceptions of invasive lobular carcinoma and inflammatory breast cancer.

REVIEW QUESTIONS

1. What is a galactocele and what is this lesion generally associated with?

2. *Peau d'orange* describes what type of breast?

3. Describe why a mass lesion is seen more clearly on fatty than on dense breast tissue.

4. The term *reduction mammoplasty* refers to what?

5. The halo is significant in the analysis of a mass on the mammogram. Describe its significance.

6. Ultrasound and MRI are especially important in diagnosing what types of implant problems?

7. Describe and give an example of a radiolucent lesion.

8. Give three mammographic characteristics of a benign circular lesion.

9. Give three causes of benign skin thickening.

10. Spicules from a lesion in the breast can cause what changes if they reach the breast skin?

11. What is the significance of asymmetric breast tissue?

12. Give two mammographic characteristics of malignant spiculated lesions.

13. Describe four characteristics of benign calcification.

14. What are the six important factors used to analyze breast calcifications?

15. What are the two main classifications of breast cancer?

CHAPTER QUIZ

1. Reduction mammoplasty often involves:
   (A) Reducing the size of individual lobules
   (B) Reducing the size of the ampule
   (C) Removal of glandular tissue only
   (D) Removal of breast tissue and relocation of the nipple

2. A patient with *peau d'orange*-type breast will present with:
   (A) Breast skin that is obviously thickened
   (B) Malignant breast calcifications
   (C) Radiopaque breast lesions
   (D) Vascular calcifications

3. Benign breast calcifications include all of the following *except:*
   (A) Vascular calcifications
   (B) Plasma cell mastitis
   (C) Milk of calcium
   (D) Casting-type calcifications

4. Which of the following describes the characteristics of benign circular or oval lesions?
   (1) Radiolucent
   (2) Halo sign
   (3) Low optical density
   (A) 1 and 2 only
   (B) 2 and 3 only
   (C) 1 and 3 only
   (D) 1, 2, and 3

5. Asymmetric breast tissue—seen as an area on one mammogram that is not reproduced on the other—is referred to as:
   (A) Stellate lesion
   (B) Spiculated lesion
   (C) Architectural distortion
   (D) Malignant lesion

6. The halo sign is often used to identify:
   (A) Benign versus malignant lesions
   (B) Tumors versus fibroadenoma lesions
   (C) Circular and oval lesions
   (D) Stellate or spiculated lesions

7. Oil cysts are:
   (A) Often malignant
   (B) Seen on the mammogram as eggshell-like calcifications
   (C) Rarely benign
   (D) Often seen after pregnancy

8. The two most common types of breast cancer are:
   (1) Lobular
   (2) Ductal
   (3) Metastasis
   (A) 1 and 2 only
   (B) 2 and 3 only
   (C) 1 and 3 only
   (D) 1, 2, and 3

9. Which of the following lesions are likely to be malignant?
   (1) Spiculated
   (2) Circular
   (3) Multiple lobulated
   (A) 1 and 2 only
   (B) 2 and 3 only
   (C) 1 and 3 only
   (D) 1, 2, and 3

10. A fibroadenoma is a:
    (A) Malignant lesion found only in older women
    (B) Fluid-filled duct
    (C) Fibrous and adenoma tissue
    (D) Round, movable, but often benign lesion

BIBLIOGRAPHY

Adem C, Reynolds C, Ingle NJ, et al. Primary breast sarcoma: Clinicopathologic series from the Mayo Clinic and review of the literature. *Br J Cancer.* 2004;91:237–241.

American College of Radiology (ACR). *Breast Imaging Reporting and Data System Atlas (BI-RADS® Atlas).* 5th ed. Reston, VA: ACR; 2013.

Anderson RL. Risks and benefits in mammography. *Radiol Technol.* 2004;75(5):215–232.

Andolina VF, Lille SL, Willison KM. *Mammographic Imaging: A Practical Guide.* 3rd ed. Philadelphia, PA: Lippincott Williams & Wilkins; 2010.

Breast arterial calcifications increase risk for coronary artery disease. *RSNA 88th Scientific Assembly.* Abstract 1234. Presented December 4, 2002.

Breast Cancer Resource Center. Non-cancerous breast conditions. American Cancer Society (ACS). Available at http://www.cancer.org/healthy/findcancerearly/womenshealth/non-cancerousbreastconditions/non-cancerous-breast-conditions-toc. Accessed August 2016.

Brinton LA. The relationship of silicone breast implants and cancer at other sites. *Plast Reconstr Surg.* 2007;120(7 Suppl 1):94S102S. Available at: http://www.ncbi.nlm.nih.gov/pubmed/18090818 DOI:10.1097/01.prs.0000286573.72187.6e. Accessed August 2016.

Harris JR, ed. *Diseases of the Breast.* Philadelphia, PA: Lippincott Williams & Wilkins; 2000.

Peart O. *Lange Q & A: Mammography Exam.* 2nd ed. New York, NY: McGraw-Hill; 2008.

Tabár L, Dean P. *Teaching Atlas of Mammography.* 3rd ed. New York, NY: Thieme Stuttgart; 2001.

Tucker AK, Ng YY. *Textbook of Mammography.* 2nd ed. Edinburgh: Churchill Livingstone; 2001.

Venes D, Biderman A, Adler E. *Taber's Cyclopedic Medical Dictionary.* 22nd ed. Philadelphia, PA: F. A. Davis; 2013.

Mammography Equipment | 5

Keywords and Phrases
X-Ray Imaging System
  The Cathode
  The Anode
  Heat Dissipation
  X-Ray Tube
Design Characteristics of Mammography Units
Electron Interaction with Matter—X-Ray Production
  Characteristic or Photoelectric Radiation
  Bremsstrahlung Radiation
X-Ray Interaction with Matter
Mammography Imaging
Mammography Tube
  General Radiography Versus Soft Tissue Imaging
  Filtration
  The Anode and Target
  Line-Focus Principle
  Focal Spot Size
  Anode Heel Effect
  Kilovoltage Range
  Exposure Time, Milliamperes, and Milliampere-Seconds
  Compression Devices
  Compression Plate
  Automatic Exposure Control
  Density Compensation and Backup Timer
  Grids
  Beam-Restricting Devices
  System Geometry (Source-to-Image Receptor Distance, Object-to-Image Receptor Distance)
Magnification
Monitors
  Acquisition Workstation
  Review Workstation

Print and Storage Options
Picture-Archiving and Communicating System
   Advantages and Disadvantages
DICOM, HIS, RIS, EMR, and HL7
Teleradiography
Lossy and Lossless Compression
Computer-Aided Detection
Image Quality: Digital and Computed Mammography
Summary
Review Questions
Chapter Quiz

Objectives

On completing this chapter, the reader will be able to:

1. Differentiate between characteristic and bremsstrahlung radiation
2. Identify different x-ray interactions with matter
3. Explain the importance of soft tissue imaging in mammography
4. Describe the basic parts of the mammography tube
5. Explain the difference between effective and actual focal spot sizes
6. Explain the anode heel effect
7. Describe how filtration can affect the emerging x-ray beam
8. Discuss the terms brightness and contrast
9. Describe the main features of the compression plate
10. Discuss the effect of grid use in mammography imaging
11. Explain the purpose of beam-restricting devices in mammography
12. Discuss how the tube "geometry" can affect the image quality
13. Explain the purpose of the magnification mammogram
14. Explain the difference between the technologist and the radiologist monitor
15. Give available print and storage options
16. Explain the meanings of PACS, DICOM, HIS, RIS, EMR, and HL7
17. Discuss teleradiography and the various compression methods
18. Explain how CAD is used to aid interpretation
19. Discuss the factors affecting image quality in digital imaging
20. Give advantages of digital imaging
21. Explain the meaning of postprocessing in digital imaging

KEYWORDS AND PHRASES

- **Anode angle** describes the angle between the surface of the target and the vertical line, perpendicular to the image plate.
- **Alternating current (AC)** refers to the current in which electrons oscillate back and forth through a conductor in alternating positive and negative directions. The rate of changing direction is called the frequency of the AC, and it is measured in hertz (Hz), which is the number of forward-backward cycles per second. Graphically, the waveform of the alternating current is represented as a sine curve.

- **Bremsstrahlung x-rays** are produced when an outer projectile electron is slowed by the electric field of the target atom nucleus. This interaction is common in tungsten targets.
- **Brightness** of the digital image refers to how dark the image is. This term replaced "density" (optical density) used to describe analog images. The *brightness* of the image is controlled not by the milliamperes (mAs) as in analog imaging but by the processing software, which will include predetermined digital algorithms. The user can also alter the brightness of the digital image after the exposure.
- **Characteristic radiation** is produced when an outer shell electron fills an inner shell void in the atom of an element. If the outer shell electron fills the void in the K shell, the x-ray emission is termed *K-characteristic x-rays*.
- **Coherent or classical scattering** describes the interaction between low-energy electrons and atoms. The x-ray loses no energy but changes direction slightly.
- **Compton scattering** occurs when moderate-energy x-rays interact with an outer shell electron and eject the electron from the atom. The x-ray loses energy and changes direction. The ejected electron is the *Compton electron*.
- **Contrast** refers to the minimum difference between two adjacent structures that can be detected in the image as separate entities. In digital imaging, contrast is dependent on pixel size. Larger pixels offer better contrast, where high contrast equals few data values between black and white, and low contrast equals many data values between black and white.
- **DICOM** (Digital Imaging and Communications in Medicine) compliant refers to a set of computer software standards that permit a wide range of digital imaging programs to understand each other.
- **Digital linear tape (DLT)** is a storage option in digital imaging.
- **Direct current (DC)** is the electron flow in one direction along a conductor. Graphically, direct current is represented by a horizontal line. The current is always positive (or always negative).
- **Effective focal spot** is the area projected onto the patient and the image plate, and it is the value given when describing the focal spot of the mammography units.
- **Film digitizers** convert a conventional radiograph to a digital version via a scanning device.
- **Grid frequency** describes the number of lead strips per centimeter (or per inch) in a grid. Grid frequencies can range from 25 to 45 lines/cm (60 to 110 lines/inch). In mammography, grid frequencies of 30 to 50 lines/cm are common.
- **Grid ratio** is the height of the lead strips divided by the distance between the strips. Generally, a common ratio in mammography is 4:1 or 5:1.
- **Halation** describes the process by which light is reflected back toward the emulsion after it passes through the emulsion and base.
- **Half-Value Layer (HVL)** is the thickness of absorbing material necessary to reduce the x-ray intensity by half of its original value.
- **Illuminance** describes the intensity of light incident on a surface.
- **Image plate (IP)**, also called imaging plate, is used in computed mammography using photostimulable phosphor (PSP) technology. The IP replaces the cassette used in analog imaging.

Reasons for dedicated mammography units:

- Breast is all soft tissue
- Similar mass densities
- Need to maximize contrast
- Need for highest possible resolution
- Need to keep radiation doses low

- **Image receptor (IR)** is also called a storage phosphor screen (SPS) or PSP. It is the device that receives the energy of the x-ray beam. This replaces the screen/film system used in analog imaging.
- **Luminance** describes the brightness of a light source.
- **Magnification factor** is calculated by dividing the source-to-image receptor distance (SID) by the source-to-object distance (SOD).
- **Matrix,** in digital imaging, is an array of pixels in which each pixel is a sample of the image represented by a numerical value.
- **Penumbra** is the unsharp shadow or "blur" surrounding the image.
- **Pixel or picture element** is the smallest component of the digital image. Each pixel has a numerical value, and pixels are normally arranged in a two-dimensional grid represented as dots or squares.
- **RAID, or redundant array of independent disks**, is a storage option in digital imaging.
- **Rectification** is the conversion of alternating current (AC) to direct current (DC).
- **Thin-film transistor (TFT) or thin-film diode (TFD)** may refer to any diode produced using thin film technology. They have amorphous silicon or amorphous selenium deposited on thin films. Amorphous silicon or amorphous selenium are photoconductors used to collect electrons.
- **Reciprocity law** states that the optical density produced on the radiograph will be equal when any combination of milliampere (mA) and exposure time is used, as long as the product of mA and time is equal. In analog imaging, the law fails at long and short exposures because the film emulsion is exposed to light from the intensifying screens. At long and short exposures, an increase in mA is required because of the reduced film speed.
- **Resolution** is the ability to image two separate objects and visibly detect them as separate entities. Resolution is categorized as *spatial resolution*—which is the ability to detect small objects with high subject contrast—and *contrast resolution*, which is the ability to distinguish anatomic structures with similar contrast.
- **Thermionic emission** is the emission of electrons from a hot cathode into a vacuum.
- **Vendor neutral archives (VANs)** are the ability to make medical images accessible across multiple modalities and between different organizations.

X-RAY IMAGING SYSTEM

There are three principal parts to any x-ray imaging system. These are the operating console, which is often digital with touch-screen technology; the high-voltage generator; and the x-ray tube. Line voltage from a wall outlet enters the x-ray circuit and is adjusted with a line voltage compensator. The line compensator measures the incoming voltage and adjusts that voltage to precisely 220 V. Incoming voltage can vary by as much as 5%, so the line voltage compensator ensures the voltage going to the x-ray tube is always constant with no variations or fluctuations. The voltage then goes to an autotransformer, which has a kilovolt (kV) meter to allow the selection of various kV values across the output terminal. The output from the autotransformer is connected to the primary coil of the step-up transformer. X-ray production requires high voltage, and the step-up transformer or high-voltage transformer coverts lower voltage into higher voltage, which is sent to the x-ray tube.

Incoming voltage is alternating current (AC); however, x-ray tubes can only operate on direct current (DC). The voltage is therefore converted using a system of rectifiers. A rectifier is a solid-state device that allows current flow in one direction only, thereby converting AC to DC. The simplest kind of rectifier circuit is the half-wave rectifier. It only allows one half of an AC waveform to pass through to the load. Modern x-ray tubes cannot operate on half-wave rectification. In full-wave rectification, both cycles of the AC sine wave must be rectified using a bridge rectifier.

Older mammography generators were three-phase generators; all modern generators are high-frequency generators. Like the older units, the incoming AC voltage is rectified to allow one-way flow of electrons using semiconductor diodes to produce a DC output. However, the modern high-frequency generators will essentially provide a constant potential with approximately 1% ripple. This is possible with the use of phase-shifted pulses that overlap each other. The high-power rectification circuits can produce a smoother DC output. This allows more efficient x-ray production and therefore produces a higher effective energy x-ray beam (Fig. 5–1). The result is higher x-ray output for a given kV and milliampere (mA) setting. After rectification, the high voltage can then be applied across the x-ray tube.

Timing circuit is used to shut off the high voltage to terminate the x-ray exposure after a selected exposure time. The timing circuit consists of an electronic device whose action will "make or break" the high voltage across the x-ray tube.

The x-ray tube has another circuit. The filament circuit is used to create thermionic emission at the cathode. The filament is located at the cathode end of the x-ray tube, and it is a coil of wire approximately 2 cm in diameter and 1 to 2 cm long. Connections from the autotransformer provide voltage for the filament circuit. A filament step-down transformer ensures that the voltage to the filament is low and the current correspondingly high. The step-down transformer coverts high voltage/low current into low voltage/high current. This high current causes the filament to glow and to emit heat and electrons. The cloud of electrons will remain in the vicinity of the filament before they are accelerated to the anode, when the high voltage is applied to the tube (Fig. 5–2).

Thermionic emission is the term used to describe the boiling off of outer shell electrons of the filament atoms.

The Cathode

The cathode, the negative side of the x-ray tube, therefore has two parts: filament and focusing cup. Typically, there are two filament sizes. The small size is used with the small focal spot

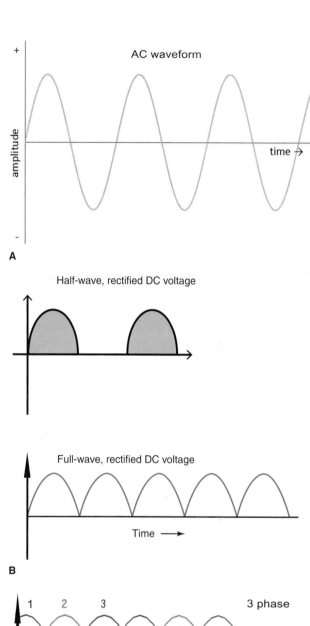

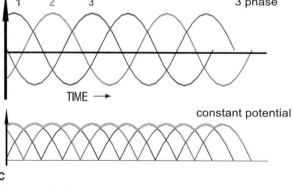

Figure 5–1. Schematic diagrams of **(A)** AC waveform; **(B)** DC waveform; and **(C)** half-wave rectification, which allows only one half-cycle of the AC voltage waveform to transmit. In a full-wave rectifier circuit, both half-cycles of the AC voltage waveform transmit to allow an unbroken series of voltage pulses of the same polarity. When a polyphase alternating current is rectified, it gives a "smoother" DC waveform (less ripple voltage).

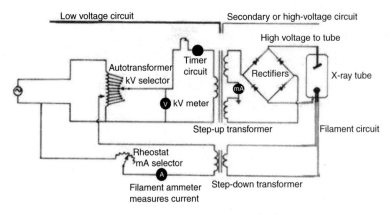

Figure 5–2. Schematic diagram of an x-ray circuit.

during magnification. The large size is used with the large focal spot during routine imaging. The focusing cup is a negatively charged cup-like device that electrostatically confines the electron beam to a small area on the anode owing to the mutual electrostatic repulsion among the electrons. The negatively charged cup controls the size and shape of the electron beam emitted from the cathode and directs the beam to the anode. The effectiveness of the focusing cup depends on the size and shape of the cup, its charge, the size and shape of the filament, and the position of the filament in the cup.

The Anode

The anode is the positive side of the x-ray tube. The anode serves as an electrical conductor to receive the electrons emitted by the filament at the cathode and conduct them through the tube to the connecting cables and back to the high-voltage generator. The anode also supports the target, which is the area on the anode struck by electrons from the cathode. Finally, the anode serves as a thermal dissipater. When projectile electrons strike the target, only 1% of the kinetic energy is converted to x-rays. The rest (approximately 99%) produces heat at the anode. Typically, the anode is rotating, consisting of a metal disk connected by a stem to the rotor. The stem is made of molybdenum, and the rotor is made of copper. A rotating anode allows the electrons to strike a different area on each exposure.

The target material can be tungsten, rhodium, or molybdenum.

The anode rotates through the principle of electromagnetic induction. This is a process whereby a paramagnetic material will become magnetized when placed in a magnetic field and lose its magnetism when removed. An induction motor consists of two parts separated from each other. The part outside the tube is called the stator, which is a series of several fixed electromagnets. The part inside is the rotor. Current is induced in the rotor by electromagnetic induction. An AC current is applied to the stator. A changing magnetic field is produced in each set of electromagnetics in the stator. A current is induced in the copper bars of the rotor. The current produced in the rotor windings will then generate a magnetic field. Because each set of the electromagnets are activated separately in the multiphase current, the rotor is forced to continually try to align with the external magnetic fields of the stator. The result is rotation of the rotors.

Heat Dissipation

The x-ray tube uses a number of methods to dissipate the heat produced at the target.

The tube is encased in an oil bath that serves as a thermal cushion to dissipate heat and an insulator to protect against electric shock. The rotating anode also helps to dissipate heat.

The tubes have a cooling fan. As the heated oil expands, it will activate a micro-switch, which will prevent an exposure if the expansion is too great.

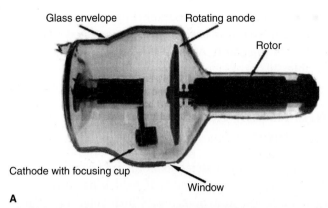

Glass envelope   Rotating anode

Rotor

Cathode with focusing cup

Window

A

B

Figure 5–3. Actual x-ray tube: **(A)** with rotating anode; **(B)** with stationary anode.

X-Ray Tube

The evacuated x-ray tube is often made of Pyrex glass or metal which can withstand heat. The vacuum allows efficient acceleration of the electrons from the cathode to the anode (Fig. 5–3).

The *protective housing* of the x-ray tube is lined with lead to reduce the levels of leakage radiation. This protects the operator from the dangers associated with high voltage and protects the tube against rough handling.

The *x-ray tube window* or *exit port* is the thin area of glass or metal enclosure approximately 5 cm or 2 inches long and allows the x-ray beam to leave the tube.

DESIGN CHARACTERISTICS OF MAMMOGRAPHY UNITS

The breast consists of all soft tissue structures with very similar mass densities and atomic numbers. Because of the similarity in tissue thickness and composition, general radiography imaging systems cannot be used to image the breast tissue. To maximize the contrast at the highest possible resolution, yet keep radiation doses low, general radiography systems have to be modified for mammography imaging. Dedicated mammography systems today have specially designed x-ray tubes, using grids and breast compression, and have an effective automatic exposure control (AEC) mechanism. These design features use soft tissue technique to enhance the differential absorption of the similar densities in the breast. In analog imaging, the cassette and film used and the development method had to be adjusted to reflect the necessary sensitivity. Digital mammography has an inherently wider latitude, which only serves to enhance the system contrast. By law, all mammography imaging in the United States must use dedicated mammography units, whether analog or digital. Mammographic imaging also involves regularly scheduled quality control checks that are mandated by the US Mammography Quality Standards Act (MQSA). These testing details are covered in Chapter 11.

ELECTRON INTERACTION WITH MATTER— X-RAY PRODUCTION

The atom is the basic unit of all matter, and atoms are present is everything. The atom consists of a central nucleus, which has protons, or positively charged particles, and neutrons, which have no charge. Orbital electrons revolve around the nucleus of the atom. Electrons are negatively charged and rotate in specific orbits or shells. The shell closest to the nucleus is called the *K shell*. Groups of atoms, bound together, are called *molecules.* If the atom has an equal amount of positive and negative particles, it is neutral. If the atom has more electrons than protons, it is a negatively charged ion, an *anion.* If the atom has more protons than electrons, it is a positively charged ion, a *cation.*

In general, when a high voltage is applied to a filament, it releases electrons in a process called *thermionic emission.* The electrons produced rapidly accelerate in the evacuated glass tube (the x-ray tube) and strike the target, which must be made of a heavy metal with a high melting point. X-rays that are produced when electrons strike a target in an x-ray tube are referred to as *ionizing radiation.* There are two types of x-ray production, characteristic or photoelectric radiation and bremsstrahlung radiation. Both of these x-ray production processes are very inefficient, with only approximately 1% of the interaction at the target resulting in x-ray production. The resulting 99% of the interaction is heat. This means that the x-ray tube has to be designed to effectively dissipate the excessive heat.

Characteristic or Photoelectric Radiation

If a projectile electron strikes the target of the x-ray tube, it can interact with an inner shell electron, knocking the electron out of its orbit. The ejection of an inner or K shell photoelectron results in a vacancy in the K shell, which is then filled by an outer shell electron. The movement of the electron from the outer to the inner shell results in the emission of an x-ray. The x-ray emitted, called *characteristic radiation*, has energy equal to the difference in the binding energies of the orbital electrons involved. A photoelectric interaction cannot occur if the energy of the incident electron is lower than the binding energy of the electron of the target (Fig. 5–4). The higher the atomic number of the target, the higher the binding energy. This means that the characteristic radiation produced by each element will be different, will depend on the atomic number of the element, and is not affected by the kilovoltage peak (kVp).

> **Types of x-ray production:**
> - Characteristics or photoelectric radiation
> - Bremsstrahlung radiation

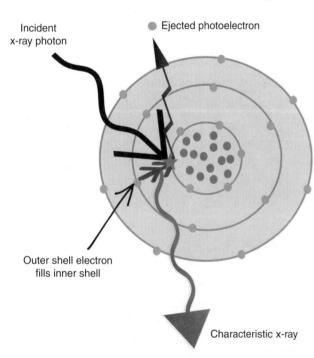

Incident
x-ray photon

Ejected photoelectron

Outer shell electron
fills inner shell

Characteristic x-ray

Figure 5–4. Characteristic radiation—an incident electron ionizes the atom by totally removing an inner shell electron. This creates a vacancy, which is filled by an outer shell electron falling into the void. This process is accompanied by a release of energy. The energy released is equal to the difference in the binding energies of the orbital electrons involved in the interaction.

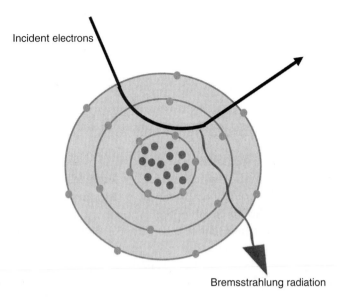

Incident electrons

Bremsstrahlung radiation

Figure 5–5. Bremsstrahlung radiation—an incident electron loses some of its kinetic energy and changes direction in an interaction with the nuclear field of an atom. This loss of energy is emitted as bremsstrahlung radiation. The electron can lose any amount of energy; therefore, the radiation emitted can have any energy up to the energy of the incident electron.

Bremsstrahlung Radiation

A projectile electron striking the target of the x-ray tube can lose its kinetic energy as it interacts with the nuclear field of the target atom. Because the electron has a negative charge and the nucleus is positive, the electron is attracted to the nucleus. It is slowed and will change course as it loses energy. The loss of kinetic energy reappears as an x-ray photon. This type of x-ray production is called *bremsstrahlung x-ray* (Fig. 5–5). Unlike characteristic radiation with specific energies, the energy of bremsstrahlung radiation can vary tremendously because the incident electron can lose some, all, or very little energy. The only certainty is that the final energy cannot exceed the initial energy.

X-RAY INTERACTION WITH MATTER

X-rays are dangerous to the human because they ionize or interact with atoms in the human body. They are therefore referred to as *ionizing radiation*. Whether they penetrate without interacting, are completely absorbed by depositing their energy, or are deflected from their original direction, these interactions can cause changes to the body at the atomic level. If the x-ray deposits energy into the tissue, this energy causes the molecules of our tissues to change or become ionized. The result is breakage of the molecule or relocation of the atom within the molecule. The most important molecule in the human body is DNA, which contains the genetic information for each cell. DNA molecules interact with each other in the form of chromosomes. If radiation damages our DNA, over time the molecules could cease to function, or function erratically because of chromosome aberrations.

Most radiosensitive tissues are:
- Stem cells
- Young cells—children
- Developing embryo or fetus
- Reproductive cells
- Cells in gastrointestinal tract
- Cells in thyroid, blood cells, or bone marrow

Least radiosensitive are:
- Mature cells
- Muscle
- Heart
- Nerve cells

The dangerous effects of radiation, therefore, take place at the cellular level in our bodies. Cells and tissues can recover from radiation injury, but sometimes the changes are irreversible, such as in genetic mutations or cell death.

Scientists also found that some organs and tissues are more radiosensitive than others. Stem cells are the most radiosensitive, whereas mature cells are more resistant to radiation. Other radiosensitive organs, tissues, or cells include younger tissues and organs in the developing embryo, the fetus (especially between 8 and 15 weeks' gestation when the rate of growth in the brain is the highest), and infants and children. Pregnant patients should therefore take precautions before any x-ray examination and should consult their physician to discuss the possible benefits versus risks before any x-ray procedure. Cells that are undergoing high metabolic activity level or rapid and continued regrowth, including the reproductive cells, cells in the thyroid, blood cells such as bone marrow, and the cells in the mucosal lining of the intestinal tract, are also more radiosensitive. The least radiosensitive cells are muscle, heart, and nerve cells because these cells undergo less cell renewal.

There are five known interactions between x-ray photons and matter at the atomic level: Compton's effect, photoelectric effect, coherent scattering, pair production, and photodisintegration. Compton's effect and photoelectric effect are the only two that play important roles in diagnostic imaging.

Compton's effect involves the interaction of moderate energy x-ray photons with the outer shell electron of the atoms (Fig. 5–6). The incident x-ray ejects an outer shell electron from the atom, ionizing the atom. This ejected electron is called *Compton's electron* or the *secondary*

X-ray interaction with matter:

- Compton's effect
- Photoelectric effect
- Coherent or classical scattering
- Pair production
- Photodisintegration

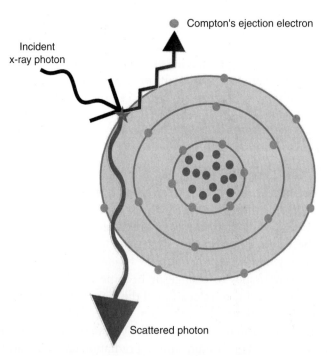

Figure 5–6. Compton's effect—a moderate-energy x-ray interacts with an outer shell electron. The x-ray is scattered and loses energy. It also causes an outer shell electron to be ejected—called *Compton's electron*.

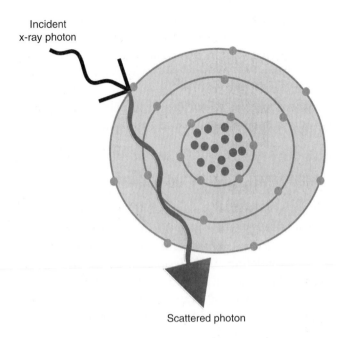

Figure 5–7. Coherent scattering—low-energy x-ray interaction with an atom. The atom releases excess energy as a scattered x-ray. Both the incident x-ray and the scattered x-ray have the same energies, but the scattered x-ray will go off in a different direction.

electron. The incident x-ray changes path but continues with reduced energy. This is Compton's scattered x-ray, and its energy is equal to the difference between the energy of the incident x-ray and the energy of the ejected electron. It will produce no useful information on the radiograph and serves only to reduce the contrast of the radiographic image.

The *photoelectric effect* involves the interaction between an incident x-ray and an inner shell or K-shell electron of the atom. The x-ray is not scattered but is totally absorbed, and the electron removed from the atom is called the *photoelectron.* This photoelectron will have energy nearly equal to the energy of the incident x-ray. This process is similar to the process of characteristic or photoelectric radiation production (see Fig. 5–4).

Coherent or *classical scattering* describes the interaction of very low energy x-rays with the electron in a target atom (Fig. 5–7). Most of these energies are too low to play a role in diagnostic imaging but could affect mammography imaging. The targeted electron becomes excited and releases its excess energy as scatter. In coherent scatter, the incident x-ray and the scattered x-ray have the same energy; however, the scattered x-ray goes in a different direction from the incident x-ray, reducing the contrast of the image.

Pair production involves x-rays at energies greater than 1.02 MeV and is not a concern in diagnostic imaging (Fig. 5–8). In pair production, the incident x-ray comes

> Compton's and photoelectric effects are the only two interactions that play important roles in diagnostic imaging.

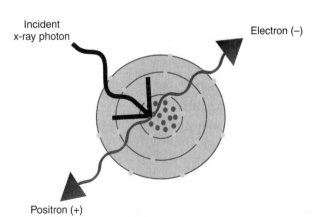

Figure 5–8. Pair production—high-energy x-ray interaction with the nucleus of the atom. The x-ray disappears, and two electrons appear: one positive and one negative.

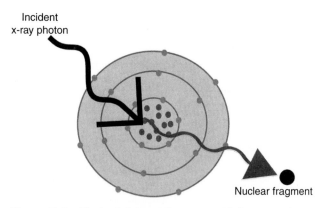

Figure 5–9. Photodisintegration—very high energy interaction with the nucleus. The nucleus becomes excited and emits a nucleon or other nuclear fragment.

Important modifications of mammography tubes:

- Compression
- Bream restriction
- Exposure factors—kV and milliampere-second (mAs)
- Filtration
- Focal spot sizes
- Grids
- Magnification
- Target material

close enough to the nucleus of an atom to cause the x-ray photon to disappear. Two electrons are formed by the disappearing photon, one positive and one negative.

Photodisintegration also involves high-energy x-rays in the region of 10 MeV (Fig. 5–9). The incident x-ray is totally absorbed by the nucleus of the atom, which then disintegrates and emits a nucleon or other nuclear fragment.

MAMMOGRAPHY IMAGING

Mammography systems involve numerous trade-offs. The aim is to produce high-quality imaging capable of visualizing microcalcifications as small as 0.1 mm or less and differentiating subtle lesions in an inherently low-contrast structure. Radiation dose in mammography must also be kept as low as possible because women will be undergoing yearly mammography, and there should not be a significant risk for developing breast cancer from the radiation involved.

Many components together play important roles in achieving the highest possible resolution. These include the use of a small focal spot, compression, and low kVp. Even the design characteristics of the x-ray tube serve to improve image quality. Today, mammography imaging in the United States is regulated by the federal government under the MQSA standard, which insists on high-quality images. These standards, which are checked with a continuous quality improvement (CQI) program, help to ensure that women in the United States will receive high-quality imaging.

Breast imaging was one of the last modalities in radiology to migrate to digital. The state-of-the-art mammography systems were originally designed using an analog method of imaging. Analog imaging produced the images in a purely mechanical way, with a final hard copy film that cannot be modified. A film was used to acquire the diagnostic information, to display the information, and to store the information. These imaging systems offered a high contrast and high spatial resolution of 11 to 20 line pairs per millimeter (lp/mm). Films, however, had drawbacks. Films could easily cost departments thousands of dollars per year. Then there are storage costs, and films were easily lost, misfiled, misplaced, or stolen. Another big disadvantage of film was the rigid exposure factors that were required to produce optimum images. There was no adjusting for overexposure or underexposure, and the fixed gradient or contrast provided a challenge for technologists because of the need to optimize contrast to aid interpretation.

In digital mammography, the acquisition, display, and storage of the image are independent processes that can be individually manipulated. The final image can be manipulated because the image is created as a series of binary digits in a process that is separate from the image display. The image can be displayed in film format, on a computer monitor, or on a flat panel. There are also many archival options available: printed film, optical disk, redundant array of independent disk (RAID), or digital linear tape (DLT) (Fig. 5–10). Digital mammography is discussed in detail in Chapter 8.

Figure 5–10. **(A)** Digital linear tape (DLT); **(B)** optical disk.

Digital imaging can be combined with a picture-archiving and communication system (PACS) enabling teleradiography and filmless libraries, which can be accessed by telephone, Internet, or any other off-site location.

MAMMOGRAPHY TUBE

General Radiography Versus Soft Tissue Imaging

Both the general x-ray tube and the mammography tube are electronic vacuum tubes with two electrodes (see Fig. 5–11). However, in general radiography, the useful kV ranges from 50 to 120. In this range, Compton's effects predominate, and there is loss of contrast with little differentiation between soft tissue structures. At lower kV, the photoelectric effect will predominate, allowing for enhanced differentiation of the soft tissue structures. Mammographic imaging is a careful balance of lowering the kV yet also trying to keep the dose to the patient and hence the mA as low as possible.

Filtration

As x-rays leave the target of the anode, they are filtered in two ways. There is inherent filtration, which is the oil in the x-ray tube, the window of the x-ray tube, the collimation assemble, and the compression plate. These structures are all a part of the mammography tube construction and will modify the beam before it enters the patient's breast. There is also added filtration, which is usually aluminum (Al) or aluminum equivalent, positioned in the path of the emerging beam, thus affecting its overall energy, defined by the half-value layer (HVL). The HVL will be affected by both the added and inherent filtration.

In general radiography, very low energy x-rays will be absorbed by the superficial tissue and will only increase patient dose without contributing to the image formation. Mammographic imaging, however, needs these low-energy x-ray photons, and it is important that any filtration does not attenuate the low-energy photons. But even in

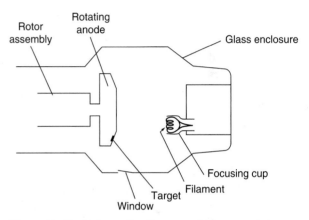

Figure 5–11. Schematic diagram of the x-ray tube showing the principal parts.

Added filtration
- Aluminum equivalent

Inherent filtration
- Oil bath
- Glass window of the x-ray tube
- Collimation assemble-mirror
- Compression plate

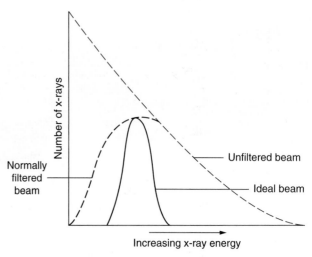

Figure 5–12. Graph of filtration versus no filtration. Filtration will remove x-ray energies that are too low and too high from the beam.

mammography, some added filtration is necessary to eliminate the extreme low-energy photons that will only contribute to patient dose and the high-energy photons that will result in a decrease in radiographic contrast (Fig. 5–12).

Inherent Filtration

The material used for the exit port or window in a mammography tube is borosilicate glass or beryllium (Be) because regular glass window would harden the emerging beam by eliminating the soft characteristic radiation. The oil bath in the x-ray tube, the mirror assemble of the collimation design, and the composition of the compression plate are all designed to attenuate very little of the x-ray beam.

Added Filtration

In mammography, added filtration shapes the emission spectrum of the x-ray beam and makes it compatible with the image receptor and breast characteristics of each patient. General radiography imaging uses approximately 1 to 2 mm Al equivalent of added filtration, but because of the effect of filtration on beam quality, mammography tubes generally use an aluminum equivalent element that is often the same element as the tube target. These filters give an equivalent filtration of only 0.1 mm of Al. This practice is designed to allow the K-characteristic x-rays of the specific target to be emitted while absorbing the higher and lower energy bremsstrahlung x-rays. Any photon energy above or below the K-shell binding energy of the filter material will be absorbed.

At the K edge, the x-ray attenuation of a material abruptly increases; as a result, the transmission through that material will decrease. Using molybdenum filtration with a molybdenum target will effectively reduce the higher energy photons in the mammography x-ray beam. When using a molybdenum target, a molybdenum filtration of 30 μm (30 μm = 0.03 mm) is used or a rhodium filtration of 50 μm (0.05 mm) is recommended. If a rhodium target is used, a filtration of 50 μm rhodium is recommended. Rhodium targets produce

Molybdenum
- Better penetration of fatty breast

Rhodium
- Atomic number higher than molybdenum
- Better penetration of dense breast

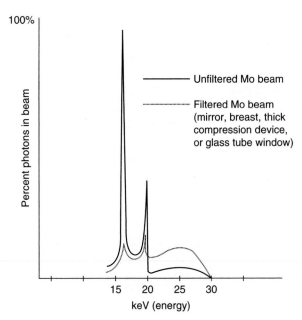

Figure 5–13. Molybdenum x-ray emission spectrum. The unfiltered molybdenum beam with prominent characteristic x-ray emissions, but also some bremsstrahlung x-ray emissions. The x-ray output below 10 keV is zero. As the energy increases, the radiation output slowly increases. At 17.5 keV, the output spikes to a high value. It then drops and begins another gradual increase. At 19.5 keV, it again spikes. These two spikes of output represent the characteristic x-ray energies for molybdenum. After the second spike, the x-ray output dips almost to zero, then begins a gradual increase, which eventually tapers off.

characteristic radiation that is 2 to 3 kiloelectronvolts (keV) higher than molybdenum targets (Fig. 5–13). A rhodium–rhodium target–filtration combination will allow a higher energy x-ray beam compared with molybdenum because of the higher K-edge x-ray attenuation for rhodium (Fig. 5–14). A rhodium–rhodium combination provides a

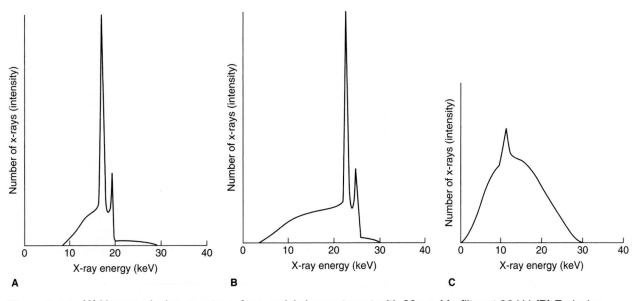

Figure 5–14. **(A)** X-ray emission spectrum for a molybdenum target with 30 μm Mo filter at 26 kV. **(B)** Emission spectrum for a rhodium target with 50 μm Rh filter at 28 kV. **(C)** X-ray emission spectrum for a tungsten target with 0.5 mm Al filter at 30 kV.

slightly more penetrating beam and is often used to image extremely dense breasts. A digital system will often use tungsten targets enabling the use of even higher tube loads; however, K-edge filters such as 0.05-mm rhodium would be used to shape the beam, therefore allowing lower exposure times at higher mA (Fig. 5–15).

Minimum HVL is specified by regulations and should not measure less than 0.30 mm Al at 30 kV or 0.25 mm at 25 kV. The HVL also should not exceed 0.40 mm Al at 30 kV to avoid filtering out the low-energy photons necessary to penetrate the breast tissue. Any HVL assessment must include total filtration, both added and inherent.

The Anode and Target

Modern mammography units are manufactured with molybdenum, rhodium, or sometimes tungsten targets, matched with the appropriate K-edge filters. These targets have a high melting point to withstand the

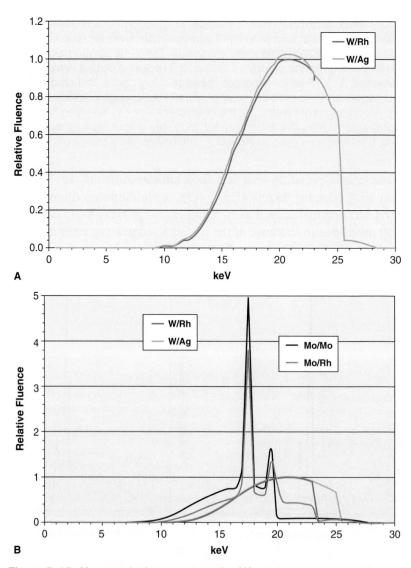

Figure 5–15. X-ray emission spectrum for **(A)** a tungsten target with rhodium or silver filter at 28 kV, and **(B)** a molybdenum target with molybdenum and rhodium filtration.

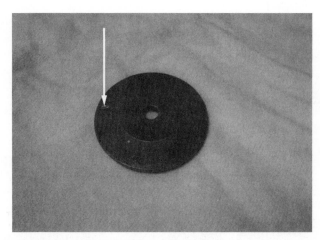

Figure 5–16. A damaged x-ray target. The *arrow* points to a "pit" in the target of the anode.

heat generated when the electrons strike the target, but they have different atomic numbers and, therefore, a different emission spectrum (Fig. 5–16). Most units have dual targets with molybdenum for imaging fatty breasts and rhodium for imaging dense breasts. Some digital units use a single target of tungsten. Single targets are more reliable and less expensive to produce. They also allow maximum anode heat loading compared with dual targets. A single target tungsten tube supports 2 to 3 times the maximum anode load compared with a dual target.

For mammography targets made with molybdenum, the most prominent x-rays are characteristic radiation, produced after a photoelectric interaction. Breast tissue consists mostly of adipose, fibrous, and glandular tissue, which all have a relatively low atomic number. The energy of the x-rays produced must be just high enough to penetrate the tissue without overpenetration, which will reduce visibility of the image detail. The x-ray range found most useful in maximizing contrast in breast tissue is in the 17 to 24 keV range. If the target is filtered with molybdenum, the characteristic energy of 19 keV from the K-shell interaction will be prominent. With proper filtration, the molybdenum anode will produce few x-ray photons above or below the preferred mammographic energy range. This makes molybdenum very effective for fatty breast tissue.

The characteristic x-rays produced using rhodium targets with rhodium filtration are similar to those using molybdenum, but because rhodium has a slightly higher atomic number, more bremsstrahlung x-rays are produced, and the K-characteristic x-rays are 2 to 3 keV higher, which provides better penetration of dense breasts (Fig. 5–17). Although better at penetrating dense breast tissue, the use of rhodium results in loss of radiographic contrast when imaging with analog technology. Rhodium is therefore not recommended for fatty breasts in analog imaging.

Tungsten targets with rhodium or silver filtration are more common in digital mammography and digital breast tomosynthesis (DBT) imaging and are not used in screen/film mammography because there the bremsstrahlung x-rays will predominate and the K-characteristic x-rays will be in ranges of 60 keV and above (Box 5–1). The higher energy

---

Box 5–1. Comparison of Target Material and Characteristic X-Ray Emission

K-Shell Characteristic X-Ray Energies
- Molybdenum, Mo (k_α = 17.48 keV; k_β = 19.61 keV)
- Rhodium, Rh (k_α = 20.22 keV; k_β = 22.72 keV)
- Tungsten, W (k_α = 59.32 keV; k_β = 67.24 keV)

Filter Material and Binding Energies
- Molybdenum, Mo (K shell = 20.0 keV)
- Rhodium, Rh (K shell = 23.3 keV)
- Silver, Ag (K shell = 25.5 keV)
- Tungsten, W (K shell = 69.5 keV)

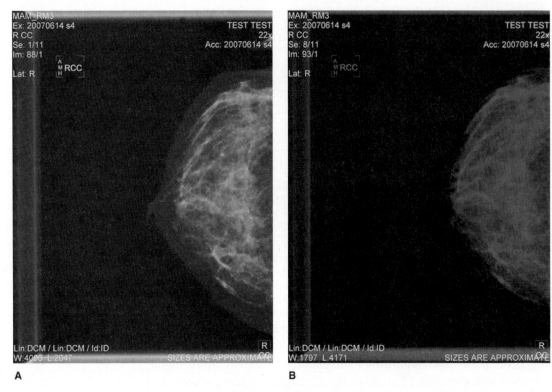

A   **B**

Figure 5–17. Images of a phantom acquired in manual mode using the same, 32 kV: **(A)** rhodium filter; **(B)** molybdenum filter. (From http://www.upstate.edu/radiology/rsna/mammography/beam/. Accessed January 2011.) (Used with permission from the Medical University of South Carolina.)

Box 5–2. Common Target Filtration Combinations

- Molybdenum target with 0.03 molybdenum filtration

- Rhodium target with 0.025 mm rhodium filtration

- Molybdenum and tungsten targets with molybdenum and rhodium filtration

- Molybdenum and rhodium targets with molybdenum and rhodium filtration

- Tungsten target with rhodium or silver filtration

x-ray photons will produce more Compton's interactions that decrease contrast in analog imaging. However, the higher energies emitted by the tungsten targets offer a wider dynamic range that can be exploited by the detectors in digital units, allowing shorter exposure times and less possibility of motion, which would degrade image quality.

In any mammography system, special target material and filter combinations will essentially shape the x-ray beam, providing the necessary kilovolt range to penetrate dense or fatty breasts (Box 5–2).

Line-Focus Principle

The focal spot is the area that electrons strike on the target. This is the actual focal spot. However, the effective or nominal focal spot is the area projection on the patient. As the size of the focal spot decreases, the heating of the target is concentrated into a smaller area. In the design known as the *line-focus principle*, the target is angled, allowing a larger area for the electrons to strike while maintaining a small effective focal spot. The angle on the target, called the *target* or *anode angle*, describes the angle between the surface of the target and a vertical line perpendicular to the image receptor (Fig. 5–18).

General radiography may use target angles of 5 to 15 degrees, but in mammography units, the target angles are approximately 20 to 24 degrees. Mammography tubes are tilted from the horizontal axis; therefore, in measuring the target angle, the angle of tilt is also taken in account. Typically, the target angle used in mammography is the sum of the angle of the anode within the tube and the tilt of the x-ray tube.

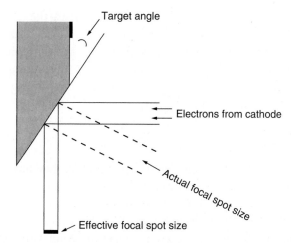

Figure 5–18. Line-focus principle. As the size of the focal spot decreases, the heating of the target will concentrate into a smaller area. Angling the target makes the effective focal spot size much smaller than the actual focal spot size. This principle will allow a large area for heating while keeping the effective focal spot small. Decrease in target angle will cause a corresponding decrease in effective focal spot size.

Both the target angle and the x-ray tube tilt are necessary so that the x-ray beam will cover a 24 × 30 cm detector at 60 to 65 cm source-to-image receptor distance (SID), while still maintaining a small focal spot size. Most manufactures will tilt the x-ray tube approximately 6 degrees from the horizontal, allowing the central ray to run parallel to the chest wall (Fig. 5–19; Boxes 5–3 and 5–4).

> **Box 5–3.** Recommended Focal Spot Sizes in Mammography
>
> - 0.4 mm or smaller for routine work. The most commonly used is 0.3 mm.
> - 0.15 mm or smaller for magnification. The most common focal spot size in magnification is 0.1 mm.

> **Box 5–4.** Determining the Size and Shape of the Focal Spot
>
> Size and shape of the focal spot are determined by:
> - Size and shape of electron beam hitting anode
> - Design and relationship of the filament coil to the focusing cup
> - Angle of the anode

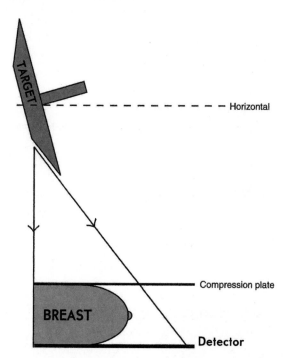

Figure 5–19. Mammography tube tilt. With the x-ray tube tilted approximately 6 degrees off the horizontal, the central rays run parallel to the chest wall so that no breast tissue is missed. Tilting the tube allows a smaller target angle and therefore a smaller effective focal spot size, while minimizing the heel effect. Tilting also allows greater anode heat capacity because the actual focal spot size is not further reduced.

Decreasing the target angle would further decrease the effective focal spot. However, the decrease in target angle is limited because of the need to cover the field size. A large anode angle should typically result in poor spatial resolution, but mammography tubes compensate by using a small actual focal spot size and limit tube damage by using lower mA and longer exposure times.

Focal Spot Size

Most mammography units are dual focus; that is, they have two focal spot sizes. The larger focal spot may be 0.4 mm but is generally 0.3 mm, and the small focal spot ranges from 0.15 to 0.1 mm. The most common size is 0.1 mm. Routine work uses the large focal spot size, whereas the small focal spot size is used for magnification work. The large object-to-image receptor distance (OID) in magnification would produce geometric unsharpness. This effect is reduced by use of a small focal spot size.

Anode Heel Effect

All x-ray tube targets are angled in the design of the line-focus principle. As a result of the angulation, some of the useful beam travels through the target material. In general, the smaller the anode angle, the greater the proportion of useful beam absorbed by the anode heel. This phenomenon is known as the *anode heel effect* (Fig. 5–20). The heel effect reduces the intensity of the useful beam at the anode end of the tube. To minimize the effects of the heel effect in breast imaging and to

> The heel effect reduces the intensity of the useful beam at the anode end of the tube.

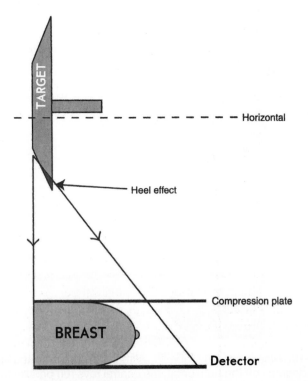

Figure 5–20. Anode heel effect. The *shaded area* indicates absorption of x-rays by the target. This results in the lower beam at the anode end of the tube. This effect is minimized in mammography by placing the anode toward the nipple (less thick) area of the breast.

allow small focal spot sizes, the cathode is positioned toward the chest wall, and the tube is tilted approximately 6 degrees off the horizontal. This allows the central ray to run parallel to the chest wall so that no breast tissue is missed. This allows for a more uniform attenuation of the x-ray beam because the greater tissue thickness is positioned toward the cathode, with the less dense nipple region at the anode.

Kilovoltage Range

A high-contrast radiograph has a lot of black and white areas compared with a low-contrast image with a lot of gray areas. Methods of controlling contrast include the use of beam restriction devices or grids to prevent scattered radiation from reaching the image receptor.

In any imaging, the kV controls the wavelength or the penetrating power of the beam. Increased kV will increase the energy of the x-ray beam. The result is greater penetrating power. Increased kV is necessary to penetrate dense breast tissue; however, at high energies, Compton's effect will predominate, producing more scatter radiation. However, as the kV is reduced, for the same tissue thickness, the penetrating ability of the beam is also reduced, resulting in the need for higher milliampere-second (mAs). Therefore, increased kV needs lower mAs and will allow lower radiation dose to the patient.

Technically, the effect of higher kV is a reduction of radiographic contrast. However, in digital imaging, the image acquisition and display are separated, and the user can alter the contrast of the image after the exposure. As a result, controlling factors of contrast are not only the kV but also the processing software and predetermined digital processing algorithms. The technical factors can be selected to minimize patient dose without compromising image contrast.

The kV selected by the technologist will depend on a number of factors, including radiologist preference, equipment calibration, manufacturer's recommendations, equipment design, processing algorithms, breast size, and breast thickness. However, the kV range on mammography units should still be kept low because the breast is all soft tissue imaging.

The actual kV range available on the mammography unit will depend on the target and filter material. Generally, the kV available on mammography units will be between 20 and 40 kV (Box 5-5).

- In analog imaging, the kV controls the subject contrast, exposure latitude, and radiographic contrast.
- In digital imaging, the controlling factor of contrast is not only the kV but also the processing software and predetermined digital processing algorithms.

Box 5–5. Kilovolt Range for Various Target Materials

- Molybdenum target tube: 24–30 kV
- Rhodium target tube: 26–32 kV
- Tungsten target tube: 23–35 kV

Exposure Time, Milliamperes, and Milliampere-Seconds

The milliampere (mA) determines the number of x-rays produced and, therefore, the radiation quantity. The mA and the exposure time are usually combined to express the milliampere-second (mAs). In digital imaging, the term *density* (optical density) has been replaced by the term *brightness*. The brightness of the image can be altered after processing; therefore, the controlling factor is not just the mAs but also the processing software and predetermined digital processing algorithms.

All dedicated mammography units will have a means to vary the selected mAs, and generators will offer selection from 20 to 600 mAs. The mAs should be kept low to reduce radiation dose to the patient,

- In analog imaging, the mA is the primary factor controlling the radiation quantity, optical density, and patient dose.
- In digital imaging, image brightness (density resolution) is controlled by processing software and predetermined digital processing algorithms.

and although exposure times must be kept as short as possible to minimize motion blur due to patient motion, extremely short times will cause grid artifacts.

The reciprocity law fails for extremely short exposure times (less than 1/100 second) and extremely long exposure (more than a few seconds), forcing the use of additional mA to compensate for the lower optical density. In routine imaging, this should not be a problem with exposure times varying from 0.4 to 1 second, but with magnification imaging, longer exposure times of 2 to 4 seconds are used; therefore, the mA must be increased to compensate for reciprocity failure. Dedicated mammography units of today will automatically compensate for reciprocity law failure.

> The primary goal of compression in mammography is to reduce breast thickness uniformly.

Compression Devices

Mammography cannot be performed without breast compression (Fig. 5–21). The primary goal of compression in mammography is to reduce breast thickness uniformly. With compression, there is also improved visualization of breast structures because the tissue is spread out over a larger area with less superimposition of underlying and overlying structures, allowing for a more uniform attenuation of the x-ray beam. This will permit uniform penetration by the x-ray beam and an optimum exposure over the entire breast. Compression will also reduce magnification of a lesion by bringing the area of interest closer to the image receptor, resulting in increased sharpness and resolution. With the reduction of tissue thickness, the compressed breast will require less kV and mAs, with the associated reduction in scattered radiation and patient dose, plus increased radiographic contrast. Motion unsharpness is also reduced because the breast is immobilized.

Compression Plate

A flat-surfaced compression plate provides compression in mammography (see Fig. 5–21). Both the height and angle of the compression plate make a difference in the final image. A compression device with a rounded or gently sloping chest wall edge does not allow uniform

Figure 5–21. Mammography compression plate.

Figure 5–22. A compression plate showing how the flex paddle angles during compression.

compression of the posterior area of the breast. The compression device should have a straight chest wall edge to allow the device to grasp the breast tissue close to the chest wall. The device should also rest flat, parallel with the image plate.

Newer mammography units have a built-in slanted compression paddle option, some called *flex paddles*. The design of these paddles is based on the principle of equal compression to all areas of the breast, especially when imaging the dense breast. With the typical compression paddle, the posterior (chest wall) edge of the breast receives the greatest force of compression, whereas the anterior portion of the breast receives little or no compression. The aim of using a sloping surface is to provide optimal compression to the entire breast. The slope of the paddle conforms to the breast contour. The compression device also has the advantage of reducing patient discomfort by not overcompressing the posterior breast tissue (Fig. 5–22).

The height of the chest wall edge of the compression paddle reduces the chance of chest tissue projecting on the mammogram. Shorter units require more care in positioning to avoid exposing the upper chest or missing posterior breast tissue. The design of the lip of the compression paddle has a lesser effect on the anterior aspect of the breast. Placement of the lip, just beyond the chest wall edge of the image plate, prevents the projection of an image of the chest wall edge of the plate on the mammogram (Box 5–6).

The material of the compression device is also important. A thin fire-retardant polycarbonate plastic is recommended. The plastic should attenuate very little of the x-ray beam and should be rigid enough to withstand the compression force.

All dedicated mammography units have compression devices with an automatic or motorized control, operated by foot pedals. All units also have a manual, hand-controlled device, controlled by the technologist. The technologist applies the force of compression; however, the maximum and minimum automatic compression force is regulated by the MQSA standards. The compression force should not reduce after final compression; and under automatic compression, the maximum force of compression should not exceed 45 lb (200 N), and the minimum

---

Compression plate:
- Material cannot attenuate low-kV photons
- Rounded posterior edge
- Straight chest wall edge

The height of the chest wall edge of the compression paddle reduces the chance of chest tissue projecting on the mammogram.

Box 5–6. Important Design Characteristics of the Compression Plate

- Flat surface must be parallel to the image plate.
- The chest wall edge may be bent upward to allow for patient comfort but shall not appear on the image.
- The chest wall edge of the compression paddle shall be straight and parallel to the edge of the image receptor.
- Systems shall be equipped with different sized compression paddles that match the sizes of all full-field image receptors provided for the system.

In digital systems, the image detector serves as the AEC device.

should be at least 25 lb (111 N). Generally, units are equipped with an automatic release, which releases the compression immediately after the exposure. A separate override feature will also automatically release the compression in cases of power failure or in situations in which the technologist needs to quickly remove an ill or extremely anxious patient from compression.

Automatic Exposure Control

Automatic exposure control (AEC) is used in all modern mammography units to control the actual exposure, therefore optimizing the image quality of the final image. With AEC on dedicated mammography machines, the chance of repeats and additional radiation dose to the patient will be significantly reduced.

Repeats due to poor exposure factors are even less common in digital imaging, which has a linear response to x-ray over a wide range of exposure values. Also, with digital imaging, the contrast, brightness, and spatial frequency of the image can be modified after the exposure.

In digital systems, the image detector serves as the AEC device. The systems can be set to automatically select the densest area of the breast for AEC detection, or the technologist can manually select the AEC detector position. Detectors used in digital units are manufacturer specific and include amorphous selenium or scintillating phosphors (such as cesium iodide). With digital imaging, because the detector covers the entire field size, the system is easily able to respond to a very small or very large breast (Box 5–7).

Density Compensation and Backup Timer

In addition to the AEC detectors, modern analog mammography systems must have a *density compensation circuit*. This refers to the optical density on the image. The density compensation circuit allows the selection of at least two steps above and two steps below the normal setting. Each step translates to a 12% to 15% increase or decrease in the mAs or a 0.15 change in the optical density.

To avoid gross overexposure, all systems are also equipped with a *backup timer*. The backup timer terminates the exposure when the backup time is reached. This avoids massive overexposure to the patient and damage to the tube. The backup timer for grid techniques is preset at 600 mAs and for nongrid at 300 mAs. The backup timer is reached during an exposure when the selected kV is too low. This will occur if low-energy photons cannot penetrate the breast because the breast is too dense or too thick.

To avoid activating the backup timer, the technologist should select a higher kV value, not higher mAs, because selection of higher mAs will not significantly change the energy of the x-ray beam. If there is insufficient breast tissue covering the detector, the backup timer is also reached when imaging breast implants. Many dedicated mammography units will automatically compensate if the selected kV is too low. This function will allow a proper exposure without reaching the backup time. With these types of units, the circuit adjusts the kV during the pre-exposure, the first 100 milliseconds (ms) of the exposure, and if the kV is too low, it is automatically increased, or the exposure is terminated quickly if the unit senses that backup timer will be reached.

Grids

The grid is basically a sheet of lead strips with a series of narrow interspace material. In typical x-ray units, the interspace material is usually lead, nickel, or aluminum. In mammography imaging, the interspace material of the grid can be carbon fiber, wood, or other low-attenuation material. X-rays needed to create a true image will travel in a straight line and will pass directly through the grid. Scattered radiation hits the grid at an angle and is absorbed by the interspace material. Grids are described by their frequency, the number of grid strips per inch or centimeter and ratio, and the height of the grid strips divided by the distance between each strip.

Grids are used to improve radiographic contrast by decreasing the amount of scattered radiation that reaches the image receptor (Fig. 5–23). Scatter increases as the energy of the photons increases and will increase the overall optical density of the radiograph but reduce

> **Grids will:**
> - Improve contrast
> - Decrease scattered radiation
> - Results in more radiation dose to the patient

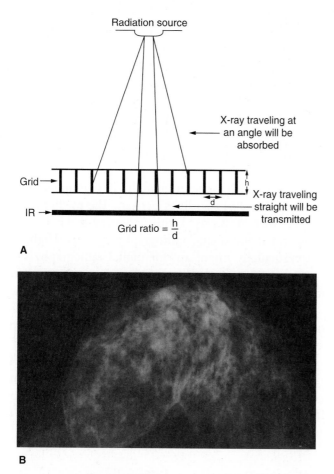

A

B

Figure 5–23. **(A)** A grid is designed to transmit radiation traveling in a straight line from the source to the image receptor and absorbs scattered radiations. **(B)** A mammogram showing the grid line—obtained if the grid is stationary during the exposure.

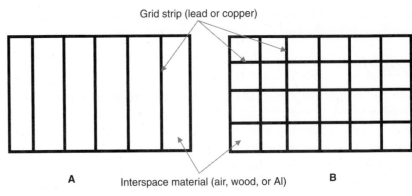

Figure 5–24. Cross section of **(A)** standard mammography; **(B)** high-transmission cellular (HTC) grid.

Characteristics of a mammography grid:

- Linear grid
- Grid ratio 3:1 to 5:1—focused to the SID
- Grid frequency 30–50 lines per cm
- Grid interspace material
  - Carbon
  - Wood

the radiographic contrast. All modern mammographic units have linear-patterned grids, which move in one direction and do not reciprocate (Fig. 5–24A). A moving grid will reduce the presence of grid artifacts. In addition to absorbing scattered radiation and improving contrast, grids will also absorb some of the primary beam. Grids will therefore result in the use of increased exposures and a corresponding increase in radiation dose to the patient. To minimize this risk, mammography grids are thinner and have a lower grid ratio than general radiography grids. The grid ratio of general radiography grids ranges from 6:1 to 16:1, but in mammography, grid ratios usually vary from 3:1 to 5:1, focused to the SID for increased contrast. On average, mammography grids have a ratio of 4:1. A higher grid ratio would require too large an increase in exposure. The grid frequency can range from 40 to 50 lines/cm (100–120 lines/inch) for low-frequency grids, 50 to 60 lines/cm (120–150 lines/inch) for medium-frequency grids, and 60 to 70+ lines/cm (150–170+ lines/inch). Low-frequency grids are generally used with a moving grid system, such as mammography. In mammography, the frequencies range from 30 to 50 lines per cm. The medium- and high-frequency grids are typically used with stationary grid holders, portable radiography, and many digital radiography systems. High-frequency grid use is particularly important for digital radiography systems to avoid aliasing artifacts. Grids are typically not used in magnification mammography and three-dimensional (3D) imaging.

Some mammography units use a high-transmission cellular (HTC) grid with characteristics of a crossed grid (Fig. 5–24B). These grids can reduce scatter in two directions rather than the one direction of a linear or focused grid. The HTC grids use copper instead of lead as the grid strip, and air rather than wood or aluminum as the interspace material. The physical design results in a grid ratio of 3.8:1, and when compared with a similar ratio linear grid, the use of an HTC grid results in equal or less radiation dose to the patient.

Beam-Restricting Devices

In general radiography, beam-restricting devices, such as cones, collimators, or diaphragms, are used to limit or collimate the size of the x-ray field to the area of interest. Collimating or decreasing field size reduces scatter production and therefore improves the image contrast, although decreasing the x-ray field will require an increase in exposure to maintain constant optical density.

Despite the benefits of collimation, in mammography, the exposure field is restricted to the size of the image plate. The entire image plate, and not just the area of the breast, is exposed. This is done to prevent extraneous light from compromising the perception of the fine detail when the radiograph is viewed. However, under MQSA standards, the collimation should not extend beyond any edge of the image plate by more than 1% of the SID.

> **Box 5–8.** Source-to-Image Receptor Distance in Mammography
>
> - Recommended source-to-image receptor distance (SID) = 50–80 cm

System Geometry (Source-to-Image Receptor Distance, Object-to-Image Receptor Distance)

The SID refers to the distance from the x-ray source to the image plate. SID is one of the many factors affecting the resolution. Lower SID will result in increased magnification. In mammography, in order to reduce unsharpness and to compensate for the relatively short SID, a small focal spot is used. The SID used is fixed generally in the range of 50 to 80 cm, with most units having a 60-cm SID (Box 5–8).

The OID refers to the distance of the object (or body part) from the image plate. The OID should be as small as possible to minimize magnification. The only exception is the microfocus magnification techniques used in mammography imaging.

MAGNIFICATION

Magnification is used in mammography to assess small lesions or microcalcifications (Fig. 5–25). Generally, magnification will reduce the resolution of the image because of geometric unsharpness. To compensate for the lost resolution, a very small focal spot is used. Magnification factors or degrees in which the image is magnified will vary. The standard is 1.5 and 1.8 times, but some machines will offer two magnification factors, with one as high as 2.0 times. To magnify the breast, the OID is increased by placing the breast on a raised platform or detector holder. This supports the breast at the higher OID. Grids are not used in magnification radiography. The large OID serves as an air gap to reduce scatter to the detector.

Magnification mammography never uses a grid. The scattered radiation is significantly reduced as a result of the large OID or air gap, and a grid will only contribute to increased radiation dose to the patient. Radiation dose to the patient is of concern with magnification. The radiation exposure to the breast, particularly to the skin of the breast, is much higher than with standard imaging because the breast is closer to the radiation source. Also, because of reciprocity failure, there is increased exposure time, further increasing the radiation dose to the patient. In general, the greater the magnification factor, the greater the skin dose to the patient (Boxes 5–9 and 5–10).

> **Focal spot size:**
>
> - 0.4 mm or smaller for routine work. The most commonly used is 0.3 mm.
> - 0.15 mm or smaller for magnification. The most common focal spot size in magnification is 0.1 mm.

> Magnification mammography never uses a grid.

> In digital imaging, the acquisition, display, and storage are separate functions that can be independently optimized.

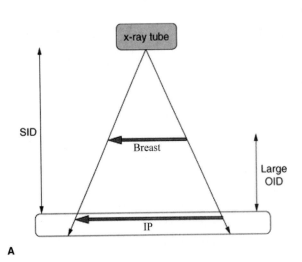

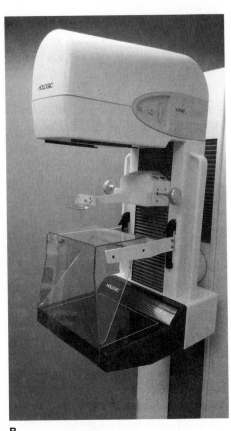

A **B**

Figure 5–25. Magnification setup: **(A)** schematic diagram; **(B)** mammography unit set up for spot magnification imaging.

Box 5–9. Common Magnification Factor

- 1.5 times
- Other factors can be 1.6, 1.7, 1.85, or 2 times

Box 5–10. Factors Affecting Resolution and Sharpness

- Motion due to long exposure times
- Increase in the focal spot size
- Increase in the object-to-image receptor distance (OID)
- Decrease in the source-to-image receptor distance (SID)
- The relationship between the OID and the SID

MONITORS

Various types of monitors are available for image display. The standard in the 1990s was the cathode ray tube (CRT). However, the liquid crystal display (LCD) and the organic light-emitting diode (OLED) display are now more common. When viewing images directly on a monitor, the ambient light should be strictly controlled. The digital image is formed as a two-dimensional (2D) matrix of square pixels of a fixed size, typically 0.04 to 0.1 mm.

The *CRT* is a specialized vacuum tube in which images are produced when an electron beam strikes a phosphorescent surface (Fig. 5–26). Most CRTs generate electromagnetic fields referred to as *extremely low frequency* (ELF). Some studies suggest that the ELF fields might have detrimental health effects on humans who are exposed to them for long periods of time. The fields are stronger to the sides rather directly in front of the tubes. Computer users should, therefore, sit 18 in. away from the CRT display.

LCD technology was originally used for notebooks or similar small computers (Fig. 5–27). This technology allows much thinner display than the CRT technology and works on the principle of blocking light rather than emitting it. The LCD display consists of two sections of polarized glass with liquid crystal between. The crystals diffuse the light and allow light of different waveforms to pass through the monitor, allowing the image to be viewed on the screen. LCD technology is

Figure 5–26. Cathode ray tube (CRT) monitor.

Figure 5–27. Liquid crystal display (LCD) monitor.

more energy efficient and safer than CRT. The newer high-quality LCDs use *thin film transistors* (TFTs). TFT-based displays have a transistor for each pixel on the screen. This allows the electrical current that illuminates the display to be turned on and off at a faster rate, which makes the display brighter and shows motion smoother. LCDs that use TFT technology are called *active-matrix* displays, which are higher-quality than older *passive-matrix* displays.

OLED systems are a new generation of organic light-emitting diode displays. They reduce power use in full-color matrix display and offer highly saturated color at a lower voltage.

OLEDs have a higher contrast and better viewing angles than the LCDs.

Both LCDs and OLEDs offer the potential for a high brightness, flat-panel display and a wide viewing angle. LCD has high luminance, but the contrast becomes lower as the viewing angle increases (see Fig. 5–27).

Comparison of CRT versus LCD:
- CRT
  - More luminescence and better color production
  - More durable
  - Faster refresh rate
- LCD
  - Less energy consumption
  - Less screen glare and better resolution
  - Lighter

Acquisition Workstation

After the exposure, a digital image will pop up, almost immediately, at the technologist workstation monitor. This workstation is also called the *acquisition workstation* (AWS). The AWS is typically low resolution, with an image resolution of 2 or 3 megapixels. The images on these monitors are sometimes referred to as the *soft-copy images* because the image data can be manipulated to accentuate or suppress various features of the image. They can be adjusted for contrast and brightness or magnified and further enhanced. Digital images viewed directly on a monitor are somewhat limited by decreased spatial resolution and luminance compared with hard-copy film in analog imaging (Fig. 5–28).

The presence of workstations also allows for rapid image viewing and serves to maintain quality assurance of the images. The technologist can immediately determine whether to repeat the examination. Workstations also allow for postprocessing of an image while viewing the changes being made.

Some of the changes that can be applied to the digital image include the following:

- Adjusting the image contrast and brightness: this is the windowing and leveling of the image. The window width controls contrast, whereas the window level controls brightness.
- The brightness value can be improved by bringing adjacent pixels in the image closer together in a process called *smoothing*.

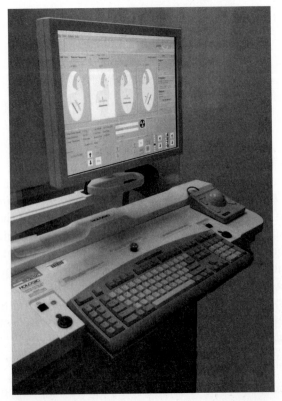

Figure 5–28. A digital unit showing the soft-copy display or acquisition workstation (AWS).

- All or parts of the image can be digitally magnified with the zoom feature, rotated, flipped, or inverted.
- The brightness of the edges of all structures can be enhanced in a process called *edge enhancement*. This process increases the visibility of the edges of structures.
- The background of the anatomy can be removed, which will allow visualization of other relevant anatomy.
- The image can be reversed. In general, dense structures image "white" on a radiograph. This is considered a negative image. In a reversal to a positive image, the dark and light pixel values in the image are reversed, and dense structures would be visualized as "black."
- In a process called *annotation*, text, rulers and measurements, and other markings can be added to the image.

Workstations enable the user to input patient data, such as name and date of birth, either by typing on an attached keyboard or by selecting from a Digital Imaging and Communications in Medicine (DICOM) work list. The AWS can be used to modify patient demographic information, then retransmit or reprint an image with the corrected information. It may also control image destination with selected transmission channels or routing codes. Typically, an AWS can hold close to 25,000 images for short-term storage on a 500-GB hard disk.

Review Workstation

The review workstation (RWS) monitors (sometime called *interpretation monitors*) used by the radiologist for interpretation are critical in digital imaging, and all diagnostic interpretation must be conducted on monitors certified by a physicist. Monitor checks include a check of the luminance to monitor the brightness; a check of uniformity to check the brightness over the surface of the entire monitor; a check of the contrast to monitor the ability of the monitor to display the full range of densities produced by the digital system; and a pattern test to detect bad pixels, spatial resolution, contrast resolution, and viewing angle cutoff. Many monitor manufacturers will provide a program or list of tests that must be conducted daily, monthly, or yearly (Fig. 5–29).

Figure 5–29. Review workstation (RWS) monitor.

In viewing the digital hard-copy printed image, a radiologist has to take into consideration the fact that ambient lighting will affect interpretation.

The mammography monitor should be a minimum of 4 megapixels, although at least a 5-megapixel monitor is recommended. Typical monitors are 21.3-inch, 5-megapixel, monochrome LCD. They are high contrast with a wide viewing angle and a luminance of up to 750 cd/m^2 (candelas per square meter). Many have luminance equalizer and calibration functions to enable smooth gray-scale display.

PRINT AND STORAGE OPTIONS

Digital mammography offers a wide variety of archival solutions, including the printed film, digital linear tapes (DLTs), optical disk, or redundant array of independent disks (RAID). Laser-printed images are lower resolution and often are not of interpretation quality. DLTs are magnetic tapes used in a cartridge tape unit. The image is encoded on the tapes based on the principle of electromagnetic induction. Because of the limited storage capacity of the tapes, they are now used primarily as backup storage. Optical disks are also called laser disks. They use laser technology to read and write to the disk. The compact disk (CD), compact disk–read-only memory (CD-ROM), and compact disk–rewritable (CD-RW) are examples of the optical disks. Optical disks can be stacked in multidisk format in drives called jukeboxes that can handle up to 100 disks. A RAID is a system of two or more disk drives within a single cabinet that collectively act as a single-storage system. The advantage of RAID is that if one disk drive is defective, the others can take its place.

Image storage or archival systems consist of a computer subsystem for storing images, patient demographic information, and a database that will allow access to this information. Various archival systems allow for a variety of storage capabilities plus variations in the rate at which data are retrieved from the system. Although in general, digital radiography decreases image archive storage space by reducing the size of the stored images with compression, mammography images uses lossyless compression. Also, the smaller pixel size of mammography digital images requires a tremendous amount of data storage. Typically, as pixels are made smaller, the image file size will increase. Images can be archived on a short-term basis by local storage on the hard drive of a workstation. They can be stored medium term by a RAID. They can also be stored long term by means of optical disks or DLTs. Other radiography images can be compressed, thereby reducing the required data storage capacity by storing data in encoded form. This allows for a more efficient use of storage space.

All storage should include a disaster recovery system, which would include backup copies of all images. Disaster recovery storage is necessary if primary or long-term storage fails; it can be configured with less expensive and often slower performing media and can be off-site or Web based. In any event, all digital systems should include duplication of data storage and redundancies to protect against system failure or even against viruses (Box 5–11).

Print & storage options
- Digital linear tapes (DLT)
- Optical disk
- Redundant array of independent disks (RAID)

Box 5–11. Example of Storage Options

- A single 18 × 24 cm image can need 16 MB of storage space
- A 24 × 30 cm image can need 28 MB of storage space
- Magnetic digital linear tape (DLT) tape holds 35 GB
- One optical disk can hold 3000 computed mammography (CM) images or 250 MB of data
- 5¼-inch single-optical disk holds 4 GB
- 5¼-inch dual disk holds 8 GB
- Multiple-platter optical jukebox (a series of optical disks) holds 5–10 TB

PICTURE-ARCHIVING AND COMMUNICATION SYSTEM

A *PACS* is a complete integrated system that offers the convenience of viewing images from multiple modalities—such as mammography, ultrasound, and magnetic resonance imaging (MRI)—all at one workstation. PACS also eliminate the need to manually file, retrieve, or transport images. The PACS image storage and retrieval system uses the DICOM standard. This allows the option of scanning nonimage data, such as patient documents and results, into the PACS for later retrieval. With PACS, the radiologist has the ability to review all the breast examinations from multimodalities, allowing comparison and integration. PACS can allow the entire database of images and reports to be accessed over the Internet, over the telephone, by satellite communication, or from cable modems. This presents the possibility of reaching remote locations across the nation or even across the world. PACS is only possible if all the systems (DICOM), hospital information system [HIS], and radiology information system [RIS]) are integrated.

Advantages and Disadvantages

After an image has been acquired onto the PACS, it cannot be lost, stolen, or misfiled. When the PACS is fully integrated, it can allow for the placement of numerous PACS terminals throughout a facility. This allows simultaneous multilocation viewing of the same image by different individuals as needed. The PACS database ensures that all images are automatically grouped with the correct examination, are chronologically ordered, correctly orientated, and correctly labeled. PACS allows the easy retrieval of patient images using a variety of criteria, such as name, hospital number, date, and referring clinician. After the exposure, all the patient's images can be sent immediately to the PACS and will be immediately available for viewing.

PACS offers the option of *prefetching* previous images of a patient to allow immediate comparison. *Prefetching* is the process whereby previous images of a patient are automatically retrieved from the long-term storage onto the short-term storage server, before the acquisition and viewing of the current imaging examination of that patient. PACS allows the user to manipulate and postprocess the image.

The biggest disadvantage of PACS is the installation expense. PACS must also be protected against system failure. All facilities using PACS should have a contingency plan in case of total failure of the system. In addition, the system should be equipped with automatic data backups to avoid data loss.

DICOM, HIS, RIS, EMR, AND HL7

Digital Imaging and COmmunications in Medicine (DICOM) is one of the most widely deployed health care messaging standards in the world. It is used for handling, storing, printing, and transmitting information in medical imaging. DICOM has a defined file format and network communication. DICOM allows communication between different medical imaging devices, such as scanners, printers, servers, workstations, network hardware, and PACS, from multiple manufactures. The National Electrical Manufacturers Association (NEMA) holds the copyright to the DICOM standard. Each device comes with DICOM Conformance Statements, which clearly state which DICOM classes they support.

Hospital information system (HIS) focuses mainly on the administrative needs of hospitals and health facilities. HIS can be a comprehensive, integrated information system. It is designed to manage all the aspects of a hospital's operation, such as medical, administrative, financial, and legal issues and the corresponding processing of services. The hospital's system would include patient's medical records, initial diagnosis, treatments, and all diagnostic tests.

Radiology information system (RIS) is the digital information storage and retrieval system used in imaging departments. The major functions of the RIS can include patient scheduling, resource management, patient registration, examination performance tracking, examination interpretation, results distribution, image tracking, and procedure billing.

An *electronic health record (EHR)*, *electronic patient record (EPR)*, or *electronic medical record (EMR)* refers to the digital collection of patient health information data. These records can be shared through network-connected systems within a facility or with other facilities. EHRs may include demographics, medical history, medication and allergies, immunization records, laboratory and diagnostic test results, radiology images, vital signs, personal statistics like age and weight, and billing information. EMRs can give patients, physicians, and other health care providers, employers, and payers or insurers access to a patient's medical records across facilities.

EMR systems can reduce the risk for data replication and are more likely to be up to date. They also decrease the risk for lost paperwork and allow searchable digital information. EMRs are also more effective when extracting medical data for the examination of possible trends and long-term changes in a patient.

In the past, the terms EHR, EPR, and EMR were used interchangeably. However, differences are evolving. The EHR is now often defined as a more longitudinal collection of electronic health information of individual patients or populations, whereas the EMR is the patient record created by providers for specific encounters in hospitals and ambulatory environments and can serve as a data source for an EHR.

Ideally, all the systems—DICOM, HIS, RIS, and EMR—should communicate with each other. The *Health Level Seven International (HL7)* organization is a not-for-profit organization that developed a system to standardize communication to allow the speedy management, delivery, and evaluation of health services. The organization was founded in 1987 and is dedicated to providing a comprehensive framework governing all standards related to health information, including the exchange, integration, sharing, and retrieval of EHRs. HL7 is able to fill the gap and allow the different computer systems within a hospital or health care setting to communicate with each other. With this system in place, clinical and administrative data between software applications can be transferred, shared, and retrieved. These include billing records, patient demographics, medical records, and appointments schedules.

Vendor Neutral Archives (VNAs) are the newest archiving solution in imaging departments. VNAs can store, manage, and share images across multiple modalities, organizations that can integrate the imaging department with other departments within a hospital, physician offices, and other off-site locations. The hope is that VNAs will eventually have the capability to store reports and insurance information on cards and non-DICAM formats, such as JPEG image files. VNAs work with an updated PACS system to tap image headers, perform lossless compression, and match new images with the most appropriate display application.

Wide area network (WAN) is a computer network that covers a large geographical area and generally involves major telecommunication or Internet links.

Local area network (LAN) is a limited computer network. A LAN network would interconnect computers to a server within a small geographical area such as within a school or hospital. Computers and other mobile devices using a LAN connection can share resources, such as a printer or network storage.

> *Wide area network (WAN)* covers a large geographical area and generally involves major telecommunication or Internet links.
>
> *Local area network (LAN)* is a limited computer network.

TELERADIOGRAPHY

Teleradiography is the transmission of the digital image from one location to another via electronic transfer such as the Internet. Transmitting digital mammography signals involves the transmission of huge data files, typically as large as 32 MB, for a single projection, and the transmission must often occur without a timeout to avoid compromising the signal. A four-series mammogram could therefore involve transmitting a 125-MB file. If the patient had prior studies, the total could easily exceed 0.5 GB per patient—just for a basic screening study. (In comparison, a typical MRI consisting of more than 100 images will need about 6.5 MB.) Computer-aided detection (CAD) can also add approximately 0.2 GB to the data load. This requires a lot of network bandwidth, which can be very expensive, especially because the American College of Radiology (ACR) requirement is that only lossless compression must be used on mammography images intended for interpretation. Fortunately, mammography compression algorithms have been developed that will allow compression of data at the site of origin and decompression at the interpretation site.

LOSSY AND LOSSLESS COMPRESSION

In information technology, *lossy compression* reduces a file by permanently eliminating certain information, especially redundant information. For example, Joint Photographic Experts Group or JPEG format is a common file format used to compress and analyze images in blocks of 8 × 8 pixels and selectively reduces the detail within each block. When the file is uncompressed, only a part of the original information is still there. The lossy compression JPEG technique is used to reduce data size for storage, handling, and transmitting content. Well-designed lossy compression technology often reduces file sizes significantly before degradation is noticed by the end user.

Lossless compression is a class of data compression algorithms that allows the original data to be perfectly reconstructed from the compressed data. Lossless compression can use a modification of the JPEG format or another file format. Lossless compression is used in mammography because it is vital that the original and the decompressed data be identical. The image is broken into smaller blocks of data at origination and put back together at the destination. With lossless compression, every single bit of data that was originally in the file remains after the file is uncompressed.

Before advancing to teleradiography, the facility will need a complete assessment of data storage capabilities and a fast, reliable, and secure transmission line. Secure lines are needed to satisfy the concerns of many patients who are expressing fears that there is insufficient data security over the public lines. A fast speed and reliable transmission of data is essential because heavy data loads could take close to an hour to transmit on slow-speed lines.

The other consideration is the issue of workstation compatibility. The PACS and all the associated components of the system must be compatible between the site of origin and the interpreting site. All quality assurances must be in compliance at both sites.

COMPUTER-AIDED DETECTION

Visually inspecting the mammogram can be a difficult, tedious, and time-consuming task because mammograms are complicated images, and screening mammography will show few abnormalities that are of concern. Some studies suggest that 5% to 10% of potentially detectable cancers are overlooked by radiologists on a screening mammogram. Studies also show that having two or more readers will significantly reduce the failure to perceive an abnormality. However, cost, which is already a major factor in mammography screening, would increase with the use of double reading, reducing the already low reimbursement rates of mammography. Computer-Aided Detection (CAD) offers a solution.

In CAD technology, digital and computed mammography are combined. The computer analyzes and prereads the mammograms, scanning every part of the radiograph and reporting any suspicious areas. The computer then displays or analyzes these suspicious areas to make the interpretation consistent from patient to patient. The computer system, in effect, can be programmed to act as a second reader.

PACS allows:

- Modality comparison
- Modality integration
- Prior image comparison

Teleradiography allows:

- Virtual consultation
- Internet access to images by telephone, cable, or satellite

---

CAD combines digital technology with computers to preread the mammograms.

- The computer displays suspicious areas.
- Normal structures are deemphasized.

CAD technology works with both digital mammography and analog imaging.

The US Food and Drug Administration (FDA) approved CAD technology in 1998 for aiding interpretation of the mammogram, and CAD now represents an efficient source of assistance to the radiologist interpreting the mammograms. The interpretation capabilities will only increase as the technology improves. CAD can be used with analog mammogram images, digital mammography, breast MRI, and ultrasound imaging of the breast.

CAD technology can also be used to convert analog mammography images into a digital signal that can be analyzed by a computer. To analyze an analog mammogram, the image must be digitized using a film digitizer. The mammogram is first fed into a digitizer. The digitizer uses a helium neon laser beam that is bounced off a series of mirrors as it scans the image. A photomultiplier tube picks up light. This is converted to an electrical signal and sent to the analog-to-digital converter (ADC), where the signal is translated to a number based on the optical density of the film and the signal received. From the digitizer, the mammogram goes to the computer reader.

In digital mammography, the digital image is sent directly to CAD. CAD works by mapping normal healthy breast tissue that is then stored as a reference. Future images or breast maps are compared with the normal reference map. Computer vision then analyses the image, detecting abnormalities by a process of enhancing the important point on the image and deemphasizing areas of little importance (e.g., if the program is detecting microcalcifications, structures that do not have a size similar to microcalcifications are subtracted). Another process separates the image into regions with similar attributes (e.g., all areas with low optical density). The program also looks for the presence or absence of spiculations at the margins of masses. The computer algorithm then estimates the likelihood that a lesion is malignant or benign. Some systems have a color-coded parametric map, with red indicating the likelihood of malignancy and blue associated with benign lesions. Green could indicate indeterminate lesions.

This technology will only be as good as the initial algorithms, and CAD cannot be used as a sole interpretation tool. CAD can offer a more consistent and logical interpretation; however, the sensitivity and specificity of the CAD system are also important. A high sensitivity may not necessarily be good because it will generate numerous false-positive alarms. At present, some CAD systems have a 90% sensitivity for detecting microcalcifications. If the specificity is high, the system will be effective in reducing the number of biopsy recommendations for benign disease. Studies show that the use of the CAD system improves the detection of subtle microcalcifications and subtle spiculated masses, halving the "miss rates" of the radiologist.

CAD for MRI was approved by the FDA in 2002 and now offers time-saving capabilities because it can be combined with MRI systems, offering computer-aided visualization and analysis. Unlike in mammography, CAD for MRI also looks at the morphology of tissue characteristics confirming irregular shapes or spiculated lesions. In MRI, a malignancy is often determined by how the tissue enhances after the injection of the gadolinium-based contrast. CAD is able to analyze the flow of contrast in and out of the tissue. In MRI, CAD can determine what the contrast does over time, how it is absorbed, and how

CAD:

High sensitivity will increase the false-positive rate.

High specificity will reduce the false-positive rate.

quickly it washes out. CAD also corrects for patient motion and realigns the magnetic resonance images automatically.

By combining this kinetic data analysis with checks of the morphology of the lesion, CAD is actually used in the initial read of a study and aids in the interpretation by performing kinetic analysis and displaying kinetic information using color coding, generally color-coding image pixels, that increase in intensity above a defined threshold for the precontrast and postcontrast images. For example, CAD can determine that the kinetics of a lesion are good, but that the lesion is irregular or spiculated and should therefore be flagged as suspicious. MRI CAD technology can automatically create subtraction images, multiplanar reformatting, maximum intensity projections, volume data, diffusion analysis, and region of interest summary series.

Studies show that initially the use of any CAD system will generally increase the recall rates and the number of biopsies; however, with experience, the radiologist soon learns to differentiate between minimally suspicious and highly suspicious marks.

IMAGE QUALITY: DIGITAL AND COMPUTED MAMMOGRAPHY

The image acquisition, the image display, and the storage of the image are separate functions in digital mammography. This means that after the image has been acquired, these functions can be modified based on the capabilities of the digital software.

In digital mammography, the term *density*, when referring to the image, has been replaced by the term *brightness*. The *brightness* of the image is controlled not by the mAs only, as in analog imaging, but also by the processing software, which will include predetermined digital algorithms. The user can alter the brightness of the digital image after the exposure. Image *contrast* is not controlled only by the kV but also by the processing software, and the user can modify the image contrast after the exposure.

Resolution is the ability to image two separate objects and visually distinguish one from the other. Resolution is directly related to focal spot size. As the focal spot size increases, resolution decreases. The aim, then, is always to minimize the effective focal spot while maximizing the actual focal spot to absorb heat. A large focal spot size will also result in focal spot blur (penumbra or geometric unsharpness). Resolution is affected by focal spot size, SID, and OID.

Spatial resolution is the ability to image small objects that have a high subject contrast (e.g., breast microcalcifications). Spatial resolution in digital imaging is also controlled by the acquisition of the image, including the pixel size, display matrix, and capabilities of the monitor. The smaller the pixel size, the better the spatial resolution of the image. In general, digital systems have a lower spatial resolution than analog systems. It is therefore important to control other geometric sources or motion. No digital system has come close to the resolution of 11 lp/mm, which is the minimum required for screen/film systems. However, the lower resolution of digital imaging systems is compensated by the higher dynamic range and higher image resolution. The greater the

> The lower resolution of digital imaging systems is compensated by the higher dynamic range and higher image resolution.
>
> In digital imaging the greater the number of pixels, the greater the image resolution.
>
> Resolution is measured in line pairs per millimeter or line pairs per inches.

number of pixels, the greater the image resolution. Resolution is measured in line pairs per millimeter or line pairs per inches.

Contrast resolution is the ability to distinguish anatomical structures of similar subject contrast. In breast imaging, it is critical to distinguish between a lesion and normal breast tissue. The more pixels there are, the greater the pixel density, which is the number of pixels in a unit area. The greater the pixel density, the higher the image resolution.

Focal spot blur is caused because the focal spot is a rectangular area on the target, not a point. The size of the focal spot varies from 0.1 to 1.5 mm. Focal spot blur is represented by a blurred region on the radiograph. Focal spot blur can be reduced by small focal spot size, large SID, and short OID.

Motion can be voluntary or involuntary and is another factor affecting resolution and sharpness of the image. Voluntary motion is under the direct control of the patient and is best reduced through patient communication, which should occur before the start of the mammographic examination to ensure patient cooperation in reducing motion. Involuntary motion is not under the conscious control of the patient and is best reduced by decreasing exposure time and immobilization. Involuntary motion is also controlled in mammography imaging by good compression, which reduces breast thickness, allowing for shorter exposure times.

Noise or *quantum noise* refers to the random disturbance that obscures or reduces image clarity and will appear as a grainy or mottled image, sometimes referred to as *snow*. In imaging, noise is controlled by the *milliamperage*, defined as the actual number of photons striking the detector. Low-exposure factors (kV and mAs) will produce few photons and therefore less signal and more noise. Higher factors will produce more signal and less noise; however, more scattered radiation will be produced. A high signal-to-noise ratio (SNR) is the best and can be controlled by the initial mA and the use of grids to reduce scattered radiation.

All images will have less resolution and sharpness than the object itself. Although changes in the SID, OID, and size of the collimated field will not affect focal spot size, in mammography, in order to produce the sharpest possible image detail, the smallest possible focal spot size is coupled with the longest SID and the smallest possible OID.

Less precise terms:
- Detail
- Recorded detail
- Visibility of detail

More precise terms:
- Spatial resolution
- Contrast resolution

Visibility of detail refers to the ability to see the detail on the radiograph. The loss of visibility can be due to any factor obscuring the image detail.

Minimize focal spot blur:
- Small focal spot
- Small OID
- Large SID

Summary

- When a fast-moving electron strikes a target, x-rays are produced. The interaction is 99% heat and 1% x-ray. Two types of x-radiation can be emitted: characteristic (photoelectric) radiation and bremsstrahlung radiation.

- X-rays will interact within the human body at the atomic level. The interaction can be Compton's effect, photoelectric effect, coherent scattering, pair production, or photodisintegration. Compton's and photoelectric effects are the only two that play a major part in diagnostic imaging.

- The characteristics of the mammography tube are all designed to maximize the imaging of the soft tissue structures of the breast. The mammographic tube is an electronic vacuum tube with a positive anode and negative cathode. The material used for the anode will affect the type of radiation produced. Analog breast imaging uses molybdenum targets for fatty breasts and rhodium for dense breasts. In digital imaging, tungsten can be used as a target material because the higher energies emitted by the tungsten targets offer a wider dynamic range that can be exploited by the detectors in digital units, allowing shorter exposure times and less possibility of motion, which would degrade image quality.

- All x-rays leaving the x-ray tube are filtered by the oil in the tube housing, the exit port window, the collimation and mirror assembly, and the compression plate. These are considered inherent filtration. To minimize the effects of inherent filtration, the exit port is made of beryllium or borosilicate glass, and the compression device is made of a low-attenuating hard plastic. Added filtration is also applied to shape the x-ray beam, removing low-energy photons that would only contribute to skin dose and high energies that would degrade the image.

- Added filtration usually matches the material of the target. Therefore, a molybdenum target would utilize a molybdenum filtration and rhodium target a rhodium filtration. Tungsten targets are filtered with rhodium or silver.

- The x-ray tube is designed to minimize the anode heel effect by placing the thickest portion to the breast, the posterior aspect toward the cathode. Tubes also utilize the line-focus principle to maximize tube loading while maintaining a small focal spot size.

- Other features of the mammography unit are as follows:
  - The kV available for use in mammography imaging will depend on the material of the target.
  - The mA is determined by the number of x-rays produced.
  - Grids are used to reduce the effects of scattered radiation.
  - Devices allow automatic exposure control (AEC).
  - Mammography tubes do not use beam-restricting devices. The collimated field is restricted to the size of the image plate.
  - The source-to-image receptor distance (SID) is fixed, often between 50 and 80 cm.
  - Magnification mammography uses a smaller focal spot size to minimize the effects of distortion.

- Use of a complete picture-archiving and communication system (PACS) with modality comparison and with teleradiography allows images to be transmitted to distant locations, such as the radiologist's home, enabling virtual consultation.

- Computer-aided detection (CAD) combines digital technology with computers to preread the mammograms. The computer will display suspicious areas, in effect acting as a second reader. CAD works by mapping the breast for a normal reference. Subsequent images are analyzed, and any abnormal microcalcifications or abnormalities are enhanced while normal structures are de-emphasized. CAD technology works with analog imaging, digital mammography, breast ultrasound, and breast magnetic resonance imaging (MRI).

- Digital imaging has separated the acquisition of the image from the image display and storage. The image quality is dependent on the processing software and predetermined digital processing algorithms.

REVIEW QUESTIONS

1. Name the five known interactions between x-rays and matter at the atomic level.

2. What type of radiation is produced when a projectile electron striking the target of the x-ray tube interacts with an inner shell electron?

3. What is the name given to the negative electrode of the x-ray tube?

4. What is the purpose of the line voltage compensator?

5. What are common tube-target materials used in mammography tubes?

6. What effect will the decreasing target angle have on the focal spot size?

7. In mammography, what type of imaging uses very small focal spot sizes?

8. To minimize the anode heel effect, the cathode is positioned at which area of the breast?

9. Why is the material of the exit port window of the mammography tube so important?

10. Which target material is often used when imaging fatty breast tissue?

11. At what exposures does the reciprocity law fail?

12. What are the maximum and minimum forces of compression allowed by the initial automatic compression in mammography imaging?

13. What is the purpose of the filament circuit?

14. What type of transformer is used in the filament circuit?

15. Why is close collimation not used in mammography imaging?

16. Generally, what is the average SID used in mammography imaging?

17. What principle is used to rotate the anode in the x-ray tube?

18. What is the photoelectron?

19. What is bremsstrahlung radiation?

20. List some items responsible for inherent filtration.

21. What is the best method of controlling voluntary motion?

22. What is main controlling factor of image brightness in digital imaging?

23. What is the main controlling factor of image contrast in digital imaging?

24. Give two possible postprocessing options available with digital imaging.

CHAPTER QUIZ

1. Added filtration in mammography imaging is:
   (A) Generally the same element as the mammography tube target
   (B) Of a higher filtration value than the filtration used in general imaging
   (C) Sometimes not recommended
   (D) Often replaced by the inherent filtration

2. Which of the following does *not* describe mammography filtration?
   (A) It shapes the emerging beam by absorbing low-energy x-rays that would only be absorbed by the superficial tissue and contribute to patient dose.
   (B) It will affect the HVL of the emerging x-ray beam.
   (C) Filtration makes the emerging beam compatible with the breast characteristics.
   (D) The filtration used in mammography imaging is usually aluminum.

3. Inherent filtration in mammography imaging implies filtration by:
   (1) Exit port window of the mammography tube
   (2) Collimator light assemble
   (3) Compression device
   (A) 1 and 2 only
   (B) 2 and 3 only
   (C) 1 and 3 only
   (D) 1, 2, and 3

4. Component parts of a mammography tube that differentiate it from a regular x-ray tube include:
   (1) Size of the focal spots
   (2) Type of added filtration
   (3) Material of the exit port window
   (A) 1 and 2 only
   (B) 2 and 3 only
   (C) 1 and 3 only
   (D) 1, 2, and 3

5. Which of the following statements are *not* true of magnification mammography?
   (A) With magnification, patient dose increases.
   (B) Magnification does not use a grid.
   (C) Magnification can be used to assess suspicious lesions.
   (D) In general, magnification can be used to image the entire breast with one exposure.

6. What is the recommended resolution of the RWS?
   (A) 2 megapixels
   (B) 3 megapixels
   (C) 4 megapixels
   (D) 5 megapixels

7. The material of the exit port is necessary to:
   (A) Harden the emerging beam
   (B) Attenuate low-energy photons
   (C) Allow the passage of low-energy photons
   (D) Increase the level of high-energy photons emerging

8. How does the material selected for the anode affect the beam quality?
   (A) The anode selection shapes the beam by produce characteristic radiation in an acceptable range.
   (B) The anode selection filters the beam by allowing only the necessary low-energy photons to emerge from the tube.
   (C) The anode selection shapes the beam by filtering out the high-energy photons.
   (D) The anode selection shapes the beam by filtering out the low-energy photons.

9. Which of the following factors will directly control the image contrast in digital imaging?
   (A) mAs
   (B) kV
   (C) Processing software
   (D) Exposure time

10. In digital imaging, a postprocessing option that will change the image brightness is called:
    (A) Edge enhancement
    (B) Windowing
    (C) Annotating
    (D) Subtraction

BIBLIOGRAPHY

Adler A, Carlton R. *Introduction to Radiologic Science and Patient Care.* 6th ed. St. Louis, MO: Elsevier Saunders; 2016.

American College of Radiology (ACR). *Mammography Quality Control Manual.* Reston, VA: The American College of Radiology; 1999.

Andolina VF, Lille SL, Willison KM. *Mammographic Imaging: A Practical Guide.* 3rd ed. Philadelphia, PA: Lippincott Williams & Wilkins; 2010.

Bushong SC. *Radiologic Science for Technologists—Physics, Biology and Protection.* 10th ed. St. Louis, MO: Mosby; 2012.

Gater L. Digital mammography: state of the art. *Radiol Technol.* 2002; 173(5):446–457.

Harvey D. SOFT Paddle. *Radiol Today.* August 30, 2004;9.

Jacobson DR. *Rad Tech's Guide to Mammography: Physics, Instrumentation, and Quality Control.* Malden, MA: Blackwell Science; 2002.

Peart O. *Lange Q & A Mammography Examination.* 3rd ed. New York, NY: McGraw-Hill; 2015.

Shepard GT. *Radiographic Image Production and Manipulation.* New York, NY: McGraw-Hill; 2003.

Special optimized full tilt: breast compression paddles for screening mammography. Available at http://www.mammospot.com. Accessed January 14, 2011.

Tucker AK, Ng YY. *Textbook of Mammography.* 2nd ed. Edinburgh, UK: Churchill Livingstone; 2001:15–24.

Venes D, Biderman A, Adler E. *Taber's Cyclopedic Medical Dictionary.* 22nd ed. Philadelphia, PA: F. A. Davis; 2013.

Breast Imaging Mammography | 6

Keywords and Phrases
Breast Imaging Mammography
  Breast Compression
  Applying Compression
Standard Projections
  Positioning Guide
  MLO—The Mediolateral Oblique Projection
  CC—The Craniocaudal Projection
Supplementary Projections
  ML—Mediolateral or 90 Degrees Lateral
  LM—Lateromedial
  LMO—Lateromedial Oblique
  TAN—Tangential Projection
  XCCL or XCCM—Exaggerated or Medial Craniocaudal
    Projection
  FB—From Below
  CV—Cleavage or Valley View
  AT—Axillary Tail
  SIO—Superior–Inferior Oblique
  Rolled Positions
  Magnification
  Spot Compression
Imaging the Nonconforming Patient
  Imaging Small Breasts
  Male Breast Imaging
  Sectional Imaging of the Extremely Large Breast or Mosaic
    Breast Imaging
  Imaging Patients with Obese Upper Arms
  Patient With an Inframammary Crease That Is Not Horizontal
  Patient With Protruding Abdomen
  Patient With Thick Axilla
  The Kyphotic Patient
  Patient With Adhesive Capsulitis (Frozen Shoulder)
  Elderly or Unstable Patients

Mentally Challenged Patient
Patients With Pectus Excavatum (Sternum Depressed) or
Pectus Carinatum (Pigeon Chest)
Barrel Chest
Delicate Inframammary Fold
Imaging Implants
Imaging the Postsurgical Site—Postlumpectomy Imaging
Postmastectomy Imaging
Mammograms of the Irradiated Breast
Imaging the Stretcher Patient
Imaging the Wheelchair Patient
Specific Imaging Problems
Nipple Not in Profile
Skin Folds or Wrinkling of the Breast
Patients With Uneven Breast Thickness
Moles and Scars
Pacemakers and Implantable Venous Access Systems
Triangulation Techniques
Before Signing Off
Summary
Review Questions
Chapter Quiz

Objectives

On completing this chapter, the reader will be able to:

1. Describe the routine four-projection series used in mammography
2. Describe proper positioning for the routine and supplementary imaging
3. Evaluate mammograms on the basis of good positioning
4. Identify the reasons for breast compression
5. Explain magnification techniques and spot compression techniques
6. Discuss labeling and radiograph identification
7. Explain the difficulty in imaging the nonconforming patient
8. Describe the Eklund or implant-displacement technique
9. Discuss how to resolve imaging problems
10. Explain breast localization methods

KEYWORDS AND PHRASES

- **Caffeine** is a drug that is essentially a central nervous system stimulant commonly used in coffee, tea, cola beverages, and certain stimulant drugs.
- **Contralateral** means on the other side of the body.
- **Detector,** also called, **image plate** (IP), is the device that receives the energy of the x-ray beam.
- **Exposure latitude** is the range of underexposure or overexposure that can occur while still producing an acceptable image. Digital receptors can correct for 50% underexposure to 100% overexposure.

- **Inframammary fold** or **crease** is the most inferior aspect of the breast where the breast attaches to the anterior chest wall.
- **Image plate (IP)** is also called a storage phosphor screen (SPS) or photostimulable storage phosphor (PSP). It is the device that receives the energy of the x-ray beam.
- **Ipsilateral** means on the same side.
- **Magnification mammography** is a technique used to improve visibility of fine detail in mammography imaging.
- **Posterior nipple line (PNL)** is the imaginary line running perpendicular from the pectoral muscle to the nipple in both the craniocaudal (CC) and mediolateral oblique (MLO) projections.
- **Spot compression** refers to the use of a small compression device to evaluate a suspicious area.
- **Subject contrast** refers to the subject or part being radiographed and the result of differential absorption of the radiation by the part. Anatomical parts with high subject contrast will have sharp differences in x-ray absorption, such as bone and soft tissue. Anatomical parts with low subject contrast will have very little difference in x-ray absorption, such as the breast.
- **Triangulation** refers to the method used to localize a lesion in the breast.

BREAST IMAGING MAMMOGRAPHY

Breast cancer screening with the mammogram has come a long way from its early beginnings. The Mammography Quality Standards Act (MQSA) of 1992 played an important role in the standardization of mammography procedure, but even before that, the voluntary accreditation adopted by the American College of Radiology (ACR) began ensuring that mammography throughout the United States was performed using recognized projections and imaging techniques. Prior chapters have dealt with the importance of maintaining a continuous quality improvement (CQI) program in mammography imaging. CQI encompasses an entire management plan or quality assurance program, the goal being to produce a high-quality image of the breast. The importance of the equipment and the process of image acquisition cannot be overemphasized, but equally important are the knowledge and skills of the technologist. Mammography screening requires dedication and commitment to produce the best possible image all of the time. It is important that technologists recognize their role in breast imaging.

The effectiveness of mammography screening is dependent on the image quality, and while there are a number of factors affecting image quality, including the inherent quality of the mammography unit and the processing equipment or methods of image acquisition, these factors are not under the immediate control of the technologist. The one factor that is totally under the technologist's control is patient positioning. It is the technologist's responsibility to take the best possible image and to include all breast tissue. A cancer cannot be detected if the area of its location is not present on the mammogram. A cancer will not be detected if the image is blurry because of motion, unsharp because of poor compression, or underpenetrated or overpenetrated because of improper technical factors.

Typically, the routine four-projection series in mammography involves imaging in the craniocaudal (CC) and mediolateral oblique

Automatic compression should never exceed 45 lb (200 N) of pressure at initial compression. The patient should not be in pain, but the breast must be taut to fingertip contact.

(MLO) projections of both breasts. The idea behind the four-projection routine in breast imaging is to image both breasts with minimal radiation dose to the patient. Positioning for these projections is relatively easy, on average, for patients in their 40s, and generally the examination will take 10 minutes or less. But patients are not clones of each other. In a diverse society, patients will be not only of different ages but also of different body shapes and sizes. Technologists will need to know how to image patients who do not fit the average profile.

With knowledge of the importance and use of each projection, the technologist will be able to adapt and modify imaging techniques in clinical situations. These include imaging the small breast, males, large-breasted patients and patients with wide breasts, patients with pectus excavatum (or sunken chest), patients with barrel chest, wheelchair patients, and stretcher patients. Technologists occasionally will need to image postmastectomy patients and patients with implants. In addition to imaging, the technologist must have knowledge of triangulation. Triangulation of a lesion is necessary whenever the technologist needs spot compression or spot magnification projections of a suspicious area.

To understand how to modify positioning, the technologist must first understand how to determine good positioning. Remember that positioning is not an exact science. There is no single technique that will work on all patients in all situations. Sometimes, the basic positioning technique will have to be modified: occasionally, the technologists will have to be creative. It will, however, remain the technologists' knowledge of the breast or lack of knowledge that will help or hinder their ability to achieve good positioning.

Breast Compression

One of the most basic principles in achieving positioning proficiency is to maximize tissue visualization on the radiographs.

The benefits versus the risks of radiation, coupled with achieving a high-contrast, high-resolution image, involved a design with numerous trade-offs. Breast compression plays a crucial role in the trade-offs that are present in mammography imaging. Compression is important in mammography by making the breast more uniform and reducing breast thickness. There are also other benefits of compression. Compression will reduce radiation dose and motion unsharpness. Compression will also separate superimposed areas of the breast tissue and bring abnormalities closer to the detector.

In general, the breast should be compressed until taut to ensure adequate compression. Often, after the automatic compression stops, manual compression must be applied to adequately immobilize and compress the breast. Although too little compression will compromise image quality, the compression should not be applied to cause the patient severe pain.

Many patients, however, fear breast compression. Because of the compression, technologists are often the focus of a variety of charges, from being uncaring to being cruel or incompetent. Patients complain of breast compression to their physicians, families, and friends. Such complaints may discourage other women from seeking a mammogram. But even worse can occur. Physicians have even been known to tell the technologist not to compress the patient's breast. These physicians try

> Compression is important in mammography to make the breast more uniform and to reduce the breast thickness.

to be responsive to their patients' needs but lack an understanding of the importance and value of breast compression during mammography.

The technologist needs to convince the patient that compression is absolutely essential in mammography. Patients can be shown images of the breast with and without compression, demonstrating how easy it is to miss a lesion, or how ill-defined and blurry a lesion without compression can appear. Technologists can explain some of the causes of a painful mammogram. One cause is breast cysts. Patients who have a history of breast cysts should be advised to schedule their mammogram 1 week after the menstrual period ends—approximately 14 days after the start of menstruation. At this time the breasts are least fluid-filled and, therefore, less sensitive. Patients with extremely sensitive breasts could be advised by their physicians to take pain medication 30 minutes before the mammogram and to use a topically applied pain gel, or the mammogram can be rescheduled.

Finally, women will be reassured if they are given some measure of control. They should be reassured that compression lasts a few seconds only, and with state-of-the-art mammography units, it is automatically released immediately after the exposure. With a thorough explanation of the need for compression, most women will be able to tolerate the compression necessary to ensure a good mammogram.

Applying Compression

Before applying compression the technologist should consider the natural mobility of the breast. The breast is more rigidly attached to the chest at its superior and medial aspects. In applying compression, the technologist should be as gentle as possible. The automatic compression should be applied with the technologist's hand between the breast and the compression plate, until the plate touches the back of the hand. The technologist should maintain firm pressure on the breast until the compression takes over. Only then should the technologist remove the hand, applying the final degrees of compression slowly (Fig. 6–1). Always apply the last degrees of compression using the manual rather than the automatic compression paddle (Box 6–1).

> **Box 6–1.** Reasons for Breast Compression
>
> Main reason:
> - Allows a more uniform attenuation of the x-ray beam by flattening the base of the breast to the same degree as the more anterior regions, permitting optimal imaging of the entire breast in one exposure
>
> Other benefits:
> - Reduces dose to the breast by reducing tissue thickness
> - Brings lesions closer to the detector for more accuracy when evaluating fine detail
> - Decreases motion unsharpness by immobilizing the breast
> - Increases contrast by reducing the amount of scattered radiation and by decreasing breast thickness
> - Separates superimposed areas of glandular tissue by spreading apart overlapping tissue, reducing confusion caused by superimposition of shadows, and allowing visualization of the borders of circumscribed lesions

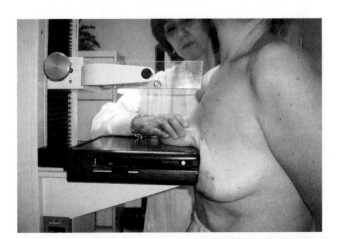

Figure 6–1. The automatic compression should be applied with technologist's hand between the breast and the compression plate (technologist standing at the lateral side of patient for demonstration only).

The MLO demonstrates the extreme posterior and upper outer quadrant. There is distortion of the anterior, central, and medial breast tissue.

The CC best demonstrates the anterior, central, medial, and posteromedial portions of the breast but is poor at visualizing the lateral and posterior lateral breast tissue.

STANDARD PROJECTIONS

In imaging, projection describes the travel of the x-ray beam through the body, whereas position describes the body's position relative to the image plate or space. The two basic mammographic projections are the CC and the MLO (Fig. 6–2). These projections are complementary. What this means is that eliminating tissue from one projection does not necessarily mean an extra projection is required (Fig. 6–3). The CC best demonstrates the anterior, central, medial, and posteromedial portions of the breast but is poor at visualizing the lateral and posterolateral breast tissue. Imaging the medial breast on the CC is important because if the medial breast is missed on the CC, it is missed from the study (Fig. 6–4).

On the other hand, the MLO demonstrates the extreme posterior and upper outer quadrant, but because it is an oblique projection, there is distortion of the anterior, central, and medial breast tissue. When imaging on the MLO projection, the detector must be parallel to the

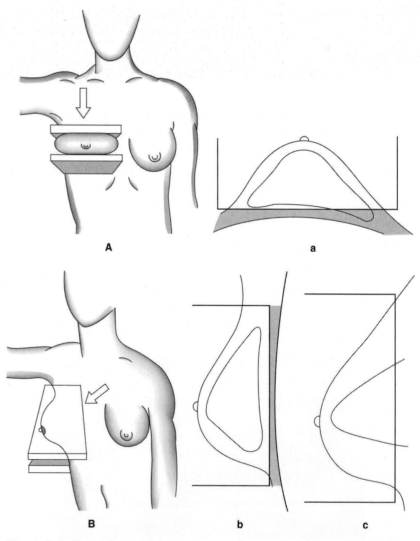

Figure 6–2. The CC and MLO projections are complementary. Schematic diagram of the **(A)** CC and **(B)** MLO projections. Shaded areas represent the posterior areas missed when compressing on each of the projections: **(a)** CC, **(b)** MLO, and **(c)** ML.

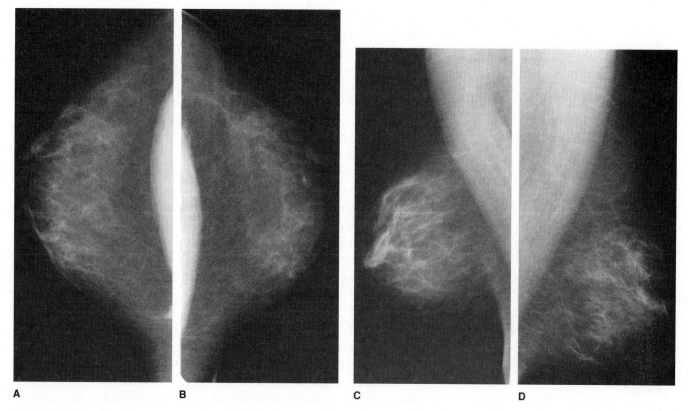

Figure 6–3. A routine mammogram showing the normal appearance of (A) RCC, (B) LCC, (C) RMLO, and (D) LMLO.

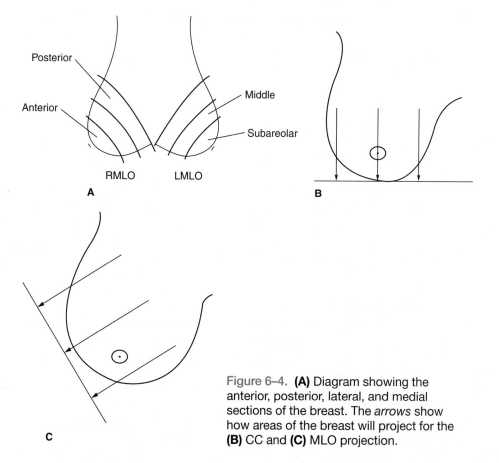

Figure 6–4. (A) Diagram showing the anterior, posterior, lateral, and medial sections of the breast. The *arrows* show how areas of the breast will project for the (B) CC and (C) MLO projection.

> When imaging on the MLO projection, the detector must be parallel to the pectoralis muscle (30 to 60 degrees for most patients). The detector should not be automatically angled at 45 degrees for every patient.

pectoralis muscle (30 to 60 degrees for most patients). The more parallel the detector is to the pectoralis muscle, the more tissue will be included in the image. The detector should not be automatically angled at 45 degrees for every patient. In such a technique, the technologist is forcing the patient to fit the machine rather than the other way around. The mammography unit is an extremely versatile unit that can adapt to most patients. It is unconscionable that the technologist should allow the patient to undergo body gyrations to fit the contours of a machine.

Regardless of the position or projection, the technologist will find it easier to manipulate the patient if the patient is relaxed and slouched forward versus rigid and erect. Patient motion is best controlled if all breast imaging is taken on suspended respiration; however, avoid telling the patient to take a deep breath before arrested respiration. A deep inspiration effort is likely to lift the shoulder or move the breast out of position. The technologist should always rehearse breathing instructions with the patient before the start of positioning.

Positioning Guide

When imaging a nonconforming patient, the technologist must first determine which area of the breast is missing before deciding which supplementary projection will best image that area. Each technologist will develop a unique sequence of steps in positioning the patient. There are, however, basic guidelines that all technologists should follow regardless of the position or projection.

Name of the Patient

The MQSA requires that all mammograms include a patient identification label. Modern mammography, ultrasound, and magnetic resonance imaging (MRI) equipment all require the user to input the patient's information before the start of the examination to avoid identifying the wrong patient data on a study. The patient identification should include the facility name; the facility location, including the city, state, and zip code; the patient's first and last name; an additional patient identification number, for example, the medical record number; and the date of the examination.

Identifying the Projection

The ACR developed a standardized projection identifier or code that is used in this text. MQSA requires that all mammographic projections include a projection label (projection and laterality) placed near the axilla. MQSA accepts the standard codes developed by the ACR.

Technologist Name

The technologist performing the examination must be identified on the image. The identification can include only the initials of the technologist as long as the facility maintains a log of the technologists and their identifying initial.

Mammography Unit Number

Facilities with more than one mammography unit must identify each unit on the image. This is an MQSA requirement.

Box 6–2. Proper Radiograph Identification and Labeling

MQSA image identification

- Name of the patient and additional patient identifier (such as hospital number or medical records number)
- Date of examination
- Projection and laterality—must be placed near the axilla using standardized codes approved by the FDA
- Facility name and location to include city, state, and zip code
- Technologist's identification
- Mammography unit identification if there is more than one unit at the facility

In addition to those previously mentioned, technical factors, including the kV, mAs, compression force, breast thickness, and degree of obliquity, are other options that can be recorded on the final image (Box 6–2).

MLO—The Mediolateral Oblique Projection

Before beginning positioning, the patient should stand slightly anterior to the detector to determine the angle of obliquity. The best imaging on the MLO is achieved when the tube angulation is correctly aligned with the patient. The breast lies anterior to and follows the line of the pectoral muscle. Positioning the detector parallel with this muscle will ensure that most of the glandular tissue is imaged. Generally, the angle should range between 30 and 60 degrees, depending on the patient's build. Taller, thinner patients generally require a steeper angulation than shorter, stockier patients.

After the angle is determined, have the patient raise the *ipsilateral* arm and drape it over the top of the detector. Next, have the patient turn both feet and body toward the detector. The upper edge of the detector is placed in the axilla. The patient should bend the elbow but should not grip the bar of the mammography unit because this will tighten the pectoral muscles, reducing the ability to compress the breast. The position of the detector is also important. If the placement is too high, generally the patient's arm is raised unnecessarily high, pulling breast tissue out of the compressed area. Too high placement will force much of the shoulder under the compression, reducing the ability to compress the lower breast. The technologist should also choose the correct size of compression plate to allow correct centering of the breast tissue and to avoid poor placement of the arm and shoulders. Using the wrong size of compression plate is similar to too high or too low placement of the detector. The arm should rest at shoulder level, relaxing the pectoral muscles and allowing the breast to fall forward. Also, the elbow should be slightly posterior, not anterior, to the detector to avoid skin folds in the axilla region.

When imaging on the MLO view, taller, thinner patients generally require a steeper angulation than shorter, stockier patients.

When the patient is in position, the technologist then places one hand on the patient's ipsilateral shoulder, further encouraging the patient to relax and drop the shoulders. With the other hand, the technologist lifts and pulls the breast gently up and across the detector, making sure to include the *inframammary fold*. The inframammary fold should be open and not overlapping the patient's abdomen. The breast is held in position on the detector as automatic compression begins. The upper chest wall edge of the compression plate should rest just anterior to the humeral head but below the clavicle. As compression begins, the anterior surface of the compression plate will pull breast tissue from the sternum. The technologist then slowly slides the supporting hand toward the nipple without completely releasing the breast until the compression plate takes over the job of holding the breast in place (Fig. 6–5).

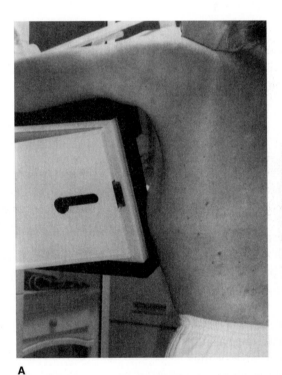

A

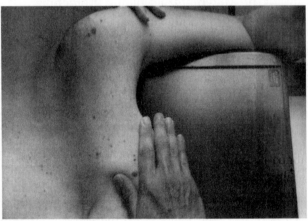

B

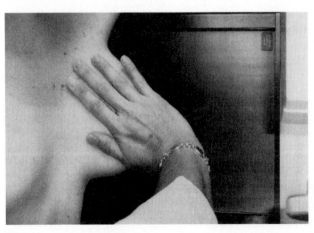

C

Figure 6–5. **(A–C)** Pictogram of MLO positioning. **(A)** Positioning the detector with the pectoral muscle. **(B)** Ensuring that the shoulder is relaxed. **(C)** Opening the inframammary fold with the "out and up" maneuver.

The hand movements necessary to avoid drooping of the anterior breast are often referred to as the "up and out" maneuver. As compression is applied, use one hand to support the anterior breast tissue to avoid skin folds—use the other hand to adjust the skin over the sternum and clavicle to reduce the "pulling sensation" that occurs as the compression plate slides over the sternum. Occasionally, the patient may have to hold the *contralateral* breast out of the field of view. Take care to avoid imaging the patient's knuckles or pulling the ipsilateral breast tissue out of the compression (Fig. 6–6).

The breast, especially the anterior breast, must be taut and should not droop. Manual compression at this point is often necessary to ensure sufficient compression.

The technologist should then check the posterior aspect of the patient to ensure that the entire posterior breast is included in the compression. The exposure is made on suspended respiration (Boxes 6–3 to 6–5).

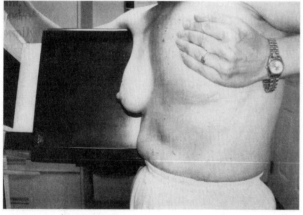
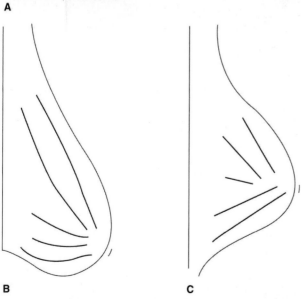

A

B   **C**

Figure 6–6. **(A)** Poor MLO positioning—the breast is drooping. Schematic diagram shows **(B)** drooping versus **(C)** correctly aligned breast tissue on the MLO.

Box 6–3. Image Evaluation of the MLO

- The pectoral muscle should be wide superiorly with a convex anterior border and should extend to or below the level of the posterior nipple line (PNL) (Figs. 6–7 and 6–8).
- The inframammary fold should be open.
- Dense areas of the breast should be adequately penetrated.

Box 6–4. Key Points When Imaging the Breast in the MLO Projection

- The degree of tube angulation will vary between 30 and 70 degrees depending on the patient size—thin patients require steeper angulation than heavier patients. Males require steeper angulation than females.
- Use the correct size of compression plate. If the compression plate is too large, the arm is raised too high—losing pectoral muscle.
- Too much pectoral muscle under the compression plate will reduce compression to the anterior portions of the breast.
- Position the arm, closest to the breast being imaged, draped over the top of detector. Place the detector in the armpit.
- Compression must adequately support the anterior breast tissue, preventing sagging and distortion of the ductal architecture.
- Appropriate markers and labeling must be used as required by MQSA.
- Expose on suspended respiration.

Box 6–5. Common Problems With the MLO Projection

- Drooping breast
- Abdominal tissue does not permit proper compression and positioning of the patient
- Missing posterior breast
- No inframammary fold
- See Fig. 6–6

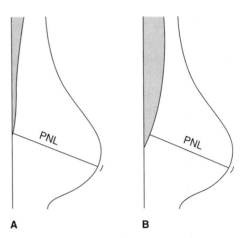

Figure 6–7. The posterior nipple line showing **(A)** concave and **(B)** convex anterior margins of the pectoral muscle.

CC—The Craniocaudal Projection

Begin positioning with the detector horizontally. The patient faces the unit with the face turned away from the breast under examination. The patient should not tilt (both shoulders should be on the same level.) Avoid having the patient resting against the unit; rather, the patient should stand slightly away from the detector, then bend a little

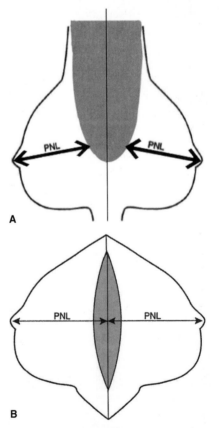

Figure 6–8. The measurement of the posterior nipple line (PNL) on the mediolateral oblique (MLO) **(A)** projection should be within 1 cm of the PNL measurement on the craniocaudal projection **(B)**.

from the waist, allowing proper imaging of the medial and posterior breast tissue.

The technologist should stand on the medial side of the breast of interest. The breast must be lifted to raise the inframammary fold, thereby positioning the detector at the level of the elevated inframammary fold. The breast is positioned in the center of the detector. Have the patient relax the ipsilateral shoulder and hold the bar of the unit with the contralateral hand. The ipsilateral arm should remain close to the patient's side. Positioning of the detector is important. Raising the detector too high will eliminate the posterior and inferior breast tissue, but if the detector is too low, the breast droops, losing superior and posterior tissue (Fig. 6–9).

With the breast in position, the patient should maintain position to ensure visualization of the medial and posterior breast tissue. It is often necessary to bring the opposite breast onto the detector to avoid losing the medial portion of the ipsilateral breast. Do not eliminate the medial breast because in the routine series only the CC will image the medial breast clearly. When the patient and detector are in place, the technologist again lifts the breast, pulling it onto the detector. One hand pulls the medial breast forward, while the other can ensure that as much of the lateral breast is imaged as possible. This must be done without rotating the patient. As the compression descends, the technologist slowly slides one hand toward the nipple, smoothing out skin folds along the way. The other hand can now move to the patient's back to prevent the normal inclination of the patient to pull back or out of the compression.

The technologist should not release the breast until the compression takes over in holding the breast in place. If the axillary fat pad or skin folds overlap the lateral aspect of the breast, the patient's ipsilateral hand can be supinated (shoulder externally rotated) or the shoulder adjusted back. The exposure is taken on suspended respiration (Boxes 6–6 and 6–7).

SUPPLEMENTARY PROJECTIONS

Breast imaging can also include supplementary projections. Each supplementary projection or position serves a main purpose. These projections/positions become useful when the standard projections are inadequate. Sometimes, the patient's history or body build is such that the standard projections are difficult to obtain. Remember, however, that ruling out an abnormality could involve imaging the breast using *any* manipulation or technique that is not a part of the routine series. Other reasons for supplementary projections include the following:

1. A suspicious area is seen in one of the routine projections, but not in the other.
2. Additional projections or positions can avoid the trauma of having an invasive procedure such as needle localization; for example, a spot compression may prove that an area showing a suspicious lesion is actually overlapping tissues.

Box 6–6. Key Points in Imaging the Breast on the CC Projection

- Position the detector at the level of the raised inframammary crease—lift the breast before positioning the detector.
- Position the patient's head away from the side being examined.
- Position the patient's feet apart with weight equally distributed.
- Patient's arm closest to the breast being examined is positioned by the patient's side.
- The contralateral arm is raised, holding the machine for support.
- Expose on suspended respiration.

Note: Always imaging the pectoral muscle on the CC can indicate missing medial breast because the patient is turned laterally (Fig. 6–10).

Box 6–7. Image Evaluation of the CC

- The nipple should be in profile.
- The nipple should be centered on the radiograph; however, do not eliminate breast tissue to center the nipple.
- The medial and lateral aspects of the breast must be included in the collimated area.
- The pectoralis major muscle is seen approximately 30% to 40% of the time. (If the pectoral muscle is seen all the time, imaging may be losing medial breast tissue.)
- Appropriate markers and labeling must be used as required by MQSA.
- The PNL line on the CC should be within 1 cm of the PNL on the MLO.
- Dense areas of the breast should be adequately penetrated.

A

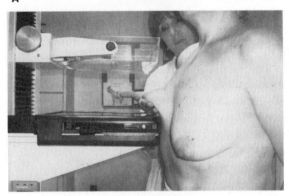

B

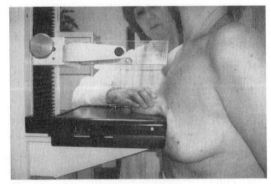

C

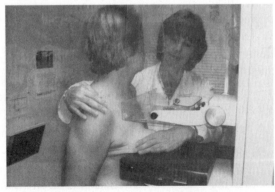

D

Figure 6–9. **(A–D)** Pictogram of CC positioning. **(A)** Lifting the inframammary crease. **(B)** Positioning the detector at the raised inframammary crease. **(C)** Pulling in medial breast. **(D)** Pulling in lateral breast.

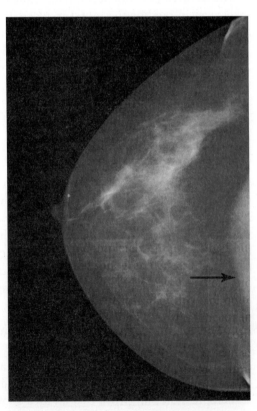

Figure 6–10. The CC projection showing pectoral muscle at the posterior.

The supplementary projections/positions recognized by the MQSA, ACR, and American Registry of Radiologic Technologists (ARRT) include the following:

1. 90-degree mediolateral (ML)
2. Lateromedial (LM)
3. Lateromedial oblique (LMO)
4. Tangential projection (TAN)
5. Exaggerated craniocaudal lateral (XCCL) and medial (XCCM)
6. Caudocranial or from below (FB)
7. Cleavage or "valley" view (CV)
8. Axillary tail (AT)
9. Superior–inferior oblique (SIO)

Other modified positions recognized by the MQSA, ACR, and ARRT:

1. Rolled medial (RM) or rolled lateral (RL)
2. Rolled inferior (RI) or rolled superior (RS)
3. Magnification (M)
4. Spot compression

ML—Mediolateral or 90 Degrees Lateral

This projection can be used to verify a finding or to localize a lesion in another dimension. It gives a true representation of the breast structures relative to the nipple. This is particularly important during needle localization. The ML can also be used to prove breast calcifications benign (Fig. 6–11). The "teacup" type of calcifications can be identified on the 90-degree lateral. The ML is poor at visualizing the posterior and lateral aspect of the breast and cannot replace the MLO. The ML is useful in locating a lesion not seen on the CC projection if the lesion is seen only on the MLO projection (Fig. 6–12). When comparing a lesion on the MLO and ML projection:

- Medial lesions move up on the lateral from their position on the MLO.
- Lateral lesions move down on the lateral from their position on the MLO.
- Central lesions will not change significantly from the MLO to the ML.

MLO and ML Comparison:
- Medial lesions move up
- Lateral lesions move down
- Central lesions will not change

Positioning

Imaging starts with the tube 90 degrees with the lateral aspect of the breast resting on the detector. The patient faces the unit with the ipsilateral arm raised to shoulder level and the arm resting on the top of the detector. With the patient leaning slightly forward, lift and pull the breast forward, opening the inframammary fold. When compressing, take care to avoid hitting the sternum.

LM—Lateromedial

The LM is ideal in improving detail of a lesion located in the medial aspect of the breast or high on the chest wall (Fig. 6–13). It is also useful to perform preoperative localization of an inferior or lateral lesion

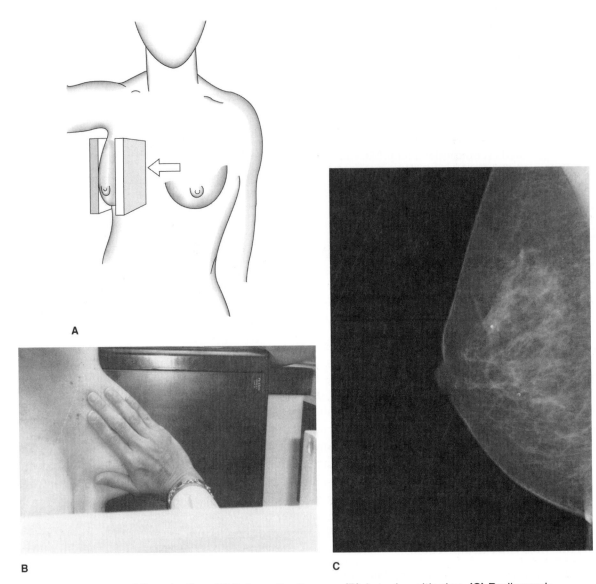

Figure 6–11. ML projection. **(A)** Schematic diagram. **(B)** Actual positioning. **(C)** Radiograph.

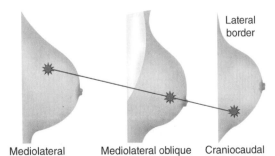

Figure 6–12. Schematic diagram showing the movement of a lesion from the ML to the MLO to the CC.

and may be more comfortable for some patients. The LM could be used instead of the LMO to image tissue missed on the MLO. The LM can replace the MLO on the nonconforming patient, and the result is similar to the MLO projection. However, the MLO places the lateral portion of the breast closer to the detector. Because most cancers occur in the upper outer quadrant of the breast, they will be visualized with sharper detail on the MLO.

Positioning

The LM begins with the tube at 90 degrees. The patient stands facing the unit with the medial breast resting on the detector. The chest wall edge of the detector rests on the sternum, with the upper edge just anterior to the humeral head and below the clavicle, on the ipsilateral side. The patient lifts the ipsilateral arm up and over the compression plate. By bending slightly, the patient will relax the shoulders. Before compression, the technologist must lift and pull

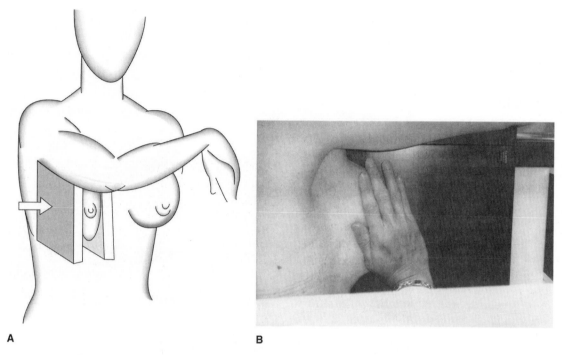

Figure 6–13. LM projection. **(A)** Schematic diagram. **(B)** Actual positioning.

the breast up and forward, rotating the patient slightly toward the detector to include the lateral breast. The breast is held in place until compression takes over in keeping the inframammary fold open. The upper arm can rest on the upper surface of the compression plate after compression is achieved. The compression plate will rest on the lateral surface of the breast.

LMO—Lateromedial Oblique

This projection is necessary when the standard MLO projection is difficult to obtain because of the patient's body build. The LMO gives improved visualization of the medial portion of the breast because that area will be closest to the detector. It can also be used on patients with prominent pacemakers or port-a-caths to avoid compressing the device, on patients with chest surgery to avoid pulling at the scar tissue and causing discomfort, with patients with prominent sternums, or to evaluate the medial aspect of the breast. The LMO is the true reverse of the MLO and results in the same projection of breast tissue (Fig. 6–14).

Positioning

The LMO is an inferolateral to superomedial projection. The tube is rotated 40 to 60 degrees parallel to the pectoral muscle. The patient then stands with the sternum and medial breast placed along the chest wall edge of the detector. The ipsilateral arm is raised over the top of the detector, and the contralateral arm will hold the bar. Compression skims the posterior rib cage with the upper border of the compression plate inferior to the humerus at the level of the axilla.

> The LMO is the true reverse of the MLO and results in the same projection of the breast tissue.

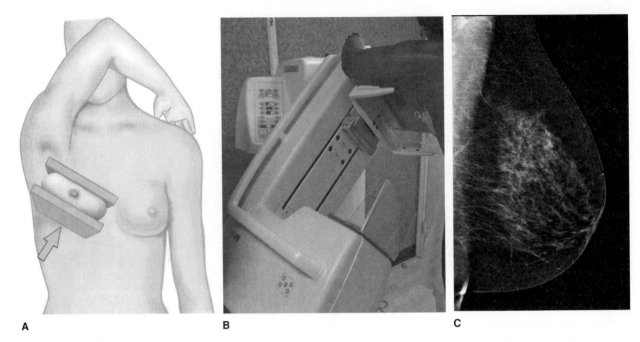

A   B   C

Figure 6–14. LMO projection. **(A)** Schematic diagram. **(B)** Actual positioning. **(C)** Radiograph.

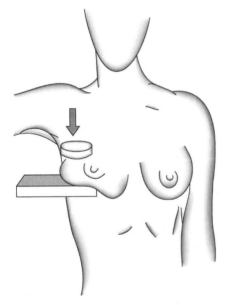

Figure 6–15. TAN projection. Schematic diagram.

TAN—Tangential Projection

This projection is most useful in projecting an area in question without superimposition of other breast tissue (Fig. 6–15). Because it skims the area of interest, it is also often used to locate skin calcifications or lesions thought to be near the skin.

Positioning

If the abnormality is palpable, a lead spot marker is placed on the abnormality during the exposure. To properly assess a nonpalpable abnormality, its exact location must be known. After the area is marked, the breast is rotated as necessary to place the marked area parallel to the direction of the central ray. The tube must be tangential to the abnormality. To aid positioning, the tube can be rotated in any direction until the shadow of the lead marker on the detector confirms the tangential placement of the marker in relation to the x-ray beam (Fig. 6–16).

XCCL or XCCM—Exaggerated Lateral or Medial Craniocaudal Projection

The XCCL is used to locate lesions in the posterior lateral aspect of the breast not seen on the CC projection (Fig. 6–17). The XCCM is use to locate lesions in the deep medial aspect of the breast not seen on the CC projection (Fig. 6–18).

XCCL Positioning

Begin positioning with the patient and unit as for the CC projection. The patient is then rotated medially to bring the outer aspect of the breast forward. The technologist then lifts the breast, pulling the posterior lateral tissue onto the compression plate. Compression must include all of the posterior lateral breast tissue. If the shoulder is still in the way, using a 5- to 10-degree tube angulation (in the MLO direction) will aid in imaging the posterior lateral breast tissue and allow the compression plate to clear the humeral head. This projection often does not include the medial breast, and to avoid distortion of the breast tissue, both shoulders should remain on the same level.

XCCM Positioning

The patient is turned laterally from the CC position. Generally, the image demonstrates the nipple off-center and more lateral. This projection often does not include the lateral breast, and to avoid distortion of the breast tissue, both shoulders should remain on the same level.

FB—From Below

This projection is ideal in imaging small breasts and can be used on kyphotic patients or on patients with pacemakers (Fig. 6–19). It can

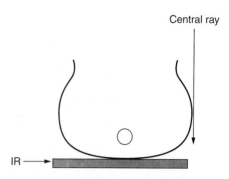

Figure 6–16. Direction of the x-ray beam in the tangential projection. The central x-ray skims the surface of the breast.

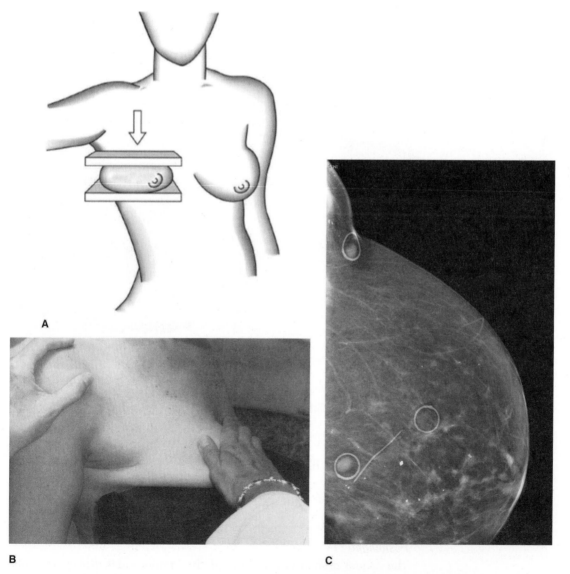

Figure 6–17. XCCL projection. **(A)** Schematic diagram. **(B)** Actual positioning. **(C)** Radiograph.

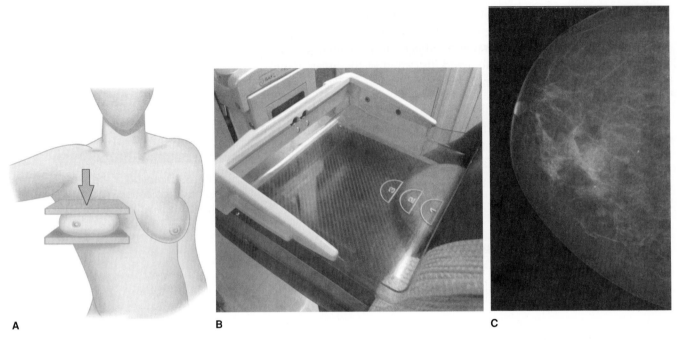

A　　**B**　　**C**

Figure 6–18. XCCM projection. **(A)** Schematic diagram. **(B)** Actual positioning. **(C)** Radiograph.

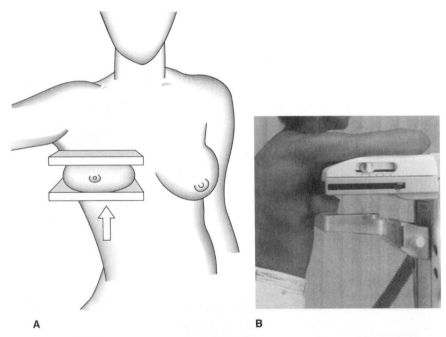

A　　　　　　**B**

Figure 6–19. FB projection. **(A)** Schematic diagram and **(B)** actual positioning. (Used with permission from GE Medical Systems.)

also be used to better visualize lesions in the superior or upper quadrants of the breast. Because the beam is direct from below, the fixed posterior surface of the superior aspect of the breast will not be included on the image.

Positioning

The caudal cranial is a reverse of the craniocaudal. Rotate the tube 180 degrees and have the patient bend over the unit. The detector is positioned at the top of the breast and the compression plate to the bottom. Similar to the CC projection, the inframammary fold must be elevated. Here the compression plate is placed at the elevated inframammary fold. Care must be taken to avoid imaging the patient's abdomen.

CV—Cleavage or Valley View

The cleavage projection, double-breast view, or medial imaging is useful in showing lesions deep in the posterior medial aspect of the breast (Fig. 6–20). This is the area closest to the chest wall.

Positioning

Begin positioning with the patient standing as for the CC projection. Lift both breasts onto the detector, making sure to first elevate the inframammary fold. Position the detector at the level of the elevated inframammary fold. The patient can hold the bar of the mammography unit with both hands for stability. With digital technology the entire

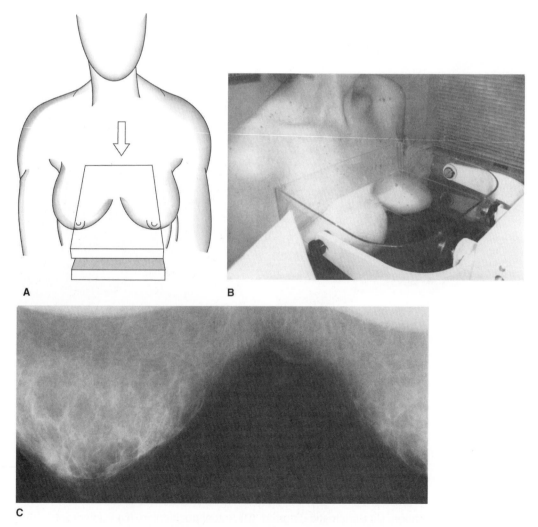

A   **B**

C

Figure 6–20. CV projection. **(A)** Schematic diagram. **(B)** Actual positioning. **(C)** Radiograph.

detector serves as an automatic exposure control (AEC) device; therefore, the placement of the breast tissue on any portion of the detector will allow automatic exposure.

AT—Axillary Tail

This projection will visualize the tail of the breast and most of the lateral breast. The AT is actually an oblique projection of the axillary area or tail of the breast (Fig. 6–21). The degree of obliquity will often depend on the radiologist's preference. It cannot replace the MLO because it does not image the far posterolateral breast tissue.

Positioning

The patient is positioned as for the MLO but with the tube rotated 20 to 40 degrees or parallel to the axillary tail, depending on the request of

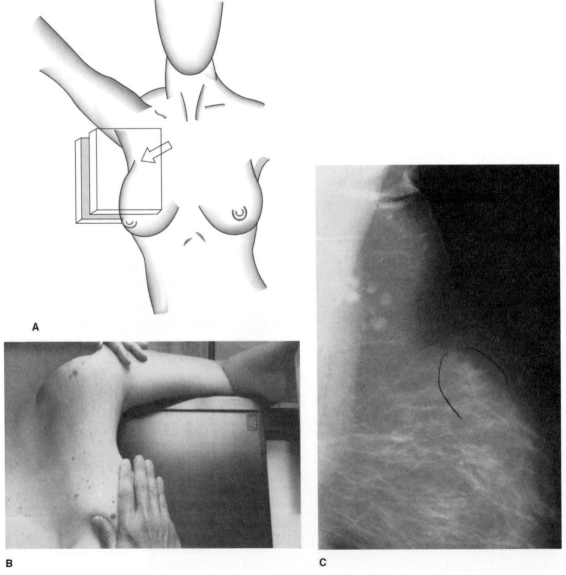

Figure 6–21. AT projection. **(A)** Schematic diagram. **(B)** Actual positioning. **(C)** Radiograph.

the radiologist. The ipsilateral breast is placed on the detector with the ipsilateral arm extending over the top of the detector. Unlike with the MLO positioning, here the posterior aspect of the shoulder rests against the detector. The arm can be bent at the elbow or kept straight. Having the patient bend at the waist will also help to pull more of the axilla area into the compression. To image the lymph nodes, some radiologists request the inclusion of portions of the humeral head and even the glenoid fossa in the image. Depending on the size of the patient, including these areas will limit compression of the breast.

SIO—Superior–Inferior Oblique

The SIO best demonstrates the upper inner quadrant and the lower outer quadrant of the breast, free of superimposition (Fig. 6–22). This projection is especially useful when imaging patients with implants using the implant-displaced positions or Eklund method, which is described later in the chapter. The 45-degree SIO can be useful in providing a projection perpendicular to the MLO and may help distinguish pseudomass from cancer. Using a 45-degree tube angulation will image the most posterior and inferior portion of the lower outer quadrant of the breast. It will demonstrate the inframammary fold or crease, especially in digital imaging. The tube can be angled 60 degrees when attempting to visualize the breast free of implant using the implant-displaced technique.

Positioning

In the SIO projection, the beam is directed from the superior lateral aspect to the inferior medial aspect of the breast. The patient faces the

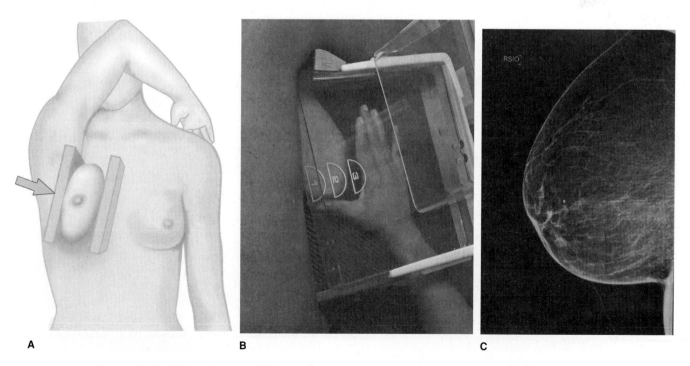

Figure 6–22. SIO projection. **(A)** Schematic diagram. **(B)** Actual positioning. **(C)** Radiograph.

unit while leaning forward from the waist. The edge of the detector is placed at the mid-sternum. The patient can hold the bar with the contralateral hand, whereas the ipsilateral arm should remain at the patient's side or lifted to rest over the top of the detector. The breast is then lifted and pulled outward and upward to bring the medial breast onto the detector. Before beginning compression, make sure the ipsilateral shoulder is not in the field of view.

Rolled Positions

> The rolled positions can be subjective and are not easy to reproduce.

These positions are useful in removing superimposed tissue when imaging dense breast (i.e., lesion is *rolled* off or away from the dense tissue) (Figs. 6–23 and 6–24).

Rolled lateral (RL)—from the CC projection, the superior portion of the breast is rolled laterally and the lower, inferior portion medially.

Rolled medial (RM)—from the CC projection, the superior portion of the breast is rolled medially and the lower, inferior portion laterally.

In these rolled positions, the breast is rolled laterally or medially from the CC projection. Some radiologists prefer a 5-degree tube angulation instead of the roll in order to aid in reproducing the image at a later date.

Rolled superior (RS) or rolled inferior (RI) positioning starts with the patient standing as for the mediolateral projection (Figs. 6–25 and 6–26).

RS—from the lateral, the portion of the breast farthest from the detector is rolled superiorly and the lower portion inferiorly.

RI—from the lateral, the portion of the breast farthest from the detector is rolled inferiorly and the lower portion superiorly.

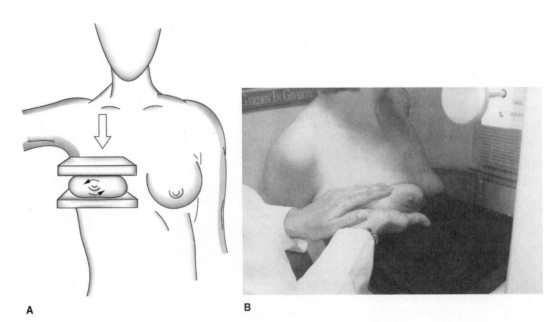

A   **B**

Figure 6–23. RL showing **(A)** schematic diagram and **(B)** actual positioning.

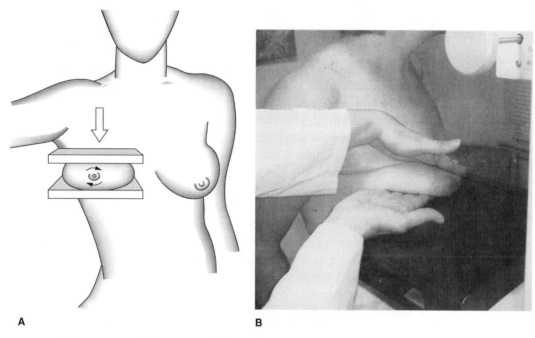

A   **B**

Figure 6–24. RM showing **(A)** schematic diagram and **(B)** actual positioning.

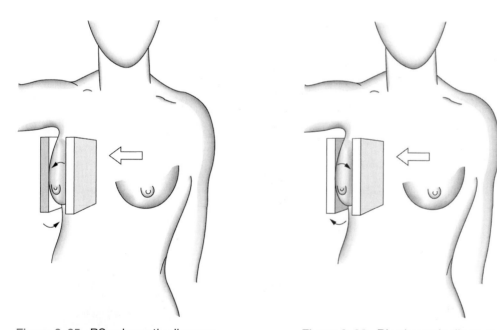

Figure 6–25. RS schematic diagram.   Figure 6–26. RI schematic diagram.

Magnification

Magnified projections are useful to visualize fine details, especially when imaging small areas of the breast. This technique is ideal in assessing the margins of lesions and can provide information on the number, distribution, and morphology of microcalcifications. Magnification can also be used to image specimen radiographs, or at surgical sites. Magnification imaging should only be performed using the correct magnification device, which does not require a grid.

The use of magnification will result in a higher skin dose to the patient because the breast is much closer to the radiation source.

Unfortunately, the use of magnification will increase patient dose because the breast is much closer to the radiation source. An advantage of digital imaging is that magnification can now be used to assess lesion size because the digital system can calculate the exact size of a lesion. To compensate for the reduced image resolution and unsharpness due to the large object-to-image receptor distance (OID), magnification utilizes a small focal spot. The large OID acts as an air gap in reducing the amount of scattered radiation reaching the detector. Grids are therefore not necessary, and their use will only increase exposure time, tube loading, and motion artifacts due to long exposure times, plus grid use will unnecessarily increase radiation dose to the patient. This is especially important because the skin dose will increase as the breast gets closer to the source of radiation. With magnification, there is a very short source/skin distance. The skin dose for a single projection during mammography screening can be as high as 1000 mrad (10 mG). However, the gonadal dose drops to 225 mrad. With magnification mammography and the shorter source-to-object distance (SOD), the skin dose is even higher (Fig. 6–27); however, the glandular dose, the dose routinely quoted, will drop to approximately 100 mrad (1 mGy).

Magnification can be accomplished in any projection, with or without spot compression.

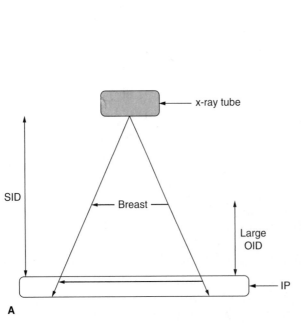

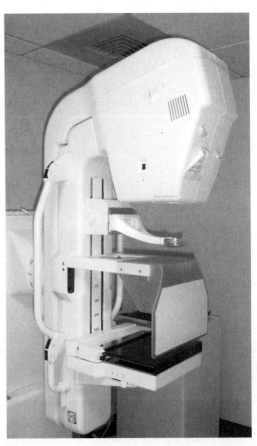

Figure 6–27. **(A)** Schematic diagram of magnification setup. **(B)** A mammography unit set up for magnification.

Spot Compression

Spot compression is used for localization of suspected abnormalities (Fig. 6–28). With spot compression, more compression is applied to a localized area of interest using a smaller compression plate. Spot compression is useful in evaluating a suspicious area and eliminating pseudomasses by spreading out the breast tissue in the area of interest and allowing for more even compression.

Spot compression can be performed in any position or projection with or without magnification. Spot compression is not recommended

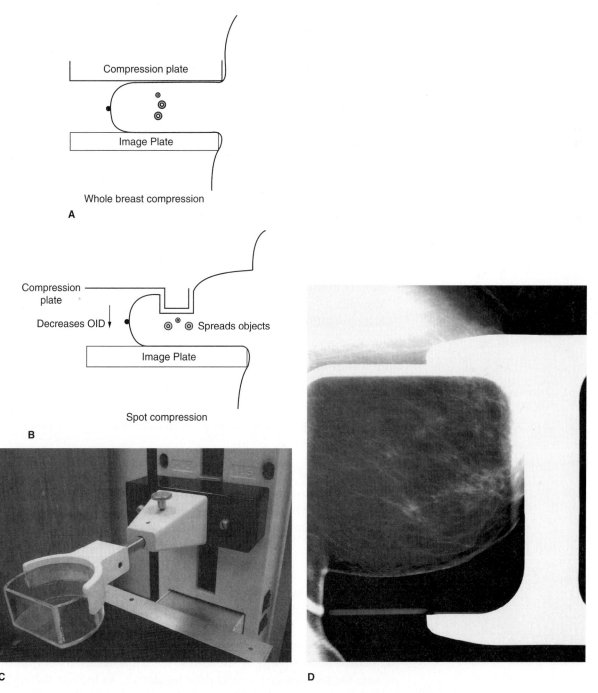

Figure 6–28. Schematic diagram and spot setup on the mammography unit. **(A)** Normal compression. **(B)** Spot compression. **(C)** Spot compression device. **(D)** Radiograph.

for imaging of lesions directly behind the nipple because the small compression plate tends to push or displace the lesion posteriorly, outside of the compression.

IMAGING THE NONCONFORMING PATIENT

Routine imaging often requires supplementary projections or positions as a further workup on a suspicious area or lesion. With the nonconforming patient, however, supplementary imaging may well be the only way to image the breast. The key to imaging is to use the mobility of the breast plus creativity in maximizing the amount of breast tissue under the compression plate. Often, imaging will require one or more supplementary projections/positions. Any extra projections/positions must demonstrate the missing breast tissue to justify the added radiation dose to the patient. A careful study of the supplementary projections/positions and breast anatomy will enable all technologists to correctly evaluate the breast and select the most suitable projection/position necessary to complete the study.

Often we will not use the basic positioning principles that state that we should never compress against a fixed margin or tissue. The movable margins of breast are the lateral and inferior margins, and the fixed margins of the breast are the medial and superior margins.

Technologists' responsibilities when imaging the nonconforming patient include the following:

- Assess safety and cooperation of the patient.
- Modify examination if necessary.
- Visualize as much breast tissue as possible.
- Note limitations on history form.

> Any extra projections/positions must demonstrate the missing breast tissue to justify the added radiation dose to the patient.

Imaging Small Breasts

It is usually easier to image a small breast for the MLO projection. Digital technology is sensitive enough to generate an automatic exposure even when imaging small breasts, so manual technique is often not necessary.

Possible Problems
In both the CC and the MLO, occasionally the technologist's fingers get squashed, yet the breast tends to slip out of the compression. The resultant image loses much of both the medial and lateral breast (i.e., all the posterior breast).

Solution
Have the patient bend laterally for the CC projection or compress while pulling the medial tissue in—remember, medial breast tissue is very important on the CC. Alternatively, roll the patient to the side being examined, bringing the elbow of the ipsilateral side forward, or have the patient bend slightly at the knees. If medial breast tissue will be missed, the technologist should do a regular CC to include the entire medial breast, then an XCCL to obtain more lateral tissue. Using a 5-degree tube angulation can also be substituted for the lateral roll.

For extremely thin breasts, a rubber spatula can be used to bring tissue away from the chest wall during compression (Fig. 6–29). The spatula is much thinner than the technologist's hand and will hold the breast in place before the compression plate takes over. On the MLO, have the patient slouch over— allowing the breast tissue to fall forward.

Although not commonly used, a breast pad can be used to provide friction, therefore allowing easier imaging of the small breast. If the pad is used occasionally, no change in quality control tests is needed. If breast pads are used on all routine work, the technologist should consult the manufacturer-specific quality control requirement.

Male Breast Imaging

The technique for imaging the male breast is very similar to imaging small breasts. Males with gynecomastia usually have enough breast tissue to image reasonably well. On the MLO, male patients usually require 65- to 70-degree tube angulation. The steeper angulation is necessary to position the detector parallel with the large pectoralis muscle.

Possible Problems

The compression tends to slide off when imaging males with hairy chests. Too much pectoral muscle prohibits compression.

Solution

If the chest is very hairy, the patient will have to be rescheduled after shaving or removing the chest hair, or the patient can shave his chest hair at the facility. Generally, the male breast can be treated as a small female breast using similar positioning techniques, for example, using a spatula to help in compression. The FB instead of the CC can also be used.

If the pectoral muscle presents a problem it may be necessary to perform an ML to image the anterior portion of the breast in addition to the routine MLO.

Sectional Imaging of the Extremely Large Breast or Mosaic Breast Imaging

Sectional imaging occurs when the breast is too large to obtain all the tissue on one detector. The first job for the technologist is to decide how to separate the breast for imaging (Fig. 6–30). If the breast is wider than it is long, then divide the imaging into three projections for each breast. Front (anterior), medial, and lateral for the CC and upper (superior), superior (upper), lower (inferior), and front (anterior) for the MLO. Always make sure there is overlapping breast tissue when imaging. Proper anatomical markers must be included.

If the breast is longer than it is wide, dividing the breast into two areas may suffice. On the CC, image the anterior and posterior potions of the breast; then for the MLO, image the upper (superior) and lower (inferior) sections. Care should be taken to image the entire breast with overlapping of the different sections. Some radiologists suggest the placement of a BB marker in the center of the breast for orientating purposes.

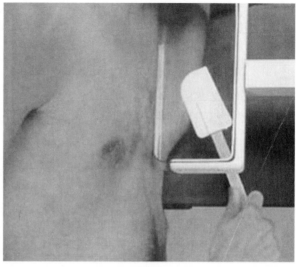

Figure 6–29. Using a spatula to image a male breast.

The technique for imaging the male breast is similar to imaging small female breasts.

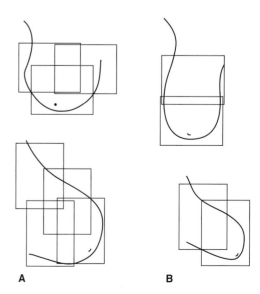

A B

Figure 6–30. Sectional imaging of extremely large breast. **(A)** If the breast is wider than it is long. **(B)** If the breast is longer than it is wide.

A similar technique used in the past was sometimes referred to as mosaic imaging or tiling of the breast.

Possible Problems

Some women can have average-sized breasts but the breast tissue wraps around to the axilla.

Solution

Generally, this would not present a problem in digital imaging. The CC, however, must be imaged using the larger compression plate. It is important not to miss the posterior breast. Sometimes, an XCCL can be used to image the necessary posterior portion of the breast. Whatever technique is used, always check positioning before making the exposure.

Imaging Patients With Obese Upper Arms

The problem generally occurs when imaging the MLO. The "fat roll" or extra tissue tends to fall into the compression area. The solution is to tape the extra tissue out of the imaging area or manipulate it so that most will fall behind the detector.

Patient With an Inframammary Crease That Is Not Horizontal

With these patients, the detector will be at the level of the inframammary crease of one side of the breast (say, the medial side), but the inframammary crease on the lateral side will still be below the image plate. Imaging will, therefore, not include all the breast tissue. The solution here is to angle the patient, tube, or both to position the inframammary crease in direct contact with the detector. This ensures that all breast tissue will be within the compression area.

Patient With Protruding Abdomen

Any patient with a protruding abdomen will present a challenge. Bearing in mind that the aim is to image the breast, not the abdomen, the idea is to keep the patient's abdomen out of the compression area. In imaging for the CC projection, have the patient step away from the unit and then lean forward. When imaging for the MLO projection, try reducing the tube angulation. The patient will, in effect, lie on the image plate, allowing easier manipulation of the protruding abdomen and ensuring an open inframammary fold.

Patient With Thick Axilla

Imaging on the CC projection is generally not a problem with these patients. The problem is most often encountered when imaging the breast on the MLO projection. The upper breast is compressed, but the compression on the lower portions of the breast is minimal to none.

Solution

One solution is to image using two exposures, the MLO for the upper breast and the ML for the lower breast. Another option is the flex paddle. Flex compression paddles are also available and have been approved by the US Food and Drug Administration (FDA). These allow

> In sectioning imaging, always make sure that there is at least 1 inch overlapping breast tissue when imaging the sections.

even compression of the breasts despite the thickness of the posterior breast or axilla region (Fig. 6–31).

The Kyphotic Patient

These patients will have exaggerated thoracic curvature and/or rounded shoulders. Problems are encountered when imaging in both the CC and MLO projections.

Solution

Use FB if possible. If it is not possible, two CC projections, one for the medial breast and the other for the lateral breast, will generally solve the imaging problem. In imaging for the MLO, try the reverse MLO, the LMO, or the LM. The ML should not be used because this projection will not image enough posterior breast. The LMO is the only projection that will give a true reverse of the MLO image. Another option is to image the patient seated. This position often helps to reduce the kyphotic curvature.

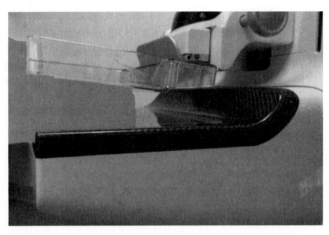

Figure 6–31. A flex paddle will apply even compression when the posterior breast is thicker than the anterior breast.

Patient With Adhesive Capsulitis (Frozen Shoulder)

A patient with a frozen shoulder is unable to lift or raise the shoulder for the MLO projection.

Solution

If the patient can lift the arm at least 90 degrees or bring the arm backward, one possibility is the reverse MLO, that is, the LMO, or the LM. With the image plate against the sternum, the edge of the compression plate will keep the raised arm out of the way.

If the shoulder is not painful, the detector can be used to lift the arm for the imaging.

> The kyphotic patients will have exaggerated thoracic curvature and/or rounded shoulders.

Elderly or Unstable Patients

As the patient population ages, older patients are coming for routine mammograms. The fear of falling can leave a patient anxious and can unnecessarily prolong the examination. It is the technologist's responsibility to assess the patient before starting the mammogram. An assessment would include a check of the patient's ability to stand unaided. Patients may need the use of walkers, canes, or other support.

Solution

Never take away any needed support from the patient. The breast can be imaged with a walker directly in front of the patient. The patient can also hold a cane or even hold onto a walker during the examination if those items provide more reassurance. The patient will certainly be more comfortable and cooperative if the fear of falling is alleviated. Another option is to have the patient transfer from holding the walker to holding the bars of the mammography unit during the mammogram.

If the patient is unstable standing, a chair examination could be the wisest choice. The technologist should document any needed modifications, especially if the modification resulted in a limited study.

> When imaging the elderly patient, the patient will be more comfortable and cooperative if the fear of falling is alleviated.

Mentally Challenged Patient

Patients who are mentally challenged will display what is generally considered to be abnormal behavior. However, the most common psychiatric disorder is anxiety disorder, which can include phobias. In imaging a mentally challenged patient, the technologist should never assume that the patient is unable to understand and should be aware that the degree of mental impairment will vary. Therefore, although the patient may travel with a caregiver, the technologist should address the patient first. The caregiver can be consulted if the technologist needs to determine what approach will work best with the patient.

In addressing the patient, the technologist should use a reassuring but firm tone of voice and should try to keep a continuous stream of conversation, explaining every step of the examination. The technologist will need to be consistent in both actions and words to avoid confusing the patient. The technologist should never interrupt the patient's outburst or argue with the patient's hallucinations or delusions. In addition, the technologist should avoid touching the patient without first asking for permission, and when talking to the patient the conversation should always be face-to-face. The use of words with double meaning should be avoided.

Irrespective of how the patient presents, the technologist should listen carefully to the patient and avoid any judgmental attitude. It is important not to attack the patient's conceptions, or give false reassurance or advice in general.

Regardless of how cooperative the patient is, the technologist should carefully monitor the patient for warning signs such as change in the volume or tone of voice, loss of eye contact, or threatening movements of the patient's upper limbs in particular. If the patient suddenly invades the personal space of the technologist, this can also indicate a problem.

A decreased stimuli environment can aid in keeping the patient calm. For example, some television programs should be avoided and the wait time for the procedure should be as short as possible.

Patients With Pectus Excavatum (Sternum Depressed) or Pectus Carinatum (Pigeon Chest)

In the condition called pectus excavatum, the sternum sometimes caves in, to the point where the ribs are protruding. It is almost impossible to image the medial breast on the CC projection. Patients with pectus carinatum (sometimes referred to as "pigeon chest") have a very prominent sternum. The bone protrudes, similar to the appearance of a pigeon's beak.

Solution
Again, two CC projections may be necessary, one for the medial breast tissue and the other for lateral. The best projection for the medial tissue would be the CV, and instead of the MLO, use the LMO. With the detector positioned against the depressed or protruding sternum, compression from the lateral side of the breast will be easily accomplished.

Barrel Chest

Here, the chest protrudes outward from the body, causing the breast tissue to extend laterally under the arms. It is usually not possible to image the entire breast using one CC projection.

Solution

Perform a CC for the medial breast, then the XCCL for the lateral breast tissue. The MLO is performed as usual because the posterior medial breast tissue missed from this projection can be imaged using the AT projection.

Delicate Inframammary Fold

Sometimes, the skin at the inframammary fold is compromised and will tear if not handled carefully.

Solution

Always practice standard precautions and infection control. The unit, in particular the compression plate, face plate, and detector, should be disinfected before each patient. If the technologist suspects that the breast skin is fragile, the examination should be rescheduled. This is particularly important with patients undergoing chemotherapy and other treatment that has compromised their immune system.

There are several over-the-counter and prescription drugs available that help to promote healing of minor skin irritations that result in fragile skin. While the technologist cannot recommend that the patient take any drug, the patient can be advised to consult with the physician.

Imaging Implants

The standard series of projections for a patient with an implant includes routine CC and MLO projections with minimal compression plus implant-displaced projections. Some facilities have the patients sign an informed consent before any implant imaging. Information about informed consents is covered in Chapter 9. An implant consent can include the following facts:

> Some facilities recommend that the patient sign a consent form before implant imaging.

1. Breast imaging with mammography is not a perfect tool and cannot detect all cancers.
2. An implant can interfere with the interpretation of the mammography study because the implant can obscure the breast tissue.
3. Compression is necessary during breast imaging and can pose potential risks. Although a rare occurrence, compression can cause implant damage, implant rupture, or implant leaks.
4. Implant imaging requires additional images, and women with implants may receive more radiation than women without implants.

Most implants can be displaced by modified compression projections. The technique was introduced in 1988 by Dr. G. W. Eklund. It is called the *Eklund* technique, implant-displaced projections, or ID projections. The result is an eight-projection series (Fig. 6–32).

The first four images will show the implant in position. These standard projections are taken to demonstrate the posterior breast tissue surrounding the margins of the implant. Compression is used solely for immobilization because vigorous compression of the implant could cause ruptures. With the implant on the detector, manual technique may be necessary to avoid an overexposure.

The other projections are called ID projections because the implant is pushed posteriorly. These projections are taken on the CC, MLO, and sometimes 90-degree lateral (ML) projections.

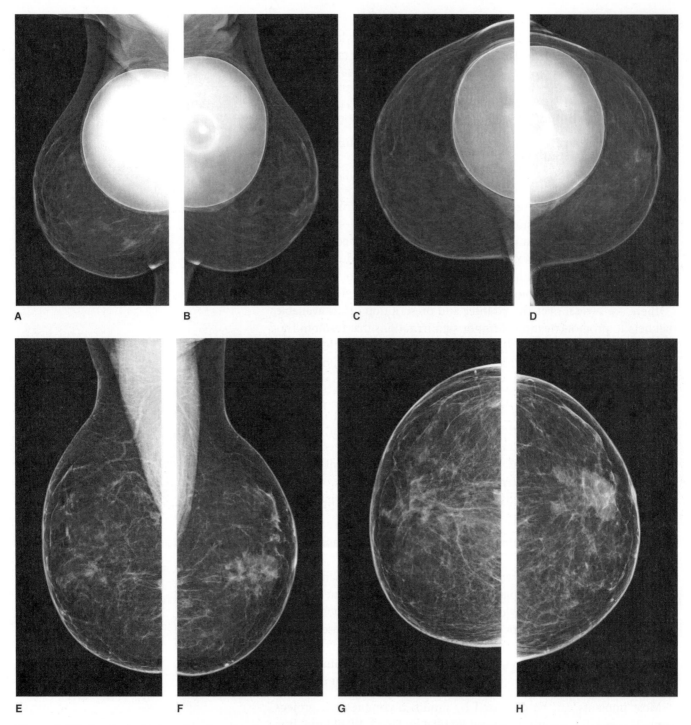

Figure 6–32. Implant-displaced series. Demonstrating modified compression **(A–D)**: RMLO and LMLO and RCC and LCC. Implant-displaced projections **(E–H)**: ID-RMLO and ID-LMLO, and ID-RCC and ID-LCC.

In imaging implants, manual technique may be necessary if the breast tissue does not cover at least the first AEC detector. However, because the entire detector is active, an automatic exposure is possible for even very small breasts. The mobility of the breast and the compression possible will depend on the type of implant, the degree of scarring and adhesion of the implant to the chest wall, and the breast size or amount of actual breast tissue. Most modern implants are placed behind the pectoral muscle, allowing flexibility in positioning.

Implant encapsulation or capsular contracture was more commonly seen with implants placed in front of the pectoral muscle. This condition produces a hard and distorted breast that is difficult to position. Implant-displaced projections are not possible with capsular contracture. The imaging with mammography is limited, and other modalities such as ultrasound and breast MRI are acceptable alternatives.

Positioning

The implant-displaced technique requires pulling the natural breast tissue forward while simultaneously pushing the implant back toward the chest wall. Compression is then applied only to the breast tissue. With the implant-displaced technique, breast tissue that would be compressed against the implant in a routine series is pulled forward and imaged free of the implant (Fig. 6–33). The implant-displaced technique will work on all implants regardless of whether the implant is placed in front of the pectoral muscle (subglandular or retromammary implants) or behind the pectoral muscle (subpectoral implants), so long as the implant is not encapsulated.

Before beginning positioning, the technologist must locate the extent of the implant by feeling for the edges to determine how large or small the implant is. Next, have the patient stand in the position for a routine

> The implant-displaced technique will work on all implants regardless of the implant placement.

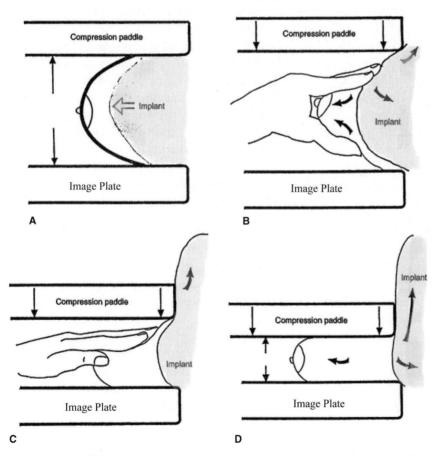

A

B

C

D

Figure 6–33. The implant-displaced method of imaging implants pushes the implant posteriorly and superiorly to image the breast tissue free of the implant. **(A)** The breast positioned for normal imaging. **(B** and **C)** Displacing the implant posteriorly and superiorly before compression. **(D)** Compression of the breast tissue free of implant.

CC or MLO. The patient then steps back, slightly away from the detector. The technologist locates the anterior edge of the implant and places the detector just posterior to the anterior edge of the implant. Use the thumb and fingers to grasp the breast anterior to the implant and pull breast tissue forward. The edge of the detector will help keep the implant pushed back. Begin compression while still holding the breast tissue. As the compression plate comes down, pull the breast tissue forward and outward while allowing the implant to displace posteriorly. The breast free of implant is compressed normally. When imaging, if there is sufficient breast tissue to cover the first AEC detector, automatic exposure control can be used. If enough breast tissue does not cover the first detector, a manual technique must be used.

Imaging the Postsurgical Site—Postlumpectomy Imaging

A recurrent tumor in the treated breast is not uncommon, and mammography follow-up of women after breast conservation is necessary to detect local recurrence at the earliest possible stage. Mammographic signs of recurrence include the development of new microcalcifications or a new mass. Enlargement of the scar and suspicious axillary nodes are also signs of recurrent tumor. Rarer signs are increased skin thickening and breast density. Failure to irradiate the original tumor can result in local recurrence within the first 6 years and sometimes even as early as 6 to 12 months after treatment.

In general, the postsurgical site can be imaged on the CC and ML or MLO projections. Scar markers are useful to identify the site of the surgical scar, although too many scar markers can be a distraction (Fig. 6–34). The wide latitude of 3D imaging now allows visualization of architectural distortion caused by surgery that was done 5 or more years in the past. It is becoming increasingly important to identify these old surgical sites with a scar marker. The marker on the area will confirm the benign finding and save further workup or even biopsy of the area. Magnification projections may be required for further evaluation of suspicious areas. The breast may be tender; compression must be applied to eliminate motion but should not be beyond what the patient can tolerate.

Postmastectomy Imaging

This imaging is controversial because not all radiologists agree with imaging the site. Most literature, however, supports the possibility of another cancer developing at the mastectomy site. Projections/positions usually include the CC, the spot position of the area of concern, an AT projection, or the MLO projection. Research suggests that if the mastectomy site is not imaged regularly, the patient should have an oncologist perform an annual visual inspection and examination of the site especially within the first 5 years of the mastectomy (Fig. 6–35).

Mammograms of the Irradiated Breast

The breast should not be imaged immediately after radiation treatment. The recommendation is 6 to 12 months after completion of radiation treatment. Radiation can cause skin thickening, trabecular thickening, increased density of the parenchymal pattern, and diffuse increase in

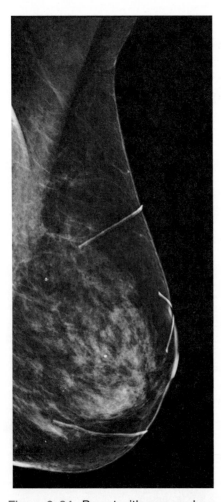

Figure 6–34. Breast with scar markers.

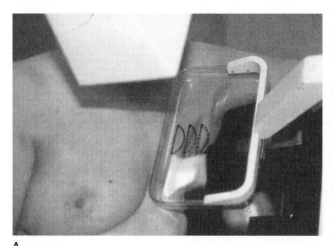
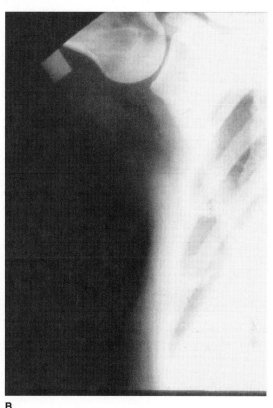

A   **B**

Figure 6–35. Mastectomy imaging using the MLO projection. **(A)** Positioning the patient. **(B)** Actual radiograph.

breast density. Most of the changes are a result of edema or coarsening of the fibrous or stromal elements of the breast, plus increased thickness and density of the ductal and glandular elements. Because of these changes, earlier mammograms will be of limited value. However, regardless of any initial change, further changes can diminish or resolve over time, but their pattern should not increase. The exception here is calcifications, which can develop up to 5 years after radiation.

When imaging the postirradiated breast at a normal interval, routine projections will be possible, but care should be taken in handling the breast. The irradiated breast is often very tender, and the skin easily broken. The purpose of the imaging is to record new mammographic patterns in the breast. If the original tumor contained microcalcifications, magnification images will fully assess the status of any calcifications at the site. If the breast does not become red, firm, or tender after treatment, there was less reaction to the radiation, and a useful mammogram can be obtained 3 to 6 months after radiation treatment.

Imaging the Stretcher Patient

The breasts can be imaged for the CC projection with the patient supine or lateral on the stretcher and the x-ray tube rotated 90 degrees (Fig. 6–36). For the MLO, the tube is positioned at 0 degrees (tube vertical), and the patient is rolled on the side being examined. The MLO is obtained, with the patient almost on the stomach, and the detector positioned under the patient's breast (Fig. 6–37). Stretcher patients can

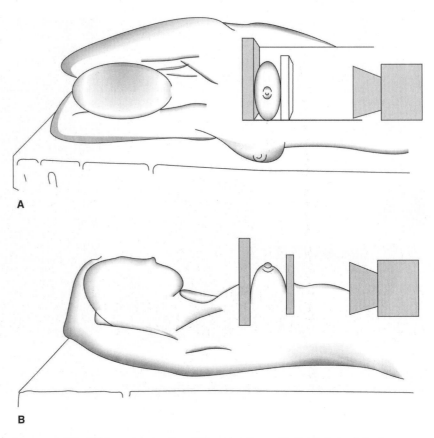

Figure 6–36. (A) Imaging on the CC projection with the tube rotated 90 degrees and the patient lateral. **(B)** Imaging on the CC projection with the tube rotated 90 degrees and the patient supine.

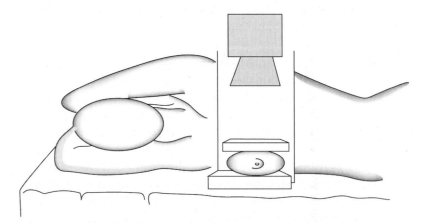

Figure 6–37. Imaging on the MLO with the tube vertical.

also be imaged with the patient recumbent, face up on the stretcher, and the x-ray tube rotated 90 degrees.

Imaging the Wheelchair Patient

The FB is especially usefully in imaging the wheelchair patient (Fig. 6–38). If imaging using the FB is not possible, the patient must be fully upright to accomplish the imaging on the CC projection. A

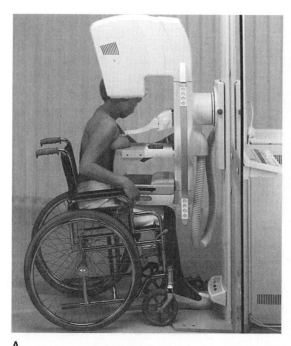

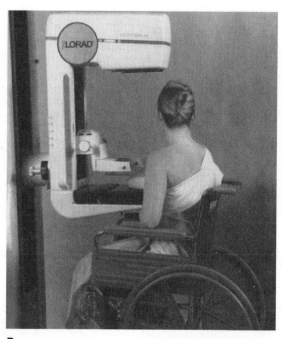

A   B

Figure 6–38. Imaging the wheelchair patient. **(A)** With spot magnification. (Used with permission from GE Medical Systems.) **(B)** Routine imaging. (Used with permission from Hologic Inc.)

pillow or cushion behind the patient's back can achieve this. The MLO will be easiest if the wheelchair has removable arm supports. If not, the patient must be "built up" so that the mammography unit will not hit the arm of the chair. Alternatively, the patient can be transferred to a chair without arms or another wheelchair with removable arms.

SPECIFIC IMAGING PROBLEMS

Nipple Not in Profile

If imaging is routinely performed using a nipple marker, nipple projections are unnecessary. The main purpose of getting the nipple in profile is to differentiate the nipple from a lesion (Fig. 6–39). In most women, with proper positioning, the nipple will automatically fall in profile on at least one projection. If the nipple is not in profile on accurate positioning, the first option should not be to reposition the breast solely to place the nipple in profile; this usually means displacing breast tissue posteriorly, which places much of the posterior breast tissue outside of the compression plate (Fig. 6–40). Always image the entire breast first. Image the nipple separately, taking nipple projections with the nipple in profile only if necessary.

Skin Folds or Wrinkling of the Breast

Wrinkles in the skin or skin folds will often produce architectural distortions or radiopaque artifacts that can obscure surrounding tissue (Fig. 6–41). To avoid excessive radiation to the patient, the

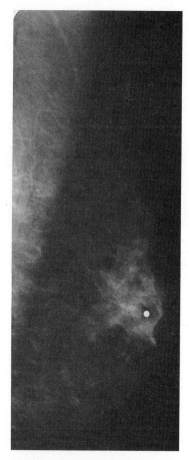

Figure 6–39. Imaging with the nipple not in profile can mimic a subareola lesion. The BB marker identifies the nipple.

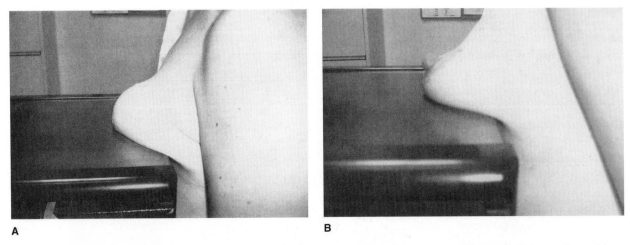

A   **B**

Figure 6–40. Patient positioning. **(A and B)** If the patient is imaged as shown, the nipple will not be projected in profile.

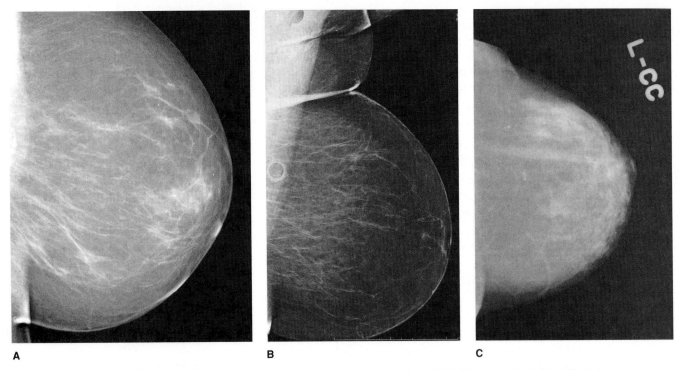

A   **B**   **C**

Figure 6–41. Skins folds can compromise the image. **(A)** The poorly positioned MLO has a skin fold at the inframammary crease. **(B)** Skin folds in the axilla area. **(C)** Skin fold in the middle of the breast can mimic pathology.

mammogram should not be repeated if the skin fold or wrinkles do not interfere with the interpretation. To remove skin folds or wrinkles, use an index finger to smooth out the breast toward the nipple as compression is applied. Sometimes, in an effort to avoid skin folds, breast tissue is eliminated from the mammogram. In such cases, always take additional projections to supplement the original. Generally, the technologist should avoid smoothing the skin away from the nipple and toward the posterior breast because this will eliminate posterior breast tissue from the imaged field.

Patients With Uneven Breast Thickness

The plate should be parallel to the pectoral muscle. This is especially important in imaging the patient with uneven breast thickness because already proper compression and imaging will be difficult. A check should also be made of the height of the detector. The upper edge should rest just in the patient's armpits. If it is impossible to compress the lower half of the breast on the CC projection, two projections may be necessary: one of the breast area closest to the chest wall (the posterior breast) and the other of the nipple area of the breast (anterior breast). The same is true for the MLO projection. The lower (anterior or nipple) area of the breast will not necessary receive sufficient compression on the routine MLO. In such cases, take an axillary tail (upper portion of the breast) with the first exposure, then image the lower (anterior) portion of the breast using the ML projection.

The other option is to use flex compression paddles. These allow even compression of the breasts despite the thickness of the posterior breast (see Fig. 6–31).

Moles and Scars

Some radiologists prefer that the technologist mark all moles and scars with radiopaque markers. Special markers are available for digital and digital tomosynthesis systems (Figs. 6–42 and 6–43). However, others feel that the radiopaque markers are a distraction and can interfere with interpretation. These radiologists prefer detailed diagrams of the breast, indicating the position of all moles or scars. Often, the patient history sheet has a schematic diagram of the breast that can be used for documentation. One problem however is that schematic diagrams are very subjective and may not provide an accurate representation of the positions of the various scars. Some studies suggest that not more than four scar or mole markers should be used on any one projection. There are also other considerations. One is that many surgeries are performed to achieve good cosmetic results. This means that, in a surgical biopsy, a superior lesion will be accessed from the inferior aspect of the breast, and a scar marker indicating the surgical site will not show the location of the original lesion. However, a scar marker may still be necessary to indicate a past surgery. The sensitivity of 3D imaging will often reveal architectural distortions caused by past surgeries. Some reports have indicated that visualization of distortion caused by surgeries over 5 years may not be seen on 2D mammogram but will be picked up on 3D. Therefore, what is seen as a new area of distortion could in fact be an old scar. A scar marker could potentially reduce the need for further work-up or a biopsy if it can be confirmed the area of architectural distortion was caused by past surgery. Fortunately, in most of the modern surgical technique, a radiopaque marker will be left in the surgical site for future reference.

> Moles or scars on the surface of the skin will project inside the breast on the two-dimensional radiograph. This can mimic a lesion within the breast.

Pacemakers and Implantable Venous Access Systems

Patients with a pacemaker or venous access port (sometimes called port-a-cath) can present challenges for the technologist. The pacemaker, if placed in the breast or even in the upper chest, can prevent or limit breast compression. The mammography should maneuver the pacemaker so that it is flat when imaging the CC projection and vertical

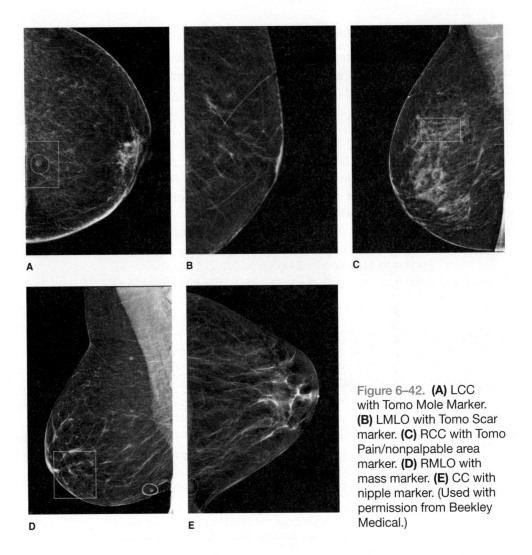

Figure 6–42. **(A)** LCC with Tomo Mole Marker. **(B)** LMLO with Tomo Scar marker. **(C)** RCC with Tomo Pain/nonpalpable area marker. **(D)** RMLO with mass marker. **(E)** CC with nipple marker. (Used with permission from Beekley Medical.)

when imaging for the MLO. With the proper manipulation and placement of the pacemaker, and depending on the thickness of the patient's breast, compression can be applied. However, the technologist should avoid vigorous compression to the pacemaker.

Venous access systems or ports can also present a problem because they are often placed just below the clavicle. The unit consists of a catheter inserted into a large vein and connected to a self-sealing silicone rubber port. In general, the port can be felt just below the skin and cannot be manipulated. Compression should be applied cautiously to avoid damage to the port.

In any examination of patients with pacemakers or ports, the technologist should note any limitation of the study in the patient records.

Triangulation Techniques

Every technologist should understand how to localize and determine the approximate positions of lesions in the breast. In breast imaging, there are three common location terms: the quadrant, the clock-face, and the region methods. In the quadrant method, the breast is divided into four quadrants—upper outer and upper inner (UOQ and OIQ) and lower outer and lower inner (LOQ and LIQ). Using the clock-face method, the breast becomes the face of a clock—the uppermost part of

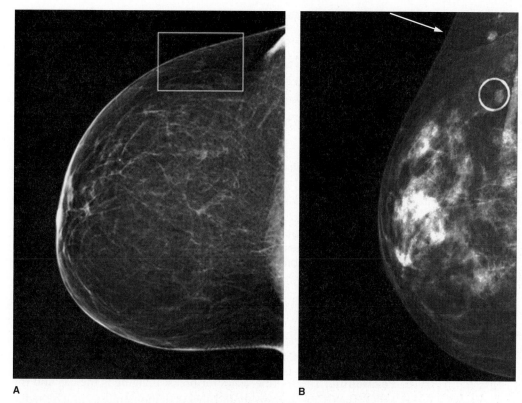

A   B

Figure 6–43. Moles and other skin lesions should be marked because they can mimic pathology. **(A)** A 78-year-old patient with screening 3D mammography. No mole marker was used and a nodule is seen in the upper outer quadrant. The technologist also charted a mole in the upper outer quadrant on a schematic diagram of the breast. However, the radiologist could not confirm the nodule was the mole and additional work-up was needed. **(B)** A modified CC taken with a mole marker in place. The mole does not correspond to the nodule seen earlier and further evaluation is required. (Used with permission from Beekley Medical.)

the breast is 12 o'clock, and the lowermost part is 6 o'clock. Note that the 5-o'clock position on the right is the lower inner quadrant, but the 5-o'clock position on the left breast is the lower outer quadrant. In the region method, the breast closest to the chest wall is the posterior region. The anterior region is located behind the nipple and the middle region between the posterior and anterior regions. Located directly behind the nipple is the subareolar region (see Fig. 6–43). Chapter 3 includes detailed information on breast location terminology.

If any two mammographic projections demonstrate a lesion, the technologist can determine the lesion's approximate location in either the MLO or CC projection by triangulation (Fig. 6–44).

1. Using fingers or a ruler, measure the distance of the line directly posterior from the nipple to the level of the lesion.
2. From the end of the first line, measure (again using fingers or a ruler) the line superior or inferior if measuring from an MLO projection, or medial-to-lateral if measuring from a CC projection to the lesion.
3. The next line measures the distance from the lesion to the skin surface.

Use your hand to mimic breast compression. Transfer the measurements to the compressed breast, again using fingers or a ruler. Use a BB marker to indicate the position of the lesion. Image using spot compression or magnification as needed.

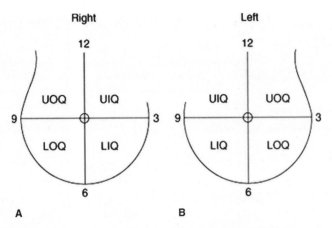

Figure 6–44. Breast localization methods used in triangulation. **(A)** Triangulation of a lesion on the MLO projection. **(B)** Triangulation of a lesion on the CC projection.

Before Signing Off

Before signing off on a mammogram, the technologist should be able to assess the following points:

1. **Positioning:** The routine series for breast imaging is the CC and the MLO. The CC projection should always demonstrate as much medial tissue as possible because this area is most likely to be missed or poorly visualized on the MLO. The MLO will visualize the maximum amount of breast tissue if the angle used is parallel to the pectoral muscle.

2. **Compression:** Improperly compressed breast tissue will result in overlapping tissue structures; nonuniform exposure, especially of the denser breast tissue; overpenetration of the thinner breast tissue; poor penetration of the thicker portion; and motion unsharpness.

3. **Exposure:** In digital imaging, the term brightness is used instead of optical density, and the image brightness can be adjusted after the exposure. Digital imaging has a much wider latitude, and it is possible to visualize the skin line and dense areas of the breast with one exposure.

4. **Contrast:** Contrast is usually highest in the thinner breast and lowest in the thicker breast due to more scattered radiation and greater tissue absorption of low-kV radiation in the thicker breast. Without contrast, there would be little differentiation between breast parenchyma and different tissue densities within the breast tissue.

5. **Sharpness:** The ability of the mammographic system to capture fine details in the image is defined as sharpness. Unsharpness may be the result of motion blur or geometric factors, such as large focal spot size, an increase in OID, or a decrease in SID.

6. **Noise:** This refers to the random background information that is due to constant flow of current in the circuit. Noise does not contribute to image quality. On the digital image noise looks like quantum mottle, and the image will appear as grainy. In digital imaging, noise is controlled by the milliamperage (mA), defined as the actual number of photons striking the detector. A high signal-to-noise ratio (SNR) is best and can be controlled by the initial mA and the use of grids to reduce scattered radiation.

7. **Artifacts:** Any radiolucent or radiopaque lesions on the image that is not a reflection of the attenuation differences in the subject can be considered artifacts. In mammography, the surface of the detector should be cleaned on a regular schedule to avoid artifacts and to provide infection control. Examples of artifacts are scratches, fingerprints, dirt, lint, or dust.

8. **Collimation:** In mammography, collimation is to the detector and not to the breast.

9. **Labeling:** Standardized labeling in mammography is important because mammograms can be legal documents, and it is important that the radiographs are not misinterpreted. In the final rules of the MQSA, labeling is divided into required, highly recommended, and recommended categories.

Summary

- Routine imaging includes the four-projection series: MLO and CC projections of both breasts.

- Sometimes supplementary projections/positions are needed if the standard projections are inadequate, if a suspicious area is seen on one projection only, or if a suspicious area needs workup to avoid the trauma of surgery.

- Supplementary projections/positions include 90-degree mediolateral—ML; lateromedial—LM; lateromedial oblique—LMO; tangential—TAN; exaggerated craniocaudal—XCCL; from below—FB; cleavage—CV; axillary tail—AT; superior–inferior oblique—SIO; rolled positions— RM, RL and RS, RI; and magnification and spot compression.

Mediolateral (ML)

- Prove "teacup" benign lesions

- Verify a finding or localize in another dimension:
  - Medial lesions move up, lateral lesions move down, and central lesions will not change.

Lateromedial (LM)

- Improve detail on medial lesions

- Image during breast localization

- Image nonconforming patients

Lateromedial Oblique (LMO)

- Replace the MLO when imaging the nonconforming patient

- Image patients with pacemakers or prominent sternum

- Give details on medial breast

Tangential Projection (TAN)

- Locate skin lesions and calcifications

Exaggerated Craniocaudal Lateral (XCCL)

- Image the extreme lateral aspect of breast on the CC

Exaggerated Craniocaudal Medial (XCCM)

- Image the extreme medial aspect of breast on the CC

Caudocranial or From Below (FB)

- Image small breasts or the kyphotic patient
- Image the superior aspect of the breast

Cleavage or Valley View (CV)

- Image the deep medial aspect of the breast

Axillary Tail (AT)

- Visualize the tail of Spence.

Superior–Inferior Oblique (SIO)

- Demonstrate the upper inner quadrant
- Image additional breast tissue in implant imaging
- Visualize the Inframammary Fold (IF), especially when using digital mammography

Rolled positions: rolled lateral (RL), rolled medial (RM), rolled superior (RS), or rolled inferior (RI)

- Remove superimposed tissue
- Localize lesions when seen only in one projection

Magnification

- Image microcalcifications
- Show the margins of lesions

Specimen Radiograph

- Confirm the removal of a lesion or calcifications

Spot Compression

- Apply more compression to an area of interest or a lesion
- Image calcifications

Imaging the Nonconforming Patient

Small Breasts

- Imaging is often easier in MLO
- Use spatula if necessary
- Roll or tilt the patient to the affected side
- Patient slouches for MLO imaging and leans away from the unit for CC imaging
- Breast pads used with caution if used for routine imaging—there are often manufacturer-specific requirements for digital imaging

Male Breast Imaging

- Similar to imaging small breasts
- FB instead of CC

- Shave chest hair if necessary
- Pectoral muscle can present a problem on the MLO
- Spatula use will avoid compressing fingers

Large Breast
- Sectional imaging (mosaic imaging) of the breast often needed
- Always overlap breast tissue sections
- Correctly label all sections

Wide Breast
- Sectional imaging of the breast or mosaic imaging on CC
- MLO projection usually images normally

Inframammary Fold Not Horizontal
- Angle patient, tube, or both to include all breast tissue for CC

Patient With Protruding Abdomen
- Keep patient away from the unit in CC imaging—patient should lean forward
- Reduce tube angulation for MLO imaging—the patient almost lies on the plate

Patient With Thick Axilla
- Image anterior and posterior breast separately—MLO for posterior breast and ML for anterior breast
  - Other option: the "flex paddle"

Kyphotic Patient
- CC imaging
  - FB or two CC projections—for medial and lateral tissue
  - Imaging the patient seated to reduce the kyphotic curvature
- MLO imaging
  - Use the LMO

Frozen Shoulder
- Use the LMO or LM—with the patient's arm moved backward
- Detector can be used to lift the arm during imaging

Elderly or Unstable Patient
- Possible chair examination
- Check with physician and document any limitations

Mentally Challenged Patient
- Use a reassuring but firm tone of voice and try to keep a continuous stream of conversation.
- Keep all conversations face-to-face.
- Never interrupt the patient's outburst.
- Never argue with the patient's hallucinations and delusions.
- Do not get angry if the patient questions your credentials/competence.
- Do not touch the patient without first asking for permission.

Pectus Excavatum (Depressed Sternum)

- CC imaging
  - Two CC projections of each breast (four-projection series)—one for medial breast, one for lateral breast
  - Other option: a three-projection series—the CV for the medial breast with an XCCL to demonstrate both lateral breasts
- MLO imaging
  - LM or LMO

Pectus Carinatum—Pigeon Chest or Barrel Chest (Prominent Rib and Sternum)

- Routine CC for medial breast tissue with XCCL for lateral breast tissue
- Routine MLO plus AT to image any missed breast tissue

Delicate Inframammary Fold

- Standard precautions when imaging
- Caution when lifting to position for the CC
  - Postpone imaging if the skin is compromised

Mastectomy Imaging

- Imaging or visual check by oncologist is recommended—always follow guidelines of facility
- Imaging can include CC or spot projections; AT or MLO projections

Lumpectomy Imaging

- Include scar markers as needed—always follow the guidelines of the facility
- Imaging can include magnification and additional projections

Irradiated Breast

- Always practice infection control
- The first mammogram is generally 6 to 12 months after completion of treatment

Stretcher Imaging

- Lock stretcher and cart
- Patient may be able to sit or assist
- CC imaging performed with tube at 90 degrees
- MLO imaging performed with the tube at 0 degrees

REVIEW QUESTIONS

1. State three benefits of breast compression.

2. Give one possible cause of painful breasts.

3. On which two margins is the breast more rigidly attached to the chest wall?

4. Why is magnification not useful in evaluating lesion size?

5. Why are the CC and MLO projections termed "complementary projections"?

6. What area of the breast (not imaged well on the MLO) is best imaged on the CC projection?

7. Why are grids not necessary in magnification mammography?

8. What factor determines tube angulation when imaging on the MLO projections?

9. What area of the breast is best imaged on the MLO?

10. Name three common problems when imaging on the MLO.

11. Which supplementary projection will best image the extreme posterior medial breast?

12. Which projection is the true reverse of the MLO?

13. Name the projection most useful in imaging skin calcification.

14. Name the projection that best images the posterior lateral aspect of the breast.

15. Why is a spatula often useful when imaging the small breast?

16. The standard Eklund technique (implant-displacement technique) will result in how many exposures when imaging a typical patient?

17. Why is imaging the irradiated breast less than 3 months after completion of radiation treatment contraindicated?

18. What is the purpose of the triangulation technique?

CHAPTER QUIZ

1. Which of the following is inappropriate or demonstrates poor positioning techniques when imaging on the CC projection?

   (A) Demonstrating the nipple in profile

   (B) Always demonstrating the posterior lateral breast

   (C) The PNL measurement is within 1 cm of the PNL on the MLO

   (D) Including both the medial and lateral breast within the collimated field

2. The projection used to demonstrate the posterior lateral aspect of the breast not seen on the CC is:

   (A) XCCL

   (B) ML

   (C) LMO

   (D) MLO

3. Key points in imaging on the MLO projection include:

   (1) Tube angulation should be between 30 and 70 degrees

   (2) The inframammary fold should be closed

   (3) Compression must support the anterior breast tissue

   (A) 1 only

   (B) 2 and 3 only

   (C) 1 and 3 only

   (D) 1, 2, and 3

4. In breast imaging, the technique used to increase compression on a small area of interest to spread out the tissue and improve resolution is called:

   (A) Cleavage

   (B) Axillary tail

   (C) Tangential

   (D) Spot compression

5. The tangential projection is often used to:

   (A) Image implants

   (B) Demonstrate skin lesions

   (C) Evaluate the margins of lesions

   (D) Remove superimposed tissue when imaging dense breast

6. When imaging implants, some of the projections taken will include an image of the implant. In these projections, the compression is used:

   (1) For immobilization only

   (2) To separate the breast tissue

   (3) To separate and spread out the implant

   (A) 1 only

   (B) 2 only

   (C) 1 and 3 only

   (D) 2 and 3 only

7. This projection best demonstrates the upper inner quadrant and the lower outer quadrant of the breast free of superimposition and can be used when imaging implants, especially encapsulated implants:

   (A) SIO

   (B) AT

   (C) TAN

   (D) LML

8. The main reason for breast compression is to

   (A) Control the radiation dose to the patient

   (B) Control patient motion

   (C) Reduce the exposure time

   (D) Ensure a uniform thickness

9. After imaging in the breast using the MLO projection, an ML projection was taken. The lesion seen on the MLO moves down on the ML. This indicates that the lesion is located:

   (A) Laterally

   (B) Medially

   (C) Centrally

10. The true lateral projection that will best demonstrate a medial lesions is:

    (A) ML

    (B) MLO

    (C) LMO

    (D) LM

BIBLIOGRAPHY

American College of Radiology (ACR). *Mammography Quality Control Manual.* Reston, VA: The American College of Radiology; 1999.

Andolina VF, Lille SL, Willison KM. *Mammographic Imaging: A Practical Guide.* 3rd ed. Philadelphia, PA: Lippincott Williams & Wilkins; 2010.

Berns EA, Baker JA, Barke LD, et al. *Digital Mammography Quality Control Manual.* Reston, VA: American College of Radiology; 2016.

Bushong SC. *Radiologic Science for Technologists—Physics, Biology and Protection.* 10th ed. St. Louis, MO: Mosby; 2012.

Gruen D. *Benign Intraparenchymal Scarring in the DBT. Beekley Medical.* Available at: http://blog.beekley.com/how-dbt-has-radiologists-coming-full-circle-back-to-marking-scars-in-breast-imaging. Accessed July 2017.

Papp J. *Quality Management in the Imaging Sciences.* St. Louis, MO: Mosby; 1998.

Peart O. *Lange Q & E Mammography Examination.* 3rd ed. New York, NY: McGraw-Hill; 2015.

Peart O. Positioning challenges in mammography. *Radiol Technol.* 2014; 85(4):417–443.

Tabár L, Dean PB. *Teaching Atlas of Mammography.* 3rd ed. New York, NY: Thieme Stuttgart; 2001.

Tucker AK, Ng YY. *Textbook of Mammography.* 2nd ed. Edinburgh, UK: Churchill Livingstone; 2001.

Venes D, Biderman A, Adler E. *Taber's Cyclopedic Medical Dictionary.* 22nd ed. Philadelphia, PA: F. A. Davis; 2013.

Breast Imaging—Digital Mammography, Ultrasound, and Magnetic Resonance Imaging | 7

Keywords and Phrases
  Ultrasound Terminology
  MRI Terminology
Analog Mammography
Digital Mammography
  Direct Flat-Panel Detector Technology
  Indirect Flat-Panel Detector Technology
  Photostimulable Storage Phosphor Detector Technology
  Photon-Counting Image Capture Technology
  Advantages of Digital Mammography
  Disadvantages of Digital Mammography
Ultrasound Imaging of the Breast
  Brief History of Ultrasound
  The Principle of Ultrasound
  The Transducer (Probe)
  Common Ultrasound Terminologies
  Uses of Ultrasound
  Management of Nonpalpable Masses
  Ultrasound Evaluation of Palpable Masses
  Doppler Effect
  Color Doppler
  Ultrasound Characteristics of Benign Lesions
  Ultrasound Characteristics of Malignant Masses
  Ultrasound Characteristics of Intermediate Lesions
  Typical Appearance of Lesions on Ultrasound
  Imaging Implants
  Ultrasound Artifacts
  Limitations of Breast Ultrasound Imaging
  Imaging Problems and Solutions
  Risks of Ultrasound
  Advantages of Breast Ultrasound
Magnetic Resonance Imaging of the Breast
  Brief History of MRI
  Electromagnetic Spectrum
  Magnetism

The Principle of MRI
Equipment
Operating Console
Computer Parameters
MRI Parameters
Use of Contrast Agents
Pulse Sequences and Techniques
Postprocessing Technique
Breast Magnetic Resonance Imaging
MRI Artifacts
Advantages and Uses of Breast MRI
MRI of the Augmented Breast
Limits of Breast MRI
Risks and Potential Complications
MRI Quality Assurance
Summary
Review Questions
Chapter Quiz

Objectives

On completing this chapter, the reader will be able to:

1. Understand basic digital terminology
2. Describe the various digital imaging technologies
3. Identify the advantages and disadvantages of digital imaging
4. Describe the principle of ultrasound
5. Discuss why high-frequency transducers will best image the breast
6. Explain the meaning of common ultrasound terminologies
7. Describe the breast ultrasound scanning techniques
8. List ultrasound recording options
9. Describe the process of Doppler and color Doppler ultrasound
10. Describe the ultrasound appearance of both malignant and benign lesions
11. State the common ultrasound imaging artifacts
12. Discuss the limits of breast ultrasound imaging
13. Give specific advantages of ultrasound imaging of the breast
14. Discuss the basic principle of magnetic resonance imaging (MRI)
15. Describe the main features of the MRI equipment
16. Identify common MRI sequencing and terminology
17. Describe how the breast MRI examination is conducted
18. State common reasons for breast MRI examination
19. Describe common MRI artifacts
20. Identify specific limits of MRI
21. Discuss the enhancement rates of benign versus malignant breast lesions in MRI
22. Identify the known risks of MRI
23. State the advantages of breast MRI

KEYWORDS AND PHRASES

- **Absorption efficiency** is a measure of the percentage of energy that strikes a receptor material that is actually absorbed by the receptor.

- **Analog-to-digital converter (ADC)** converts analog signal from the image receptor/detector to a digital signal for the computer to manipulate for processing, display, and storage.
- **Bannayan–Riley–Ruvalcaba syndrome (BRR)** is a rare genetic condition characterized by the growth of hamartomatous polyps of the small and large intestine. The condition, which manifests in childhood, is also associated with macrocephaly, benign fatty tumors (lipomas), hemangiomas, and thyroid problems.
- **Bytes** are a group of eight bits, where a bit represents the smallest unit of measure of computer storage. One kilobyte (kB) 10^3 equals 1024 bytes, 1 megabyte (MB) 10^6 equals 1 million bytes, 1 gigabyte (GB) 10^9 equals 1 billion bytes, and 1 terabyte (TB) 10^{12} equals a 1000 billion bytes. The higher the bit value, the higher the contrast.
- **Computer reader** or **CR** scans the image receptor (photostimulable storage phosphor, or PSP) with a laser beam to initiate the emission of light. This emission is called photostimulated luminescence (PSL).
- **Contralateral** refers to the opposite from the side being examined.
- **Conversion efficiency** is the percentage of energy absorbed by a receptor that is converted to usable output.
- **Cyst** is the term used to describe a spherical, fluid-filled structure. In ultrasound, the cyst has few or no internal echoes. Cystic structures are generally circumscribed, have smooth borders, and are usually round or oval in shape with acoustic enhancement posterior to the lesion. On an ultrasound image, a cyst appears completely echo free. This means it is filled with fluid—serous fluid, blood, or bile. Any cyst can be subjected to at least one or more of the following three complications: rupture, infection, and bleeding.
- **Cowden's syndrome** is a rare genetic disorder characterized by the development of skin lesions, multiple benign hamartomas, and neoplastic growth throughout the body.
- **Detective quantum efficiency (DQE)** is the product of absorption and conversion efficiency. It is the percentage of energy that strikes a receptor and results in a useful output signal. DQE measures how efficiently a digital detector can convert the remnant beam to useful data. High DQE means lower patient dose.
- **DICOM**, Digital Imaging and Communications in Medicine, is the industry standard for transferring radiologic images and other medical information between computers. It is based on an open-system international standard and allows digital communication between diagnostic and therapeutic equipment and systems from multiply manufacturers.
- **Digital-to-analog converter (DAC)** converts digitally manipulated data back to an analog (shades of gray) signal to be sent to the display monitor.
- **Diodes** are vacuum tubes with two electrodes: a cathode and an anode.
- **Display contrast** in digital imaging is determined by the "window" and "level" of the displayed region of the breast. The window is the range of intensities that are displayed. The digital image has a wide latitude—called the dynamic range. The contrast in any display image can be changed or modified by the technologist.
- **Dynamic range** is the range of values over which a system can respond and is known as the gray-scale range. Dynamic range will,

therefore, refer to the number of shades of gray that can be represented in each pixel and represents the receptor's ability to respond to different exposure levels.

- **Exposure latitude** is the range of underexposure or overexposure that can occur in producing an acceptable image. Analog receptors can produce an acceptable image, but not an ideal image, within a range of 30% underexposure to 50% overexposure. Digital receptors can produce an image that appears acceptable on the display monitor, but which was produced at 50% underexposure to as much as 100% overexposure.
- **Field of view (FOV)** describes how much of the patient is imaged in the matrix. The FOV is equal to the pixel size times the matrix size. FOV is important in mammography because to image most of the average female population, the FOV should be the size of the largest detector or cassette. The maximum FOV available is 24 cm × 31 cm.
- **Half-life** refers to the time taken for a quantity of radioactivity to be reduced to half of its original value. The half-life of a radioisotope can range from minutes to hours. Fluorodeoxyglucose (FDG) has a relatively long half-life of 2 hours.
- **Image receptor (IR)** also called the storage phosphor screen (SPS) or photostimulable storage phosphor (PSP). It is the device that receives the energy of the x-ray beam and forms the image of the body part in digital mammography (DM) and computed mammography (CM).
- **Ipsilateral** refers to the same side as the side being examined.
- **Li–Fraumeni syndrome (LFS)** is a rare hereditary condition in which individuals are predisposed to multiple cancers, caused by an alteration in the *p53* tumor suppressor gene. The common cancers associated with this syndrome are adrenocortical carcinoma, breast cancer, brain cancer, leukemia, and sarcoma.
- **Matrix** refers to the layout of cells in rows and columns. Each cell corresponds to a specific location in the image.
- **Modulation transfer function (MTF)** is the ability of the detector system to transfer its spatial resolution characteristics to the image (i.e., record) of the available spatial frequencies to produce an image that is exactly like the object.
- **Noise** is the random background information that is detected but does not contribute to the image quality. All digital imaging has noise. It can look like quantum mottle.
- **PACS**, picture archiving and communication system, is essentially a networking system in which images acquired from different modalities at multiple locations, such as ultrasound or magnetic resonance imaging (MRI), can be viewed at one location or shared and archived.
- **Parallax** refers to the apparent displacement of an observed object when it is imaged from two or more different points.
- **Photoconductor** is a material that, when irradiated, produces free electrons. These electrons can be collected if a voltage is applied to the photoconductor.
- **Photostimulated luminescence (PSL)** is the release of stored energy within a phosphor or photostimulable plate by stimulation with visible light.
- **Pixel,** or picture element, is the smallest component part of an image. Each pixel has a numerical value, and they are normally arranged in a two-dimensional grid represented as dots or squares.

- **Prone** is lying face down, on the stomach.
- **Scintillation** is the process by which certain materials scintillate or emit a flash of light in response to the absorption of ionization radiation. The light emitted is directly proportional to the amount of energy absorbed by the material. Only certain crystalline structures will scintillate.
- **Scintillation phosphors** are inorganic crystals that respond to ionization radiation by scintillation. They include thallium-activated sodium iodide (NaI: T1) or thallium-activated cesium iodide (CsI: T1).
- **Sensitivity** measures the ability to respond to or register small changes or differences. An imaging system that is highly sensitive will detect all changes in the breast, whether normal or abnormal.
- **Signal-to-noise ratio (SNR)** is a comparison of the strength of the information in the digital image to the strength of the noise in the image. Most digital images have more signal strength at low spatial frequencies and less at high spatial frequencies. In MRI, signals will measure the voltage induced in the receiver coil by the precession of the vector in the transverse plane. The noise can be background electrical interference inherent to the system, and it also depends on patient size and area under examination. High SNR has little noise interference. Increased noise results in decreased contrast.
- **Spatial frequency** is measured in line pairs per centimeter (1 p/cm) or line pairs per millimeter (1 p/mm). The ability to image high spatial frequency means the ability to image very small objects. Higher spatial frequencies also give better spatial resolution. **Spatial resolution** is the ability to distinguish and separate between two adjacent structures in the image. In MRI, reducing slice thickness increases spatial resolution, whereas increasing slice thickness reduces spatial resolution.
- **Specificity** is the quality of being precise rather than general. An imaging system that is highly specific can effectively differentiate between normal and abnormal changes within the breast.
- **Supine** is lying face up or on the back.
- **Windowing** is the process whereby the technologist postprocesses the image, allowing only a "window" of the entire dynamic range to be viewed on the computer monitor. Window level controls the brightness display. Window width controls the brightness difference displayed, or contrast.

Ultrasound Terminology

- **Acoustic enhancement** refers to the fact that sound traveling through a fluid-filled lesion such as a cyst is barely attenuated. This means that there is *through transmission* and there is an appearance of white (hyperechoic) area distal to the lesion. That is, the area immediately distal to the lesion appears to have more echoes than the surrounding areas.
- **Acoustic impedance** is the resistance encountered by the ultrasound wave as it passes through the medium. The speed of sound in body tissue is relatively constant at 1540 m/s.
- **Anechoic,** or echo free, means without internal echoes. An echo-free structure is sonolucent.
- **Carcinomatosis** or **carcinosis** is the development of many carcinomas throughout the body.

- **Color Doppler ultrasound** is a technique utilizing the shift in frequency when ultrasound is used to visualize a part in motion. A Doppler ultrasound technique can estimate how fast blood flows by measuring the rate of change in its pitch (frequency). In color doppler, the computer changes the Doppler sounds into colors that are overlaid on the image of the blood vessel
- **Complex** is used to describe a structure that is combined cystic and solid. The structure can appear as a cyst with one or more septations or internal debris. An example is a hematoma. When a cyst contains only fluid, it is called simple cyst. When a cyst contains any other materials along with fluid, it is called complex cyst.
- **Doppler effect** or **shift** is the apparent change in the frequency of sound wave if there is a relative motion between the sources of the sound reflector. Doppler effect can be used to measure the velocity of the blood flowing in a vessel. The reflector moving toward the sound source will show an apparent increase in the observed frequency compared with the actual frequency emitted by the sound source. Reflector moving away from the sound source will show a decrease in the observed frequency compared with the actual frequency emitted by the sound source. These differences can be measured in terms of a phase shift and recorded as the "Doppler frequency." This can then process to produce either a color flow display or a Doppler sonogram. In vascular imaging, erythrocytes are the moving reflectors.
- **Enhancement,** posterior enhancement, acoustic enhancement, posterior acoustic enhancement, or enhanced through transmission are terms referring to the bright white artifact distal to a fluid-filled structure. It is the result of lack of impedance when the sound waves pass through fluid and is then less attenuated by the underlying structures. The result is an increase in echogenicity as the sound waves strike underlying structures. Posterior enhancement is a characteristic of fluid-filled structures such as cysts.
- **Echogenic or hyperechoic** in ultrasound means having more echoes than surrounding organs or brighter echoes than other organs. In ultrasound imaging, echogenicity of the liver is considered standard. Echoes brighter than those in liver are considered hyperechoic and darker than those in liver are considered hypoechoic. Gallstones is an example of an echogenic lesion.
- **Echopenic or hypoechoic** is used to represent structures that have fewer echoes or are less echogenic on an ultrasound. The term is relative, and changes in the normal echogenicity of a structure could represent pathology.
- **Focal zone** is the area of the sound beam where the resolution is highest. For a disk-shaped single-element probe, the sound beam has an hourglass shape. The narrowest point is the focus. Area on either side of the focus has better resolution and is called focal zone.
- **Frequency** is the number of waves emitted per second. Instead of using the phrase per second, scientists use the word hertz (Hz), which is synonymous with "per second." The unit of frequency is therefore the Hz. Higher frequency creates better images but cannot penetrate deeply, and vice versa.
- **Fresnel zone (near field)** describes the area of the hourglass-shaped sound beam between transducer face and focus.

- **Gain** is the strength of the echoes throughout the image and is one of the receiver functions of the ultrasound system that adjusts the brightness of an image. The gain can be changed by varying the power output from the system.
- **Hertz** is the standard unit of frequency used in ultrasound. It is equal to 1 cycle/second.
- **Homogenous** means uniform in composition.
- **Hyperechoic** means having more echoes than surrounding organs or being brighter than other organs because there is little attenuation of sound traveling through a fluid-filled structure. Structures distal to the lesion will appear to have more echoes than the surrounding areas (e.g., cysts). In ultrasound imaging, echogenicity of the liver is considered standard. Brighter than liver is hyperechoic, and darker than liver is hypoechoic.
- **Hypoechoic** means having fewer echoes than surrounding organs, that is, not as bright.
- **Isoechoic** means similar echogenicity to the liver.
- **Piezoelectric effect** is the conversion of reflected ultrasound beam to an electrical signal.
- **Piezoelectric** is ability of crystals to change shape and emit sound waves when an electric current or mechanical stress is applied, or vice versa.
- **Real time** refers to the many imaging frames that run together to create a cinematic view of the tissue that is similar to watching a movie versus looking at a photograph.
- **Reverberations** are artifacts resulting in the multiple reflections of sounds returning from a large acoustic interface to the transducers. The initial echoes return to the tissue again and again, causing additional echoes parallel to the first.
- **Scan** is used in ultrasound as a verb, meaning to perform an ultrasound scan.
- **Shadowing** is the failure of a sound beam to pass through an object. The sound is either partially reflected, completely reflected, or absorbed. This blockage is caused by reflection or absorption of the sound and may be partial or complete. Shadowing can indicate pathology. There is less shadowing in benign lesions, such as fibroadenoma.
- **Solid** is used to describe something that is not cystic, such as a structure that does not contain fluid, and on the ultrasound image will appear gray or white depending on its content. The margins may be smooth (likely benign) or irregular (likely malignant).
- **Through transmission** refers to the amount of sound passing through a structure.
- **Transducer** is the probe containing crystals that are capable of converting energy from one form to another (e.g., conversion of sound energy to electric energy, and vice versa).
- **Velocity** or speed in ultrasound is a measure of how fast the ultrasound waves travel through a particular medium. In physics, **velocity** has both magnitude and direction, whereas **speed** has only magnitude.
- **Wavelength** measures the distance a wave travels in a single cycle. As the frequency increases, the wavelength decreases.

MRI Terminology

- **Bandwidth** is the range of radiofrequencies in a pulse. The bandwidth generally matches the processional frequency of the nuclei in the middle of the slice excited by the radiofrequency (RF) pulse.
- **Catenary** refers to a deep curve, taking the appearance of having a chain hanging free from two points or hanging from around the neck.
- **Dipole** refers to the north and south poles of a magnetic field.
- **Echo time (TE)** (time-to-echo or echo delay time) is the time interval from one pulse cycle or a series of pulse cycles to the measurement of the magnetic resonance (MR) signal (spin–echo signal) and is usually measured in milliseconds.
- **Flip angle** describes the angle by which the net magnetization vector rotates after an RF pulse.
- **Free induction decay** refers to the reduction in the induced signal or voltage in the receiver coil. This results as the transverse magnetization decreases after the RF pulse is switched off.
- **Gradient spoiling** is the use of gradients to dephase or rephase the residual magnetization. Often, it is used to dephase the residual magnetization so that it is jumbled at the beginning of the next repetition.
- **Inversion time (TI)** is the time between a 180-degree inverting pulse and a 90-degree excitation pulse in an inversion recovery pulse sequence. It is usually between 200 and 2000 ms.
- **Longitudinal and transverse magnetization** describes the strength of the net magnetization in the longitudinal and transverse direction. This represents a convenient way of describing two different aspects of the net magnetization of an element by representing the combined magnetic forces of the individual nuclei of the element.
- **Longitudinal relaxation** is the return of longitudinal magnetization to equilibrium and transverse magnetization to zero. It is also known as T1 recovery or spin–lattice relaxation because it is a process by which hydrogen nuclei give up energy to their environment or lattice.
- **Nephrogenic systemic fibrosis and nephrogenic fibrosing dermopathy (NSF/NFD)** are rare conditions associated with some gadolinium-based contrast used in MRI. It was first identified in 1997 and most often will occur in patients with impaired renal function. The condition can develop within 2 days to 18 months after receiving the contrast injection and is characterized by tight, rigid skin that will eventually make any movement at the joints difficult or impossible and may lead to multiorgan failure and death.
- **Precession** is a type of motion that can be described as a wobble. This motion is seen in a spinning top. As the top spins, the spinning gets slower and slower, and just before the top falls it begins to wobble.
- **Proton density (PD)** is a measure of the concentration of mobile hydrogen available to produce the MR signal. The proton density of a tissue is the number of protons per unit volume of that tissue. The higher the concentration of mobile hydrogen nuclei, the stronger the net magnetization at equilibrium (Mo) and, therefore, the more intense the MR signal. A strong MR signal results in a better image. Mobile hydrogen nuclei are loosely bound hydrogen nuclei such as those found in liquids.
- **Proton density weighted imaging** is achieved by using a long repetition time (TR) between pulses and a short time to echo (TE). The TR must be longer than the T1 recovery time of most tissues (more than

2000 ms), and the TE shorter than the T2 decay time of most tissues (less than 20 ms), to reduce if not eliminate the effects of both T1 and T2 contrasts. This permits the RF signal from a larger number of protons to be detected, creating the image from a greater number or densities of spinning protons. Proton density images have relatively low contrast because of the slight differences between the percentages of hydrogen present in various tissues; but areas with high proton density are seen as bright, and areas of low proton density are seen as dark.

- **Relaxation time** is the time taken for the physical changes that were caused by the RF pulse to disappear, returning the atoms to the state they were in before the application of the RF pulse. Relaxation will, therefore, result in recovery of magnetization in the longitudinal plane and loss of magnetization in the transverse plane.
- **Repetition time (TR)** or **time to repetition** is the time interval between two successive pulse cycles and is usually measured in milliseconds.
- **Spin** is the term used in MRI to describe the rotation of hydrogen nuclei.
- **Spin–lattice** or **longitudinal relaxation time (T1)** is sometimes called thermal relaxation time. T1 is the time taken for the nuclei to give up their energy to the surrounding environment (lattice) and return to their original longitudinal magnetization or magnetization in the longitudinal plane (or recover 63% of their maximum strength). This energy release follows a 90-degree RF pulse. T1 time will vary according to the protons involved and the magnetic field. The value of T1 will increase with the strength of the magnetic field.
- **Spin–spin** or **transverse relaxation time (T2)** is the time taken for the nuclei to exchange energy with neighboring nuclei or the time taken for the nuclei to lose 63% of their coherence or alignment with each other after an RF pulse. It is called the transverse relaxation time because it is the loss of the transverse magnetization that determines T2 relaxation time. (The transverse magnetization returns to equilibrium.) It is also known as spin–spin relaxation because it describes the relation between individual hydrogen nuclei or spins. This time depends on the tissue. T2 relaxation time never exceeds T1 relaxation time.
- **T1-weighted images** will measure the intensity contrast between any tissues in the image due mainly to the T1 relaxation properties of fat and water. To produce a T1-weighted image, a short TE is used to eliminate the effects of T2 and a short TR is used in order not to eliminate the effects of T1. The flip angles are long, and because TR controls how far each vector can recover before it is excited by the next RF pulse, the TR is kept short so that neither fat nor water has a chance to recover full longitudinal magnetization before the next RF pulse is applied.
- **T2-weighted images** are achieved when the TE is long enough to give both fat and water time to decay. The contrast in the tissue will, therefore, depend on the difference in T2 times of fat and water.
- **T2*** is a time constant indicating the reduction in the voltage induced in the receiver coil following an RF excitation pulse. It is due to transverse relaxation time (T2) and inconsistencies in the magnetic field that do not exactly match the external field strength. T2* is always shorter than T2.
- **Vector** describes a mathematical quantity. It is usually symbolized by arrows and can represent both magnitude and direction.

ANALOG MAMMOGRAPHY

Analog mammography is a method of imaging that produce the images in a purely mechanical way with a final hard copy that cannot be modified. A film is used to acquire the diagnostic information, to display the information, and to store the information (Box 7–1).

Films, however, have drawbacks. They can easily cost departments thousands of dollars per year. There are also storage costs, and films are easily lost or stolen. Another big disadvantage of films are the rigid techniques required to produce optimum exposures. There is no adjusting for overexposure or underexposure, and the fixed gradient or contrast can still provide a challenge for the mammographer because there is a critical need to optimize contrast to aid interpretation. Analog mammography imaging is rarely used in the United States. This method of imaging has been replaced by digital mammography.

DIGITAL MAMMOGRAPHY

Digital technology provides ease of use plus enhanced imaging quality and performance. Radiation passes through the breast to strike a digital detector (Fig. 7–1). This specialized detection system generally offers reduced radiation dose to the patient. Positioning, however, is just as important in digital. Digital mammography (DM) does have the advantage of ease of access to a picture archiving and communication system (PACS) and digital storage. Digital also enables teleradiography, allowing radiologists to access images from remote locations via the Internet or telephone. With DM comes digital imaging terminology. In DM some of the terminology is still in flux. The old term *cassette* is still being used, although it is sometimes replaced by the correct terminology, *image plate* (IP). In some digital systems, the film was replaced by a storage phosphor, also called the *photostimulable storage phosphor* (PSP) or *image receptor* (IR). More recently, the term *detector* is being used to represent the area struck by x-rays leaving the breast, and this term is slowly replacing the previous terminology.

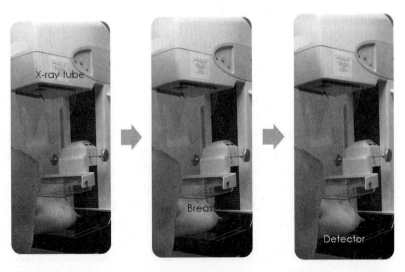

Figure 7–1. X-ray passes through the breast and strikes the detector.

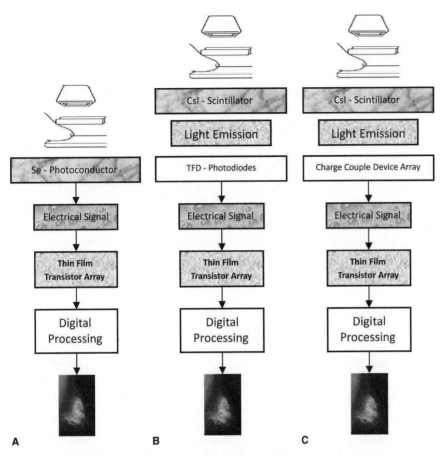

Figure 7–2. **(A)** Direct flat-panel digital detector. **(B)** Indirect flat-panel digital detector using thin-film diodes. **(C)** Indirect flat-panel digital detector using charge-coupled devices.

Currently, three different DM detectors are available in the United States (Fig. 7–2):

1. Flat-panel detectors systems (cassetteless), sometimes referred to as full-field digital mammography (FFDM). There are two types of flat-panel detectors: direct and indirect.
2. Computed mammography (CM) systems using PSP technology
3. Photon-counting technology using a crystalline silicon detector and photon-counting electronics in a multislot system

Box 7–2 provides a list of units that are approved by the US Food and Drug Administration (FDA).

DM technology uses a built-in non-removable detector system (so-called cassetteless system), whereas CM adapts the analog system with digital technology by integrating a removable detector. Both DM and CM have the ability to provide ease of use, plus enhanced imaging quality and performance. DM offers hands-free imaging without the need to carry a cassette. With CM technology, the removable IPs are designed to fit into the Bucky device of standard analog mammography unit. In CM technology the IP contains the storage phosphor screen (SPS) or PSP. It is the PSP that receives the energy of the x-ray beam and forms the latent image. The PSP is sensitive to radiation, not light, and can be used repeatedly. After the exposure, the IP is removed from the x-ray unit and inserted into a

Box 7–2. The Following Units Are Approved by the US Food and Drug Administration

- GE Senographe Pristina with Digital Breast Tomosynthesis (DBT) Option—03/3/2017
- Fujifilm ASPIRE Cristalle with Digital Breast Tomosynthesis (DBT) Option—01/10/2017
- Siemens Mammomat Fusion on 09/14/15
- Siemens Mammomat Inspiration with Tomosynthesis Option (DBT) System on 4/21/15
- GE SenoClaire Digital Breast Tomosynthesis (DBT) System on 8/26/14
- Fuji Aspire Cristalle Full-Field Digital Mammography (FFDM) System on 03/25/14
- Siemens Mammomat Inspiration Prime Full-Field Digital Mammography (FFDM) System on 06/11/13
- iCRco 3600M Mammography Computed Radiography (CR) System on 04/26/13
- Philips MicroDose SI Model L50 Full-Field Digital Mammography (FFDM) System on 02/01/13
- Fuji Aspire HD Plus Full-Field Digital Mammography (FFDM) System on 09/21/12
- Fuji Aspire HD-s Full-Field Digital Mammography (FFDM) System on 09/21/12
- Konica Minolta Xpress Digital Mammography Computed Radiography (CR) System on 12/23/11
- Agfa Computed Radiography (CR) Mammography System on 12/22/11
- Fuji Aspire Computed Radiography for Mammography (CRM) System on 12/8/11
- Giotto Image 3D-3DL Full-Field Digital Mammography (FFDM) System on 10/27/11
- Fuji Aspire HD Full-Field Digital Mammography (FFDM) System on 9/1/11
- GE Senographe Care Full-Field Digital Mammography (FFDM) System on 10/7/11
- Planmed Nuance Excel Full-Field Digital Mammography (FFDM) System on 9/23/11
- Planmed Nuance Full-Field Digital Mammography (FFDM) System on 9/23/11
- Siemens Mammomat Inspiration Pure Full-Field Digital Mammography (FFDM) System on 8/16/11
- Hologic Selenia Encore Full-Field Digital Mammography (FFDM) System on 6/15/11
- Philips (Sectra) MicroDose L30 Full-Field Digital Mammography (FFDM) System on 4/28/11
- Hologic Selenia Dimensions Digital Breast Tomosynthesis (DBT) System on 2/11/11
- Siemens Mammomat Inspiration Full-Field Digital Mammography (FFDM) System on 2/11/11
- Carestream Directview Computed Radiography (CR) Mammography System on 11/3/10

DBT SYSTEMS

- GE Senographe Pristina with Digital Breast Tomosynthesis (DBT) Option—03/03/17
- Fujifilm ASPIRE Cristalle with Digital Breast Tomosynthesis (DBT) Option—01/10/17
- Siemens MAMMOMAT Inspiration with Digital Breast Tomosynthesis (DBT) Option—04/21/15
- GE SenoClaire Digital Breast Tomosynthesis (DBT) System—08/26/14
- Hologic Selenia Dimensions Digital Breast Tomosynthesis (DBT) System—02/11/11

Box 7–2. (Continued)

- Hologic Selenia Dimensions 2D Full-Field Digital Mammography (FFDM) System on 2/11/09
- Hologic Selenia S Full-Field Digital Mammography (FFDM) System on 2/11/09
- Siemens Mammomat Novation S Full-Field Digital Mammography (FFDM) System on 2/11/09
- Hologic Selenia Full-Field Digital Mammography (FFDM) System with a Tungsten target in 11/2007
- Fuji Computed Radiography Mammography Suite (FCRMS) on 07/10/06
- GE Senographe Essential Full-Field Digital Mammography (FFDM) System on 04/11/06
- Siemens Mammomat Novation DR Full-Field Digital Mammography (FFDM) System on 08/20/04
- GE Senographe DS Full-Field Digital Mammography (FFDM) System on 02/19/04
- Lorad/Hologic Selenia Full-Field Digital Mammography (FFDM) System on 10/2/02
- Lorad Digital Breast Imager Full-Field Digital Mammography (FFDM) System on 03/15/02
- Fischer Imaging SenoScan Full-Field Digital Mammography (FFDM) System on 09/25/01
- GE Senographe 2000D Full-Field Digital Mammography (FFDM) System on 01/28/00

Source: http://www.fda.gov/Radiation-EmittingProducts/MammographyQualityStandardsActand-Program/FacilityCertificationandInspection/ucm114148.htm.

computer reader (CR) that removes the PSP. The PSP is then scanned with a laser-scanning device. The PSP will release light that is output to an electrical, then digital, format. The PSP is reusable because, after scanning, it can be erased using a powerful beam of light and then reloaded into the IP. These steps all take place within the CR (Fig. 7–3).

Photon-counting image capture detects and captures individual x-ray photons leaving the breast and uses a crystalline silicon detector and photon-counting electronics to capture the image. A precollimator removes scatter not directed to the detector, and a postcollimator removes scatter leaving the breast.

In all imaging methods the first step of DM is the acquisition of the image. In digital, the acquisition of the image is totally separated from the image display. This means that to optimize the image, both the method of acquiring the image and the method of display must be optimized. If a high-resolution image is acquired and the monitor resolution is poor, the viewer will not have any of the benefits of the high-resolution image acquisition. However, the digitally acquired data are not lost, so just by switching to a high-resolution monitor, the viewer will obtain all the benefits of the high-resolution acquisition.

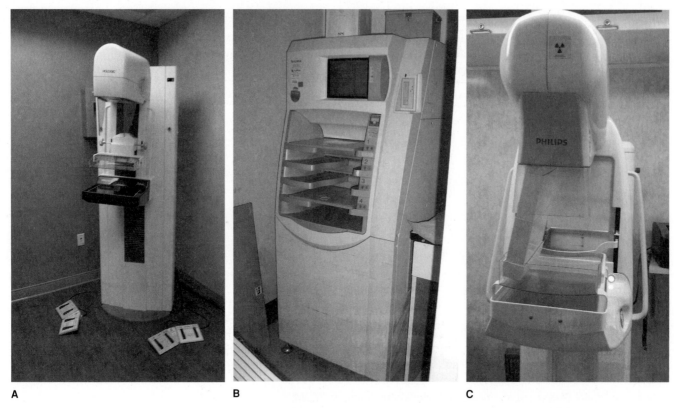

Figure 7–3. **(A)** Flat-panel Hologic unit. **(B)** Computer reader processor. **(C)** Philips MicroDose Unit.

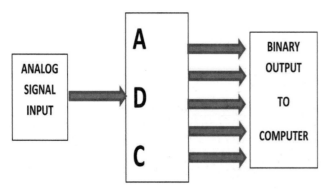

Figure 7–4. Schematic representation of analog-to-digital conversion.

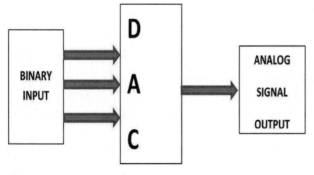

Figure 7–5. Schematic representation of digital-to-analog conversion.

Regardless of the method used to acquire the image, the digital image is formed as a two-dimensional matrix of pixels and is viewed on the display monitor. Digital data can be copied and modified; distributed worldwide; or archived for storage, retrieval, and display (Fig. 7–4).

To acquire the image, the x-rays pass through the breast and strike the detector. The information from the detector is then sent to an *analog-to-digital converter* (ADC). The image undergoes *quantization* in the ADC. This converts analog signal from the detector to a digital signal for the computer to manipulate. Binary language is the only language that the computer will understand; therefore, various shades of gray in an image must be converted to binary language.

The computer receives a digital signal from the detector and applies selected algorithms to the data received. This is the process of *preprocessing*. However, the computer outputs a signal in computer language. This digital signal must be converted to an analog signal to display on a monitor. The preprocessed digital data is sent to a *digital-to-analog converter* (DAC). The DAC converts the digitally manipulated data back to an analog signal. The image will appear on the display monitor as various shades of gray or sent for storage (Fig. 7–5).

A digital image is a *matrix* of picture elements (*pixels*). A matrix is a box of cells with numeric values arranged in rows and columns. Each cell corresponds to a specific location in the image. The matrix size is determined by the number of pixels in the rows and columns and expressed by listing the number of pixels in each dimension (length and width). Common sizes are 256 × 256, 512 × 512, and 1024 × 1024 (Fig. 7–6).

The individual cell in the pixel matrix can be measured in two ways: by *pixel size,* which is the side-to-side measurement within an individual pixel; or by *pixel pitch,* which measures from the center of one pixel to the center of an adjacent pixel, including any gaps between.

Pixel size is important because it is a measure of the spatial resolution. As the pixel size decreases, the amount of data contained in the images rapidly increases, but the electronic noise will also increase. Therefore, the greater the number of pixels per inch, the greater is the spatial resolution. Although the detection of microcalcifications, for example, is dependent on the system resolution, an increase in pixel size will increase resolution but will also increase noise that can deteriorate image quality. To image a 100-μm microcalcification, the pixel must be equal to or smaller than 100 μm (0.1 mm). Pixel size will also affect the storage, transmission time, and archival quality of the image (Fig. 7–7).

The *pixel pitch* will determine the spatial resolution of a CR system or PSP systems. In the flat-panel systems the spatial resolution is determined by the fixed size of the thin-film transistor (TFT) and the detector element (DEL) size, which will be discussed later (Fig. 7–8).

> Analog signals are electrons or light.
>
> Digital signals are numeric data in the form of binary language.

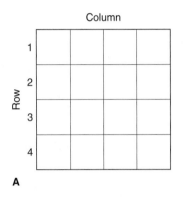
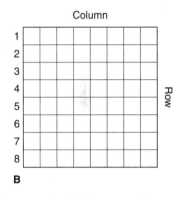

Figure 7–6. Cell matrix refers to the amount of columns and rows. **(A)** 4 × 4 matrix. **(B)** 8 × 8 matrix. Each cell in the matrix is called a pixel.

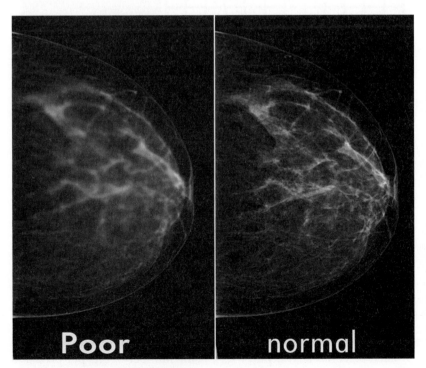

Figure 7–7. Decreased pixel size = increased spatial resolution; increased pixel number = increased image resolution.

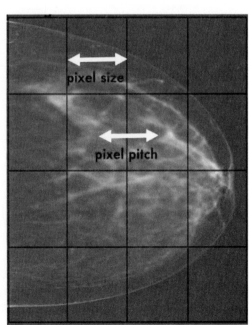

Figure 7–8. Schematic representation of the pixel pitch.

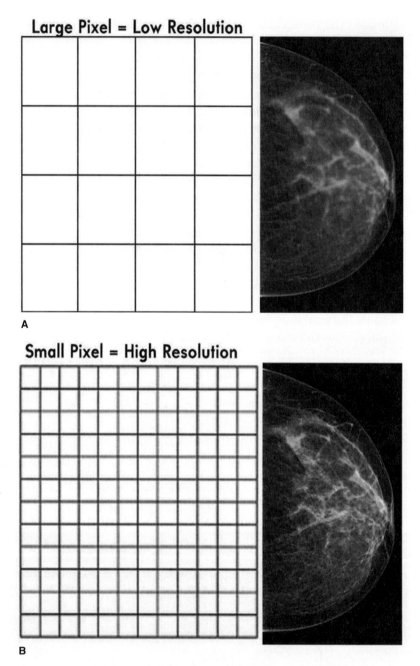

Figure 7–9. (A) Low pixel density = low resolution. **(B)** High pixel density = high resolution.

Pixel density measures the number of pixels in a unit area. The more pixels there are, the greater the pixel density. As the pixel density increases, the *image resolution* will also increase. Each digital image has between 25,000 and 1 million pixels (Fig. 7–9).

Each pixel contains *bits* of information. A bit is a single binary digit, a zero or one (0 or 1) that is used to express a decimal digit. A bunch of 8 bits equals a *byte,* and 16-bit value is called a *word.*

Each pixel has a *bit depth,* which is the number of bits per pixel and determines the shade of gray demonstrated. It is represented by the range of numbers assigned to the pixel and the number of gray shades that a pixel can produce. The pixel could be assigned any value between 0 and 2056. The value of the pixel will determine the image brightness. Also, the higher the bit value, the higher the contrast.

The smaller the pixel size, the better the spatial resolution. The spatial resolution of a digital system is the minimum separation between two objects at which they can be distinguished as two separate objects in the image. In digital, the spatial resolution describes the ability of the system to accurately display objects in two dimensions. (In analog imaging, the spatial resolution is determined by phosphor crystal in film.) The unit of spatial resolution is line pairs per millimeter (lp/mm).

Image contrast is also dependent on pixel size. As the pixel size increases, the contrast will also increase, where high contrast represents few data values between black and white and low contrast has many data values between black and white (Fig. 7-10).

The biggest advantage of digital system is the wide *latitude*, basically a linear response to the intensity of x-ray exposure that enables optimization of contrast and brightness. The digital exposure latitude, which represents the range of underexposure or overexposure that can occur while still producing an acceptable image allows the system to correct for 50% underexposure to 100% overexposure. In contrast, analog receptors can correct for 30% underexposure to 50% overexposure. It also gives the system a wide *dynamic range,* which refers to the receptors' ability to respond to different exposure levels. Figure 7–11 shows the wide range of exposure response value for the digital receptors. For the technologist, quantum noise is the only visual cue that indicates when the digital receptor is underexposed. When an image is overexposed, the image contrast is decreased because of the reduced signal difference reaching the IR. This is totally different from the visual cue

> Gray scale = 2^k, where k = bits available.
> Most radiography systems use an 8-, 10-, or 12-bit depth.

> Unit of spatial resolution is line pairs per millimeter (lp/mm).

> In digital imaging, quantum noise is the visual cue indicating an underexposer. Lower image contrast is the visual cue indicating an overexposer.

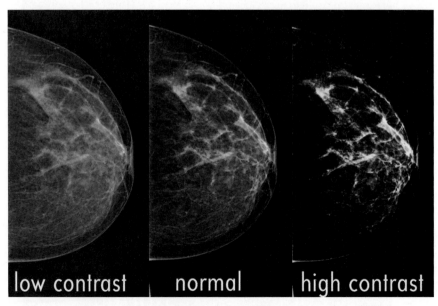

low contrast   normal   high contrast

Figure 7–10. Contrast is dependent on pixel size. Increased pixel size = increased contrast. High contrast = few data values between black and white. Low contrast = many data values between black and white.

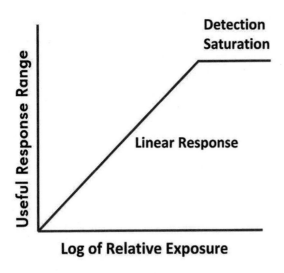

Figure 7–11. Dynamic range is the receptor's ability to respond to different exposure levels.

- Increased noise = decreased contrast
- Increased image contrast = decreased noise

Window level adjusts brightness.

Window width adjusts contrast.

for underexposure—a white or light film—and the visual cue for overexposure—a black film—in the analog world.

Most digital systems will quote the *detective quantum efficiency* (DQE). The DQE represents the percentage of energy that strikes a receptor and results in a useful output signal and measures how efficiently a digital detector can convert the remnant beam to useful data. Systems with a high DQE will offer much lower patient dose.

Noise in digital imaging is something to be avoided. Noise represents the random background information due to the constant flow of current in the circuit. It will not contribute to image quality, and on the digital image noise looks like quantum mottle. Increased noise results in decreased contrast. The *signal-to-noise ratio* (SNR) is a measurement of noise in the image. A high SNR has little noise interference (Fig. 7–12).

Another measure of the contrast in digital imaging is the *contrast-to-noise ratio* (CNR). CNR measures contrast as it relates to noise.

Preprocessing options are controls by the system's manufactures. However, postprocess is under the control of the technologist. Windowing and leveling are options available at the technologist monitor or acquisition workstation (AWS). *Window level* controls the brightness of the image displayed, and *window width* controls the brightness difference or range of pixel values displayed, or contrast.

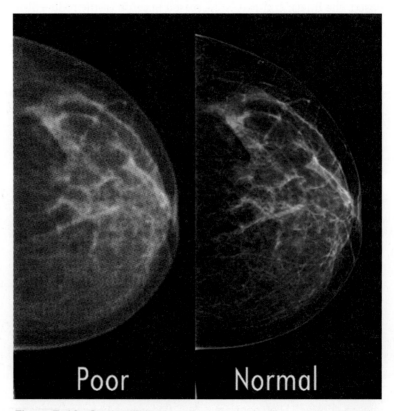

Figure 7–12. On the digital image, noise looks like quantum mottle.

Direct Flat-Panel Detector Technology

Non-removable digital detectors are categorized as direct or indirect. The electrical signal on both the direct and indirect detectors can be erased and the detector reused.

In the direct system, x-rays strike a photoconductor, and an electrical signal is created in a one-step process. Selenium (Se) is the common photoconductor used in direct flat-panel detector systems. Selenium is ideal for mammographic systems because of its high x-ray absorption efficiency, high resolution, low SNR, and dose efficiency (Fig. 7–13).

There is no light scatter, which is a common problem in the indirect systems, and the spatial resolution is limited only by the pixel size. Of course, the optimal pixel size is determined by limitation of the display system, the information system, and the manufacturing process. The electronic signal or charge in the direct system is collected with little lateral spread or blur, but to achieve this, the selenium layer has to be relatively thick, which can result in other forms of blurring. X-ray photons may travel from within the selenium layer, away from the site of absorption, causing blurring. An x-ray not striking perpendicular to the detector can also result in blur (Fig. 7–14).

X-ray beam strikes the photoconductor (e.g., amorphous selenium [aSe]). A high-voltage charge applied to the amorphous selenium just before the exposure allows the amorphous selenium to convert the x-ray to electrons. The electrons are collected by a TFT array, and the components of the TFT are arranged on amorphous selenium as the semiconductor material. Electrons then migrate to the DELs on the TFT. The signal from the TFT is sent to ADC and then to a computer for display.

TFT or *thin-film diode* (TFD) may refer to any diode produced using thin-film technology. They have amorphous silicon or amorphous selenium deposited on thin films. Amorphous silicon or amorphous selenium are photoconductors used to collect electrons; that is, it accepts and stores freed electrons from the ionized selenium layer (the latent image). The TFT is a complex circuit device that collects electrons emitted from either amorphous selenium or amorphous silicon. The TFT generates an electric charge from each stored electron. The electrons are arranged in a matrix array, allowing the charge pattern to be read pixel by pixel. A variation of the TFT is used in high-quality flat-panel liquid-crystal displays (LCDs).

Detector elements (DELs), or dexels, are areas located within the TFT. Each square in the matrix is a DEL. DELs collect the electrons given off by the amorphous selenium or the amorphous silicon representing individual components of the digital image. After the exposure, the DELs read those electrons in a sequential pattern that matches their location within the

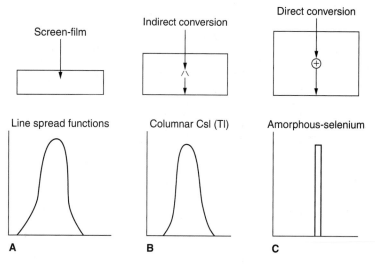

Figure 7–13. Line spread comparison in the screen/film system. **(A)** Indirect system. **(B)** Direct-conversion system. **(C)** The screen/film and indirect-conversion detectors have a broader line spread function than the direct-conversions systems.

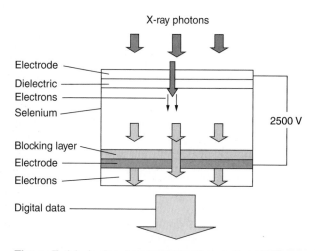

Figure 7–14. In the direct-conversion system the x-rays strike the selenium photoconductor, generating electrons that are collected on a pixel array and sent to computer as a digital signal.

DEL is the sensitive component of the TFT.

DEL size controls the recorded detail, or spatial resolution, for the flat-panel device.

DEL size contributes to the image blur present in a flat-panel detector receptor.

The larger DELs in a flat-panel detector cause more image blur.

The technologist cannot change DEL size.

There are two types of flat-panel digital systems:
- Direct flat-panel digital systems
- Indirect flat-panel digital systems

detector matrix. As the electrons are extracted off the TFT array, they are sent to the ADC, which sends the digital signal to the computer. The size of the DEL controls the spatial resolution for the flat-panel device.

DEL size also contributes to the image blur present in a flat-panel detector receptor. The larger DELs in a flat-panel detector cause more image blur. The technologist cannot change the size of the DEL, which is fixed for that piece of equipment. After the DELs are read, the flat-panel detector automatically erases and is ready for another exposure.

Fill Factor

Each square in the TFT matrix has sensitive and nonsensitive areas. It is the sensitive areas on the DELs that collect the electron charge. The fill factor is the ratio of the sensitive area to the entire detector area and is usually expressed as a percentage. Fill factor will affect the spatial resolution and signal-to-noise ratio. A typical fill factor is 80%.

Hologic, Fujifilm, and Siemens have also developed a direct-conversion system, using a TFT flat panel with an amorphous selenium (a-Se) photoconductor. With direct digital detection, a photoconductor absorbs the x-rays and directly generates the electronic signal without any intermediate steps. A high voltage of approximately 2500 V is applied across the detector. This external electrical field causes electron-hole pairs to drift toward a pixel electrode. There, they are collected on a pixel capacitor. The electron hole travels along the straight line of the electric field, and there is no lateral movement of the charge. The result is an extremely narrow point spread of approximately 1 μm. The photoconductor is sufficiently thick to stop most of the incident x-rays; however, if the photoconductor material is too thick, this will result in parallax and decrease resolution.

Indirect Flat-Panel Detector Technology

Indirect flat-panel digital systems use a two-step process. A scintillator, such as cesium iodide (CsI) doped with thallium (Tl), absorbs the x-rays and generates a light scintillation that is then detected by an array of TFT or TFD. In the indirect system, the components of the TFT are arranged on amorphous silicon. The TFTs convert the light photons to electronic signals that are captured using TFTs. Some systems use charge-coupled devices (CCDs) as an alternative to the TFTs light-collecting array and readout method. Electrons then migrate to the DELs on the TFT. The signal from the TFT is sent to ADC and then to a computer for display. Indirect systems can suffer from poor resolution caused by light spread in thick scintillators and poor quantum efficiency within thin scintillators. This effect is minimized by the columnar structure of the CsI (TI).

The *GEMS General Electric Senographe* method is indirect (Fig. 7–15). It uses a large area matrix of phosphor diodes on an amorphous silicone phosphor that is coated with thallium-activated cesium iodide. The CsI is used to absorb x-rays, and the silicone chips will detect the x-ray photons. The CsI phosphor layer is grown in crystalline columns directly on the image panel.

This results in excellent coupling of the phosphor and the flat panel. The CsI channels light produced from the absorption of x-rays directly

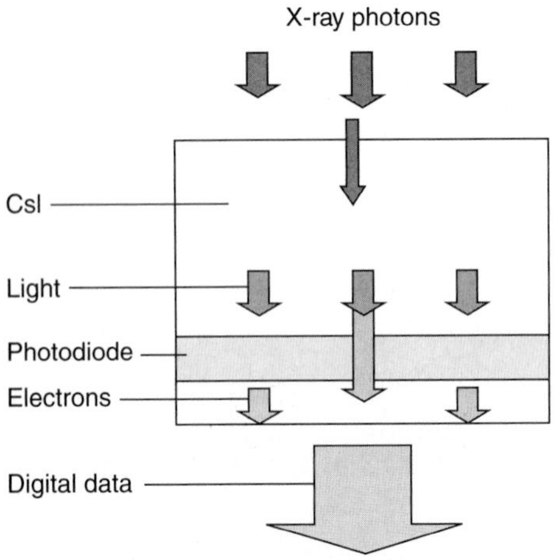

Figure 7–15. The Senographe 2000 D detectors have independent stages that can be optimized independently. The x-rays striking the cesium iodide phosphor layer are converted to light that is transmitted to the photodiodes, where the light is converted into electronic signals.

to the detectors. A photo diode/transistor flat panel converts the light image into electronic charge and stores the charge until the image readout occurs. The detector components can be optimized separately for absorption of the signal, whereas the photodiode, which converts light to electrical charge, can remain thin to minimize lag or ghosting and reduce electronic noise. Some of these systems use a 1-mm-thick sheet of glass coated with CsI plus an amorphous silicone detector. Such a system can be retrofitted on an analog mammography unit.

Photostimulable Storage Phosphor Technology

CM is a less expensive method of digital technology. CM is based on the original Fuji computed radiography system and uses another form of indirect conversion. One problem facing digital systems is the size of the detector. A large detector will allow imaging of nearly all women, but because the detector is built-in, a larger detector will result in poor positioning of small-breasted patients. In CM the detectors are placed in IPs and can be loaded in standard-sized analog Bucky trays. This system, like the analog system, has two detector sizes, 18 × 24 cm and 24 × 30 cm. Instead of a film/screen as the image detector, CM uses a flexible IR, which is coated with storage phosphors. The IR is sometimes called a PSP. The PSP is not sensitive to light, and the IP serves to protect the PSP, not to keep out light (Fig. 7–16). Lead backing in the IP absorbs backscatter radiation.

The PSP is less than 1 mm thick. It absorbs x-ray energy within the crystalline material of the phosphor and forms a latent image in much the same way a film does. The phosphor layer can be Ba (barium), F (fluorine), Cl (chloride), Br (bromine), I (iodine), or Eu (europium). The PSP is constructed much like an analog screen. Both have a protective laminate to protect the phosphor layer. The PSP has a conductive layer to conduct away static, a flexible support layer that allows flexibility of the PSP, and a backing layer to prevent damage to the PSP during transportation (Fig. 7–17). With such a similar structure, the PSP can be placed in the IPs, which are really modified cassettes, and exposed with standard x-ray equipment. The x-ray passes through the patient and interacts with the electrons in the PSP. Electrons in crystals move from their normal orbital location to a higher level. The number of electrons affected is directly proportional to the amount of energy absorbed by the PSP. This is the latent image formation. The image formed is in proportion to the amount of radiation incident on the phosphor and will represent a pattern of transmission through the breast.

After exposure, the IP is placed in a CR, where the PSP is removed (see Fig. 7–18). A roller system transports the PSP to the laser scanning area of the reader.

A rotation polygonal mirror system uses a point scanner to direct a laser beam to the PSP. The finely focused laser beam of infrared light scans the phosphor in a raster pattern, meaning that the beam sweeps horizontally left to right at a steady rate, then rapidly moves back to the left. This provides energy to the electrons trapped in the phosphor center. The energy provided allows these metastable electrons (electrons at high energy) to return to their ground state and to release energy absorbed from

Figure 7–16. Computer reader.

The intensity of light emitted from the PSP is proportional to the amount of radiation absorbed by the storage phosphor.

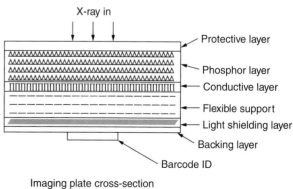

Imaging plate cross-section

Figure 7–17. A cross section of an imaging plate.

Computer reader system

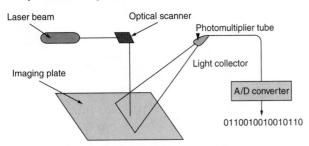

Figure 7–18. Photostimulable storage phosphor scanning.

the x-ray beam. The release of energy is in the form of a visible emission of blue-purple light. This emission is called *photostimulated luminescence* (PSL). The intensity of light emitted from the PSP is proportional to the amount of radiation absorbed by the storage phosphor. The laser light is red, whereas the PSP emits a blue-purple light—to allow the photodetector to recognize the light used for plate scanning versus image formation.

During the reading process, light that is emitted from the PSP is collected by a light guide and sent to a photomultiplier tube (PMT), photodetector (PD), or charge-coupled device (CCD). The PMT detects the blue light given off by the trapped electrons as they return to their normal neutral state. Signal from the PMT is sent to an ADC and then to a computer for display. The resultant digital information can now be electronically transmitted, manipulated, and more efficiently stored.

Some excited electron will always remain; if not removed, they will result in ghosting on subsequent PSPs. Any residual latent image is removed by flooding the phosphor with very intense light. The high-intensity light erases the plate by releasing any remaining electron energy and allows the PSP to be reusable (Fig. 7–18).

In CM, the barcode reader on the IP allows for the patient's information and identification. It is during the scanning process that data from the PSP, as well as the anatomical menus, are analyzed. Each image is processed under a specific anatomical menu that correlates with the examination that was done. Image characteristics will determine the reading sensitivity and the exposure latitude. Scanning the bar code for each patient is therefore extremely important to ensure that the image is linked to the correct patient. This is also the step that tells the computer software what type of processing codes to use for an image of a particular anatomical structure. The PSP bar code may be scanned before or after an exam is performed. If the IP is scanned for a CC projection and is used for a mediolateral oblique (MLO) radiograph, the image will not display properly because of incorrect processing codes.

After exposure, the PSPs should be processed immediately because light is actually being emitted in a slow steady emission all the time. The longer the interval between the exposure and the reading, the smaller will be the resultant signal and therefore the poorer the signal. However, the primary image is in digital format, and therefore the image data can be manipulated to accentuate or suppress various features of the image. The Fuji FCR system will display the image on the console within 50 seconds. The pixel size is approximately 50 μm. The typical computer reader can hold 400 GB of data.

The Fuji Computed Radiography Mammography Suite (FCRm) and Carestream Directview Computed Radiography (CR) Mammography System both use CM technology. Images from these systems can also output to a laser printer in a typical print time of 140 seconds.

An advantage of the PSP system is that it has the wide latitude of digital imaging. The disadvantage is that the image on the PSP loses 25% of its energy in 8 hours. It is also noisy at low exposures and is very

PSP advantages and disadvantages

- The PSP system has a wide latitude.
- Image on the PSP loses 25% of its energy in 8 hours,
- Low exposures can create a noisy image.
- The PSP is very sensitive to radiation, which can contribute to a noise image.

sensitive to radiation, which can contribute to a noise image. The PSP should not be left in the x-ray room during an exposure and must be erased every 24 hours if not used.

Photon-Counting Image Capture Technology

The photon-counting MicroDose mammography system detects and captures individual x-ray photons leaving the breast using a crystalline silicon detector and photon-counting electronics, and therefore eliminating the signal loss that occurs in digital systems when ADCs are used. The system uses tungsten anode and aluminum as the added filter, and with the use of a multislot precollimation and postcollimation system that virtually eliminate scatter, a grid is not necessary. The precollimator serves to remove scatter not directed to the detector, and the postcollimator removes scatter from the breast. Both the multislot system and the lack of a grid allow further reduction in radiation dose. Images are acquired with a resolution of 25 megapixels (Fig. 7–19).

Advantages of the Photon-Counting System

The photon-counting system offers an ergonomic design. The detector is curved. It can also withstand temperatures to 10° to 50° C (50°–122° F). No ADC is needed; therefore there are no lost signals and no ghost images. This allows 100% fill factor. Studies show that the photon-counting system gives 40% less radiation dose per projection than FFDM systems (Box 7–3).

Advantages of Digital Mammography

In DM the patient information is easily included on the mammogram. Methods used can include selecting the patient's name from a work list, using a magnetic strip—similar to a credit card—that is swiped through a card reader to transfer the patient information encoded on it to the reader, or by scanning a barcode on the patient's requisition.

Digital systems have separated the individual processes of image acquisition, image display, and image archiving, thereby providing the opportunity to independently optimize each process. The biggest advantage of digital systems is associated with the wide latitude of the digital signals. The higher sensitivity of the detectors demonstrates a linear response (input to output) to the intensity of x-ray exposure over a broad range. This allows better penetration of denser breast tissue; in addition, because the response to exposure is linear, optimum contrast can be obtained for both fibroglandular and fatty regions of the breast. A comparison of the film response on the characteristic H and D curves of the radiographic film shows that digital systems will provide far more information in the low- and high-exposure regions of the image (Fig. 7–20).

Digital imaging provides a wider dynamic range, and the limited spatial resolution is outweighed by improved contrast resolution. This will allow imaging at higher kV values with targets of higher atomic number and could therefore allow a reduction in patient dose. Newer digital units now use tungsten targets with filtrations of silver for dense breast and rhodium for fatty breast. With specially designed x-ray tubes

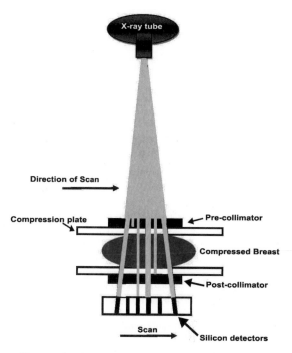

Figure 7–19. Photon-counting image capture.

Patient identification options:

Work list

Barcode reader

Magnetic card swipe

Box 7–3. Multislit Scanning

- Scan time 3–15 seconds
- Tungsten anode, 0.5-mm Al filter, no grid used
- Small detector and tube moves across the breast

Average exposure for a 46-mm-thick breast = 29 kVp, 11.27 mAs, 0.48 mGy

Average Glandular Dose (AGD)

Sample exposure factors include:

- CC projection: 26 kVp, 13.0 mAs, 0.42mGy AGD, 33-mm thickness
- MLO projection: 26 kVp, 12.9 mAs, 0.44 mGy AGD, 30-mm thickness

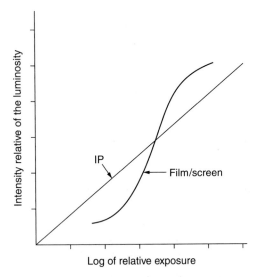

Figure 7–20. The characteristic curve for digital imaging versus film/screen. Digital systems will have a linear response whereas film/screen systems will have a curvilinear response.

and filter systems, the x-ray energies can be tuned or customized for each woman's individual breast type, which is important because breasts vary dramatically in composition and size. Using the optimal x-ray energy for each woman's breasts will improve the overall quality of the resulting image while keeping the radiation dose to a minimum. The wider dynamic range of the digital systems allows a more efficient use of x-ray photons incident on the detector, and there is the potential to even further reduce the dose to the patient during a screening mammogram.

DM can improve the facility's workflow by reducing the number of repeated examinations because it will compensate for exposure errors (within a certain magnitude). The ability of DM to reduce repeats also reduces the cost associated with those repeats. Included in the cost is a more efficient use of time. DM will, therefore, streamline the department, and because the image is in digital format, the storage space is also reduced.

With DM, the medicolegal consequences associated with lost films will be minimized. With electronic storage of the digital data, there is no deterioration of the image, and an image copied at a later date is an exact duplication of the initial image. In addition, any number of copies can be made.

Images can be viewed quickly and easily from any monitor within the facility, and digital images can also be transmitted to workstations outside the facility. Because the rapid viewing can drastically reduce imaging time, DM is commonly used in stereotactic imaging to guide breast biopsies.

The one important consideration in DM is that patient positioning remains critical; therefore, breast compression and patient communication remain essential to achieving quality imaging.

Disadvantages of Digital Mammography

With DM, technologists have lost the visual cues associated with overexposure or underexposure. This often results in *dose creep*, whereby the average technical factors needed to image a part slowly increases, and can also deemphasize radiation safety.

Digital systems must have a reliable and secure backup because sooner or later any system will fail or be subjected to natural disaster or virus attacks.

A heavy mammography workload can also use up a lot of disk space on archive servers.

Some radiologists fear that DM will lead to the outsourcing of reports, thus eliminating the need for a group of radiologists at any one site. There is also a concern that the quality of interpretation provided by radiologists hired at private teleradiology companies could compromise patient care. Interpretation depends on the quality of the image transmitted and the credentials of the interpreting radiologist. Outsourcing is not a problem in breast imaging. According to the American College of Radiology (ACR), radiologists are required to be licensed in the state in which they read the breast images and the state in which the transmitting facility is located. In addition, the radiologist must be credentialed at the transmitting facility.

Advantages of digital imaging:
- Optimization of contrast
- Reduce patient dose
- Improve workflow by reducing repeats
- Reduce cost associated with repeats
- Reduction in lost films
- Reduced storage cost and space
- Multiple storage options—short, medium, and long term
- Compression storage

The use of digital imaging is resulting in increased reliance on postmanipulation of the image and thus decreasing the ability of technologists to set correct technical factors.

Despite the listed drawbacks, nothing will detract from the single greatest advantage of DM—the elimination of virtually all repeats due to technical factors.

> **Disadvantages of digital:**
> - De-emphasizes radiation safety
> - Risks for data loss
> - Effects of ambient on interpretation
> - Reliance on postprocessing

ULTRASOUND IMAGING OF THE BREAST

Brief History of Ultrasound

Scientists were always aware that sound waves could travel through various mediums; however, the real breakthrough in echo-sounding techniques came with the discovery of the piezoelectric effect in certain crystals by Pierre Curie and his brother Jacques Curie in Paris, France, in 1880. Later, underwater sonar detection systems were developed for the purpose of underwater navigation by submarines in World War I. The term SONAR actually refers to **SO**und **N**avigation **a**nd **R**anging, and the idea of measuring distances under water became particularly important when trying to locate the Titanic after the ship sank in 1912.

The use of ultrasound in medicine started initially as a therapy tool rather than as a diagnostic tool because high-intensity sound waves were found to create heat and be particularly disruptive on animal tissue. High-intensity ultrasound progressively evolved to become a neurosurgical tool. William Fry at the University of Illinois, and Russell Meyers at the University of Iowa, performed craniotomies and used ultrasound to destroy parts of the basal ganglia in patients with Parkinson's disease. Peter Lindstrom in San Francisco reported ablation of frontal lobe tissue in moribund patients to alleviate their pain from carcinomatosis. Ultrasonic energy was also extensively used in physical and rehabilitation medicine. In 1953, Jerome Gersten at the University of Colorado reported the use of ultrasound in the treatment of patients with rheumatic arthritis. Karl Theodore Dussik, a neurologist/psychiatrist from the University of Vienna, Austria was regarded as the first physician to have employed ultrasound in medical diagnosis.

It was not until the 1940s that ultrasound was put in use as a diagnostic tool. A physician at the Naval Research Institute in Bethesda, Maryland, George Ludwig, began experiments on animal tissue using pulse-echo ultrasound. In 1954, smaller and better transducers, using the newer piezoelectric crystals, lead zirconate titanate (PZT), were discovered. PZT had a high electromechanical coupling factor and more superior frequency-temperature characteristics. The newer transducers had better overall sensitivity, frequency handling, coupling efficiency, and output.

> Ultrasound was put to use as a diagnostic tool in the 1940s.

John Julian Wild, an English surgeon who migrated to the United States, and Donald Neal, an engineer, noted that malignant tissue was more echogenic than benign tissue, and the former could be diagnosed because of their higher tissue density and failure to contract and relax under pressure. In May 1953 they produced real-time images at 15 MHz of a 7-mm cancerous growth of the breast. Since then, the use of ultrasound has expanded extensively, and today it is used both as a diagnostic and treatment tool.

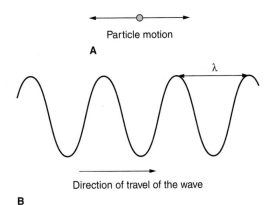

Figure 7–21. Longitudinal wave. Sound is a longitudinal wave with velocity (v), frequency (Hz), and wavelength (λ). **(A)** Particle motion. **(B)** Direction of travel of the wave.

> The basic principle of ultrasound is based on the piezoelectric effect.

The Principle of Ultrasound

Sound is a mechanical, longitudinal wave that propagates through a medium. Longitudinal means that the waves vibrate back and forth in the same direction that the wave is traveling (Fig. 7–21). Sound waves can be expressed in the form of a mathematical formula called the wave equation, with the unit of frequency expressed as hertz (Hz). The velocity in which the sound travels depends on the tissue density of the material through which it passes. Ultrasound uses frequencies above 20,000 Hz and can go as far as 30 MHz, which is beyond human hearing. For abdominopelvic imaging the routine is 3 to 5 MHz, and for vascular and superficial imaging an 8- to 12-MHz frequency is used. Audible sound waves are at frequencies between 20 and 20,000 Hz. Frequencies of less than 20 Hz are called infrasound. Both ultrasound and infrasound are not audible to humans (audible sound: 20–20 kHz; ultrasound: >20 kHz). This high-frequency ultrasound wave is used to create images of the soft tissue organs and blood vessels in the body:

$$Velocity = frequency \times wavelength.$$

Sound cannot travel through a vacuum and travels poorly through gases because the gas molecules are widely separated (Fig. 7–22). The closer the molecules of a medium, the faster sound waves will travel through that medium. In general, ultrasound is not ideal for imaging structures containing air, such as the lung, stomach, or bowel. Bone and metal will therefore conduct sound very well, and when compared with lung and fat, sound will travel faster through bone. The average speed of sound waves through soft tissue is 1540 m/s, and ultrasound machines have based their measurements and calculations on this assumed average. Also, bone conducts sound at much faster speeds than soft tissue; therefore ultrasound machines cannot accommodate for the difference in the speed of soft tissue and bone and will not clearly image bone or structures covered by bone. If the difference of speed in soft tissue and bone is large, the entire ultrasound will bounce back from soft tissue–bone interface to the transducer, creating no image beyond bone.

The basic principle of ultrasound is based on the piezoelectric effect. Ultrasound testing is performed using a small handheld device called a transducer or probe. The ultrasound transducers contain many piezoelectric crystals, which are capable of converting energy from one form to another (e.g., electric energy to sound energy, and vice versa). As the transducer is moved slowly over the part to be tested, crystals in the transducer vibrate when activated by an electrical pulse. The vibrations of the crystals produce sound waves that are

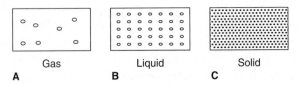

Figure 7–22. Sound travels in gas **(A)**, liquid **(B)**, and solid **(C)**. Sound travels or propagates worst in a gas because the molecules are widely separated. Sound travel is better in liquids but best in solids.

sent through the tissues. Returning echoes are reflected from the tissue structures back to the transducer. When the echo is received the crystals vibrate again, generating an electrical voltage (piezoelectric effect), which is proportional to the strength of the returning echo.

Since sound travels through soft tissue at 1540 m/s, the distance between the transducer and the structure can be calculated by calculating the time elapsing from the transmission to the return of the signal, and converting that value into distance. A computer analyzes the strength of the returning echo, and an image of the structure is displayed on a video monitor. A radiologist or sonographer can then view the image on the monitor.

The strength of the retuning echo is related to the angle at which the beam strikes the structure being imaged (Figs. 7–23 and 7–24). The more perpendicular the beam, the stronger the returning echoes. If the echo strikes at an angle, it is likely to scatter—resulting in a poor return echo. In imaging a structure, the sonographer should, therefore, image from varying angles to produce the best image. Imaging from varying angles will also send smaller echoes to help define the borders and contents of the structure.

The strength of the returning echoes also depends on the difference in acoustic impedance between the various tissues in the body. *Acoustic impedance* is determined by the tissue density. It is resistance faced by the traversing sound wave. The greater is the tissue density difference between two structures, the stronger are the returning echoes defining the borders between the structures. A highly radiopaque malignant lesion in the breast will therefore be easily characterized because it is easy to distinguish it from the surrounding radiolucent structures in the breast.

The Transducer (Probe)

The main part of the ultrasound transducer consists of crystals made up of piezoelectric material. Lead zirconate titanate,

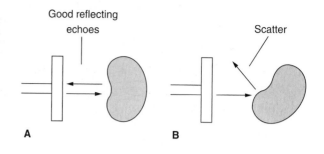

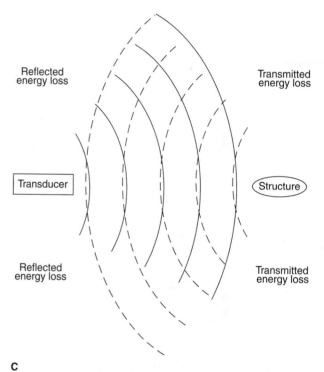

Figure 7–23. Strength of returning echoes. **(A)** The sound beam is stronger when the beam is perpendicular to the part. **(B)** If the beam is not perpendicular to the part, scatter is produced. **(C)** Energy is lost in both the transmitted and reflected directions.

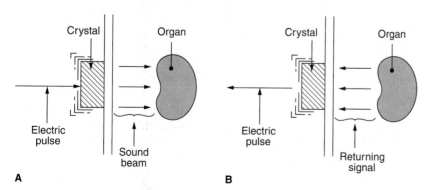

Figure 7–24. Echoes striking a structure. **(A)** The electrical pulse strikes the crystals and produces a sound beam that travels through tissue. **(B)** Echoes from the structure are reflected back to the crystal. The crystal vibrates, producing an electrical impulse in proportion to the strength of the returning echo.

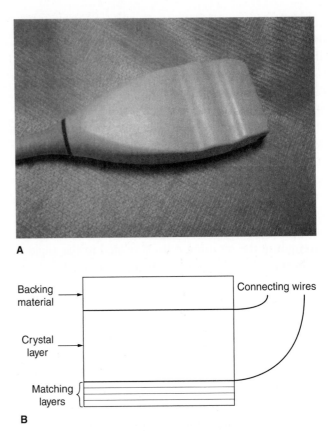

Figure 7–25. **(A)** Actual ultrasound transducer. **(B)** Schematic diagram of the transducer showing the position of the backing material, crystals, and matching layers.

Box 7–4. In Diagnostic Ultrasound, the Transducers Serve Two Functions

- Converts electric energy to acoustic pulses that are sent into the patient
- Receives the reflected echoes and converts their weak pressure changes into electrical signals so that the computer can process them

---

popularly known as PZT, is the most commonly used crystal. The crystals convert electrical voltage into acoustic energy on transmission, and acoustic energy into electrical energy on reception. The transducer also has matching layers in front of the transducer crystals that provide an acoustic connection between the transducer crystals and the skin and that decrease reflection, allowing more of the sound waves to propagate into the body (Fig. 7–25). Behind the crystals is damping material, such as rubber, to decrease secondary reverberations of the crystals with the returning signals. An insulated housing encloses the component parts of the transducer including two wire electrodes, which enter and leave the transducer housing (Box 7–4).

Transducers used in ultrasound range from 2.5 to 16 MHz. They come in different shapes and sizes and are used for different organs. The lower transducer frequencies (2.5, 3, or 4 MHz) are used to image deep structures in the abdomen and pelvis, and for transcranial Doppler. They have greater penetrating abilities but less resolution. The higher frequency transducers (5, 7, 12, 15, or 16 MHz) have greater resolution but less penetrating power. These are perfect for breast imaging where the transducers need only penetrate to a depth of approximately 5 cm. It is common practice to use 14 MHz or higher for most breast imaging and 12 to 10 MHz for a very dense or large breasts. Only occasionally will a large breast need a 5-MHz transducer to assess a deep lesion.

There are several types of transducers. Breast imaging uses the linear array transducers, in which multiple transducer elements are arranged in a straight or curved bar. Groups of element crystals in the transducer are electronically pulsed at once to act as a single larger element. Pulsing occurs sequentially down the length of the transducer face, moving the sound beam from end to end. Because the element crystals are pulsed as a group, there is a delay time difference in the returning signals, thus allowing focusing at different depths. Using the linear array transducers will provide a large rectangular image on the computer screen.

Common Ultrasound Terminologies

Echogenicity

In ultrasound imaging, smooth gray color of the liver parenchyma is considered standard and labeled as isoechoic; any structure brighter than that is labeled as hyperechoic or echogenic. An echogenic structure looks white on the ultrasound image. Structures darker than the liver parenchyma are labeled as hypoechoic or echo poor. If a structure is completely dark, it is labeled as anechoic or sonolucent, or echolucent or echo free. Fluids are echo free.

Calipers

Calipers in ultrasound are used to measure distances (e.g., length, width, and depth). A dotted line can be created around the outline of a structure to calculate the circumference or the area (Fig. 7–26). When imaging, it is always recommended to image the same area with and without the caliper's dotted line because the dotted lines can sometimes obscure pertinent details on the margins of a lesion.

Gain

The system gain is a regulation of the brightness of the image or degree of echo amplification. The gain is measured in decibels and is controlled by the output power or transmitted voltage to the transducer. As

> In breast imaging, it is a common practice to use a 14-MHz or higher transducer and 10- to 12-MHz transducers for very dense or large breasts.

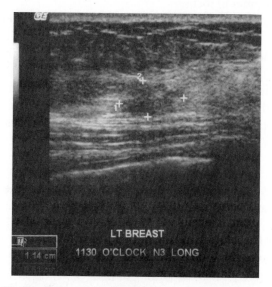

Figure 7–26. The plus signs representing caliper markings on an ultrasound image are used to measure distances.

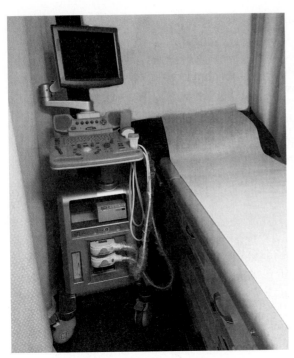

Figure 7–27. The ultrasound unit.

the sound travels through the tissue it becomes weaker in a process known as attenuation. Thus returning echo is extremely weak. Gain compensates for that weakening. Increased gain produces a stronger echo, but too much gain can create artifactual echoes in fluid-filled structures such as cysts. Too little gain, on the other hand, can cancel real echo information in a cyst. The gain setting should be sufficient to differentiate between a simple cyst and a solid mass. In ultrasound, often, internal echoes will be seen within a cyst because of improper gain or power adjustment. Sonographers can control overall gain or depth-specific gain. Depth-specific gain control is known as depth gain compensation (DGC) or time gain compensation (TGC).

Real-Time Scan

All modern ultrasound machines provide real-time imaging to produce a two-dimensional image on a computer monitor. Real-time scanning provides constant feedback of the structures under investigation, with the results displayed on a screen. This ability allows ultrasound to easily image parts without the rigid criteria of stillness that is necessary in mammographic imaging (Fig. 7–27).

Transducer Focal Zone

The sound beam will vary in shape and resolution, with structures close to the skin surface that are within the focal zone of the transducer distorted and difficult to see. Also, the divergence of the transmitted ultrasound beam causes a decrease in the intensity of the signal with increasing distance from the transducer. The result is that the energy is spread over a larger area. The reflected beam will also diverge, returning to the transducer with a loss of intensity. Focusing the beam concentrates the acoustic energy over a smaller area at a specific distance from the transducer. Compared with a nonfocused beam, the focused beam produces a stronger echo, reflected to the transducer. Usually focus is placed at the depth of interest. The sonographer can change the position and number of focus.

To achieve good resolution when imaging, the sonographer must use a transducer that is focused to the right depth. If imaging a structure close to the skin or superficial tissue within the breast, such as within the first 1.5 cm of the skin surface, it is usually necessary to use an offset, also called a "stand-off" pad, or a lot of gel between the transducer face and the surface of the skin. Most systems use electronic focusing that permits the transducer to be focused at one or more variable depths that can be altered by the sonographer.

Zoom

Zoom allows for magnification of the image by increasing the pixel size. With magnification, the resolution of the image deteriorates.

The Ultrasound Image

The breast is seen on the ultrasound as a multilayered structure, with skin and fibroglandular structures being relatively echogenic, while subcutaneous and retromammary fat layers are echo poor. The skin line is seen as a high echogenic border. The chest wall is also echo poor in younger patients but may become more echogenic as the patient ages and the breast parenchyma is replaced by fatty tissue. Except in adolescents, young women, pregnant women, and during lactation, the glandular structures of the breast are mostly surrounded by fat, which will appear less echogenic than the surrounding parenchyma.

On ultrasound, the fibroglandular layer has a flat posterior surface, whereas the anterior surface is cone shaped toward the nipple. Usually, the gland is thicker in the upper outer quadrant, with Cooper's ligaments seen as fine echogenic bands passing from the anterior surface of the gland to the superficial layers of the superficial fascia, where they attach to the skin. Ligaments tend to tent slightly, with the anterior surface assuming a catenary (curved) outline. If imaging with 15-MHz probes, ducts as small as 150 μm (150 μm = 0.15 mm) can be visualized. This means that the lactiferous sinuses and ducts behind the areola are sometimes seen as echolucent tubular structures, tracing outward to the periphery. Lymph nodes are seen as bean-shaped, echo-poor lesions with an echogenic central line extending part way into the parenchyma from one side—this represents the fatty hilum.

Overall, the sonographic appearance of both breasts should be consistent bilaterally, and a lack of consistency can indicate pathology.

Preparing for the Procedure

Ultrasound examinations can be performed by a trained ultrasound technologist, also called a sonographer or radiologist. The preparation is simple. Patients are often examined supine and will be asked to remove all clothing and put on a hospital gown.

For some pelvic ultrasound procedures, the patient has to drink plenty of water—up to 32 oz—before the examination, but this is not necessary for breast ultrasound imaging. Water is used in pelvic imaging because sound waves do not exist in a vacuum and sound does not travel well through air. A full urinary bladder serves as an acoustic window for imaging pelvic organs. Air-filled lungs and bowels containing air conduct sound so poorly that these structures cannot be imaged with ultrasound instruments. Also, any structures behind them cannot be seen. An adjacent soft tissue or fluid-filled organ is, therefore, used as a window through which to image a structure that is obscured by air. The fluid-filled organ most often used is the bladder.

Before beginning the examination, the sonographer should carefully examine the breast and identify the location of any pathology or mass. The technologist should document skin thickening, unusual lumps, moles, eczema or ulcers, dimpling or puckering and nipple changes. If mammograms are available, they should be reviewed before the ultrasound. The breast can then be exposed, and an ultrasound gel is applied. The gel must fill the space between the transducer and the patient's skin; otherwise sound will not be transmitted across the air-filled gap. If the patient is supine, the ipsilateral arm of the breast under examination is raised, and occasionally the contralateral or ipsilateral shoulder is raised to flatten the breast on the chest wall and reduce the tissue distance through which the ultrasound will have to travel.

Scanning Technique

Scanning can be a comprehensive scanning of the entire breast; a screening ultrasound, limited scanning to correlate mammography

> **Two basic types of ultrasound examination:**
>
> Screening
>
> Diagnostic

findings; or a diagnostic ultrasound. Radial (along a line drawn from nipple to the outer margin of the breast) and antiradial (perpendicular to radial) scan planes can also provide additional diagnostic information on the shape of the lesion. A radial plane is along any line connecting the nipple to the peripheral margin of the breast, and the antiradial plane is perpendicular to the radial plane.

> Ultrasound scanning should always be done in two planes: longitudinal (sagittal) and transverse.

Scanning should always be done in two planes: longitudinal (sagittal) and transverse. The longitudinal and transverse axis of the structure being visualized may not necessarily be the longitudinal and transverse axis of the body because the structure could lie obliquely in the body. The ultrasonographer should watch the monitor carefully as the gain setting on the unit is increased. If the lesion is a cyst, the echoes will appear first in the part of the cyst closest to the transducer. At least two sets of images of a lesion should be obtained, one set without measuring calipers and one set with calipers. The lesion should be measured in at least two or more dimensions, and the dimensions should be recorded on the image. All measurements should be in the same unit, millimeters or centimeters, and the reference points used for measurements should be shown on the images of the lesion.

Screening ultrasound is a comprehensive scan, most often bilateral, that is often reserved for young women, or women with dense breasts. A clockwise approach is used with the nipple as a reference point. The patient lies supine with the ipsilateral arm raised. For large breast imaging, the patient can be turned slightly to the contralateral side, especially if the breast tends to fall laterally. In any imaging, labeling is important. The sonographer should start at the top 12-o'clock position, with 12 o'clock described as the portion of the breast from the nipple up to the clavicle. From the nipple down would then be the 6-o'clock position. Scanning should continue in a clockwise direction with the transducer kept as nearly perpendicular to the skin as possible in order to accurately reproduce suspicious areas (Fig. 7–28).

A diagnostic ultrasound can be performed on a palpable mass or a mammographic lesion. The sonographer will palpate the mass and fix it

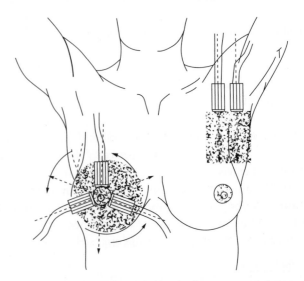

Figure 7–28. Scanning technique. A clockwise approach is used to scan the breast tissue. Parallel scanning can be used in the axilla area.

between two fingers. The examination is done with the patient supine unless the mass is only felt in the erect position. Large breasts are often imaged erect because in this position it may be easier to immobilize the mass. Small breasts may require a stand-off pad to bring area of interest in the focus of the field of view. A partially filled saline bag may be placed between the skin and the transducer to act as a stand-off pad.

The *axilla* is scanned using parallel sweeps, but before scanning the axilla, a check should be made for enlarged nodes. Often, when scanning the axilla, the arm is slightly lowered, and if a suspicious area is visualized, gentle compression with the transducers can be used to confirm the presence of a mass (see Fig. 7–28).

Sonographer and Patient Interactions

Patient care is very important in ultrasound because the sonographer spends a long time in close contact with the patient. Imaging, therefore, requires the development of a good patient and sonographer rapport. Often, the sonographer will find that conversation with the patient yields clues to help focus the study.

Patients also often solicit information from the sonographer. Sonographers should be familiar with the policy of their institution. Some facilities allow sonographers to answer questions only related to the bioeffects of ultrasound or the procedure.

Sonographer and Radiologist Interactions

The sonographer must work as a team with the radiologist. Ultrasound examinations are extremely operator dependent, and the information the sonographer conveys to the radiologist can affect the final diagnosis. The sonographer can report unusual artifacts and confusing scans and can also rescan at the discretion of the radiologist, whenever there is a need to clarify a difficult area. The radiologist should also be informed of any problems encountered during the scanning process or changes the sonographer made to produce the scan. If the radiologist is not available for consultation immediately after the scan, the sonographer should make detailed notes of the entire imaging technique.

Documentation

Both the location and plane of the scan should be clearly labeled. Images should be labeled either right or left, with the examination date, patient name, and additional patient identifier such as a medical record number. The image and any lesions should be identified using a clock notation or labeled schematic diagram of the breast. The position of the lesion relative to the distance from the pectoral muscle or nipple should be noted along with the transverse or longitudinal orientation of the probe with respect to the lesion. This is necessary for any restudies of the same area.

Documentation of the sonographer's name or signature and of any clinical finding is essential. Documentation must also include the clinical history, such as date of onset of any symptoms, recent pregnancy, whether the lesion is painful, and skin changes or nipple discharge.

> Ultrasound examinations are extremely operator dependent, and the information the sonographer conveys to the radiologist can affect the final diagnosis.

Recording the Image

The two-dimensional scan produces an image that is normally displayed on an LCD monitor, but the screen is erased and updated as each new scan is acquired. To save important clinical findings, the sonographer can generate a hard copy, or the information can be sent for digital storage. An integrated PACS system will allow the radiologist or physician to view the images at a later date. If digital transmission is not available, a printed hard copy is required because the physician will not necessarily be present during the scan, and images from follow-up examinations can only be compared with a hard copy of those obtained from a previous examination.

Ultrasound images can be recorded using a variety of formats including magnetic tape recorders, which record a series of images for subsequent playback in real time. With digital integration, the most common means of recording and viewing the ultrasound images are no longer in film format. However, images can be printed in paper format directly to a laser or desktop printer. This method gives the patient a quick image of the ultrasound but is not recommended for interpretation. The ultrasound image can also print directly to a digital dry laser printer. Usual ultrasound printers do not use an ink cartridge but instead use thermal papers.

> Ultrasound is almost 96% to 100% accurate in the diagnosis of cysts.

Uses of Ultrasound

Conventional ultrasound imaging is most often used in determining whether the mass seen on the mammogram is solid or cystic. Ultrasound is almost 96% to 100% accurate in the diagnosis of cysts.

Many physicians are now recommending a mammogram plus a breast ultrasound for patients with dense breasts because the dense breast tissue of younger women, between ages 35 and 45 years, often does not image well on the mammogram. Dense breast notification laws are in place in a number of states that give women the option to take further action, including having an ultrasound. Ultrasound of dense breasts has been found to detect cancers missed on the mammogram, especially when the lesion is palpable.

Ultrasound is a valuable adjunctive tool in determining whether a lesion seen on one projection of the mammogram but not the other is really a mass versus a pseudomass. Many radiologists will request an ultrasound if a suspicious mass is seen on the mammogram because ultrasound is often the first biopsy choice. There is no radiation, lower cost, and real-time imaging when compared with stereotactic mammography or magnetic resonance imaging (MRI) biopsies.

Ultrasound can save the anxiety and discomfort associated with surgery because a simple cyst if diagnosed on ultrasound may not require aspiration or biopsy. The technology can be used during a needle-guided location for excisional biopsy, cyst aspiration, or core-needle biopsy.

Ultrasound can be used to delineate intraductal extension toward the nipple and can be useful to assess implants for leaks or rupture.

Management of Nonpalpable Masses

Nonpalpable masses are often identified at mammography screening, and the workup of these masses is based on the type of mass. They can be oval, rounded, gently lobulated, or circumscribed, or they can have partially circumscribed and obscured margins or be irregular masses with ill-defined or spiculated margins. Most often, if the mass is first seen on the mammogram, it is further evaluated with spot compression or magnification images; however, ultrasound can provide useful information on whether the mass needs to be biopsied. If the mass is irregular or ill defined, an ultrasound can still be performed because ultrasound-guided biopsy is easier, faster, and less costly and does not involve radiation. If, however, the option is an ultrasound-guided needle biopsy, prebiopsy ultrasound imaging will be necessary.

Some radiologists advocate the use of ultrasound as an alternative screening tool for multiple circumscribed masses visualized mammographically, to prevent further surgery or biopsy. In examining any mass, the final determination should be based on a combination of factors, including mammographic, ultrasonographic, and clinical findings.

On ultrasound, a simple cyst has circumscribed margins, sharp anterior and posterior walls, no internal echoes, and posterior enhancement. When true echoes (not artifactual) are seen within a cyst, they are usually because of proteinaceous material or cellular debris.

Ultrasound Evaluation of Palpable Masses

If the mass is palpable but is not seen on the mammogram, ultrasound is still a useful tool. The simple cyst will be easily identified on ultrasound. A painful cyst can be aspirated. If the mass seems solid at ultrasound, a needle biopsy is recommended. Masses with oval, round, lobular, circumscribed, or partially obscured margins should be further assessed mammographically with spot compression or magnification, but ultrasound assessment is useful in ruling out a simple cyst.

An irregular mass or a mass with an ill-defined or spiculated margin that is palpable should be biopsied. Ultrasound-guided needle biopsy can be used.

Doppler Effect

When a high-frequency sound beam strikes a moving structure, such as blood flowing in a vessel, the reflected sound returns at a different frequency. This change in frequency is called Doppler shift. Ultrasound is inaudible, but Doppler shift is audible to the human ear. The velocity of the moving structure can be calculated from this frequency change. The returning frequency will be increased if flow is toward the sound source (transducer) and decreased if flow is away from the sound source. Changes in the speed of the blood can indicate narrowing of an area of the blood vessels (stenosis) or other pathology. This ultrasound technique is known as the Doppler technique, and it is used to evaluate blood flow through the arteries or veins and the vascularity of a mass.

The frequency of the returning wave can also be converted into an audible signal. Typically, veins have a low-pitched hum, whereas arteries have an alternating pattern with a high-pitched systolic component and a low-pitched diastolic component. Optimal imaging in Doppler is achieved only when the flow is at an oblique angle to the transducer. If the angle is increased, the detected Doppler shift decreases. If the sound beam becomes perpendicular to the direction of motion, no shift will be detected. The angle should be between 45 and 60 degrees to avoid creating incorrect velocities, and it definitely should not exceed 60 degrees.

Color Doppler ultrasound assigns a different color to the red blood cells in a vessel depending on their velocities and the direction of the blood flow

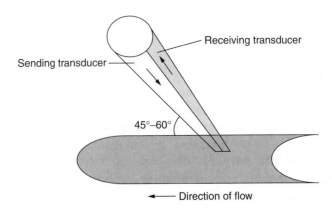

Figure 7–29. Doppler and color-flow principle. Sound will only be detected if the angle between the transducer and the vessel is between 45 and 60 degrees.

Color Doppler

Color Doppler ultrasound assigns a different color to a vessel depending on its velocity and direction of the blood flow (Fig. 7–29). The color assignment is performed very rapidly, so a real-time imaging is still generated. In most systems, the flow of blood away from the transducer is given the color blue and the flow toward the transducer is assigned the color red. Brighter colors indicate faster velocities, and darker colors, slower velocities. The fastest velocities may be yellow or white. Turbulent flow will show as a mixture of colors.

Normal vessels of the breast are seen on Doppler as colored lines running along the normal structure, similar to Cooper's ligament. On a gray-scale ultrasound image, Cooper's ligaments appear as echogenic lines. They have a linear or smooth, curved shape and taper smoothly with regular smooth branches. Generally, two sets of lines are seen— branches from the lateral thoracic artery running in the anterior part of the breast and branches from the internal thoracic artery (internal mammary artery) passing laterally into the breast from the parasternal region. Occasionally, branches of the intercostal arteries are seen in the posterior surface of the breast.

The absence of vascularity of a lesion on color Doppler favors a benign lesion, but vascularity can be associated with malignancy or with inflammation. If color Doppler demonstrates flow within the lesion, the mass is usually solid.

Ultrasound Characteristics of Benign Lesions

Generally, benign masses have smooth borders and uniform internal echoes with sharp anterior and posterior borders. They do not disrupt the surrounding architecture. They move freely through the tissues and are compressible with good posterior acoustic enhancement. Most cysts are compressible. But posterior enhancement can occur behind both solid masses and cysts (Box 7–5).

Ultrasound Characteristics of Malignant Masses

Malignancies generally are irregular with ill-defined borders and an infiltrating edge with alternating echopenic and echogenic straight lines

---

Box 7–5. Ultrasound Characteristics of Benign Lesions

- No internal echoes—avascular—in simple cyst. (Benign lesions can be solid as well and will have internal echoes.)
- Posterior enhancement
- Homogeneous or hyperechogenicity
- Thin echogenic pseudocapsule around the lesion
- Ellipsoid shape (width greater than anterior–posterior diameter)
- Fewer than four gentle lobulations
- Compressibility—they are more easily compressed than malignant lesions
- Acoustic shadowing is even and less marked in benign lesion
- Calcifications are larger and more discrete

radiating from the surface of the mass. In an echopenic fatty breast, the nodular strands are relatively hyperechoic, whereas in echogenic fibrous tissue, the strands appear hypoechoic. Their internal echoes are heterogeneous, and they may have irregular acoustic shadows because they infiltrate and drag the surrounding tissue with them as they move. They tend to be incompressible and are usually vascular with tortuous and irregular vessels. Generally, benign masses spread in a horizontal line because they are limited by the tissue plane, but cancerous tumors are not confined by tissue planes and, therefore, often spread vertically, becoming taller than they are wide. Cancers also tend to expand toward the nipple within the ducts, a process called duct extension. They can have multiple small lobulations on the border of the mass. Calcifications are sometimes seen as punctuate areas within the mass (Box 7–6).

Ultrasound Characteristics of Intermediate Lesions

Occasionally, the ultrasound will generate false-negative or false-positive results because some malignancies may mimic a benign mass on ultrasound, or vice versa. Ultrasound features should be incorporated with an overall assessment of the mass, including mammograms and diagnostic workup, such as spot compression. If any ultrasound or mammographic features suggest malignancy, a biopsy is necessary (Box 7–7).

Typical Appearance of Lesions on Ultrasound

Abscesses have the same general ultrasound characteristics as hematomas. They have a cavity with irregular walls, and although the borders are well seen, they will be thick and irregular. Abscesses often occur in the periareolar region. They are very tender, and the breast is red and hot over the mass. They often have internal echoes. The clinical history is very important when evaluating abscesses or hematomas because the internal appearance can evolve over time.

Microcalcifications are not well seen on the ultrasound. Occasionally, they are seen as intense echogenic foci with distal shadowing. They are especially difficult to detect in tissue that is itself echogenic, especially the fibroglandular parts of the breasts. Macrocalcifications can appear as a "shadowing" area (Fig. 7–30).

> **Box 7–6.** Ultrasound Characteristics of Malignant Lesions
>
> - Irregular shape or spiculated outline
> - Angular or greater than three lobulations
> - Microlobulated—there may be small lobulations on the borders
> - Ill-defined
> - Spiculated margins—alternating echopenic and echogenic straight lines radiate from the lesion
> - Anterior-to-posterior diameter (height) greater than width (taller than wide)
> - Marked hypoechogenicity
> - Attenuating distal echoes—shadowing because of the failure of the sound beam to pass through the lesion
> - Duct extension
> - Punctuate calcifications

> **Box 7–7.** Ultrasound Characteristics of an Intermediate Lesion
>
> - Echo texture (homogeneous or heterogeneous)
> - Echogenicity similar to fat
> - Normal or enhanced posterior echoes

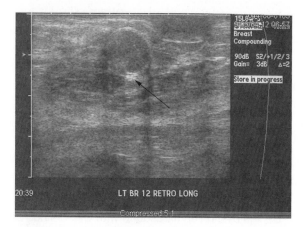

Figure 7–30. Calcifications in an oil cyst (*arrow*).

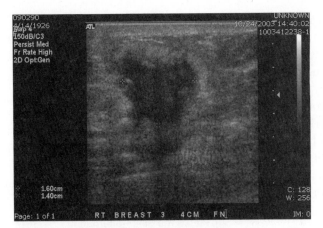

Figure 7–31. Cancer on an ultrasound showing irregular shadowing.

Cancers typically have an echo-poor central area and are often somewhat heterogeneous with distal shadowing. The lesions tend to be longer than wide because they grow across the boundaries of the breast and disrupt the adjacent fibro-glandular layers. Cancers are noncompressible and are relatively fixed. Along with the cancerous changes, the Cooper's ligaments become thickened and distorted. Cancer can also cause dilated ducts and skin thickening leading to enlarged intercostal or axillary nodes. The dilated ducts are usually clearly visualized on the ultrasound, and the nodes are seen as echo-poor spherical masses approximately 1 cm in diameter. Malignancies are often highly vascular, with vessels penetrating the lesion itself. Lesions smaller than 5 to 8 cm are difficult to visualize on the ultrasound and will often mimic the appearance of a fibroadenoma. Lesions such as infiltrating lobular or inflammatory carcinoma, which do not form a discrete mass, are also difficult to detect on ultrasound (Fig. 7–31).

Carcinoma in situ is noninvasive and difficult to detect on the ultrasound.

Cysts are well-defined echo-free spaces with good posterior acoustic enhancements. They can be solitary or multiple, unilateral or bilateral, round or oval. They have smooth, well-defined walls and can occasionally have thin and uniform septations. Multiple cysts adjacent to each other can mimic a septation on ultrasound. The result is that a simple cyst may appear complex. They are spherical when tense, flattened when lax, and often form in bunches. Cysts generally form from the ductal and glandular structures, and they are attached to the surrounding parenchyma. This makes them only partially mobile. The best time to evaluate a patient with cystic breast would be 10 to 12 days after the first day of the menstrual cycle.

Generally, any cyst that does not have all the characteristics of a benign mass should be biopsied or aspirated (Fig. 7–32). Benign cysts are always avascular. Thickening or irregular walls are suspicious characteristics, and some cysts have low-level echoes due to sebaceous contents. Calcification in a cyst wall can also prevent "through

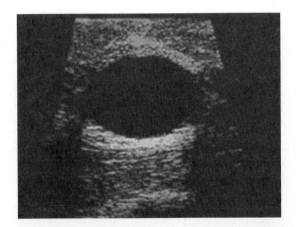

Figure 7–32. Benign cyst on an ultrasound with well-defined echo-free space and acoustic enhancement.

transmission." It is often difficult to differentiate these cysts from a malignant tumor on the ultrasound.

Fibroadenomas are seen as well-defined lesions with an oval or multilobulated contour. They are often seen in women taking hormone replacement therapy or women between 10 and 30 years old, and may vary in size. Fibroadenomas can have round or oval, smooth or lobulated contours (Fig. 7–33). They are sometimes well-defined and easily differentiated from the surrounding tissue, but often the internal echoes are isoechoic. Occasionally, the fibroadenoma will have uniform low-level internal echoes with the same intensity as the subcutaneous fat. There is often through transmission because the fibroadenoma will not attenuate sound unless it is calcified or has undergone hyaline necrosis. They are very mobile but only slightly compressible. Large fibroadenomas tend to be more vascular than a normal benign lesion, but on color Doppler the vessels are usually seen running around the lesion rather than through it. It is often difficult to distinguish small fibroadenomas from a malignancy.

Fibrocystic change is a general term including conditions such as cystic diseases, adenosis, epithelial dysplasia, fibrous disease hyperplasia, and fibrosis. Normal cystic conditions can include having multiple cysts of various sizes, or multifocal hyperplasia of the epithelial or stromal tissues of the terminal lobules, which can mimic malignancies. Often, the breast is lumpy and painful with regional thickening of the gland, especially the upper outer quadrant. Ultrasound usually reveals generalized segmental ductal changes with prominent stromal and epithelial tissue changes. The ducts are surrounded by fine echo-poor lines, giving the ultrasound a mottled or sponge-like look.

Galactocele presents as a firm movable mass. On ultrasound, its appearance can depend on its fat and water content. It can be oval or elongated with well-defined walls, low-level internal echoes, and posterior acoustic enhancement. A galactocele is a cyst containing milk.

Hematomas are seen as irregular-walled cavities with echogenic fluid.

Lymph nodes are found throughout the breast, often in the upper outer quadrant (Fig. 7–34). They are well-defined, rounded, oval, or kidney-shaped masses with an echogenic fatty center.

Lymphoma will produce a large echo-poor mass with well-defined borders.

Metastasis to the breast is seen as a mass with low-level heterogeneous echoes. These are noncompressible and highly vascular.

Phyllodes tumors are lesions similar in appearance to a large fibroadenoma with well-defined margins and distal enhancement. They tend to have multiple cystic spaces within the mass and are vascular on color Doppler.

Scarring in the glandular layers of the breast will not necessarily be directly under a scar on the skin surface. One reason is that often, in order to produce good cosmetic results, the surgical incision is not made directly over the mass to be removed. Scars show a fine linear echo-poor shadowing. Scars older than 3 to 6 months are avascular on color Doppler.

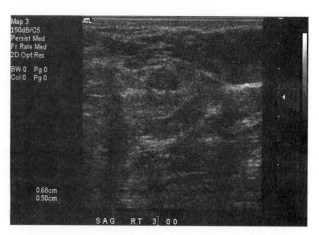

Figure 7–33. Fibroadenoma on an ultrasound. A well-defined lesion with uniform internal echoes.

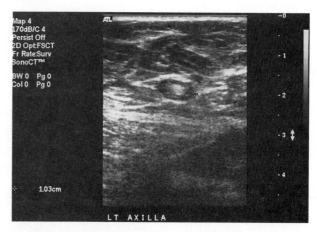

Figure 7–34. Lymph nodes on an ultrasound seen with characteristic vascular hilum.

Imaging Implants

In general, older implants have a single lumen, whereas the newer ones are double lined. Implants can have an inner bag of silicone and an outer bag of saline. The material used for implants is echo free and appears as a black space surrounded by a smooth echogenic capsule. Silicone implants are generally seen with a depth of approximately twice the actual depth because the speed of sound in silicone is approximately half of that of body tissue. Because the delay in return echo from the back wall of the prosthesis is used by the scanner to calculate its depth, and the calculations are based on the speed of sound in tissue, the longer delay translates and is interpreted by the scanner as a greater depth. The lateral dimensions are not affected by this error because it is measured by the span of the echoes along the transducer surface. Ultrasound can easily detect masses outside of the implant.

In the extracapsular ruptured implant, the silicone granulomas will be seen in the periphery of the implants, or within the breast parenchyma as a "snowstorm" appearance compared with the echo-free "space" represented by the actual implant. These high echoes are very characteristic of silicone and are sometimes even seen in the lymph nodes if the free silicone is taken into the lymphatic system. Focal collecting of silicone is sometimes seen as an echolucent appearance.

Intracapsular ruptures are difficult to detect because the silicone is confined to the pseudocapsule. There may be an increase in echogenicity within the implant, and globules of silicone may be visible in the edge of the bag. These represent silicone, which has escaped the bag and is now surrounded by it. A bulging contour could indicate rupture, with multiple, parallel, horizontal echogenic lines within the bag representing layers of collapsed bag. These horizontal patterns are referred to as the "stepladder" pattern and are similar in appearance to echogenic bands that can extend all the way across the implant, outside and within the silicone shell. The silicone nodules seen outside the implant appear hypoechoic (grayish) and add to this stepladder effect.

Ultrasound Artifacts

All artifacts are fake, transient, and not present in all images. Artifacts in ultrasound can be related to equipment malfunction, operator dependent, or due to the effects of sound interacting with the tissues. Some of the main causes are outlined next.

Equipment Artifacts

Calibration problems result when the ultrasound unit is improperly calibrated, causing incorrect measurement when using the calipers.

Artifactual noise can be caused by nearby equipment, which could produce electrical interference. Noise caused by nearby equipment will appear as a repetitive pattern throughout the image.

Operator-Dependent Artifacts

Noise is an operator-dependent artifact caused by excessive gain. Excess gain can cause a simple cyst to appear like a solid lesion. The choice of the wrong transducer or improper placement of the transducer

---

Artifacts in ultrasound can be related to equipment malfunction, operator dependent, or due to the effects of sound interacting with the tissues.

(e.g., poor contact with the skin) can result in missed masses or masking of actual problems.

Motion artifacts can be caused by breathing or movement during the scanning process. The resultant image is blurred or distorted. The sonographer should give clear instructions to the patient or have the patient suspend respiration during the scan.

Operator pressure is a result of too much pressure applied by the sonographer. The pressure should be just sufficient to keep the transducer in contact with the skin.

Sound-Tissue Artifacts

Reverberation artifacts occur when the sound is repeatedly reflected within the tissue and produces repeated parallel bands in the image. They are often seen in silicone implants parallel to the anterior implant surface. They also occur if a stand-off pad is used to image structures close to the surface.

Linear artifacts are small linear echoes near the border of the bag of an implant representing bag folds. They can mimic implant rupture.

Shadowing on a mass favors malignancy. Edge shadows extend deep from the margins of any clearly defined structure (e.g., a cyst or a fibroadenoma). They are caused by fibrous bands of Cooper's ligaments.

Limitations of Breast Ultrasound Imaging

A major disadvantage of breast ultrasound imaging is that it is only as good as the person holding the transducer. The examination is highly operator dependent. The sonographer must be able to identify all organs in the body and differentiate between real and artifactual echoes. If enough force is not applied to the breast, normal breast tissue can look like a mass. Also, if the lump is not palpable and the technologist is not experienced, the area of interest can be missed entirely. Applying the incorrect gain (amount of sound) can result in false echoes, making a simple cyst look solid.

Ultrasound requires dedicated training to interpret the images. Internal echoes in a cyst can be the result of artifactual reverberation or other artifacts. It is easy to miss a mass with ultrasound. Small lesions are especially easy to miss, but larger lesions will also be missed if their echogenicity and texture are the same as the surrounding tissue. Ultrasound does not display all solid masses, even in the mammographically dense breast. If a mammographic lesion is not visible on ultrasound, the lesion must be assumed to be solid and, based on clinical findings, should be biopsied.

Another serious limitation of ultrasound is that ultrasound waves are almost 100% reflected in bone because bone conducts sound at a much faster speed than soft tissue and because the current ultrasound instruments cannot tell the difference in speed between soft tissue and bone.

This means that bone or structures covered by bone will not be imaged well.

Although ultrasound imaging is not affected by the patient's age or the tissue density of the patient's breasts as in mammography, the patient's body size can affect the ultrasound image. Breathing can be a problem if Doppler is utilized. A high-frequency, high-resolution transducer will penetrate only to a depth of about 5 cm, which means

---

Ultrasound artifacts

Equipment artifacts
- Calibration
- Artifactual noise

Operator-dependent artifacts
- Noise
- Motion
- Operator pressure

Sound-tissue artifacts
- Reverberation
- Linear
- Shadowing

Bone or structures covered by bone will not imaged well on ultrasound; therefore the technology will not image microcalcifications.

that more penetrating power will necessitate a lower frequency transducer, therefore sacrificing image resolution. In breast ultrasound imaging, obese patients are very difficult to scan and may limit visualization of certain structures. Locating a cyst deep within a large fatty breast may be difficult. But even in a normally sized breast, cysts less than 5 mm in diameter are difficult to differentiate. Large breasts may be easier to scan in the erect position where they can be easily stabilized for scanning. The lactating breast will also present a challenge because it generally contains a range of tubular structures representing milk-filled ducts. This usually increases the difficulty in interpreting the ultrasound.

Imaging Problems and Solutions

Sound cannot travel in a vacuum and needs a medium to travel through. A gel must be applied to the transducer and patient before scanning.

Cartilaginous portions of ribs can be confused with a fibroadenoma if the scan cuts across the rib in a longitudinal section. This can be overcome by always visualizing a lesion in two planes.

Intramammary lymph nodes can be mistaken for masses. The location and appearance of the nodes—echogenic with a vascular hilum—can, however, help in differentiation.

Edge shadowing from Cooper's ligaments can be mistaken for pathology. This can be avoided by noting the origination of the point of the shadow and shadow characteristic. Shadows are fine lines and will be obliterated by probe pressure.

In color Doppler, too much pressure will obliterate small vessels.

Risks of Ultrasound

Ultrasound uses no ionizing radiation, and there are no documented risks of harmful bioeffects or side effects of the medical use of ultrasound to image the breast; however, the frequency levels used in ultrasound have been steadily increasing, and studies have shown that high-frequency sound waves can be harmful, especially to the fetus, because they create heat and microcavitations in the tissues. Heat is noted more in bones and calcified tissues because they absorb ultrasound energy and convert it into heat energy. This can be especially problematic for fetal head scanning. Although no harmful effect has been reported in breast ultrasound, some authorities recommend shorter scanning time.

Advantages of Breast Ultrasound

The most important advantage is that ultrasound is free from radiation! With state-of-the-art equipment, ultrasound can detect cysts as small as 2 to 3 mm in diameter within a small breast. The dense breast tissue of women younger than 30 years does not image well on the mammogram. These women are, therefore, ideal candidates for ultrasound evaluation of suspicious palpable lesions. Breast ultrasound is often used as a screening tool for women with dense breast tissue, as an important adjunct to mammography. Ultrasound can be used to determine whether a mass is solid or fluid filled, and the technique is invaluable during needle-guided excision biopsy, cyst aspiration, or core-needle biopsy. Ultrasound is also useful to assess implants to detect leaks or rupture.

Sound cannot travel in a vacuum and needs a medium to travel through.

A gel must be applied to the transducer and the patient before scanning.

MAGNETIC RESONANCE IMAGING OF THE BREAST

Brief History of MRI

Radio waves have been around since the late 19th century when Edison and other scientists developed radio communications. While investigating the behavior of solids and liquids in a strong magnetic field, the scientists found that the nuclei of these materials absorbed radio waves at specific frequencies. The technique was then called nuclear magnetic resonance (NMR). It was soon discovered that the spectrum of absorbed frequencies from a chemical substance reveals its molecular structure. In the 1940s, Felix Bloch and Edward Purcell continued to work independently, experimenting with the nucleus of atoms in a magnetic field. Their experiments led to the NMR spectrometers, which were used to determine the molecular configuration of a material from the analysis of its NMR spectrum. The two scientists, Bloch and Purcell, were awarded the Nobel Prize in Physics in 1952. Also, in 1991, the Nobel Prize for Chemistry was awarded to Richard R. Ernst of Zurich for his contributions to the field of magnetic resonance spectroscopy.

By the late 1960s, other scientists, notably Raymond Damadian and Paul Lauterbur, discovered that not only did the nuclei of materials in a strong magnetic field absorb radio waves at specific frequencies, but also the exact frequencies absorbed depended on the chemical environment of each atomic nucleus. By modifying and adapting NMR techniques, they were able to first produce images of animals and then humans. The "nuclear" in NMR was dropped because of the patient's fears about radiation, and this new imaging technique, now called magnetic resonance imaging (MRI), was used to show subtle difference in contrast between body tissues.

As early as 1978, even before head images and years before the first commercial MRI machines appeared on the market, researchers began imaging the breast. By the late 1980s, MRI of the breast was still thought to be of little value, but soon research with contrast agents revealed that most breast cancers were enhanced and could be differentiated from some benign conditions. Today, breast MRI has found its place and is used as an important adjunctive tool in detecting breast cancer.

MR imaging uses radiofrequency (RF) waves.

Electromagnetic Spectrum

There are many types of electromagnetic radiation, and they can be grouped together to make what is called the electromagnetic spectrum. In the middle of the spectrum is visible light (red, orange, yellow, green, blue, and violet). Visible light is the smallest part of the spectrum and the only portion that we can see. At the long-wavelength, low-energy end of the spectrum are radio and television waves. There are also radiofrequency (RF) waves. RF waves are commonly used in MRI and to broadcast signals to and from television and radio stations. The short wavelength RF waves are known as microwaves (Fig. 7–35). The only differences between the photons of the various portions of the electromagnetic spectrum are frequency and wavelength, and there is usually considerable overlap with the wavelength of the various sections.

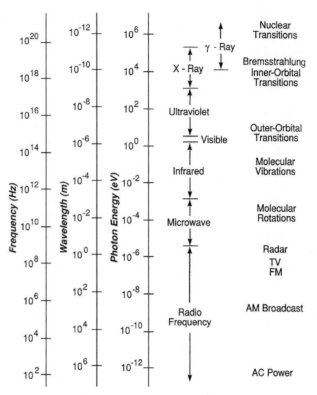

Figure 7–35. The electric magnetic spectrum. There is considerable overlap of the wavelengths of the various sections.

Magnetism

Magnetism is a characteristic property of certain materials to generate imaginary lines of a magnetic field around the material (Fig. 7–36). All magnets will attract iron and have dipoles, meaning two poles—north and south. Any charged particle in motion will create a magnetic field. The magnetic field of a charged particle is perpendicular to the motion of that particle.

The largest natural magnetic field is the earth. Earth has a magnetic field because it spins on an axis. Many materials also exhibit magnetic properties. These are called ferromagnetic materials and are strongly attracted by a magnet and can be permanently magnetized by exposure to magnetic fields, for example, iron and nickel. Materials that are unaffected by a magnetic field are called diamagnetic or nonmagnetic. Examples would be wood or glass. Some materials are intermediate; these are paramagnetic materials and are loosely attracted to a magnet. A paramagnetic material becomes magnetic when it is placed in a magnetic field but lose its magnetism as soon as it is removed from the field. The contrast agent used in MRI is made of a paramagnetic material. Another type of magnet is the electromagnet. The electromagnetism is induced magnetism due to an electron current. The material is only magnetized while the current is flowing. The Tesla (T) is the standard unit for measuring magnetic fields. The Tesla replaces the older unit, the Gauss (G); 1 T equals 10,000 G.

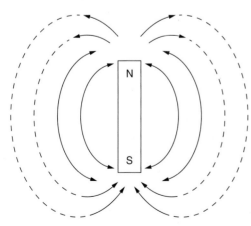

Figure 7–36. Magnetic field around a particle. Ferromagnetic materials such as iron have imaginary lines leaving the North Pole and entering the South Pole.

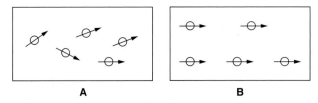

Figure 7–37. **(A)** Hydrogen atoms randomly aligned in the body in its natural state. **(B)** The hydrogen atoms aligned in a magnetic field.

The Principle of MRI

MRI is essentially an image of the interaction of body tissues with radio waves in a magnetic field. These interactions result in high- or low-intensity signals that appear as bright or dark areas on the image.

Certain nuclei in the periodic system table of elements, such as hydrogen, phosphorus, and sodium, possess magnetic properties and will behave like small bar magnets and align with an outside magnetic field like the needle of a compass. These elements have very specific and predictable behavior when exposed to a magnetic field. In the human body, hydrogen (symbol, H) was the element of choice because the body is more than 70% water. Water is found in the body as fluids, or bound to larger molecules as protein or as hydrogen atoms within fat. Each water molecule consists of two hydrogen atoms bound to an atom of oxygen (symbol, H_2O—where O is the symbol for oxygen). Hydrogen has a very simple atomic structure and consists of one proton and one electron. Because the hydrogen nucleus has only one proton, it is often referred to as a "proton" in MRI. Also, the hydrogen nuclei are bound with their atoms to form a molecule. This binding arrangement is sometimes called a lattice.

Normally, the hydrogen nuclei in the body are randomly oriented, with their magnetic poles pointing in varying directions. As the body is subjected to a magnetic field, the high-energy hydrogen nuclei align themselves against the magnetic field (Fig. 7–37); however, the low-energy hydrogen nuclei in the body essentially become magnetized and are aligned with the external magnetic field. Because the low-energy protons outnumber the high-energy protons, this leaves a net magnetization.

Not only are the hydrogen atoms aligned to the external magnetic field, they also rotate around their axis as much as the earth rotates on its axis. This rotation is called *precession*. Precession describes a wobble motion similar to the last motions of a spinning top just before it falls. Within the magnetic field, all protons precess at the same frequency but not necessarily in coherence or in phase. The frequency is determined by the strength of the magnetic field; it is a constant known as the Larmor frequency (Fig. 7–38).

If the body part remains under the influence of a magnetic field for a sufficiently long period of time, the individual magnetic fields or proton spins will stabilize to a steady value. This is referred to as the net magnetization at equilibrium and is the sum of the individual magnetic moments of each hydrogen nucleus. In MRI, vector diagrams are used to describe the MRI phenomena. A vector has both direction and magnitude. Arrows in a vector diagram will indicate direction, and the size of the arrows will indicate magnitude. Using the conventional system, the external magnetic field is parallel to the Z axis; therefore, the net

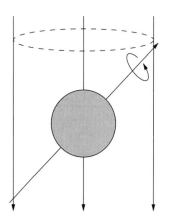

Figure 7–38. When a nuclear magnetic moment, represented by a hydrogen nucleus, is placed in a magnetic field its axis of rotation precesses about the magnetic field.

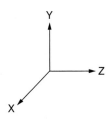

Figure 7–39. Longitudinal and transverse magnetic fields. The Z axis represents net magnetism at equilibrium.

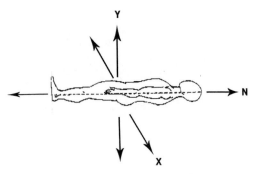

Figure 7–40. Magnetic field acting on the patient. The long axis of the patient lies on the Z axis when the patient is in the permanent magnet. The image shows how the vector diagram would line up.

magnetism can be described as the equilibrium magnetization along the Z axis (Figs. 7–39 and 7–40).

Precession will change the net magnetic field. The net magnetic field will flip off the Z axis and will point somewhere else. The result is net magnetization in the XY plane. Transverse magnetization describes the net magnetization in the transverse plane (XY plane), and longitudinal magnetization describes loss of transverse XY magnetization and the regrowth of net magnetization in the Z plane.

If an RF pulse is applied to the body, the protons in the nuclei absorb energy and become excited. They go from a low-energy to a high-energy state and try to realign themselves against the magnetic field. This reorientation results in a flip; they are flipped transversely on their axis. This distance or angle is known as the flip angle, and RF excitation pulses are named after the flip angle they induce. Therefore, a 90-degree pulse will produce a 90-degree flip angle, and a 180-degree pulse will lead to a 180-degree flip angle (Fig. 7–41). The flip angle is determined by the duration of the pulse. Therefore, creating a 180-degree pulse means inducing twice as much energy as a 90-degree pulse. The RF applied must be the same frequency as the precessing hydrogen nuclei; that is, the RF applied must be at the Larmor frequency of hydrogen.

As long as the RF field is applied, the hydrogen nuclei will precess in phase or in unison and are said to have phase coherence. After the RF is removed, the protons relax, lose energy, and lose phase coherence as they return to their original orientation. To return to equilibrium, the nuclei must give up the energy gained from the transmitted RF pulse. This loss of energy is termed relaxation. The rate of relaxation varies with the individual proton in the body, and the relaxation times can be measured. Also, since the relative time for each element of tissue differs, and relaxation results in an RF emission or echo, the differences can be measured and will give information about the physical and chemical tissue under examination. The information can therefore be used to create a three-dimensional image of the part.

As relaxation occurs, the amount of magnetism in the longitudinal plane gradually increases. This is called the T1 recovery. Also, the amount of magnetism in the transverse plane gradually decreases. This is called T2 decay (or T2 relaxation).

T1 relaxation or recovery, therefore, results in longitudinal magnetism due to energy moving to the surrounding environment or lattice. T2 relaxation or decay is due to interactions between the magnetic fields of adjacent nuclei and results in loss of transverse magnetization.

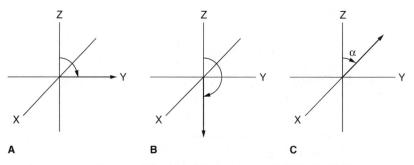

Figure 7–41. In MRI, rotation can be by a 90-degree pulse **(A)**, by a 180-degree pulse **(B)**, or using smaller flip angles for fast imaging **(C)**.

As the strength of the transverse magnetism decreases, so does the strength of the signal in the receiver coil. This reduction in induced signal is called the free induction decay (FID) signal.

The intensity and duration of RF signals from different body tissues will vary depending on how quickly the net magnetization for each tissue reorientates itself longitudinally (T1 relaxation) and loses its transverse orientation (T2 relaxation). The difference in RF signals is recorded and used to generate images. A large value for T1 or T2 will indicate a long, gradual relaxation time; a small value will indicate a rapid relaxation time. T1 relaxation time will be represented by an increasing value and T2 relaxation time by a decreasing value because T2 represents the relaxation from a state of excitement to a state of equilibrium where the relaxation is zero. T2 will always be less than T1.

A very simple pulse sequence in MRI is a combination of RF pulse, the MRI signal, and a period of recovery.

Equipment

Like a typical computed tomography (CT) scanner, the MRI scanner has three main sections—the table, the scanner, and the computer (Fig. 7–42). The table resembles a bed or stretcher, which slides in or out of the scanner (depending on what part of the body is being scanned). The table is sometimes slightly curved to conform to the circular bore of the scanner and uses a hydraulic system to move the patient in and out of the bore. The bore is donut shaped, generally approximately 50 to 60 cm in diameter, and contains the primary magnet assembly and various secondary magnets. The secondary magnets are at room temperature, but the primary magnet must be supercooled and is immersed in liquid helium.

At room temperature, all materials are resistant to the flow of electric current, but below a critical temperature, superconductivity will be achieved, and once an electric current starts to flow, it will flow indefinitely. This eliminates the need to provide a current source to keep the primary magnet magnetized. The primary magnet is kept at supercooled temperatures near absolute zero (0° K) using special cooling

> The primary magnet is kept at supercooled temperatures, near absolute zero (0°K), by using special cooling agents called cryogens, to achieve superconductivity.

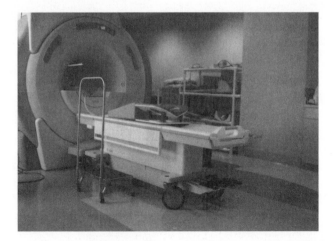

Figure 7–42. The MRI unit with a breast coil in place on the table.

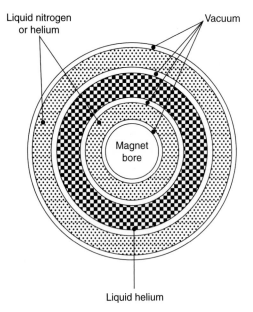

Liquid nitrogen
or helium

Vacuum

Magnet
bore

Liquid helium

Figure 7–43. Superconducting magnet is surrounded by several layers of insulation.

agents called cryogens, to achieve superconductivity (Fig. 7–43). The cryogens used can be liquid helium and/or liquid nitrogen, and the containers for supercooling the coils are called cryostats. At these low temperatures there is virtually no resistance to an electrical current, making very high magnetic field strengths possible. With superconductivity, the magnetic resonance (MR) signals will achieve the desired SNR, resulting in shorter examination times and reduced motion artifacts, also producing higher resolution images. With superconductivity, however, the primary magnet will always be on.

The overall large size of the MRI scanner is necessary to hold the primary and secondary magnets, which are not visible. Closest to the patient are the first secondary magnets or RF coils or RF probes. These are actually radiofrequency antennae. The RF coil or probe produces a radio signal, and both transmit and receive the MR signals. The RF coil must completely surround the part under examination. The basic body coil is used as an RF transmitter for all scans and an RF receiver for body imaging. Other body parts coils or surface coils are connected with cables to the MRI scanner and are used only as receiver coils. The signal detected by the surface RF coils comes from all tissues within the sensitive volume of the coil.

Coils improve SNR by reducing the detected noise; surface coils will not detect signals from tissue not under examination. The closer the coil is to an area of interest, the less RF energy will be needed to create transverse magnetization. The closer the receiver coil is to the area excited, the more signal volume will be detected. For example, the breast coil will register signals from the breast but will not register signals (noise) from the abdomen because the abdomen is far from the coil's sensitive volume. For this reason, individual coils are made to accommodate various body parts (e.g., head, extremity, and breast) (Fig. 7–44).

The next coils are called gradient coils. They are large electrical conductors that produce fast gradient magnetic fields by switching on and

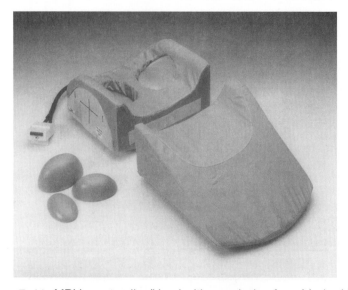

Figure 7–44. MRI breast coils. (Used with permission from Medrad Inc.)

off. There are three pairs of gradient coils to produce magnetic fields in one of three directions, or obliquely. These directions correspond to the *X*, *Y*, and *Z* axes. Gradient coils are broad copper–conducting bands referred to as conductors. They are typically 10 mm wide and 4 mm thick to carry the intense electric currents of up to 30 A that are required to generate the magnetic field gradients. When a gradient coil is switched on, the electric current causes the conductor to expand as a result of resistive heating; when the gradient is switched off, the conductor cools and contracts. The alternating contraction and expansion produces sound.

These gradient coils must be capable of switching on and off very rapidly. Switching times of the gradient coil can be less than 500 μs (0.5 ms). It is the switching on and off of the gradient coils, necessary to measure the MRI signals reflecting from the patient's body, that produces the thump-thump-thump sounds heard during the MRI examination.

Next are the shim coils, which are closest to the primary magnet, and in some powerful magnets they are incorporated within the super-cooled primary magnet. In preparing for a slice, the shim coils will further refine the homogeneity of the magnetic field by adjusting the power to each individual shim coil after the patient is positioned (Box 7–8).

> **Box 7–8.** Summary of the Component Parts of the MRI Unit
>
> - Magnet
> - Patient table
> - Shim coils
> - Three sets of gradient coils
> - Body RF coil (transmitter and receiver) and surface coils
> - Computer
> - Image processor
> - Main console with video monitor and keyboard
> - Image-storage unit
> - Power unit

Operating Console

Generally, there are two consoles, one for operating the MRI unit to acquire data and the other for manipulation of the images after they have been collected. The MRI operator controls the scanning by specifying the RF pulse sequence, pulsing time interval, matrix size, field of view (FOV), and number of excitation (NEX) or number of times image acquisitions are repeated for each slice. The operator will not have to initiate the various pulses but must choose the appropriate sequencing method for the particular examination. The most common variable in operating the MRI is the time interval (T1, TE, and TR) used to enhance the image quality.

Computer Parameters

The *matrix size* will define the number of pixels on the resultant MRI. Increasing the size of the matrix means increasing the number of data points, but not increasing the size of the pixel. This will increase spatial resolution but will also result in decreased SNR and increased scan time. An image with more pixels will have better differentiation between separate areas.

The *FOV* describes the amount of body area to be included in the image. Decreased FOV improves spatial resolution because the same number of pixels is seen in a smaller body area. Decreasing FOV will also decrease SNR.

The *SNR* is the proportion of RF signal to random background noise used to construct the image. High *SNR* increases image quality, whereas low SNR decreases image quality. An image with a low SNR will have a grainy appearance.

The *spatial resolution* will depend on the size of the matrix used for imaging. A larger matrix has more pixels, allowing for more details to be depicted and increasing the spatial resolution. The spatial resolution is the clarity with which different areas of the image are distinguished. An image with low spatial resolution will look blurry.

MRI Parameters

MR images are made with RF in the range of approximately 1 to 80 MHz. This technique differs from visible images, in which radiation is reflected from the body, or x-ray images, in which radiation is transmitted through the body. The signals emitted from the body during the MR examination are detected, interpreted, and used to produce an image on a computer monitor. There is no ionizing radiation from MRI examinations. During the MRI examination, the RF pulses are turned on or off. This pulse cycle is repeated in sequence to produce the MRI. The MRI signal is termed a spin echo.

The most common MRIs performed are TI-weighted and T2-weighted images.

T1 and T2 are related to the rotational and tumbling frequency of molecules. Small molecules reconnect more rapidly than larger ones. Smaller molecules are found in liquid and have shorter T1 and longer T2. Larger molecules found in lipids (fat) have longer T1 and shorter T2. T1 weighting is performed to produce images in which the contrast between tissues reflects difference in T1; similarly, T2 weighting will reflect image difference in T2.

T1 and T2 imaging is based on other common MRI parameters: the TR and TE. Both are measured in milliseconds. TR is the time between the applications of each RF pulse. TR will, therefore, determine the amount of T1 relaxation allowed. TE is the time interval between a 90-degree RF pulse and the measurement of the MRI signal and is, therefore, a measure of how much T2 relaxation is allowed before the signal is read. The MRI technologist can manipulate both TR and TE.

Another parameter that influences the amount of signal obtained from body tissue is the number of hydrogen nuclei within a specific volume of tissue or the proton density (also called spin density). MRI signals depend on the presence or absence of hydrogen nuclei and are also sensitive to the environment of the hydrogen nuclei. Tissues containing large numbers of hydrogen nuclei emit high-intensity signals, whereas tissues with less hydrogen emit fewer signals. How the hydrogen is bound within a molecule also determines the strength of the MRI signal. Although bone is visualized on MRI, because bone has little water and fat content and the hydrogen nuclei are bound so rigidly it is harder to produce a detectable signal from small bony structures. MRI is therefore not the best technology to image microcalcifications. Air will look black on MRI because it has no fluid content and has little hydrogen. The hydrogen atoms in the body are contained within different tissue types. In a cyst, the atoms are loosely bound and will relax quickly. Atoms in fat or calcium will hardly move. These different results will produce various contrasts in the image.

MRI is not the best technology to visualize microcalcifications, because like bone, microcalcifications have little water and fat content and the hydrogen nuclei are rigidly bound.

Although fat and water are the two extremes of contrast in MRI, there are actually no absolutes. An image may be heavily weighted for T1 by keeping TR and TE as short as possible, or the image may be heavily weighted for T2 by keeping TE and TR as long as possible, but TR is never long enough nor TE short enough to totally eliminate intermediate weightings, and a wide variety of images will be generated with these intermediate weightings. The actual values for TR and TE will depend on the type of MRI scanner plus other factors, such as the facility's preference (Boxes 7–9 and 7–10).

Slice selection will determine the slab of tissue to be imaged in the body. The gradient coils are used to select the section or slice of body to be imaged. The magnetic field gradient is placed perpendicular to the plane of the slice, and the body is exposed to the increasing or decreasing magnetic field strengths. The slice thickness is determined by the bandwidth of the RF pulse and/or the range gradient magnetic field strength across the FOV. For a thin slice, a narrow bandwidth is selected; for a thick slice, a broad bandwidth is selected.

When a transverse slice is produced, the Z gradient is the slice-selection gradient. The stronger the Z gradient currents, the stronger will be the Z gradient magnetization, and this will result in thinner transverse slices. The X gradient is the horizontal axis across the patient from side to side (right to left). This can be used for slice selections in the sagittal images. The *Y* axis is the vertical axis through the patient. It is used to perform slice selection in the coronal plane (Fig. 7–45). All three gradients are energized simultaneously to obtain an oblique image. In MRI, the operator selects the slice thickness required, and the unit will automatically apply the appropriate bandwidth.

Use of Contrast Agents

Image contrast is described as the brightness between two regions of an image. In MR images, this represents the difference in RF signal amplitude or intensity. To enhance image contrast, most breast imaging is performed with a gadolinium-based contrast. Gadolinium is a

Box 7–9. Summary of Common MRI Parameters

- Fat has a short T1 and T2 time
- Water has a long T1 and T2 time
- On a T1-weighted image, fat will appear bright
- On a T2-weighted image, water will appear bright

Box 7–10. Echo Pulse Sequence

- T1 weighting
  1. Short TE (10–20 ms)
  2. Short TR (300–600 ms)
  3. Scan time 4 to 6 minutes
- T2 weighting
  1. Long TE (more than 80 ms)
  2. Long TR (more than 2000 ms)
  3. Scan time 3 to 6 minutes
- Proton weighting
  1. Short TE (20 ms)
  2. Long TR (more than 2000 ms)
  3. Scan time 3 to 6 minutes

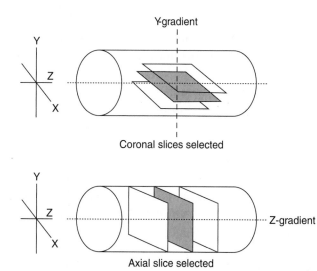

Figure 7–45. Axial and coronal slice selections.

To enhance image contrast, most breast imaging is performed using a gadolinium-based contrast. Gadolinium is a paramagnetic element.

paramagnetic agent designed to enhance the T1 relaxation time and permit the visualization of lesions with shorter TR, thus decreasing scan time. The gadolinium must be bounded or chelated to another compound because gadolinium by itself is extremely toxic. At present, most paramagnetic contrast agents are administered intravenously, although some are administered orally or by inhalation. Gadolinium-chelated contrast is excreted by the kidneys within 3 hours of the injection.

Pulse Sequences and Techniques

Fast spin echo (FSE) is a method used to shorten the conventional spin echo by using multiple 180-degree RF pulses to produce up to 16 spin echoes within a single time to repeat (TR) interval. MRI time is reduced because data are acquired from additional sections while the first section is still proceeding through the pulse sequence. FSE requires hardware that is capable of producing faster RF pulses. FSE produces better spatial resolution in less time, resulting in less motion artifacts. FSE will also reduce artifacts from metal implants.

Fat suppression technique, or fat saturation, is used to distinguish between fat and nonfatty tumors. It is the inversion recovery technique that can be used to suppress the signal from fat. A 180-degree RF inversion pulse is followed by a 90-degree pulse, applied at an inversion time T1 that just matches the time when fat longitudinal magnetization is relaxing. The fat saturation pulse sequence will improve contrast of the remaining tissue.

Gradient-echo pulse sequences were developed as a faster alternative to the standard spin-echo pulse sequence. Gradient-echo sequences permit very rapid scanning, with the data acquisition for a single slice taking place in seconds. The sequence uses only one RF excitation pulse at less than or equal to 90 degrees. The signal of the echo is produced (refocus) by using the gradient magnetic fields rather than 180-degree RF pulses. The gradient pulse is created by reversing the direction of the magnetic field gradient or by applying balanced pulses before and after the RF pulse to cancel out the position-dependent phase shifts that have accumulated because of the direction of the magnetic field gradient. This canceling or rephasing generates an echo that is recorded and measured like another RF signal. Because the RF pulse only slightly decreases the longitudinal magnetization, the recovery time is much quicker. This allows for a shorter TR and, therefore, more rapid imaging.

The image contrast in gradient echo is controlled largely by flip angles. A larger flip angle (>45 degrees) increases T1 weighting, while smaller flip angles (<20 degrees) increase proton density weighting.

Gradient spoiling uses a spoiler gradient pulse at the end of the pulse sequence. The spoiler pulse is applied along the slice-selection direction after data acquisition, to destroy residual transverse magnetization before another RF excitation pulse is transmitted. As a result, images containing minimal T2 effects and maximum T1 weighting are generated. Spoiling techniques are used to show lesions with a high fluid content.

Inversion recovery pulse sequence takes a lot of time and is considered a two-pulse sequence. The net magnetization is inverted by applying a 180-degree RF pulse. The inversion rotates the vector about its base, but as the vector begins to relax another 90-degree pulse is applied. The time between the initial 180-degree RF pulse and the following 90-degree RF pulse is the inversion time. Inversion recovery produces high-contrast imaging but will take a long time unless used with FSE.

Short T1 inversion recovery (STIR) is a version of the inversion recovery pulse sequence that is used to enhance the T1-weighted contrast imaging, especially for tissues with a short T1 relaxation time such as fat. A short inversion time allows the elimination of the high signals from fat on a T1-weighted image, therefore improving the image of true pathology surrounded by fat. STIR is often used with FSE. It uses a T1 that corresponds to the time it takes for fat to recover from full inversion to transverse plane so that there is no corresponding longitudinal magnetization of fat.

Spin-echo pulse sequencing is a popular pulse sequence because it reduces scan time. Spin echo uses a 90-degree pulse followed by one or more 180-degree rephrasing pulses to generate a spin echo. To produce a T1-weighted image, a short TE and a short TR are used, and only one echo is generated. To produce proton-density weighted and T2-weighted images, two rephrasing pulses are used. For proton density, a short TE and long TR are used. Spin-echo pulses are weighted for T2 by using long TR to eliminate the effects of T1 and long TE in order not to eliminate the effects of T2.

Steady state is a condition in which the TR is shorter than the T1 and T2 times of the tissue. The steady state is achieved by repeatedly applying RF pulses at time intervals less than the T1 and T2 times of all tissue. There is, therefore, no time for the transverse magnetization to decay before the pulse sequence is repeated again. There is coexistence of both longitudinal and transverse magnetization during the data acquisition period. Flip angles of approximately 30 to 35 degrees in conjunction with a TR of 20 to 50 ms are used to maintain the steady state. Most gradient-echo sequences use the steady state to achieve the shortest TR and scan time.

Three-dimensional imaging is a method of acquiring images in volume. The entire tissue volume is excited by the RF pulse instead of a single slice. The volume can be divided into slices at any slice thickness and in any plane. This provides high-resolution images of small slice thickness and also allows manual manipulation of images and a three-dimensional object. The technique is invaluable during surgical planning.

Two-dimensional imaging is a method of acquiring images slice by slice. There is minimal time lapse between slices, and by exciting tissue at certain frequencies the slices are created at predetermined locations. Like three-dimensional imaging, two-dimensional imaging is a volumetric method of acquisition that increases scan time but allows extremely good resolution.

Postprocessing Technique

Subtraction

The subtraction postprocessing technique is used to enhance one tissue type while suppressing the background tissue. The final set of pictures

Three-dimensional imaging is a method of acquiring images in volume.

Two-dimensional imaging is a method of acquiring images slice by slice.

could show the postcontrast images subtracted from the precontrast images in order to display regions that are highlighted by the contrast. Subtracted images can be used to highlight a spiculated area showing invasive lobular carcinoma.

Breast Magnetic Resonance Imaging

Breast MRI examinations are best conducted with 1.5-T or higher magnet and dedicated breast coils. Precontrast and postcontrast breast imaging is performed after giving the patient a gadolinium chelated contrast, administered intravenously. There are no moving parts in the MRI bore, and everything is electronically controlled. Various magnetic fields are activated, and both breasts are imaged simultaneously. The way the body's atoms respond to those fields and how they relax when the magnetic field is removed are noted and sent to the computer, along with information about where the interactions occurred. These echoes are continuously measured by the MRI scanner, and numerous points are sampled and fed into the computer. MRI can easily acquire direct images of breasts in almost any orientation without moving the patient. Each MRI sequence is an acquisition of data with specific image orientation and specific type of image appearance.

This information is processed to compose a computer-generated image of the particular slice, representing the different areas of the breast. Each image will show a thin slice of the breast tissue. All the images are then compiled by the computer to allow studying and viewing from different angles.

MR scans can generate 1000 to 2000 images that the radiologist reads and interprets. All data from the scan are stored electronically on magnetic or optical tapes or disks.

Patient Preparation

Patients scheduled for a breast MRI need a through explanation of the examination. The examination can be very traumatic if the patient is not properly prepared.

The patient will lie face down on the table with the breasts suspended in special breast coils. The breast must be properly positioned in the breast coils to achieve an optimal SNR. The patient should be told that the examination can last 35 to 45 minutes and that they will have to remain still throughout the examination to avoid motion artifacts. They also need to be made aware of the fact that they will be lying prone on a table that will slide into the bore of the magnet, and they also must be made aware of the noise during the examination. Some departments give patients earphones with music or audio books to help block the noise and distract them. There are a few patients who find the MRI bore too claustrophobic, and they are unable to undergo this procedure. However, newer MRI units have a much shorter and somewhat wider bore and are much less confining.

The actual examination is totally painless and comprises a series of two to six sequences, with each sequence lasting between 2 and 7 minutes as the radio signals are turned on and off. Typically, the examination can last 35 to 45 minutes.

> The MRI examination is totally painless and comprises a series of two to six sequences, with each sequence lasting between 2 and 7 minutes.
>
> Typically, the examination can last 35 to 45 minutes.

Imaging Standards

Transaxial and sagittal images are most often done. Most tumors demonstrate rapid contrast enhancement within 5 minutes of contrast injection and can, therefore, be differentiated from normal breast parenchyma. Within 10 minutes of contrast injection, tumor enhancement is equal to the breast parenchyma. The breast appearance on the MRI will depend on the amount of parenchymal fatty replacement. Fat has an intense signal on T1-weighted images and moderate signal on T2-weighted images. The central parenchymal tissue is usually gray on both T1- and T2-weighted images. Cysts are identified on T2-weighted images as an area of intense signal. If the tumor were located in fat, a T1-weighted image would provide the contrast to permit its detection. Tumor would be hidden in T2-weighted images because T2 relaxation times of tumor and fat are identical, making them appear the same on the image. T1 relaxation times of a tumor and muscle would produce identical images; therefore, a tumor in muscle would be hidden in a T1-weighted image. RF spoiling techniques can be used to demonstrate a tumor with a high fat content. Fat suppression will enhance lesions on T1-weighted images when the surrounding, normal fat will obscure the lesion.

Most types of MR imaging systems offer a wide variety of imaging options designed to improve image quality, reduce scan time, and effectively differentiate between various tissue types. To produce high-quality MR images, the technologist should move beyond fixed imaging parameters and know how to completely evaluate the MR scan. Every high-quality imaging involves trade-offs among time, SNR, and spatial resolution, and by manipulating these factors the technologist should be able to produce high-quality work without compromising the image quality. It is also important for the technologist to understand the benefits and limitations of the various imaging options.

Imaging Sequence

The standard protocol in breast MR imaging includes fat-saturated T2-weighted FSE images, or sagittal STIR, non–fat-saturated T2 images and T1-weighted images, sagittal STIR images, and fat-saturated spoiled gradient-echo (FSPGR) images. Other imaging can include three-dimensional axial, three-dimensional spoiled gradient, and FSE T1 and T2 imaging of implants. Imaging is always acquired before and post-contrast injection. Dynamic postcontrast imaging is usually performed at multiple times after the intravenous administration of contrast (Figs. 7–46 to 7–49).

All imaging is done bilaterally using a 1-T or higher field strength, dedicated breast coil, a slice thickness no greater than 3 mm, high spatial resolution (1 mm all planes), and high temporal resolution (maximum 5 minutes) (Box 7–11).

MRI Artifacts

Aliasing Artifact

If the part is outside of the field of view but within the RF excitation range, the RF signals cannot be properly interpreted. Signals outside

---

Imaging is almost always acquired before and after contrast injection, and the postcontrast imaging is obtained immediately after the bolus of injection is administered.

Box 7–11. Sample Imaging Sequence

- Before contrast: sagittal short T1 inversion recovery, T2-weighted non–fat saturated, and three-dimensional fat-saturated spoiled gradient-echo

- After contrast: three-dimensional fat-saturated spoiled gradient-echo and three-dimensional axial fat-saturated spoiled gradient-echo (Figs. 7–46 to 7–47)

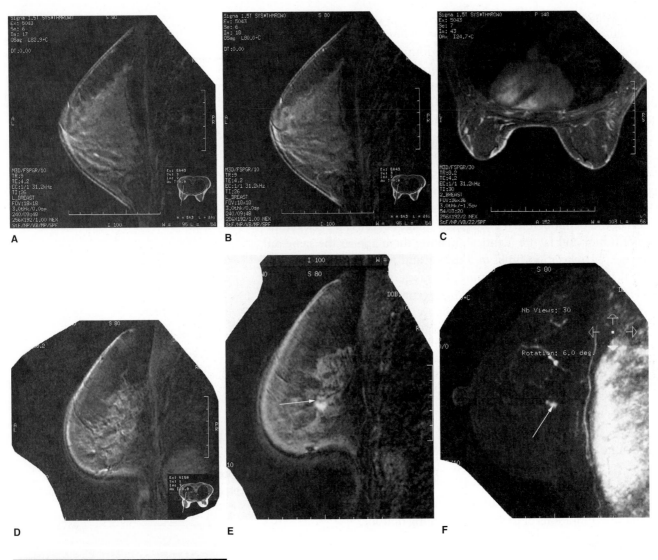

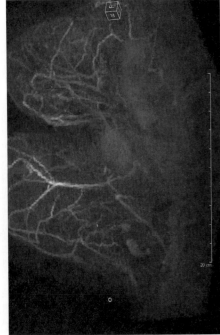

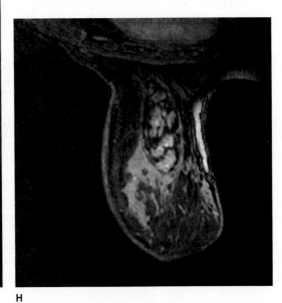

Figure 7–46. Normal and diagnostic MRIs. Normal **(A)** three-dimensional fat-saturated spoiled gradient-echo (FSPGR) precontrast, **(B)** three-dimensional FSPGR postcontrast, and **(C)** three-dimensional axial postcontrast. Diagnostic **(D)** three-dimensional FSPGR precontrast, **(E)** three-dimensional FSPGR postcontrast, and **(F)** three-dimensional rendering of the series of precontrast and postcontrast images. The enhanced areas on images **(E)** and **(F)** are possibly malignant. **(G)** A three-dimensional image of blood vessels to breast **(H)** fibroadenolipoma on MRI.

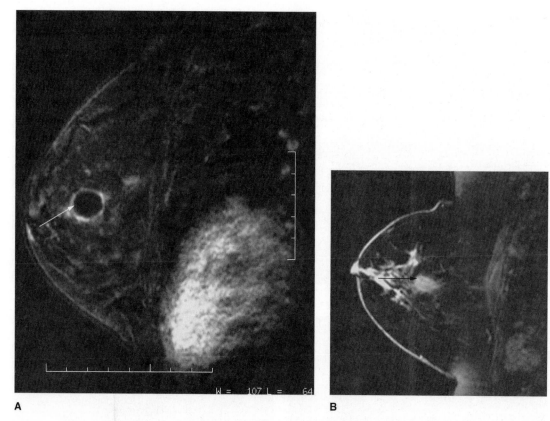

A

B

Figure 7–47. **(A)** Breast cyst on a postcontrast three-dimensional fat-saturated spoiled gradient-echo MRI. **(B)** Cancer on a breast MR image (*arrows*).

the FOV will be projected over the real area of interest but at the opposite end of its true location, resulting in aliasing or *wrap-around* artifacts.

Body Shape Conductivity and Extension

Extension of body parts outside of the coil will result in metallic-like artifacts at the edge of the area of interest. The curvilinear artifact conforms to the shape of the magnetic field at the edges and may have a characteristic pattern for individual magnet system.

Chemical Shift Artifact

This artifact is a result of slight differences between the resonating frequencies (*Larmor frequencies*) of similar protons, for example, fat and water. The hydrogen in water is bound more tightly to oxygen than the hydrogen in fat is bound to carbon. As a result, fat precesses at a slightly lower frequency than water. This produces a bright rim of signal at one interface and a dark rim on the opposite side of the particular organ (e.g., between fat and water in other body tissues). Chemical shift artifact intensifies with increasing magnetic fields but can be reduced or eliminated by increasing the strength of the gradient magnetic fields and reducing the field of view, which produces larger bandwidths.

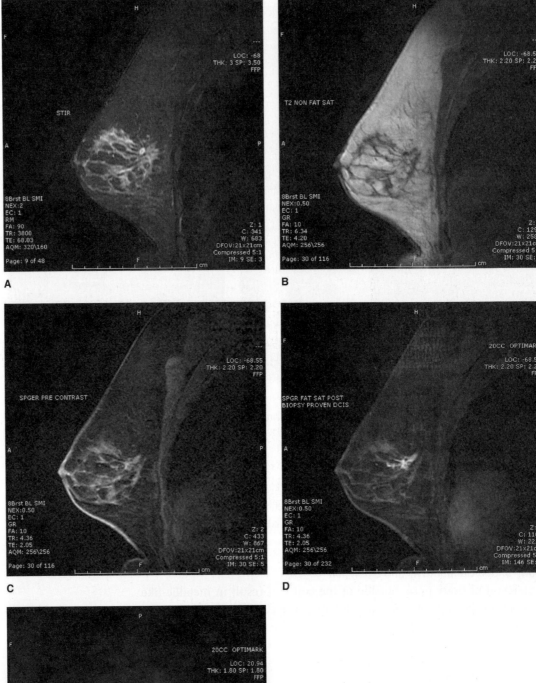

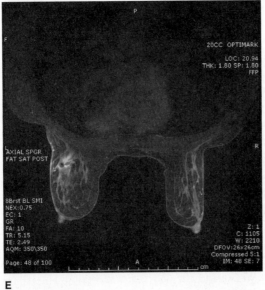

Figure 7–48. MR imaging series showing a large area of enhancement with irregular margins— malignant lesion on left breast. (A) Sagittal short T1 inversion recovery, (B) sagittal T2 weighted non–fat saturation, (C) sagittal vibrant (spoiled gradient-echo) precontrast, (D) sagittal vibrant (spoiled gradient-echo) postcontrast, and (E) axial spoiled gradient-echo postcontrast.

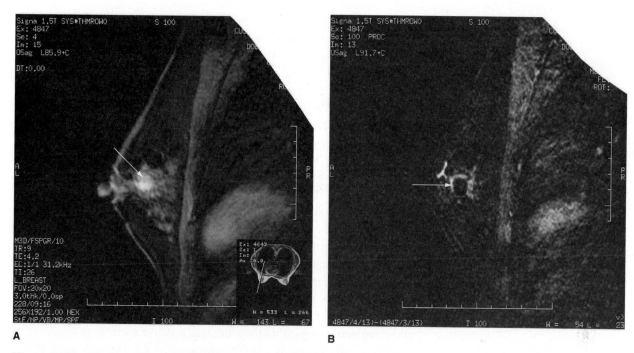

Figure 7–49. (A) Three-dimensional fat-saturated spoiled gradient-echo. **(B)** Subtracted image showing an ill-defined area of enhancement, which was recommended for biopsy (*arrows*). The enhancement characteristic of the breast carcinomas will vary. Often, there is homogeneous enhancement, but occasionally there is only rim enhancement.

Ferromagnetic Material Artifact

Ferromagnetic material will result in partial or complete loss of signal at the site of the metal plus a warping distortion of the surrounding area (Fig. 7–50). Metallic artifacts can be caused by belts, buckles, keys, or even mascara or other makeup and certain types of nylon clothing, such as gym shorts.

Motion Artifacts

Motion artifacts will be caused by movement of the patient, whether voluntary or involuntary. The MR image is degraded, especially if the

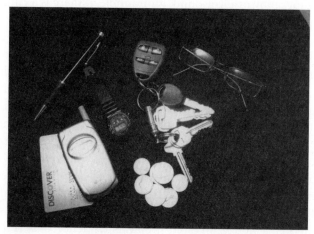

Figure 7–50. Ferromagnetic items to avoid bringing into the MRI suite.

Magnetic resonance imaging characteristics of malignant spiculated or stellate lesions

- Rapid lesion enhancement after intravenous contrast agent administration
- Peripheral enhancement ("ring" enhancement)
- Partially irregular margins

motion occurs during the middle of the acquisition phase. The image will be poorly defined or smeared in appearance. Voluntary motion can be eliminated by immobilization or making the patient as comfortable as possible.

System-Related Artifact

A system-related artifact is any artifact due to the magnetic properties or RF field in the primary coil or each secondary coil.

Advantages and Uses of Breast MRI

MRI was originally used to image joints and soft tissue structures, but since 1991 MRI of the breast has been approved by the FDA as a supplemental tool in the screening of breast cancer. Contrast-enhanced MRI can demonstrate sensitivity to lesions smaller than 1 cm. Breast MRI, in fact, can report sensitivity well into the 90% range in some studies and is proving extremely useful in imaging dense breast tissue and checking silicone leaks. Breast MRI is also considered cost effective when replacing surgical biopsy for false-positive mammograms.

MRI of the breast is now recommended for patients with *BRCA* mutations; patients with a first-degree relative with breast cancer, especially if the cancer developed when the patient was younger than 40 years; patients with a lifetime breast cancer risk of at least 20%; patients who received chest wall radiation between ages 10 and 30 years; and patients with high risk related to Li-Fraumeni, Cowden's, or Bannayan-Riley-Ruvalcaba syndrome.

The American Cancer Society (ACS) recommends that women with a greater than 20% lifetime risk for breast cancer should have a breast MRI and a mammogram every year. For patients with a lifetime risk between 15% and 20%, the decision to have an MRI is made on a case-by-case basis.

A negative MRI will virtually exclude a cancer because MRI shows a sensitivity to breast cancer of close to 96%. MRI can therefore reduce the number of biopsies resulting from false-positive mammograms or ultrasound examinations on women with dense breasts, or can be used just as a screening for high-risk patients who have dense breasts on mammography. MRI can also be used to evaluate lesions not identified on mammography or ultrasound, for example, if there is a clinical finding of a palpable lump that is not identified on mammography or ultrasound.

MRI of the breast can be used to map tumor extent in newly diagnosed breast cancer patients and can detect multifocal or multicentric disease in the ipsilateral breast, breast cancer in the contralateral breast, or lesion invasion to the chest wall. MRI can also detect cancer reoccurrence after lumpectomy or can determine whether the cancer has spread to the lymph nodes or chest wall. By tracking the possible spread of breast cancer, MRI can be used as a staging tool to evaluate treatment options, that is, to determine the feasibility of breast conservation versus mastectomy and the potential need for radiation (see Fig. 7–48), chemotherapy, or chest wall resection. MRI can also determine whether a newly inverted nipple is evidence of a retroareolar cancer.

Because MRI can detect and analyze vascular enhancement patterns of breast tissue and nodules, it can be used to evaluate patients with positive surgical margins for residual cancer either for re-excision of margins or for mastectomy. This makes MRI extremely useful when the results of other modality imaging are indeterminate. Also, malignant lesions have mostly irregular margins, whereas smooth lobulated borders of benign lesions are easily identified on the MRI. MRI can even detect the presence of any suspicious lymph nodes not identified on the physical examination, the mammogram, or the ultrasound.

After radiation therapy, MRI can be used to distinguish postoperative and postirradiation therapy scarring from recurrent cancer. MRI can evaluate the effects of the neoadjuvant chemotherapy response, the extent of residual disease in patients undergoing chemopreventive therapy, and the effectiveness of chemotherapy, all allowing better management of the chemotherapy treatments.

For patients with breast implants, MRI can determine implant rupture or leakage because MRI images are not affected by the presence of silicone or saline implants. *Spaghetti strings* are the typically appearance of a ruptured silicone on the MRI. Even patients with silicone injections directly into the breast can benefit from MRI screening, and MRI will image breast tissue that is compressed by an implant.

> Magnetic resonance imaging characteristic of benign circular or oval lesions
> - No enhancement or slow enhancement after contrast agent administration

MRI of the Augmented Breast

The augmented breast consists mostly of water, fat, and silicone, each of which has distinctive MRI characteristics (Figs. 7–49 and 7–51). Silicone produces only minimal signals because silicone has a unique MRI resonance frequency and long T1 and T2 relaxation times. This allows the easy differentiation between silicone and surrounding fat and water. Fat and silicone have similar resonance frequencies; therefore, fat suppression technique will partially suppress the silicone signals, limiting the technique selection. To produce a silicone-only image, the signals from water protons can be suppressed and a short inversion recovery time used to exploit the difference in T1 relaxation times of fat, water, and silicone. Alternatively, by exploiting the phase-shift differential, precessional frequencies from the signal intensities of fat, water, and silicone will be noted. The use of relatively long TE with FSE images will emphasize T2 weighting and aid in giving a greater contrast between silicone and surrounding breast tissue.

Limits of Breast MRI

Some studies suggest that after the intravenous injection of contrast, benign breast lesions either do not enhance at all or demonstrate slow uptake of contrast. The phenomena led to the suggestion of charting the dynamic measurement of contrast uptake in each lesion and attempting to differentiate malignant from benign based on the rapidity of initial uptake and washout of contrast. Breast cancer does tend to enhance faster and wash out faster than benign breast tissue. But further studies have demonstrated too much overlap in enhancement characteristic of benign and malignant lesions. Generally, peripheral enhancement is seen mostly in malignant lesions, but fat necrosis can also demonstrate this pattern.

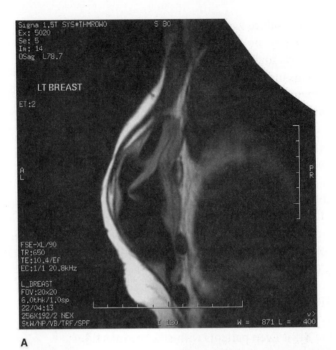

A

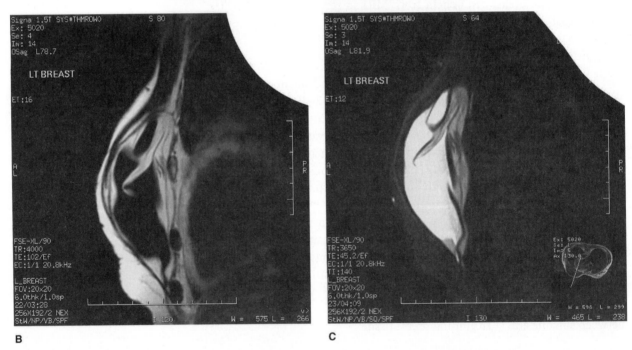

B
C

Figure 7–51. MR images of the augmented breast. T1 weighted **(A)**, T2 weighted **(B)**, and sagittal short T1 inversion recovery **(C)** of the augmented breast demonstrating intracapsular and extracapsular rupture. The fluid is seen posterior to the implant, with smaller amounts anterior and lateral to the implant.

MRI with computer-aided design (CAD) has offered improved interpretation efficiencies, helping radiologists to reduce the time spent with a manual analysis of the thousands of MR images. CAD units featuring motion correction have also greatly improved study analysis.

Specificity of MRI of the breast can also be improved by making all prior images available at the time of the interpretation and scanning the patient at the appropriate time. Premenopausal patients should be scheduled at days 7 to 15 of the menstrual cycle, and postmenopausal patients should discontinue hormone replacement therapy at least 30 days before the MRI. Also, postirradiation patients should wait at least 6 months after treatment before a breast MRI.

Contrast-enhanced breast MRI can demonstrate sensitivity to lesions significantly smaller than 1 cm in the breast; however, the specificity of MRI scans of suspicious breasts is as low as 30% to 70% because MRI cannot always distinguish between a cancerous and a noncancerous abnormality. MRI relies on the inherent vascularity of malignant lesions. The problem here is that benign lesions can also be vascular and will demonstrate enhancement. The result is a high false-positive rate for MRI, causing overdiagnosing and overtreatment, including unnecessary additional biopsy or workup on the patients. Fibroadenomas, recent scars, fibrocystic changes, radial scars, proliferating benign disorders, areas of inflammation, and active glandular tissue can all show enhancement on the MRI.

Because fibroadenomas, proliferating benign disorders, areas of inflammation, and active glandular tissue can all show enhancement, a method used by some radiologists to improve the specificity of the MRI is to measure the rate of contrast uptake during the initial minutes after contrast injection. This specificity increase, however, comes at the expense of spatial resolution and/or coverage of the breast. Also, MRI poorly visualizes the axillary nodes mainly because contrast enhancement occurs for both normal and pathologic nodes. Plus, a contrast injection poses the associated risks, and all breast MRI, except those done for implant rupture, must be done with contrast.

MRI is unable to image microcalcifications clearly, particularly lesions smaller than 5 mm, which are often associated with early-stage ductal carcinoma *in situ* (DCIS). MRI detection of microcalcifications has a sensitivity range of 45% and up. The technology is very expensive, costing sometimes 10 times more than a mammogram and is, therefore, not cost-effective as a screening tool. Also, because of the limitations of the MR suite and the longer examination time, patients with physical disabilities who are unable to lie flat for the breast MRI cannot be scanned. This examination is especially difficult for patients who are claustrophobic.

The size and the weight of the MRI unit make placement and location critical. Another aspect of MRI limitations is related to the dangers of ferromagnetic materials to patients and also to the imaging. Any internal ferromagnetic material can pose the danger of shifting or failing to work if exposed to the MRI magnetic fields. On the image, ferromagnetic elements will cause susceptibility artifacts, resulting in the loss of signal and spatial distortion. This can occur even when the ferromagnetic materials are too small to be seen with the naked eye. Although many devices are now MR compatible, details of the manufacturer and the date of manufacture are important in determining if the internal device is MR compatible. Patients with cardiac pacemakers, defibrillators, neurostimulators, cochlear implants or internal hearing aids, implanted infusion pumps, aneurysm clips in the brain, and metallic orbital foreign bodies should be carefully screened before any MR examination and may not be able to have the MR examination. Other devices such as surgical clips, prostheses, shrapnel, and metal implants should be checked for ferromagnetic properties. Patients should be properly assessed if they have been exposed to metallic flakes or silver as a result of working around metal finishing, welding, or grinding equipment. Patients must also remove coins, watches, keys, bobby pins, credit cards, and pocket knives before entering the MRI suite.

Risks and Potential Complications

During the MRI, the patient is placed within a powerful magnetic field. Patients are required to provide the technologist with a detailed medical history before the MRI examination because any metal objects can potentially be extremely dangerous inside or around the magnet of the MRI unit.

Because the magnet is always on, the major risks of MRI are associated with the magnetic properties of the MRI machine. Even metal zippers or buttons and belt and bra fasteners may need to be removed, and some electrocardiogram leads have been known to cause localized burns during MRI examinations.

MRI has been reported to cause burns associated with the presence of ferromagnetic objects inside or outside the body. Also, a rise in body temperature has been noted in the RF field used by MRI. Although none of these changes has produced any adverse effects, the long-term effects of MRI are still not fully understood, and pregnant women, especially those in early pregnancy, are usually advised to avoid MRI examinations.

Research documents a connection between gadolinium-based MRI contrast agents and nephrogenic systemic fibrosis (NSF). Gadolinium-based MRI agents are safe for most patients; however, studies have shown a causative link between the gadolinium contrast and NSF in patients with compromised kidney function. All patients scheduled to have an MRI should be screened for kidney disease or decreased kidney function. The exact cause of NSF is still under investigation, but the current theory is that gadolinium, which is very toxic, is normally bound (or chelated) to a ligand molecule, which makes it safe for humans. If the renal functions are impaired, the gadolinium remains in the body longer and is more likely to separate from its ligand molecule in a transmetallation process. The result is that the gadolinium ion is distributed to the skin and other tissues, where it activates cells to start the fibrotic process.

Many radiologists have changed MRI protocols, including reducing the dosage of gadolinium-based contrast to avoid the occurrence of NSF.

MRI Quality Assurance

Basic quality assurance procedures for the MRI unit consist of measuring the SNR and other data on phantoms daily or weekly. Weekly logs of RF transmitter voltage should also be maintained. Other testing includes checks of the image uniformity spatial resolution, gradient pulse flip angles, and Larmor frequencies of known nuclei and T1 and T2 relaxation times.

> The major risks of MRI are associated with the magnetic properties of the MRI machine.

Summary

Digital Technology

There are three types of digital technologies: flat-panel technology, photostimulable storage phosphor (PSP) technology, and photon-counting technology. Flat-panel technology (digital mammography, or DM) uses a non-removable detector system, whereas PSP technology (computer mammography, or CM) adapts the analog system with digital technology by integrating a removable detector. DM offers hands-free imaging without the need to carry a cassette. In CM, the removable detector systems work similarly to an analog system, whereby the cassette is replaced by an image plate (IP) and the film is replaced with the image receptor (IR). In CM, the latent image forms on the IR (PSP), which is in a removable detector. The PSP is removed and scanned, and the image is digitized. Photon-counting technology was recently approved in the United States. It detects and captures individual x-ray photons leaving the breast using a crystalline silicon detector and photon-counting electronics. The system has a precollimator to remove scatter not directed to the detector and a postcollimator to remove scatter from the breast.

- In digital technology, the image is formed as a two-dimensional matrix of pixels, in which the greater the number of pixels per inch produces the greater the resolution.

Advantages of Digital Imaging
- Wide exposure latitude
- Optimization of contrast
- Reduced patient dose with best practice
- Improved workflow by reducing repeats
- Reduced cost associated with repeats
- Reduction in lost images
- Every reprint is an original
- Electronic image storage or archival in encoded form to reduce space
- Reduced storage cost and space
- Multiple storage options—short, medium, and long term
- Compression storage

Disadvantages of Digital Imaging
- Loss of experience with manipulating technical factors
- Loss of ability to control patient dose
- De-emphasis on thinking of radiation safety and patient protection
- Risks for data loss due to insufficient backup system or disasters
- Concern about outsourcing of reports
- Increased reliance on postprocessing of image

Ultrasound imaging, using a linear array transducer with frequencies of 10 MHz and higher, is employed as an adjunctive imaging tool to screen dense breast, to determine whether a mass seen on the mammogram is fluid filled or solid, or to assess implants for leaks. Breast ultrasound uses high-frequency sound waves based on the piezoelectric effect.

- Sound is a mechanical longitudinal wave measured in units of hertz (Hz).
- Breast imaging is performed with the patient lying comfortably on a bed. The technology uses no ionizing radiation, and there are very few documented risks or harmful bioeffects. In imaging, because sound cannot travel through a vacuum, a gel must be applied to the skin to act as a conductor.

Magnetic resonance imaging (MRI) of the breast uses complex magnetic properties of elements and the interaction of body tissue with radio waves in a magnetic field. No x-rays are involved. Echoes or signals from the body are continuously measured by the MRI scanner, and a digital computer reconstructs the echoes into images of the breast. MR imaging of the breast is generally performed after the injection of a paramagnetic compound, which includes gadolinium. The contrast helps differentiate malignant lesions, which enhances and washes out rapidly, whereas benign lesions enhance and wash out slowly. For the MRI procedure, the patient lies prone on the table with the breasts falling into specialized breast coils. The table slides into the bore of the magnet, and numerous points are sampled in an examination lasting up to 45 minutes. MRI can be used in the following:

- Map tumor extent—as a staging tool to evaluate treatment options
  - Locate retroareolar cancer
  - Detect multifocal/multicentric diseases
  - Detect recurrence
  - Evaluate dense breasts
  - Evaluate uses of breast MRI
  - Evaluate positive surgical margins for residual cancer
  - Evaluate the effects of chemotherapy response
  - Distinguish postoperative or postirradiation scarring from cancer
  - Evaluate implants
- Most of the risks and complications of MRI are associated with metallic dangers. There is a very low risk for nephrogenic systemic fibrosis (NSF), mainly for patients with renal dysfunction. MRI technology is very expensive. It poorly visualizes the axillary nodes and cannot image calcification.
- Contraindications to MRI are as follows:
  - Cardiac pacemakers
  - Aneurysm clips (intracranial)
  - Intraocular ferrous foreign bodies
  - Pregnant patients should consult their physician before imaging
- Ultrasound and MRI increasingly are being used to image the dense breast of younger women. Unlike mammography, the accuracy of both is independent of the age of the patient. Ultrasound is very reliable in demonstrating a simple cyst, but complex cysts and other breast lesions are often problematic and result in

indeterminate findings. MRI is proving a valuable asset in detecting multifocality of lesions. One major drawback of both these imaging techniques is their inability to detect microcalcifications clearly. For breast ultrasound and MRI to compete with mammography as a successful screening tool, they should be able to visualize all cancers that can be seen mammographically plus offer additional benefits. Ultrasound and MRI can play a valuable role in improving the sensitivity of mammography by reducing the amount of false-negative results. At present, however, neither can stand alone as a screening tool.

REVIEW QUESTIONS

1. The range of values over which a digital system can respond, also known as the gray-scale range, is called what?

2. In digital imaging, which matrix of cells will have less information: a 5 × 5 or a 10 × 10?

3. What is the name of the photoconductor used in direct digital systems?

4. What is the name given to the device that replaces the film in the computed mammography systems?

5. In digital imaging, name the process whereby a phosphor crystal will emit light when exposed to radiation.

6. Name two methods used to display the digital image.

7. What is described as the biggest advantage of digital mammography?

8. On what effect is the basic principle of ultrasound based?

9. Higher frequency ultrasound transducers have less penetration power. What feature makes them ideal for breast imaging?

10. What is the effect of applying too much gain when imaging a simple cyst using ultrasound?

11. Why is it is recommended that a lesion be imaged with and without a caliper in ultrasound?

12. What information can ultrasound give about a cyst?

13. Describe the borders of a benign lesion.

14. On ultrasound, if a lesion is irregularly shaped with spiculated margins, what is it likely to be?

15. What type of lesion tends to show "posterior enhancement" on the ultrasound image?

16. What type of ultrasound lesion generally has spiculated margins?

17. How many protons and electrons does the hydrogen atom have?

18. In MRI, how would you describe the wobble motion similar to the motion of a spinning top just before it falls?

19. In MRI, why is the primary magnet kept super-cooled?

20. Identify the coil closest to the patient during an MRI examination.

21. What action/s create/s the noise heard during the MRI examination?

22. In MRI, describe the relation of T1 to T2 in larger molecules, such as those found in fat.

23. What describes the time interval between a 90-degree pulse and the measurement of the MRI signal?

24. Slice selection is determined by the selection of which coils in MRI?

25. Give the term describing an imaging sequence using one RF excitation pulse at less than or equal to 90 degrees.

26. How soon after injection of contrast are the MR images taken?

27. What type of artifact is created if the part is outside the RF coil during MR scanning?

28. Name two common benign disorders that can mimic malignancy on the MR image.

29. What factor/s is/are associated with the single major risk during an MRI examination?

30. Can a patient with an aneurysm clip have a breast MRI?

31. Give two uses of MRI of the breast.

CHAPTER QUIZ

1. As pixel size decreases, the amount of data contained in the images will:

    (A) Increase

    (B) Remain the same

    (C) Decrease

    (D) Pixel size has no effect on image data

2. In digital imaging, the dynamic range refers to:

    (A) Range of intensities that are displayed

    (B) Random background information that is detected but does not contribute to image quality

    (C) Range of value over which a system can respond

    (D) Layout of the cells in rows and columns

3. The detector system in digital mammography is categorized as direct if:

    (A) The x-rays are absorbed by the detector and an electrical signal is created in one step

    (B) The x-rays are absorbed and converted to light, which is then detected by TFD

    (C) The system has a wide dynamic range

    (D) The spatial frequency is high

4. During the reading process, the light emitted from the PSP is collected by

    (A) TFT

    (B) CCD

    (C) ADC

    (D) PMT

5. Most of the risks of MRI of the breast are associated with:

    (A) The magnetic properties of the patient

    (B) Ferromagnetic metals

    (C) Technologist errors

    (D) Radiologist error

6. What is the name of the scintillator used in the flat-panel technology?

    (A) Amorphous silicon

    (B) Cesium iodide

    (C) Amorphous selenium

    (D) Thin-film transistor

7. In MRI of the breast, a large value for the T1 or T2 will indicate:

    (A) A long, gradual relaxation time

    (B) A rapid relaxation time

    (C) Hydrogen atoms are not aligned to the external magnetic field

    (D) The RF field was not applied

8. Sound waves cannot travel through:

    (A) A medium

    (B) Vacuum

    (C) Gasses

    (D) Liquids

9. Typically, a lesion with no internal echoes, no spiculations, and fewer than four gentle lobulations would represent what type of lesion on ultrasound?

    (A) Malignant

    (B) Intermediate

    (C) Carcinoma

    (D) Benign

10. The ultrasound transducer is used to:

    (1) Convert electric energy to acoustic pulses

    (2) Convert acoustic pulses to electric energy

    (3) Change the speed of the sound waves

    (A) 1 and 2 only

    (B) 2 and 3 only

    (C) 1 and 3 only

    (D) 1, 2, and 3

BIBLIOGRAPHY

Adler A, Carlton R. *Introduction to Radiologic Science and Patient Care.* 6th ed. St. Louis, MO: Elsevier Saunders; 2016.

American Cancer Society. Learn About Cancer. Detailed Guide. http://www.cancer.org/cancer/. Accessed August 2016.

Andolina VF, Lille SL, Willison KM. *Mammographic Imaging: A Practical Guide.* 3rd ed. Philadelphia, PA: Lippincott Williams & Wilkins; 2010.

Balu-Maestro C, Chapellier C, Bleuse A, et al. To compare the value of conventional imaging modalities and MRI for determination of response to neoadjuvant chemotherapy for breast cancer. *Breast Cancer Res Treat.* 2002;72(2):145–152.

Bassett LW. Imaging of breast masses: Breast imaging. *Radiol Clin North Am.* 2000;38(4):669–690.

Broome DR, Girguis MS, Baron PW, et al. Gadodiamide-associated nephrogenic systemic fibrosis: Why radiologists should be concerned. *AJR Am J Roentgenol.* 2007;188:586–592.

Bushong SC. *Radiologic Science for Technologists: Physics, Biology and Protection.* 10th ed. St. Louis, MO: Mosby; 2012.

Digital Mammography: Technology Review. Milwaukee, WI: GE Medical System; 2002.

Digital Mammography. http://www.fda.gov/radiation-emittingproducts/mammographyqualitystandardsactandprogram/facilitycertificationandinspection/ucm114148.htm. Accessed August 2016.

Fundamental of Digital: Review Physics, Technology and Practical Considerations. LORAD Hologic. Danbury, CT: Women's Health; 2002.

Greenstein OS. MR imaging of the breast: Breast imaging. *Radiol Clin North Am.* 2000;38(4):899–1160.

Greenstein C, Manasseh DME. *Breast MRI: A Case-Based Approach.* Philadelphia, PA: Lippincott Williams & Wilkins; 2011.

Haus AG, Yaffe MJ. Screen film and digital mammography. Breast imaging. *Radiol Clin North Am.* 2000;38(4):871–896.

Peart O. *Lange Q & A Mammography Examination.* New York, NY: McGraw-Hill; 2015.

Rotten D, Levaillant JM. The value of ultrasound examination to detect and diagnose breast carcinomas: Analysis of the results obtained in 125 tumors using radiographic and ultrasound mammography. *Ultrasound Obstet Gynecol.* 1992;2(3):203–214.

Seerm E. *Digital Radiography: An Introduction.* Clifton Park, NY: Delmar; 2011.

Sundararajan S, Tohno E, Kamma H, et al. Role of ultrasonography and MRI in detection of wide intraductal components of invasive breast cancer: A prospective study. *Clin Radiol.* 2007;62:252–261.

Tucker AK, Ng YY. *Textbook of Mammography.* 2nd ed. London, UK: Churchill Livingstone; 2001:217–235.

Velex N, Earnest DE, Staren ED. Diagnostic and interventional ultrasound for breast disease. *J Surg.* 2000;180(4):284–287.

Venes D, Biderman A, Adler E. *Taber's Cyclopedic Medical Dictionary.* 22nd ed. Philadelphia, PA: F. A. Davis; 2013.

Vyborny CJ, Giger ML, Nishikawa RM. Computer-aided detection and diagnosis of breast cancer breast imaging. *Radiol Clin North Am.* 2000;38(4):725–740.

Warner E, Plewes DB, Shumak RS, et al. Comparison of breast magnetic resonance imaging, mammography, and ultrasound for surveillance of women at high risk for hereditary breast cancer. *J Clin Oncol.* 2001;19(15):3524–3531.

Woo J. A short history of the development of ultrasound in obstetrics and gynecology. http://www.ob-ultra-sound.net/history.html. Accessed February 2011.

Yun-Chung C, Shih-Cheh C, Mi-Ying S, et al. Monitoring the size and response of locally advanced breast cancers to neoadjuavant chemotherapy with serial enhanced MRI. *Breast Cancer Res Treat.* 2003;78(1):51–58.

Breast Imaging—Adjunctive Modalities | 8

Keywords and Phrases
Diagnostic Options
Digital Mammography Technologies
   Digital Breast Tomosynthesis or Three-Dimensional
     Mammography
   Digital Subtraction Mammogram
   Dual-Energy Mammography
Ultrasound Technologies
   Automated Ultrasound and Three-Dimensional
     Breast Ultrasound
   Full-Field Breast Ultrasound
   Elastography
   Ultrasound Imaging Using Capacitive Microfabricated
     Ultrasonic Transducers
   Contrast-Enhanced Ultrasound
Magnetic Resonance Imaging Technologies
   Magnetic Resonance Spectroscopy
Molecular Imaging Technologies
   Nuclear Medicine
   Fluorodeoxyglucose–Positron Emission Tomography
   Positron Emission Mammography
   Radionuclide Imaging with Single-Photon Emission
     Computed Tomography
   Breast Scintigraphy, Scintimammography or Breast
     Molecular Imaging
   Breast-Specific Gamma Imaging
   Lymphoscintigraphy—Sentinel Node Mapping
Radiation Dose Related to Nuclear Imaging
Other Imaging Modalities
   Optical Imaging
   Cone-Beam Computed Tomography
   Computed Tomographic Laser Mammography
   Transscan or T Scan

Microwave Imaging Spectroscopy
Neutron-Simulated Emission Computed Tomography
Summary
Review Questions
Chapter Quiz

Objectives

On completing this chapter, the reader will be able to:

1. Identify the main difference between digital mammography and digital tomosynthesis
2. Understand the concept of three-dimensional ultrasound
3. Discuss similarities and differences between magnetic resonance imaging and magnetic resonance spectroscopy
4. Differentiate between anatomy imaging technologies and molecular imaging technologies
5. Describe how positron emission mammography differs from positron emission tomography
6. Describe breast-specific gamma imaging and how it differs from scintimammography
7. Explain the advantage of the dual-head gamma system used in breast molecular imaging
8. Describe lymphoscintigraphy
9. Describe the following emerging technologies and state their main advantages and disadvantages in breast cancer screening: computed tomographic laser imaging, optical imaging, transscan, and elastography

KEYWORDS AND PHRASES

- **Angiogenesis** is the process of forming new blood vessels and is a common property of a malignant tumor.
- **Breast-specific gamma imaging (BSGI)** is a functional imaging study of the breast. A radioactive pharmaceutical that will concentrate in malignant lesions is injected into an arm vein, and the patient is scanned using a high-resolution, small-field-of-view gamma camera.
- **Color Doppler ultrasound** is the shift in frequency when ultrasound is used to visualize a part in motion. Doppler can be used to measure the velocity of the blood flowing in a vessel.
- **Contrast enhanced mammography** is based on the association of cancer with increased vascularity. During the procedure, iodinated contrast agents are injected into an arm vein.
- **Contrast-enhanced ultrasound** injects microbubbles, gas-filled spheres, into the vascular system. The technology can be used to measure blood flow to structures and organs.
- **Digital breast tomosynthesis (DBT)** or **three-dimensional (3D) mammography** creates a 3D reconstruction of the tomographic planes of the breast after a sequence of projection images are acquired.
- **Digital subtraction mammogram** exploits the angiogenesis present in breast cancers. After a digital imaging of the breast, the patient is

given an intravenous injection of contrast. The postcontrast image is digitally subtracted from the precontrast.

- **Dysesthesia** refers to an abnormal or unpleasant sense of touch and can involve sensations such as numbness, tingling, burning, or pins and needles.
- **Dual-energy mammography**, also called contrast media mammography (CMM), uses a contrast agent that is injected immediately before the imaging. Two exposures of the breast are made at different energies, one between 20 and 30 kV and the other between 40 and 80 kV. The tissue characterization of the breast is obtained from a weighted subtraction of one image from another.
- **Half-life** refers to the time taken for a quantity of radioactivity to be reduced to half of its original value. The half-life of a radioisotope can range from minutes to hours. Fluorodeoxyglucose (FDG) has a relatively long half-life of 2 hours.
- **Hypesthesia** is a condition that involves an abnormal increase in sensitivity to stimuli of the senses, such as sounds, tastes, textures, and touch.
- **Lymphoscintigraphy,** or sentinel node mapping, also uses a radiopharmaceutical. The tracer is injected into subareaola lymphatic plexus and leaves the breast through the lymph nodes. The procedure is based on identifying the sentinel node or first node in the chain of nodes draining the breast. This node is identified use a Geiger counter. The absence of cancer in the sentinel node is thought to be a strong indication of the absence of metastatic disease.
- **Optical imaging** uses infrared light to penetrate through the tissue. Extra blood vessels that accompany a tumor will be highlighted by the pattern of light absorption because the presence of a tumor will distort light in a characteristic way.
- **Parallax** refers to the apparent displacement of an observed object when it is imaged from two or more different points. The technology works on the principle that most cancerous tissues use vast amounts of sugar. The technology uses fluorine-18 fluorodeoxyglucose (^{18}F)FDG, which is a short-lived radioactive isotope. A special gamma camera detects radiation emitted from radioactive isotopes injected into the vein. Because the isotope is carried in the blood and will attach to glucose molecules of breast cancer, the metabolic activity and therefore the incidence of cancer can be assessed.
- **Radiopharmaceuticals** are radioactive drugs, sometimes referred to as radioactive tracers or radioactive iodine, used in nuclear imaging studies for diagnostic or therapeutic purposed.
- **Scintillation** phosphors are inorganic crystals that respond to ionization radiation by scintillation. They include thallium-activated sodium iodide (NaI: T1) and thallium-activated cesium iodide (CsI: T1).
- **Scintigraphy** is the injection of a radioactive isotope into the body. The detection of the isotope will reveal body functions and diseases.
- **Scintimammography** or scintigraphy is a functional or molecular imaging procedure. The procedure is sometimes called **breast molecular imaging (BMI).** During the procedure, 20 to 25 mCi of the radiopharmaceutical technetium-99m (^{99m}Tc)-sestamibi is injected into the contralateral arm vein of the patient. The tracer accumulates in malignant lesions and can be imaged with a gamma camera.

- **Sensitivity** measures the ability to respond to or register small changes or differences. An imaging system that is highly sensitive will detect all changes in the breast, whether normal or abnormal.
- **Seroma** refers to fluid buildup in tissues.
- **Specificity** is the quality of being precise rather than general. An imaging system that is highly specific can effectively differentiate between normal and abnormal changes within the breast.

DIAGNOSTIC OPTIONS

Although recent studies have questioned the effectiveness of mammography in screening younger women, in the fight against breast cancer the mammogram remains the number one screening tool. For older women, there is no question of the effectiveness of a mammogram. Mammography in general demonstrates increased sensitivity and specificity in women older than 50 years but has a lower sensitivity among younger women. This lack of sensitivity is driving the need for a more perfect imaging system or better detection methods. Today, there are numerous imaging modalities, and researchers are constantly seeking to improve on the available technology. They cannot yet replace the mammogram, but each new modality can be used to provide important diagnostic information and can, therefore, be considered an important adjunctive tool that should be used in conjunction with mammography.

DIGITAL MAMMOGRAPHY TECHNOLOGIES

Within a few short years, digital mammography has come to dominate the field of breast imaging. The technology offers significant advantages over the analog system of imaging, yet despite its numerous advantages, the final image is still in two-dimensional form in which overlapping tissue can mask important details. This means that the inherent disadvantages of mammography have not disappeared. In mammography a cancer is visualized as a white area within the background density of the breast. The sensitivity of the system will, therefore, depend on breast density, patient age, and hormone status of the patient. Mammography also tends to understate the multifocality of a lesion, with inadequate compression and poor positioning affecting interpretation. Two-dimensional (2D) imaging is therefore not 100% effective as an imaging tool. It is limited by tissue superimposition. Overlapping tissues can mask tumors or mimic tumors, leading to false-negative or false-positive reports. To minimize the disadvantage of 2D imaging, the concept of three-dimensional (3D) imaging, or digital breast tomosynthesis (DBT), was incorporated in breast screening.

Digital Breast Tomosynthesis or Three-Dimensional Mammography

Digital tomosynthesis is a modification of digital imaging, and although it has been around for more than 30 years, it is now combined with digital mammography. During imaging, the breast is compressed and positioned for any of the standard projections. The x-ray tube rotates

> Digital tomosynthesis offers 3D reconstruction of tomographic planes of the breast.

around the patient's breast and takes a sequence of projection images. The technology then creates a 3D reconstruction of the tomographic planes of the breast after exposure. The result is a 3D reconstruction of the breast from the multiple images. Tomosynthesis takes advantage of an effect called parallax in which an object that is closer appears to move a greater distance against the background than an object that is farther away when viewed from various angles. The parallax principle allows the visualization of a lesion by blurring out planes above and below the lesion. This allows structures located at different depths to project at different locations, removing the major disadvantage of 2D imaging systems. DBT removes tissue superimposition that hides or mimics pathologies in 2D images, in addition to improving visibility of lesions in the breast by reducing tissue superimposition. Other advantages to the patient include a reduction in the recalls needed to work up suspicious lesions and an increase in the breast cancer detection rate; after all, without tissue overlap, benign and malignant lesions are seen more clearly (Figs. 8–1 and 8–2).

To date, the US Food and Drug Administration (FDA) has approved, cleared, or accepted the following full-field digital mammography and DBT units for use in mammography facilities in the United States, as indicated by date. In addition, the Philips photon-counting 3D system is pending FDA approval.

- GE Senographe Pristina with Digital Breast Tomosynthesis (DBT) Option on March 3, 2017
- Fujifilm ASPIRE Cristalle with Digital Breast Tomosynthesis (DBT) Option on January 10, 2017
- Siemens Mammomat Inspiration with Tomosynthesis Option (DBT) system approved on April 21, 2015
- GE SenoClaire Digital Breast Tomosynthesis (DBT) system approved on August 26, 2014
- Hologic Selenia Dimensions Digital Breast Tomosynthesis (DBT) system approved on February 11, 2011.

Although the FDA approved the units, the FDA does not accredit 3D tomography because no standards have been developed. The FDA instead has developed a certificate extension program to cover situations in which a mammography device has been approved for marketing, but in which accreditation standards have not yet been established. The certificate extension program allows facilities to legally use 3D tomography. It does, however, specify that the facility must have an FDA-approved accreditation body to accredit the 2D imaging component of the 3D tomography unit. The 2D portion of DBT must be accredited with the accreditation body.

Hologic Digital Tomosynthesis

The Hologic 3D mammography unit was the first to achieve FDA approval. This unit has the x-ray tube moving in an arc across the breast during a DBT exposure. The tube takes 15

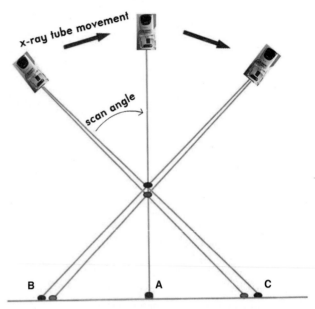

Figure 8–1. Digital breast tomosynthesis. **(A)** In two-dimensional mammography, the image generated is a summation of objects above the detector. **(B & C)** With tomosynthesis imaging, images acquired from different angles separate structures at differing heights. (Used with permission from Hologic Inc.)

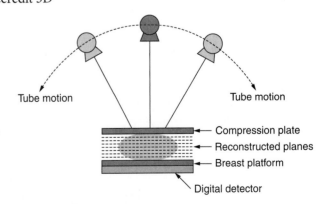

Figure 8–2. Tomosynthesis acquisition geometry showing the direction of motion of the x-ray source and the orientation of the reconstructed planes. (Used with permission from Hologic Inc.)

low-dose images in four seconds. The x-ray beam is continuously "on," and the images are acquired from different angles because the tube sweeps from −7 to 0 to +7. The patient does not move during exposure, and the projection images are reconstructed into 1-mm slices. The number of reconstructed images is based on the breast thickness in millimeters. The first slice is closest to the detector, whereas the highest slice number is closest to the compression paddle.

Reconstruction is always done in 1-mm-thick slices so that a 4-cm-thick breast translates to 40 mm. The system adds 6 mm to ensure inclusion of all breast tissue, which equals 46 slices. A 5-cm-thick breast coverts to 50 mm plus 6 mm—therefore 56 slices. To visualize calcifications, slices can be slabbed for greater thickness.

To aid in reduced radiation dose to the breast, the Hologic DBT does not use a grid. The system offers four imaging options.

1. Normal 2D imaging
2. 3D imaging (tomosynthesis)
3. COMBO imaging, which takes a 2D and 3D exposure with one compression. The 3D image is taken first, immediately followed by the 2D image. The patient's breast remains compressed during the two-image sequence.
4. C-view imaging, which takes a 3D exposure and reconstructs a 2D image without the added radiation exposure. The result is a 3D image plus a reconstructed image of the same projection.

Motion in the Hologic 3D imaging can be problematic when compared to 2D imaging, because of the longer acquisition time and multiple images acquired during the exposure. The acquisition time for the COMBO imaging is 12 seconds versus 4 seconds for 2D imaging. Motion can occur at one point, at multiple points, or throughout the duration of the projection series. Motion can also manifest at different areas of the breast, which may or may not impact breast tissue and detail sharpness.

In the Hologic system, radiologists do not routinely review the projection dataset, where motion can be easily confirmed or ruled out. The projection dataset is available only at the technologist acquisition workstation. It therefore becomes the responsibility of the technologist to repeat images when motion is identified. To appreciate motion, the technologist must carefully review the 15 projection images in cine mode before sending them to the radiologist. Hologic suggests that the technologist should check the following:

- Nonlinear movement of calcification—calcifications should move in a straight line parallel to the chest wall. Any anterior-to-posterior movement would then suggest motion
- Objects or lesions looking sharper in one projection, but not the opposing projection
- Any anterior/posterior movement of breast lesions, calcifications, or objects

The technologist should focus on the axilla and inframammary fold. In the axilla, checks should be made for anterior-to-posterior movement of the pectoral muscle, for lymph nodes that are shifting back and forth or out of view, and for structures that shift in and out of view (any anterior-to-posterior movement). At the inframammary fold the

Radiologists do not routinely review the projection dataset, where motion can be easily confirmed or ruled out.

The projection dataset is available only at the AWS. It therefore becomes the responsibility of the technologist to repeat images when motion is identified.

Any anterior-to-posterior movement of the breast images or objects within the breast will indicate motion.

technologist can check for abdomen movement and check for structures that shift in and out of view.

During the projection series of exposures, the x-ray tube moves in a path parallel to the chest wall. The resulting breast images and objects should also move smoothly along this same pathway in a medial-to-lateral or lateral-to-medial direction; therefore, any anterior-to-posterior movement of the breast images or objects within the breast will indicate motion.

Breathing Technique

Hologic has specific recommendations for minimizing patient motion during the COMBO imaging sequence. These suggestion are not mandatory, nor will they be effective for every facility.

- Engage the patient in the technique by informing them about the 3D and 2D technology.
- Describe the C-arm movement.
- Explain that motion can affect the image.
- Instruct the patient in the breathing technique. Explain that "*stop breathing*" means just that. The patient *should not* take in a breath and hold it because if he or she is unable to breath-hold for the length of the exposure, a deep breath will result in motion.
  - Breathing Technique Steps
    - Depress the exposure controls
    - While the x-ray tube is moving into position to start the tomosynthesis:
      - Instruct patient to *stop* breathing for the 3D acquisition.
      - At the conclusion of the tomosynthesis sweep, instruct the patient to breathe.
      - As the tube moves to center, listen for the completion of the grid movement. Then, instruct the patient to *stop* breathing for the 2D acquisition.

Hologic Tomosynthesis With Generated Two-Dimensional Images: The C-View

The C-view offers a reconstructed 2D image of the breast without additional radiation. Referred to as *s2D imaging*, these images are reconstructed 2D images generated from 3D DBT data set. The process is to perform the standard DBT of both breasts (craniocaudal [CC] and mediolateral oblique [MLO] projections). The computer then generates a synthesized 2D image for each projection. This process offers a significant radiation dose reduction for patients, as shown in the table here:

> Hologic systems offer a C-view reconstructed 2D image of the breast without additional radiation.

Imaging Options	Radiation Dose
2D	1.2 mGy
3D	1.45 mGy
COMBO (2D + 3D)	2.65 mGy

This is important because with a large or dense breast, the American College of Radiology (ACR) recommended dose of 3 mGy for routine imaging could be easily exceeded.

Drawbacks of Hologic Digital Breast Tomosynthesis

A major disadvantage of the Hologic DBT system is that motion is appreciated best at the technologist's workstation, which holds the projection images. Reconstructed images sent to the radiologist will not easily depict motion—and the projection images are not available at the radiologist's workstation. Because motion is incorporated into the reconstruction, it is harder to confirm it on a reconstructed image.

Other disadvantages are as follows:

- Imaging implants in 3D gives patients more radiation because of the lower kVp used.
- The large number of images produced requires longer reading times.
- Calcifications can be degraded with thinner slices, although slabbing will help.
- DBT is not possible for certain specialized projections, such as from below (FB) magnifications, and if the breast is more than 24.5 cm thick.

General Electric SenoCare

In 2015, the FDA approved General Electric (GE) SenoCare—a DBT unit that makes nine exposures to acquire the 3D MLO projection. The unit was approved for a routine sequence of CC and MLO projections of both breasts, plus 3D imaging of both breasts in the MLO projections *only*. This imaging sequence emits a radiation dose similar to the 2D mode.

The unit uses a 25-degree scan angle with a step-and-shoot system that is marketed to eliminate focal spot and motion blur because the tube completely stops for each exposure. The x-ray beam is pulsed while the gantry moves, which results in a 10-second exposure time for the DBT imaging. Although there is a greater chance of motion with the longer exposure time, it is cancelled by the step-and-shoot pulsed imaging system and the unit's lack of continuous motion.

SenoCare also includes GE's adaptive statistical iterative reconstruction (ASiR) dose-reduction technology, marketed to allow reconstructed images with lower dose to the patient and better visualization of microcalcifications. An additional feature of the SenoCare system is dual-track tubes, which uses molybdenum and rhodium that optimize exposures based on breast density and compressed breast thickness. The device's automatic optimization of parameters (AOP) technology identifies the densest breast regions and selects the appropriate anode, filter, kV, and mAs to allow for consistent image quality at the most optimal radiation dose.

> **FDA-approved imaging sequence for GE SenoCare**
>
> 1. 2D routing sequence of CC and MLO projections
> 2. 3D MLO projections only.

GE SenoCare Stats	
Image size	24 cm × 30 cm
Angular range	±12.5
Target/filter	Mo/Mo, Mo/Rh, & Rh/Rh
Filter material/thickness	Mo: 0.03 mm
	Rh: 0.025 mm
kVp	Mo/Mo: 24–30
	Mo/Rh: 26–32
	Rh/Rh: 26–40

The GE DBT units offer a "V-preview" option. Similar to the "C-view" of the Hologic, acquiring the V-preview does not result in additional radiation to the patient. After taking the 3D projection, the V-preview 2D image is generated from the raw DBT projection dataset.

A disadvantage of the GE system is that the face shield moves during exposure, which means the patient cannot rest her face against the shield during the DBT exposure, especially for the CC projection. The face shield is used during regular 2D imaging to prevent the patient's face from projecting in the exposure field. Another disadvantage is the recommendation of 3D in the MLO only. Although 3D is possible in the CC projections—and is often used in the routine sequence of imaging—its use runs counter to FDA approval and will result in more radiation to the patient.

Siemens Mammomat Inspiration

The FDA approved the Siemens DBT unit in April 2015. The unit, Mammomat Inspiration, offers the largest angular range: ±50 degrees, with the highest number of projection images. This unit uses 25 projection images for 3D reconstruction. Siemens markets its unit as offering a 50% dose reduction by using a tungsten/rhodium anode target/filtration system. It also offers high detective quantum efficiency using five different dose levels: Mo/Mo, Mo/Rh, or W/Rh with direct-to-digital amorphous selenium (aSe) capture. The unit also has a calming MoodLight—an integrated LED panel with a broad range of colors to choose from. Technologists can change the colors of the unit with the season—for instance, using pink each October to celebrate Breast Cancer Awareness Month. This soothing light can provide a relaxing examination.

Like the Hologic unit, the Siemens unit includes a sliding grid that is removed during the DBT exposure. This removal eliminates an extra radiation-absorbing material from between the breast and detector, offering 100% of the primary radiation beam to the breast and consequently using less radiation. The other feature of the Siemens unit is called Progressive Reconstruction Intelligently Minimizing Exposure—"PRIME"—which allows the operator to use less dose without compromising image quality. Siemens units offer contrast enhancement and an individualized compression feature, which compresses only as long as the woman's breast is soft and pliable. The compression will also stop when it reaches optimal compression for the best image quality. A disadvantage of the Siemens is the longer exposure time associated with the angular range.

Fujifilm ASPIRE Cristalle with Digital Breast Tomosynthesis

Fujifilm ASPIRE Cristalle recently achieved FDA approval in the United States. It has been available in Europe as AMULET Innovality. The system offers two tomosynthesis modes. Standard mode (ST) has an angular range of ±7 degrees (15 degrees total) and high resolution (HR) has an angular range of ±20 degrees (40 degrees total). HR mode image can visualize spiculation and marginal structure of tumor clearly due to higher depth resolution and higher plane resolution. Images are acquired using Hexagonal Close Pattern (HCP) Direct Conversion Technology. This technology uses hexagonal-shaped pixels in the detector, designed to obtain images with maximum image quality at reduced radiation dose. Fujifilm markets this technology as offering a 50 μm output and improved detector sensitivity when compared to conventional square pixel FFDM detectors (Fig. 8-3). This in turn allows reduced exposure

FujiFilm Aspire Stats	
Target/filter:	W/Rh and W/Al
kVp range:	22–49
Tube current:	max 200 mA
mAs range:	Large focus: 2–600
Detector:	aSe + TFT (hexagon pixels)
Detector size:	24 × 30 cm
Output pixel size:	50 μm
Exposure time:	ST = less than 4 seconds; HR = less than 10 seconds
X-ray tube motion:	Continuous

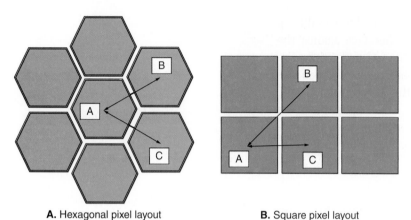

A. Hexagonal pixel layout   **B.** Square pixel layout

Figure 8–3. Hexagonal close pattern (HCP) direct conversion technology. **(A)** Hexagonal pixel layout. **(B)** Square pixel layout.

dose without degrading the image quality. The product is marketed as offering 20% reduction in dose when compared to conventional detectors that use square pixels. Another feature of the unit is the Intelligent AEC (iAEC). This considers the breast composition and uses the data to calculate the optimal technical factors for each examination. Another unique feature of the unit is the Ompression plate compression that offers not only back-to-front flexion as in other units, but also side-to-side flexion of the compression paddle.

The acquisition time is less than 4 seconds in ST mode and less than 10 seconds in HR mode. The acquired images are reconstructed into a series of high-resolution 1-mm slices that can be displayed individually or dynamically in a cine mode. ST mode offers less depth resolution when compared with an HR mode image. However, the depth directional structure of lesions such as microcalcification is superior in ST mode.

The system also offers two types of processing. Pattern 1 enhances spiculations and calcifications while keeping maximum contrast for the viewing of masses within the glandular tissue. Pattern 2 maximizes the visualization of fine spiculations and calcifications.

The system uses tungsten target with either rhodium or aluminum filters and uses a continuous tube motion to complete the exposure.

Philips MicroDose Tomosynthesis

Although it is available in Europe, the MicroDose unit has not yet been approved for use in the United States. The x-ray tube has the same multislit precollimator and postcollimator features as the regular 2D unit. The image receptor rotates around the detector in one arc using a tomographic angle of 11 degrees, and the unit uses a continuous slit-scanning imaging method.

This system offers the lowest patient dose of all the DBT units. The higher kVp used in imaging can be associated with lower contrast images.

MicroDose Stats	
Image size	21 lines × 24 cm
Target/filter	W/Al
Filter material/thickness	0.5-mm Al
Tube voltage	26–38

Important Consideration

Positioning in DBT imaging is just as important as in digital or analog imaging, and DBT can only legally be performed at accredited and certified facilities in the United States. In any tomosynthesis examination, gross or subtle motion unsharpness must be minimized. Repeats due to motion will increase radiation dose to the patient. Some of the contributing factors include inadequate compression, movement of the patient during the exposure, and poor positioning. If the patient is uncomfortable, she will make subtle adjustments, which can contribute to motion.

Although regular breathing is not generally a problem during the exposure, deeply breathing can result in motion on the final image.

Specific Training

The Mammography Quality Standards Act (MQSA) has set rules on the equipment used for breast imaging, the technologist performing the examination, the radiologist reading the examination, and the physicist inspecting the equipment. Although there are technological differences between these DBT systems and differences in their FDA-approved indications for use (IFU), all DBT system are currently treated as a single mammographic modality under the MQSA definition. This means that under MQSA rules, in addition to any state licensing or training, all personnel are required to complete an initial 8 hours of training before independently using any new mammographic modality such as DBT. New mammographic modality training does not need to be in the form of CE credits and can be obtained through a variety of formats, including, but not limited to, professional training by the manufacturer's application specialist, seminars, online or self-study, or continuing medical education. Training can also be provided by another technologist or medical physicist who has been trained in DBT and has experience with the modality.

Documentation of training can be an attestation letter, a certificate from a lecture, a continuing education seminar or the manufacturer.

This rule also applies to interpreting physicians who are qualified to interpret DBT mammograms and physicists who are qualified to perform equipment evaluations or surveys of DBT mammography units. Technologists must also perform regular manufacturer-recommended quality assurance tests.

Personnel must also be active in the field (with yearly documentation of a specific number of mammograms performed or interpreted and the number of equipment units inspected). All personnel are also required to document regular continuing education activities. The FDA requires each facility to submit exact documentation on personnel and equipment before the unit can be used for screening or diagnostic mammograms. The physicist must also complete a detailed report on the mammography unit, called the Mammography Equipment Evaluation (MEE).

Medicare Reimbursement

Although there is still no comprehensive payment for DBT under the Medicare model, the agency began reimbursing for DBT beginning in 2015. Medicare DBT documentation must be included in the report with either a screening or diagnostic digital mammogram because the Centers for Medicare and Medicaid Services have deemed the codes as *add-ons*. Consequently, there is no way to receive reimbursement from Medicare for a standalone DBT examination that is performed separately from a screening or diagnostic 2D mammogram.

Potential Uses

Studies have shown that DBT can be extremely effective for imaging women with dense breasts. The technology also reduces the callback rate and is ideal for diagnostic workup or problem solving. Routine use

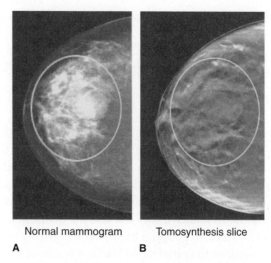

Normal mammogram   Tomosynthesis slice
A   **B**

Figure 8–4. A comparison of normal mammogram **(A)** and tomosynthesis slice **(B)**. (Used with permission from Hologic Inc.)

of DBT results in an 8.7 to 30% reduction in callbacks at some facilities, in addition to reducing false-positive rates by 13%.

Digital tomosynthesis has been approved for use as a diagnostic tool in the United States, Europe, Canada, and Asia and does not replace the 2D mammography.

Currently, tomosynthesis machines are similar to digital mammography systems, and patient positioning for both the CC and the MLO projections is similar to the mammogram. In one system, a COMBO-mode technology allows acquisition of the traditional 2D digital mammogram and a 3D breast tomosynthesis in the same compression. This enables correlation between the two imaging systems. DBT removes superimposed tissue, especially in dense fibroglandular breast, which will reduce the callback rate, identify true versus pseudo lesions, visualize clear mass margins, and allow better edge definition of lesions (Fig. 8–4).

Disadvantages

Manufacturers are working hard to reduce the radiation dose for DBT. Currently, all units operate under ACR guidelines, with the average dose less than 3 mGy per projection. Still, the dose can record higher than the average 2D imaging dose in some systems, especially when imaging dense or larger breast.

Digital tomosynthesis requires a tremendous amount of data storage, more than 2.5 to 4 GB per breast. A typical digital mammogram uses 80 to 100 MB per study. Many facilities must ensure that they have the supporting transmission speeds and the necessary data storage capacity before adopting DBT.

Another notable drawback to the procedure is the large number of images produced, which extends the radiologist reading time. Although computed-aided detection (CAD) is available—digital technology that combines with computers to preread mammograms—the significant increase in the number of images per breast mandates an increase in reading time; this affects the overall time of a screening or diagnostic mammogram. With CAD, the computer displays suspicious areas—acting as a second reader.

The technology is less effective when imaging fatty breast; there is less tissue overlap. DBT can be difficult—or even impossible—to achieve when imaging large breasts. If a breast has to be sectioned for imaging, this can increase a routine four series of images to eight or more.

Reports also indicate that without proper immobilization the system produces motion artifacts, and some systems have shown difficulties in delineating the morphology of microcalcification. However, taut compression, slabbing in some cases (displaying tomo data in slices larger than 1 mm), and clear communication of breathing instructions to the patient could eliminate this problem.

The final disadvantage that is unique to the Hologic system is that motion must be identified and resolved at the technologist's workstation.

> **DBT Units: Target/Filtration Combinations**
>
> **Digital Mode**
>
> Tungsten target with rhodium filtration (fatty breast)
>
> Tungsten target with silver filtration (dense breast)
>
> **Tomography Mode**
>
> Tungsten target with aluminum filtration

Digital Subtraction Mammogram

The digital subtraction mammogram exploits the angiogenesis present in breast cancers. After digital imaging of the breast, the patient is given an intravenous injection of contrast. The postcontrast image is digitally subtracted from the precontrast. The technique is being used to explore the subtle relationship between tumors and background tissue and might allow the detection of cancerous lesions.

Dual-Energy Mammography

In dual-energy mammography, also called contrast media mammography (CMM), the contrast agent is injected immediately before the imaging. Two exposures of the breast are made at different energies, one between 20 and 30 kV and the other between 40 and 80 kV. The tissue characterization of the breast is obtained from a weighted subtraction of one image from another. If performed with a contrast injection, the images are acquired after the injection of contrast. The normal background is removed to produce an image of the contrast agent. Another method is to obtain a mask image, before the injection of the contrast. The contrast is then injected, and these images are subtracted from the mask image so that the normal anatomy is removed, leaving an image of the contrast. The technology is still under investigation; however, it is hoped that the technique can someday be used to highlight calcification within the breast.

ULTRASOUND TECHNOLOGIES

Ultrasound technology offers a fast and worry-free method of imaging. There is no radiation, and the images are acquired with the patient lying comfortably on a table. However, ultrasound imaging can be extremely wearing for the technologist's wrist, and some ultrasound technologists now suffer from carpal tunnel syndrome and other related conditions from the repeated scanning motion needed to acquire the ultrasound images. Newer technologies hope to provide a better image with less stress on the carpals.

Automated Ultrasound and Three-Dimensional Breast Ultrasound

Automatic ultrasound offers systematic breast scanning. The technologist positions a cup-like transducer over the breast. The cup holds a 14.5-cm transducer instead of the 4-cm transducer of typically ultrasound units. This allows the scanner to capture 14 × 17 × 5 cm volume for 3D diagnostic flexibility. A single button automatically centers the device and performs the scan, which can take 10 to 15 minutes, and the images are acquired in longitudinal rows, with overlapping to ensure complete coverage. A mechanical arm of the unit controls transducer speed and position, and the technologist maintains appropriate contact pressure and orientation, vertical to the skin.

After acquisition, postprocessing algorithms can be applied based on the nipple location. The technology can generate 400 to 500 images, with data capture in 2D format and reviewed from any angle, or 3D reconstruction can be applied. Some units have a shorter scanning time (60 seconds) and offer the ability of capturing 2000 to 3000 images of the breast, which can be viewed in a continuous ciné playback loop (Fig. 8–5).

The Automated Breast Ultrasound SomoVu

The Automated Breast Ultrasound (ABUS) SomoVu software displays slices of each volume from three orthogonal planes: coronal, axial, and sagittal. The radiologist can adjust how the planes are displayed by selecting various options on the tools menu.

To image, the technologist places a single-use mesh membrane over the surface of the transducer to aid acoustic coupling. Three per breast, in anteroposterior, lateral, and medial projections, are taken, giving a total of six volumes to complete a breast examination. Each scan captures 350 images. Typically, the slices are 2 mm thick with 1 mm of overlap. However, the thickness is adjustable. Slice thickness and overlap can be customized according to the radiologist's individual workflow preferences to as thin as 0.02 mm to as thick as 1 cm. The total examination time of a bilateral breast screening is about 15 minutes.

> Ultrasound technologies involve no radiation, and images are acquired with the patient lying comfortably on a table.

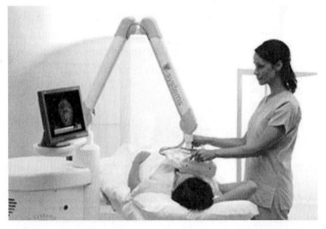

Figure 8–5. SomoVu Automated Breast Ultrasound System. Copyright © Siemens Healthineers 2017. Used with permission. https://www.siemens.com/press/en/pressrelease/?press=/en/pressrelease/2009/imaging_it/him200902016.htm

The ABUS uses a location system to report the position of lesions in three coordinates: clock position, distance from nipple, and depth beneath the skin surface. An image of the lesion from each of the three orthogonal planes is automatically captured and attached to the report with the location information to facilitate fast and accurate identification of the region of interest during diagnostic workup. The images can be annotated, and all the images along with the report can be automatically saved to a picture archiving and communication system (PACS) or sent to DICOM (Digital Imaging and Communications in Medicine) as PDF reports. The SomoVu generates a report for each examination reviewed by the radiologist. These reports are similar to the worksheets currently filled out by the ultrasound technologist while performing handheld studies.

The FDA has specified that special training is required by all technologists and radiologist before using the ABUS technology for screening. Technologists must hold a current American Registry for Diagnostic Medical Sonography (ARDMS) license or an American Registry of Radiologic Technologists (ARRT) license with current registration in breast ultrasound, or an ARRT license (or unrestricted state license) in mammography, and must also meet MQSA requirements. All technologists must have five continuing education units specific to breast ultrasound.

ABUS examination is used as an adjunct to mammography, and the patient must first complete her routine digital mammogram. ABUS imaging requires a dedicated workstation, with proprietary software for interpretation of ABUS images. The workstation converts the ABUS image data into a 3D volume that can be controlled by the radiologist. Unfortunately, this means that the ABUS images cannot be interpreted with the existing workstations.

Advantages of this procedure include the following:

- Minimal compression
- Patient can breathe normally
- No ionizing radiation
- Noninvasive test—no contrast
- Large field of view helps reduce tissue exclusion
- Can visualize dense breast tissue and abnormalities
- Provides a global representation of the entire breast
- Provides easy correlation between other imaging modalities

Disadvantages of this procedure include the following:

- The need to frequently reimage the patient

An advance on this technology is 3D Doppler ultrasound, which offers the ability to visualize the blood flow in and around malignant tumors. Malignant tumors generally have a more extensive and higher and faster velocity flow than benign tumors. This information would be captured by the 3D Doppler unit and used to differentiate malignant versus benign lesions.

Full-Field Breast Ultrasound

Whole-breast sonography is a technology using units similar to the mammography unit. The system uses three components: a special scan station, an image-processing computer, and a view station. The breast is positioned in a unit similar to positioning for a mammogram. In one sweep, 400 to 800 sagittal and perpendicular ultrasound images are acquired of the breast. The images convert into six to eight somograms—tomographic ultrasound slices—which can be viewed next to their mammographic counterpart. Each somogram represents a section of breast tissue approximately 5 to 10 mm thick. The scan lasts 60 seconds and uses light compression. It is proving particularly effective in imaging dense breast.

Elastography

Elastography can be used to both detect and classify a breast tumor. Currently, it is too time-consuming and the equipment too cumbersome to be practical as a scan of the entire breast. It is being used to study lesions seen on ultrasound, and investigations are under way to reduce the size of the equipment or to develop handheld devices. It is hoped that faster computers will reduce the time of the study.

The principle of elastography is that a cancer is relatively hard compared with surrounding breast tissue, and the more a cancer is compressed, the harder it gets. Cancers will also compress differently from normal tissue. Imaging involves a two-step process. Strain elastogram—or the rate of displacement—will reveal information about the stiffness of the tissue and its cellular structures, and modulus images then offer a visual presentation of the spatial distribution of properties of the material. Modulus is a mathematical quantity that relates to the load and deformation in the material. Modulus is also related to stiffness. The technique can detect minute levels of hardness.

Images are taken before and after light compression. The compression applied is similar to the pressure from palpation with the fingertips. After compressing the breast, it takes approximately 20 seconds to get an image.

Ultrasound Imaging Using Capacitive Microfabricated Ultrasonic Transducers

Current ultrasound technology uses piezoelectric crystals to vibrate and create electrical signals. These crystals, however, lack the bandwidth and beam-focusing capabilities required to take advantage of high-frequency ultrasound units and three-dimensional imaging. Contrast capacitive microfabricated ultrasonic transducer (cMUT) technology will improve contrast agent imaging and enable volumetric scanning.

The technique is based on the use of microscopic silicone drums about the width of a human hair and operates similar to a stereo speaker and microphone. An electrical signal to the drum creates an electrostatic force on the membrane, causing it to vibrate and emit ultrasound. Echoes returning from the body also cause the drum to vibrate and produce an electrical signal. These signals are used to create a visual image. It is hoped that the technique will revolutionize the use of ultrasound in the diagnostic setting and reduce the need for more expensive and invasive technology, such as magnetic resonance imaging (MRI) or surgical biopsies, in evaluating women with dense breast or scar tissue.

Volumetric imaging will also reduce scan time and enable the adoption of whole-breast ultrasound examinations.

Contrast-Enhanced Ultrasound

Microbubble contrast agent introduces gas-filled bubbles into the circulation system. The bubbles are typically 3 µm in diameter. A volume of less than 200 µL of air or gas is injected intravenously to improve conventional ultrasound scanning by enhancing both gray-scale images and flow-mediated Doppler signals.

Microbubbles work by resonating in an ultrasound beam, rapidly contracting and expanding in response to the pressure changes of the sound wave. They vibrate particularly strongly at the high frequencies used for diagnostic ultrasound imaging. This makes them several thousand times more reflective than normal body tissues.

Specific characteristics of the oxygen, nutrients, and vascular supply to the cells are often an indication of an active breast cancer. The microbubbles can be designed to attach to an integrin, which is a well-characterized molecular marker that is an indication of tumor growth. Ligands that bind to receptors characteristic of malignant diseases are identified and bound to the microbubbles, thus enabling the microbubble to accumulate selectively in areas of interest.

Early studies suggest that this technology could substantially increase the sensitivity of ultrasound to metastatic disease and identify lesions less than 1 cm in diameter, for which all imaging methods lack sensitivity. However, this technology is still experimental.

MAGNETIC RESONANCE IMAGING TECHNOLOGIES

MRI of the breast is playing an increasing role in breast cancer screening. However, the lack of specificity of MRI has led to a large increase in the number of false-positive interpretations. Studies estimate that 80% of breast biopsies under MRI are proved to be benign.

Magnetic Resonance Spectroscopy

Proton magnetic resonance (MR) spectroscopy could help reduce the high false-positive rates of the breast MRI by providing chemical information about a lesion. Proton MR spectroscopy works by measuring the levels of choline compounds in a lesion.

By adding approximately 10 minutes to the typical breast MRI study, the radiologist can see the choline makeup of a lesion. This is particularly important when evaluating non–mass-enhancing lesions, which frequently pose a dilemma to the radiologist. Studies have shown that the choline is a by-product present in cancerous lesions and is often a mark of an active tumor.

MOLECULAR IMAGING TECHNOLOGIES

Molecular imaging looks at the biological activity and shows what cells are doing and how they are functioning. This is opposed to the anatomic imaging provided by modalities such as mammography, ultrasound,

> Molecular imaging shows changes at the cellular level long before these changes can be detected structurally.

computed tomography (CT), or MRI. Molecular imaging can show changes at the cellular level long before these changes can be detected structurally.

The aim of current imaging is the detection of breast cancer in its earliest stages. But the ultimate aim is to find a means of preventing the disease rather than curing it. With the exciting possibility of molecular imaging, this aim is that much closer to reality. By targeting specific enzymes or cell structures in the body, it could theoretically be possible to detect, measure, and target specific cancer cells. At present the technique lacks specificity, but it is showing significant promise for the future, and once achieved, perhaps, breast cancer will become just another chronic but treatable disease.

Molecular imaging shows promise toward personalizing medicine based on each individual's unique genetic and metabolic makeup. In the near future molecular imaging hopes to revolutionize the understanding and approach to breast cancer care by diagnosing and staging of tumors with accurate noninvasive diagnostic and screening tools, by designing more effective drug treatment and monitoring the therapeutic treatment or therapy, and by evaluating responses and effectively adding personal designs to treatment plans.

Molecular imaging includes a diverse field of modalities. The main molecular imaging modality is nuclear imaging; however, optical imaging in which light-producing proteins attach to specific molecules, such as those at the surface of cancer cells, is becoming increasingly important. The molecules will then emit a low level of light from inside the body, and special goggles will be used to see the fluorescence caused by the infrared light source. This technology, called "near-infrared fluorescence (NIRF)," could eventually allow the visualization of lymphatic channels and lymph nodes in real time. Fluoroestradiol (FES) is another experimental technology that measures the estrogen receptor markers of a lesion and could be used to track the receptor status of tumors. Another technology, fluorothymidine (FLT), looks at cellular growth and proliferation. Both of these technologies could be used to determine and monitor a patient's treatment.

Nuclear Medicine

Nuclear medicine uses artificially produced radioactive isotopes. These are also called radiopharmaceuticals, radioisotopes, radionuclides, or radioactive tracers. Radiopharmaceutical can be introduced into various organs of the body, either orally or intravenously. The radioisotopes emit gamma rays and will travel through the body, exposing all organs; however, they generally will have an affinity to a specific organ or tissue type. Nuclear medicine procedures must be completed within a given time period because the radiopharmaceuticals have relatively short half-lives. This half-life factor is what makes the radiopharmaceutical possible for human diagnostic purposes. Isotopes are absorbed by or taken up at varying rates or in different concentrations by different organs or tissue types. A diseased or poorly functioning tissue will emit amounts of radiation that are different from healthy tissue, and a gamma camera can be used to record the emitted radiation (Fig. 8–6). Areas of reduced radioactivity are called cold spots, whereas areas of

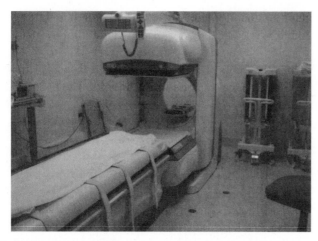

Figure 8–6. Gamma camera used for positron emission tomography imaging of the breast.

increased activity are called hot spots, indicating the presence of disease or injury.

The traditional gamma camera has a large crystal detector called a scintillation crystal. These crystals detect the emitted radiation signal and convert that signal into light. The light is then converted to an electric signal, which is digitized into an image by a computer.

At present, nuclear imaging technologies are not used as screening technologies; however, the technology is less expensive than breast MRI and offers comparable sensitivity. In addition, by combining new molecular agents with traditional imaging tools, nuclear imaging is becoming an increasing important component of what is now being referred to as molecular imaging. In recent years, positron emission tomography (PET) imaging, breast scintigraphy or scintimammography, and lymphoscintigraphy have proved useful for both diagnostic and therapeutic purposes. With newer breast gamma cameras, positron emission mammography (PEM) and breast-specific gamma imaging (BSGI) have evolved. These are advanced molecular breast imaging (MBI) techniques that show promise of increased sensitivity and specificity.

The main clinical drawback to MBI is that it uses 8 to 10 times the radiation of a standard mammogram. This dose would need to be significantly lower before the technologies can be implemented as routine screening tools.

Fluorodeoxyglucose–Positron Emission Tomography

PET imaging uses extremely short-lived nuclides in diagnostic imaging (Fig. 8–7). Fluorine-18 fluorodeoxyglucose (^{18}F-FDG) is the most common radiopharmaceutical used for PET imaging. It is used because breast cancer has a particular affinity to it. This nuclide has a short half-life and is quickly broken down by the liver and excreted. Its complete name is fluorodeoxyglucose, methionine, tyrosine, fluoroestradiol, norprogesterone, 2-deoxy-2-fluoro-d-glucose labeled with fluorine.

Patients undergoing a PET scan must fast for 4 to 6 hours. About 1 hour before the appointment, the technologist will take a drop of blood from the patient's finger to check the blood sugar level. If it is

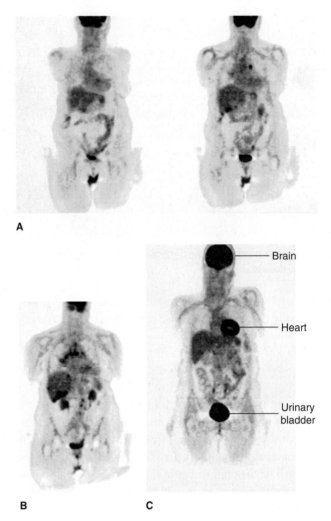

Figure 8–7. (A–C) Whole-body positron emission tomography imaging for restaging of breast cancer, showing metastatic disease to the thoracic and lumbar vertebra, the mediastinum, and the liver. Uptake in the heart, brain, and urinary bladder is normal (*arrows*).

> PET and PEM technologies work on the principle that most cancerous tissue uses vast amounts of sugar.

within the acceptable range at the appointment time, a total of 10 to 12 mCi of the radiopharmaceutical is injected into an arm vein. The patient then lies quietly for up to 1 hour before the imaging, usually in a separate "quiet room," to allow the radiopharmaceutical to circulate and be absorbed by the body. Immediately before the imaging the patient is required to void. Next, the patient lies on a table in a unit somewhat similar to the MRI or CT units with a pancake-shaped gamma scanner. The table slides in and out of the scanner during the examination while the gamma scanner is used to detect the radiation emitted.

The technology works on the principle that because of the increase in metabolic activity, most cancerous tissue uses vast amounts of sugar (glucose) and at a much higher rate than benign tissue. The radioactive substance used, FDG, decays by positron emission with a half-life of 110 minutes and is metabolized in the body like sugar. It will, therefore, go to the tissues that are most active. During the process, there is an annihilation of the positron by an electron, creating two 511-keV gamma rays emitted in opposite directions, which are then detected by the gamma camera and can be used to generate an image.

Positron Emission Mammography

The recent development of organ-specific units now offers advanced breast imaging options. The size of the gamma cameras in the past and the lack of breast compression would inhibit the resolution of the final image and did not allow reliable detection of subcentimeter lesions. Also, because the patients were lying on a table for the examination, the older technology does not allow correlation with a mammogram when the patient is imaged in the upright position. The newer PEM units are high-resolution units with a greater sensitivity and specificity.

The technology uses two gamma cameras specifically designed for breast imaging. One technology uses a detector made of 2000 lutetium-containing photo detection crystals mounted in the compression paddles. During the acquisition phase the detectors move within the paddles, acquiring coincident counts. The user has the ability to select the travel distance of the detector to customize the breast coverage. The patient is seated during the study and positioned similar to the MLO and CC projections for mammography. The breast is gently compressed between the two cameras during the detection process. The images data can then be collected in the standard MLO or CC positions as used with regular mammograms. The unit is also capable of imaging the axillary, cleavage, and mediolateral positions as needed. The PEM units also allow limited-angle tomographic reconstructions, perhaps approximately 12, with each slice thickness about 4 mm, corresponding to the travel distance divided by 12. The total image-acquired range can be approximately 48 mm.

The small units allow closer breast positioning, making acquisition more efficient. Two images of each breast are acquired, with each image acquisition taking approximately 10 minutes. However, a typical PEM examination includes two scans per breast, and the entire procedure, including time in the quiet room, could take up to 2 hours. Often a PEM study will be followed by a whole-body PET, and there is no added radiation because both scans are completed with a single injection.

Use of FDG-PET and PEM Imaging

For patients with an ambiguous mammogram, FDG-PET can give valuable clarifications. FDG-PET can be used to stage lymph node involvement before surgery and is very accurate in detecting lymph node spread in the mediastinum area around the heart, great vessels, and internal mammary nodes on either side of the sternum. If a preoperative PET scan is positive in the axillary node, this is usually a strong indication of lymph node involvement.

Whole-body PET imaging will provide useful information on metastases to both soft tissue and bone and can detect recurrent metastases. PET has an accuracy of 87% to 90% in detecting sites of metastatic disease and appears to be more reliable than mammography for identifying relapsed tumor. PET and PEM imaging are very accurate in identifying a relapsed tumor, even in patients with negative serum tumor markers but with suspicious clinical findings, and they are more specific than MRI in discriminating between fibrotic scar tissue, necrosis, or tumor. PET and PEM imaging are also less affected by hormonal changes and breast density than MRI.

> The small size of the PEM unit allows data collection in the standard MLO and CC projections.

PEM can give important information on both the staging and restaging of cancerous breast lesions. Accuracy in staging and early identification of recurrent tumors is critical for therapy choice. If the chosen therapy is effective, for example, if the radiation therapy or chemotherapy will kill the cancer cells. PEM can determine whether cell death has occurred. Also, the uptake of FDG in tissue correlates with the histologic grade and potential aggressiveness of breast cancer. Studies show that tumors with a high uptake are generally more sensitive to chemotherapy. This fact can be used when evaluating staging and treatment options.

The value of mammography is often limited after breast augmentation, and in this situation PEM might offer some advantage in screening for cancer. PEM has a sensitivity of 93% for invasive cancers and 93% for ductal carcinoma *in situ* (DCIS) for tumors larger than 2 cm. It is not affected by metal in the patient's body, breast density, hormone replacement, or menopausal status and generates fewer images than MRI, making it easier to interpret.

Scanners are now available that can perform both PET and CT scans in the same imaging session. The images that are obtained can be overlaid or fused so that the functional information of the PET scan can be accurately localized in the high-resolution anatomic images of the CT scan. Fusion imaging performed on these hybrid scanners has been shown to improve whole-body scanning for diagnostic accuracy.

Combining PET and dynamic MRI information in assessment of breast lesions found on mammography can decrease the number of biopsies of benign lesions because the high specificity of PET complements the high sensitivity of dynamic breast MRI.

Dedicated breast PET/CT or PEM units show promise in improved detection in primary breast cancer, while also providing a method for image-guided biopsy. PEM biopsies using a vacuum-assisted device have been approved by the FDA in the United States; ongoing research in the molecular basis of cancer will likely result in new agents that may better identify tumors currently not well imaged by FDG. Potentially, molecular therapies that target specific cell receptors or processes may be tagged with a positron and imaged, thereby allowing not only imaging but also treatment of breast cancer.

Analysis of FDG-PET

Although the basic mechanism of uptake is through glucose metabolism, other factors that result in variations in FDG uptake in breast tumors have not been clearly identified. Care must be taken in interpreting PET or PEM scans of patients after biopsy or surgery because sites of inflammation or infection will also display increased uptake of FDG.

Blood sugar level affects tumor uptake on any PET or PEM scans. Whenever there is high blood glucose, the tumor uptake of FDG may be decreased. Some facilities will check the patient's blood levels before giving the radiopharmaceutical injection. Also, because muscles use glucose for metabolism, and the recovering muscles will continue to take up glucose for approximately 24 hours, patients must curtail any vigorous exercise, such as jogging or weight lifting, for at least 48 hours before a PET or PEM scan.

- Patients must fast and curtail vigorous exercise before the scan.
- After the injection and before scanning, the patient must lie quietly for approximately 60 to 90 minutes.

Tumor size and cell type are factors that affect PET scan accuracy. The accuracy in detecting tumors larger than 2 cm is high; however, PET may miss approximately one third of invasive cancers smaller than 1 cm. Although the accuracy of the technology has improved with the use of the dedicated gamma camera, as used in PEM, PEM and PET technologies are still more likely to identify invasive ductal carcinomas yet miss invasive lobular carcinomas. Also, these technologies are not helpful for identification of noninvasive tumors. Because PEM cannot detect small breast tumors or microscopic tumors, it cannot be used as a replacement for sentinel node mapping, although positron imaging is very useful in identifying recurrent metastatic disease.

PET or PEM imaging is not currently marketed as a primary breast cancer detection tool.

PET and PEM results can be affected by:
- Blood sugar levels
- Vigorous exercise
- Breast biopsy or surgery
- Inflammation or infection
- Tumor size or type

Radionuclide Imaging With Single-Photon Emission Computed Tomography

Researchers are experimenting with using a limited-angle pinhole single-photon emission computed tomography (SPECT) study and adapting the existing equipment in a technique termed "PICO-SPECT" to improve the detection of breast lesions. A recent patient imaging study showed higher contrast and more spatial detail of the tumor tracer uptake with PICO-SPECT. The study demonstrated the potential of this dedicated breast imaging tool to detect smaller lesions when they have sufficiently higher uptake than the surrounding normal breast and illustrates a successful and necessary first step in the development of a new imaging approach.

Breast Scintigraphy, Scintimammography, or Breast Molecular Imaging

Scintimammography, or scintigraphy, is a functional or molecular imaging procedure. The procedure is sometimes called breast molecular imaging (BMI). During the procedure, 20 to 25 mCi of the radiopharmaceutical ^{99m}Tc-sestamibi is injected into the contralateral arm vein of the patient. The injection should be made with a butterfly needle and followed by a 10-mL saline flush to minimize extravasation and streaking artifacts. If both breasts are under investigation, the injection should be made in the dorsalis pedis vein. ^{99m}Tc is a cationic lipophilic agent that is transferred across the cell membrane and absorbed by the mitochondria. The FDA has approved ^{99m}Tc, marketed under the trade name of Miraluma for use in scintimammography. ^{99m}Tc is a gamma-emitting isotope that accumulates in malignant lesions in the breast with a higher metabolic rate than benign lesions, which do not take up the isotope.

The breast is not compressed in scintimammography. Imaging is performed 60 to 90 minutes after the injection. Each projection takes approximately 10 minutes to image. The patient is imaged in the prone position using a table with a cutout insert designed specifically for scintimammography. This allows the breast to hang freely during the imaging process. After injected into the bloodstream, ^{99m}Tc will concentrate in breast cancer cells. The examination generally consists of three planar images: lateral images in the prone position, posterior oblique images also prone, and anterior images in the upright or supine position. The prone position allows optimal visualization of the entire breast and relaxation of the pectoralis muscle, allowing separation of the breast tissue from the chest or abdominal organs. In the supine position, the natural contours of the breast are seen and can be used as easy anatomic landmarks. Both breasts are imaged for comparison.

Breast-Specific Gamma Imaging

In a newer version of scintimammography, emissions from the breast are detected using a very small, high-resolution breast-optimized gamma camera. The technology, BSGI, uses the same radiopharmaceuticals; however, after the injection, the patient is seated and the breast placed between the gamma cameras (Fig. 8–8). Gentle compression is applied during the detection process, which is generally 6 to 10 minutes per image.

The latest version of the BMI technology uses a cadmium zinc telluride gamma camera incorporated into a new dual-head breast-imaging system. In a single-head gamma camera the fall-off of spatial resolution with distance means that low-contrast lesions located opposite the single detector could be lost in the noise. With a dual-head system the second opposing detector should detect this lost lesion. In addition, recording two facing views provides a means to localize a lesion's depth and estimate its size. This system reports an 88% accuracy in detecting lesions 10 mm in diameter. Lesions 2 to 4 mm will not be detected with this technology.

During the scan each breast is lightly compressed for 5 to 10 minutes between the gamma cameras, with just enough pressure to prevent

- BSGI uses a small, high-resolution breast-optimized gamma camera.
- The radiopharmaceutical must be given in the contralateral arm vein of the patient.

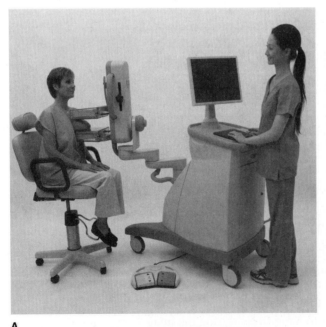

A

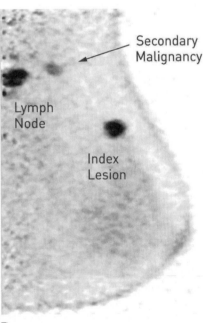

B

Figure 8–8. **(A)** Organ-specific positron emission tomography scanner (Used with permission of CMR Naviscan Corporation.) **(B)** Positron emission mammography showing a secondary lesion.

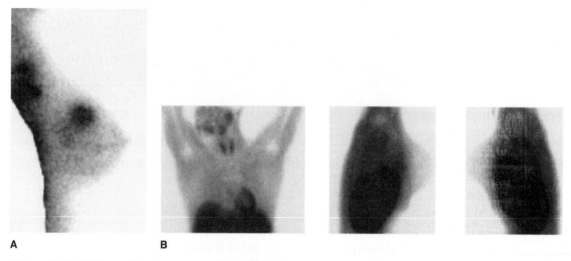

Figure 8–9. **(A)** A positive breast-specific gamma imaging study. **(B)** A negative scintimammography study.

motion. The compression force is typically about 15 lb, which is a lot less than the force required for regular mammograms.

In both scintimammography and BSGI, the higher metabolic cancerous cells will absorb a greater amount of the radiopharmaceuticals, and this will be revealed as uptake areas on the images. A positive image is seen as an area of increased uptake. A negative scan is defined as an uptake on the 10-minute image and complete washout of activity on the 60- to 90-minute images (Fig. 8–9).

Uses of Scintimammography or BSGI

Scintimammography reports a sensitivity of 84% and a specificity of 80% for the detection of malignant lesions ranging from 5 to 80 mm (with a median of 20 mm). It can be used to assess the level of axillary node involvement and to determine the true extent of a disease in order to optimize surgical planning. ^{99m}Tc-sestamibi is especially useful in patients with dense breasts, implants, diffuse calcifications, and breast tissue scarred by radiation or surgery or on patients with indeterminate mammograms who are not being referred for biopsy.

BSGI is less expensive than breast MRI, with a better specificity which will result in fewer false-positive biopsies. It has the added advantage of a short examination time. BSGI has proved to be twice as effective in finding cancers when compared to breast ultrasound.

BSGI is also proving invaluable for breast cancer staging to show multifocal disease or axillary lymph involvement. If the patient is scheduled for a lumpectomy, the presence of a multifocal lesion can change the choice of therapy, and a slow tumor clearance of Tc99m can predict a good response to chemotherapy in patients with non-Hodgkin's lymphoma. Tc99m can also differentiate between responding and nonresponding tumors early in the course of chemotherapy, which is important in treatment planning. It has also been found that patients with a multidrug resistant (MDR) gene are predisposed to a poor chemotherapy response. These patients clear Tc99m more rapidly than other patients. This correlation can perhaps be used to identify these patients in advance of treatment (Fig. 8–10).

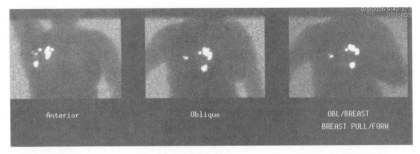

Figure 8–10. A positive scintimammography study.

Analysis of Scintimammography and BSGI

Scintimammography does not detect small breast cancers as effectively as mammography screening but is useful in detecting tumors in the dense breasts of younger women. However, some benign diseases such as fibroadenomas give false-positive results, and skin folds or muscle folds in the upper arm can mimic axillary uptake. Lesions in the upper pole of the breast are easier to detect, whereas lesions smaller than 1 cm are difficult to detect, especially if they lie deep within the tissue.

The injection should be made in the contralateral arm or foot and should be a good nonextravasating injection to avoid nonspecific axillary node uptake. Injecting the radiopharmaceutical into the subcutaneous axillary region often results in a false-positive reading.

BSGI reduces many of the drawbacks of the older technology. The cameras encircle the breast, enabling imaging from many angles. The units can image in the mediolateral oblique, craniocaudal, lateromedial, and mediolateral projections, therefore obtaining the same projections as a mammogram. The systems can also rotate 360 degrees around the breast. They are user-friendly, and the design allows correlation of the images to the mammograms. The improved spatial resolution using a dual-head gamma camera can detect 10-mm lesions and lesions as small as 3 to 3.3 mm, compared with the 4- to 5-mm lesions detected by the single-head gamma cameras technology. An even further modification is the production of high-resolution three-dimensional images of the breast. The three-dimensional images obtained in this way are called emission or function mammotomograms.

Lymphoscintigraphy—Sentinel Node Mapping

The lymphatic system receives cell waste products, and generally breast cancer leaves the breast through the lymphatic system. The idea behind lymphoscintigraphy or sentinel node mapping is that the first lymph node, called the sentinel node, that receives drainage from a tumor can be used to predict the presence or absence of tumor in the remaining nodes. If the sentinel node is negative, it is likely that the cancer is still contained within the breast, and the absence of cancer in the sentinel node is thought to be a strong indication of the absence of metastatic disease to other nodes. In cases of conservation breast surgery, identifying the sentinel node will avoid a random blind dissection of the axillary nodes, which carries a higher degree of complications, including lymphedema, decreased mobilization, numbness, infection, permanent

A positive sentinel node would indicate that the cancer is no longer contained within the breast.

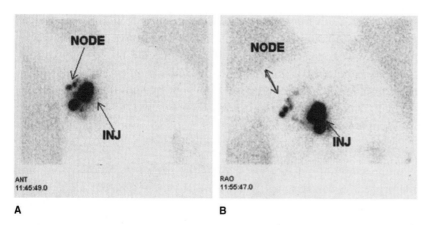

Figure 8–11. Anterior **(A)** and right anterior oblique **(B)** projections showing positive uptake on a lymphoscintigraphy study.

hypesthesia, dysesthesia, or painful seroma. Women can be spared the considerable psychological distress involved in the removal of the axillary nodes if the presence or absence of axillary lymph node metastasis from the breast can be ruled out. A positive sentinel node would indicate that the cancer is no longer localized to the breast (Fig. 8–11).

Indications for Lymphoscintigraphy

Candidates for this procedure are patients considered for axillary node dissection who do not have nodes that are clearly involved or patients with a high-grade tumor. Lymphoscintigraphy is not recommended in patients who have had prior surgical biopsies on the breast because surgery can disrupt the lymphatic pathways and could affect the examination results. The procedure employs a technique similar to scintimammography in that a small amount of radioactive substance is injected into the patient. The site of choice here is often the subareola lymphatic plexus or directly in the region of the lesion.

Often, 3 to 5 mL of a blue dye—isosulfan blue vital dye—is also injected into the region of the tumor. The dye is then carried to the sentinel node. The dye is generally used along with the radioactive tracer. Use of the dye alone increases the amount of dissection and tissue disruption, and often it is impossible to localize the sentinel nodes in the internal mammary or other node chains.

For lymphoscintigraphy, a radioactive substance—^{99m}Tc-sulfur colloid (^{99m}Tc-SC)—is injected into the tumor bed 1 to 9 hours before surgery. ^{99m}Tc-SC is bound by disulfide bonds to the protein constituents of gelatin and is rapidly taken up by the lymph system and travels to the lymph nodes under the arm. Generally, 1 mCi of ^{99m}Tc-SC is mixed with 4 to 8 mL of saline. Patients may receive an initial injection of local anesthesia. If the ^{99m}Tc–SC is used along with the blue dye, the protocol is to inject 0.4 mCi of the technetium-labeled sulfur colloid approximately 2 to 4 hours before surgery. Then, 10 minutes before surgery, the 1% isosulfan blue dye is injected. Both are injected at the same site, around the primary tumor. After the injection, the site must be carefully wiped and all gloves and wipes placed in a radiation bag.

During surgery, the surgeon uses a Geiger counter to locate and biopsy these radioactive lymph nodes. If the blue dye is also used, all

> During surgery the surgeon uses a Geiger counter to locate, identify, and biopsy radioactive lymph nodes.

blue-stained nodes are also located and removed. The procedure can provide prognostic information based on the presence or absence of regional lymph node metastases. If the nodes are negative for tumor, radial axillary node dissection is not needed. Even if a cancerous node is identified, fewer lymph nodes are removed. By identifying the sentinel node, patients are spared an extensive node dissection surgery with its resultant side effects.

Analysis of Lymphoscintigraphy

Lymphoscintigraphy poorly visualizes the deep lymphatic system, and the procedure can have a 10% false-negative rate. The particular size of the sulfur colloid is important, with optimal results achieved if filtered sulfur colloid is used. Filtered ^{99m}Tc-SC will eliminate all particles larger than 0.22 µm in diameter (1 µm [1 micron or micrometer] = 0.001 mm). Most early failures were the result of using unfiltered ^{99m}Tc-SC because of the fear that the unfiltered colloid has a greater tendency to migrate through the axillary node chain. Unfiltered ^{99m}Tc-SC migrates slowly and easily becomes trapped in the interstitial tissues, rather than migrating to the lymphatics. This problem, however, can be solved by carefully coordinating the time of injection with the time of surgery.

Another limitation of lymphoscintigraphy is associated with the site of the injection. Studies have shown that if the injection is made in the dense fibrous tissue surrounding a tumor, it may never be taken up by the lymphatics. The best site of the injection is still under debate, with some studies suggesting that the injection should be made in the sub-areola lymphatic plexus, and others stating that the injection must be made directly into the tumor site for best results. Studies also found that if deep anesthetic is used before the injection of the radioactive substance, there is no migration toward the axilla. The injection must, therefore, be made with local skin anesthetic, which unfortunately will result in discomfort to the patient.

> In addition to the radioactive substance, often 3 to 5 mL of a blue dye is injected in the region of the lesion.

RADIATION DOSE RELATED TO NUCLEAR IMAGING

BSGI and PEM imaging are not suitable for screening because of the high doses of radiation used. The radiation involved in a single BSGI examination poses a risk that is 20 to 30 times greater than that from digital mammography in a woman 40 years of age, and the risk from a single PEM screen is 23 times higher.

The average "effective dose" from mammography is about 0.44 to 0.56 mSv, which is equal to about 2 months of natural background radiation; the effective doses from BSGI and PEM (6.2–9.4 mSv) are equal to approximately 2 to 3 years of natural background radiation exposure.

The major difference in radiation is because BSGI and PEM imaging involve intravenous injections of radioisotopes, which expose all organs of the body to gamma ray emissions, whereas mammography with lower energy x-rays exposes only the breasts.

The distribution of these radionuclides in the bloodstream, their uptake into tissues, and their clearance in the liver (for ^{99m}Tc used in BSGI) and kidney (for FDG used in PEM) result in radiation exposure of organs in the chest, abdomen, and pelvis. The highest risk is to the colon in BSGI and the bladder in PEM. The organs receiving the highest doses and therefore at greatest risk for cancer induction from radionuclide administration are the colon, lungs, and bladder. Cancers occurring in these organs are less curable than breast cancer, so the risk ratios for BSGI and PEM, compared with mammography, are greater for cancer mortality than for cancer incidence.

However, newer techniques using a gamma camera system with dual detectors and special collimators offer a much lower radiation dose (2–4 mCi), which is comparable to the radiation dose receiving during one or two mammograms.

OTHER IMAGING MODALITIES

Optical Imaging

Optical imaging uses infrared light to penetrate through the tissue. Research shows that some structures absorb light and some scatter it. Extra blood vessels that accompany a tumor will be highlighted by the pattern of light absorption because the presence of a tumor will distort light in a characteristic way. It is hoped that the technology will draw attention to areas beneath dense breast tissue, especially in younger women. The technology is still being investigated, but a digital infrared imaging system, by Infrared Sciences Corp. (ISC), recently obtained approval from the FDA in the United States to assist in the early detection of breast cancer.

The technique can be combined with ultrasound, MRI, or DBT in evaluating lesions found in any of these modalities. With ultrasound, the use of diffuse optical tomography (DOT) involves placing the ultrasound transducer in the middle of an optical source. After the ultrasound locates the lesion, the DOT is performed by shining infrared light into the area and measuring light absorption at two optical wavelengths. Infrared light can penetrate deep into the tissues up to 4 cm.

A modification of this technique is the use of near-infrared spectroscopy (NIRS), which is a combination of optical spectroscopy, and breast magnetic resonance NIRS, and DOT. Both the NIRS and DOT technologies enable analysis of hemoglobin and deoxyhemoglobin concentrations to provide information on tissue vascularity, oxygenation water percentages, and lipid concentration. The technique employs an array of 16 fiberoptic bundles, with one fiber acting as a light source and the remaining fibers acting as light detectors. This array is moved across the breast, recording light signals at locations around the breast. The difference in the transmission and detection of light through the breast is related to the total hemoglobin concentration and oxygen saturation of each section of breast tissue. The process of angiogenesis makes a malignancy more vascular because it has a higher hemoglobin concentration. It is also less oxygenated, possibly because of the increased metabolic activity or inefficient oxygen distribution system. Because the distinct tissue elements that absorb wavelength-specific light in the breast are known, this information is used to generate a map of the tissue elements of the breast. When used after a contrast MRI, NIRS will focus on areas enhanced by the gadolinium chelated contrast. The

- Average radiation dose with cone-beam CT is 9.9 mGy.
- Average radiation dose from a four-projection mammogram is 10 mGy.

information from the MRI is therefore used as a guide to the region of interest. If used with ultrasound, the ultrasound locates the lesion, and DOT focuses the area of interest.

The sensitivity of this modality is 92%, with a specificity of 93%, even with lesions smaller than 2 cm.

Cone-Beam Computed Tomography

Researchers at several sites in the United States have been studying a proposed scanner design that produces highly detailed CT images of the breast yet uses radiation levels that are comparable to or less than mammography. A recent study of the cone-beam CT technology found that on average patients received a dose of 9.9 mGy (0.99 rad), which is actually lower than 10 mGy (1 rad), the average dose for the four-projection mammography series.

For the imaging, the patient lies facedown on a specially built table with cutouts for the breast. The x-ray tube rotates 360 degrees around the table, taking images and collecting data. It will generate approximately 300 to 500 images within 10 to 17 minutes. The information is sent to a computer, which then creates 3D images of the breast that can be viewed from any angle.

Advantages
The breast is not compressed, and the patient lies comfortably during the study.

Analysis of the Technology
Mammography is still better at detecting and imaging microcalcifications than cone-beam CT.

Computed Tomographic Laser Mammography

Computed tomographic laser mammography (CTLM) involves no ionizing radiation (Fig. 8–12). The breast is scanned for temperature differences that might indicate a growing tumor. This is based on the principle that malignant tumors develop numerous new blood vessels in a process called *angiogenesis*. Hemoglobin in the blood absorbs the CTLM laser light much more than the surrounding tissue. The more blood flows, the whiter the image, and normal vessels will have a different pattern appearance than cancerous vessels.

During the examination, the patient lies facedown on a table with the breast suspended through a hole in the table. Inside the console, a low-wavelength laser sweeps around the entire breast, making a 360-degree circle. Several passes are made, with each pass taking approximately 45 seconds to complete. The process takes 15 minutes or longer depending on the breast size. The system then calculates how the light is absorbed and uses the information to create an image.

CTLM gives a colorful three-dimensional, cross-sectional view of each breast. The images are bright green and show the breast from the front, sides, and back.

> CTLM is based on the principle of angiogenesis.

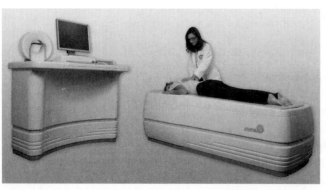

Figure 8–12. Computed tomographic laser mammography unit. (Used with permission from Imaging Diagnostic Systems, Inc.)

Currently, CTLM has not yet been approved for general use by the FDA. It has, however, been approved for use in Canada, Europe, and Asia.

Advantages of CTLM
- There is no breast compression.
- There is no radiation.
- It can be useful in imaging dense breast.
- The technology can differentiate cystic versus solid lesions.
- It can be used to image breast implants.

Analysis of CTLM
- It cannot detect microcalcifications.
- There are reports of high false-positive rates.
- It is difficult to image small breasts with this system because not much breast tissue will go into the machine.

Transscan or T Scan

T scan uses low-level electric currents to map the breast (Fig. 8–13). It is also called electrical impedance imaging or electrical impedance spectroscopy (EI). The currents are applied either through an electrode patch attached to the arm or by a handheld cylinder, similar to the ultrasound transducer. The breast is not compressed, and there is no radiation, injection, or radionuclides. Although still experimental, the test has been approved by the FDA since 1999 as an adjunctive breast cancer screening tool. T scan is less expensive than MRI or even nuclear imaging studies, such as sentinel node mapping or scintimammography.

The technique works by creating an image "map" of the breast using a small electrical current. One volt of continuous electricity (about the same as a flashlight battery) is transmitted into the body through either an electrode patch attached to the arm or a handheld cylinder. The electric current travels through the breast, where it is measured at skin level by a probe placed on the breast. The electrical properties of tissue have been found to be associated with tissue cellularity, and cancer is more cellular than benign breast disease or normal breast tissue. The cancerous tissue, which conducts electricity differently from healthy tissue, will appear on the resulting images as bright white spots.

Microwave Imaging Spectroscopy

Microwave imaging involves exposing a cross section of the breast tissue to nonionizing radiation and recovering data on the electromagnetic properties of the different tissue components in the path of transmission. In this imaging modality, an array of antennas is positioned at a fixed diameter of 15.2 cm from the breast. The breast and the array can also be immersed in a saline solution to increase the range of transmission frequencies and limit stray microwave radiation. During the image acquisition stage, the array moves vertically to scan the breast in 1-cm increments from top to bottom. A single antenna will emit microwaves. These waves are detected by nine antennas forming an arc around the opposite side of the breast. Data are collected and used to generate an image of the electromagnetic properties of the

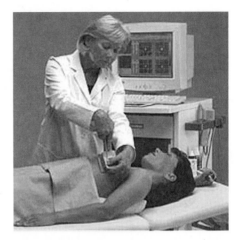

Figure 8–13. T scan using a device similar to the ultrasound transducer. (Copyright © Siemens Healthineers 2017. Used with permission.)

different tissues. Breast cancer, benign breast conditions, and normal breast tissue will all generate specific tissue differences because breast cancer has higher water content than benign or normal breast tissue.

Neutron-Simulated Emission Computed Tomography

Neutron-simulated emission computed tomography (NSECT) is the noninvasive mapping of the concentration of stable or radioactive isotopes in any 3D volume of the breast. By imaging the relative concentrations of trace elements in breast tissue, the subtle difference in chemical composition between benign and malignant tissue will be detected. NSECT uses fast neutrons with energies between 1 and 10 MeV to bombard the body. They deposit excess energy in the nuclei of atoms and molecules. To release this energy, the nuclei release gamma ray photons. The photons released will depend on the specific nuclei of the atom or molecule. A detector can then record the emitted gamma rays from various angles, creating a 3D map of the specific chemical isotopes. The advantage of this technology is the possibility of detecting cancers even before they begin to cause anatomic changes, potentially at the precancerous stage.

This technology is still in the early stages of development, but preliminary results indicate that the system would be less expensive than a CT or MRI system and would give radiation doses similar to the screen/film mammography.

Summary

Breast imaging is an exciting and dynamic field. The technology is evolving, and detection tools are constantly emerging. Many alternative technologies attempt to address issues such as radiation concerns, patient comfort, and specificity, and although the mammogram is still the gold standard, mammographers and other imaging specialists need to be prepared and to be informed. Any new technology will involve a learning curve, but that curve can be minimized if technologists are prepared and willing to integrate new technology into present imaging techniques.

Digital Tomosynthesis (DBT)

This technology has the x-ray tube rotating around the patient. There is no movement of the patient, and the results are high-resolution cross-sectional three-dimensional (3D) images, which eliminate overlapping structures. The images can be reconstructed with slice separations of 1 mm. s2D images are reconstructed 2D images generated from the 3D DBT data set. The use of s2D imaging does not require additional radiation exposure after the initial DBT exposure.

Three-Dimensional and Color Doppler Ultrasound

The technology works by assigning different colors to blood flow depending on velocity and direction of blood flow, for example, to and away from the transducer. Cancers generally are highly vascular, and the use of high-frequency sound to image blood flowing to a structure could improve the detection rates for breast cancer. The images can be captured in two-dimensional (2D) format or 3D reconstruction can be applied.

Volumetric Imaging

Another advance in ultrasound is the concept of volumetric imaging. Instead of multiple sweeps across the breast as in regular ultrasound, a single probe covers the breast, automatically sweeping to create 3D volume image of the breast. After acquisition, postprocessing algorithms can be applied based on the nipple location.

Elastography

This imaging technology uses information from the ultrasound signal to produce an image display of the elastic properties of breast tissue. It is based on the theory that cancerous tissues are hard compared with normal breast tissue, and this difference in tissue stiffness can be imaged.

Magnetic Resonance Spectroscopy (MRS)

This is a noninvasive imaging similar to magnetic resonance imaging (MRI). In MRS, however, the functional breast cancer by-product, choline, is measured. The hope is that this technology will improve the detection rate of MRI and prevent unnecessary breast biopsies.

Molecular Imaging

Mammography, ultrasound, and MRI image the anatomical characteristics of breast cancer, whereas molecular imaging is based on how cancerous cells function. These technologies include:

- Positron emission tomography (PET) or positron emission mammography (PEM)
- Breast-specific gamma imaging (BSGI), breast scintigraphy, or scintimammography
- Lymphoscintigraphy or sentinel node mapping

PET or PEM imaging

PET imaging uses fluorodeoxyglucose (FDG), a radioactive tracer that is injected into the arm vein. The technology is based on the fact that cancerous tissue uses vast amounts of sugar. The radioactive substance is metabolized in the body like sugar and will go to tissues that are most active. Special gamma scanners detect the radiation emitted. The main difference between PET and PEM technology is the higher resolution images obtained with PEM technology because small gamma cameras are used.

BSGI, Scintimammography, or Breast Molecular Imaging

This technology uses the radiopharmaceutical technetium-99m (^{99m}Tc)-sestamibi, which accumulates in malignant lesions. When a small specialized gamma camera is used, the technology is called BSGI. BSGI is more sensitive than scintimammography, in which a regular gamma camera is used.

Lymphoscintigraphy or Sentinel Node Mapping

This technology is used to identify the lymphatic spread of breast cancer. A radiopharmaceutical is injected into the subareolar lymphatic plexus closest to the lesion or into the lesion. The radiopharmaceutical travels to the sentinel node, allowing identification of that node for dissection and eliminating the need for extensive lymph node dissection. This procedure is often performed using a blue dye in addition to the radiopharmaceutical to allow better visualization of the deep lymphatic.

OTHER IMAGING MODALITIES

Computed Tomographic Laser Mammography (CTLM)

This technology is based on the properties of the hemoglobin in a vascular tumor, which will absorb the CTLM laser light. The patient lies prone with the breast suspended in a hole in the table. The breast is scanned 360 degrees for temperature differences, which would indicate the presence of a tumor. Imaging takes 15 minutes per breast.

Optical Imaging

This technology uses infrared light to penetrate breast tissue. The extra blood vessels clustering a tumor distort light in a characteristic way.

Cone-Beam Breast Computed Tomography (CT)

This technology images the breast using radiation, similar to a CT scan. The patient lies on a cushioned table with a cutout in the middle and the breast suspended through an opening.

The scanner takes 360-degree scans without compression, approximately 300 images in 10 seconds.

Bioelectric Imaging or Electrical Impedance Scanning (EIS)

This technology uses low-level electric currents to "map" the breast. There is no compression, radiation, injections, or radionuclides. It is less expensive than nuclear imaging and MRI and ideal for imaging young women, but is contraindicated for pregnant women or those with electrically implanted devices.

REVIEW QUESTIONS

1. What is the technique whereby 11 to 15 low-dose tomography images are taken using one or two compressions of the breast?

2. What ultrasound technology identifies cancer by the relative hard properties of cancerous tissue?

3. What technology identifies the chemical choline, present in cancerous lesions?

4. Name three different nuclear imaging technologies.

5. Give three general terms used to describe the radioactive substance used in nuclear medicine studies.

6. What technology is based on the principle that most cancerous tissue uses a vast amount of glucose?

7. What nuclear imaging modality sees exercise as a contraindication?

8. Name the radioactive substance used during a scintimammography scan.

9. What major factor differentiates the scintimammography scan from the BSGI?

10. What syndrome is related to the repeated scanning motion used by ultrasound technologists?

11. In addition to the radiopharmaceutical, what other substance is often injected when trying to determine whether a cancer has left the breast?

12. What term is used to describe the process of forming new blood vessels?

13. Which technique uses infrared light to penetrate breast tissue?

14. What modality uses x-rays to image the breast without compression, giving an average dose of 9.9 mGy?

15. Give three limitations of CTLM.

16. What technique employs a low-level electric current to map the breast?

CHAPTER QUIZ

1. Scanning the breast to locate a cancer based on the vast amount of glucose/sugar used by cancer cell is called:

   (A) Scintigraphy

   (B) PEM imaging

   (C) MRI

   (D) Lymphoscintigraphy

2. Which of the following technologies is based on the principle of angiogenesis?

   (A) Scintigraphy

   (B) PEM imaging

   (C) MRI

   (D) CTLM

3. Which of the following types of breast imaging method will fall under the category of nuclear imaging?

   (1) PEM

   (2) MRI

   (3) Scintimammography

   (A) 1 and 2 only

   (B) 2 and 3 only

   (C) 1 and 3 only

   (D) 1, 2, and 3 only

4. Which of the following procedures requires the patient to lie still for 60 to 90 minutes after an injection and before the actual procedure?

   (A) Magnetic resonance spectroscopy (MRS)

   (B) PEM

   (C) Elastography

   (D) BSGI

5. A technique of injecting a radioisotope directly into the tumor bed to tract the extent or spread of a cancer, describes:

   (A) Scintigraphy

   (B) Scintimammography

   (C) Lymphoscintigraphy

   (D) Positron emission tomography

6. A digital method of scanning the breast to create three-dimensional reconstruction of tomography planes of the breast is:

   (A) Lymphoscintigraphy

   (B) BSGI

   (C) Digital breast tomosynthesis

   (D) Optical imaging

7. The principle of elastography is that:

   (A) Cancers are relatively hard compared with surrounding breast tissue

   (B) Electric current can travel easily thought the breast tissue

   (C) Malignant tumors develop numerous new blood vessels

   (D) Radioactive substances have an affinity to malignant cells

8. The radioactive compound used in BSGI is:

   (A) Fluorodeoxyglucose

   (B) Gadolinium

   (C) ^{99m}Tc-sestamibi

   (D) Radiation

9. An imaging technology that measures choline, the functional breast cancer byproduct, is:

   (A) Magnetic resonance spectroscopy (MRS)

   (B) Lymphoscintigraphy

   (C) Elastography

   (D) BSGI

10. A technology that uses low-level electric currents to "map" the breast is:

    (A) Magnetic resonance spectroscopy

    (B) Digital breast tomosynthesis

    (C) Bioelectric imaging or electrical impedance scanning (EIS)

    (D) Positron emission mammography

BIBLIOGRAPHY

Adler A, Carlton R. *Introduction to Radiologic Science and Patient Care.* 6th ed. St. Louis, MO: Elsevier Saunders; 2016.

Baltzer PAT, Benndorf M, Dietzel M, et al. False-positive findings at contrast-enhanced breast MRI: A BI-RADS Descriptor Study. *Am J Roentgenol.* 2010;194:1658–1663.

Bushong SC. *Radiologic Science for Technologists—Physics, Biology and Protection.* 10th ed. St. Louis, MO: Mosby; 2012.

Cryotherapy—Researchers look to ice breast cancer. *Radiology Today.* 2010; 11(5):26. http://www.radiologytoday.net/archive/rt0510 p26.shtml.

Digital Breast Tomosynthesis Is Now FDA Approved. http://www.hologic.com/en/medical-professionals/online-home/breast-tomosynthesis. Accessed October 2011.

Helvie M. Digital mammography imaging: Breast tomosynthesis and advanced applications. *Radiol Clin North Am.* 2010 Sep; 48(5): 917–929. doi: 10.1016/j.rcl.2010.06.009. Available at https://www.ncbi.nlm.nih.gov/pmc/articles/PMC3118307/ Accessed May 2017.

Imaginis. T-scan Impedance Imaging of the Breast is Safe and Effective. http://www.imaginis.com/t-scan/t-scan-impedance-imaging-of-the-breast-is-safe-and-effective. Accessed September 2016.

Imaging Diagnostic Systems Inc. CTLM® Computed Tomography Laser Breast Imaging system. http://imds.com/about-computed-tomography-laser-mammography. Accessed May 2017.

Kopans B. Digital breast tomosynthesis from concept to clinical care. *American Journal of Roentgenology.* 2014;202: 299-308. 10.2214/AJR.13.11520. Available at http://www.ajronline.org/doi/abs/10.2214/AJR.13.11520. Accessed May 2017.

RSNA Press Release. MR Spectroscopy Reduces Unnecessary Breast Biopsies: Technology Identifies Chemical Make-up of Breast Tumors. September 25, 2007. http://www.prweb.com/releases/RSNA/MR_ Spectroscopy/prweb556 016.htm. Accessed February 12, 2010.

RSNA News Release. Combined Imaging Technologies May Better Identify Cancerous Breast Lesions. Released: November 9, 2010. http://www.rsna.org/media/pressreleases/pr_target.cfm?ID=522. Accessed February 12, 2011.

Schilling K. Breast imaging with a positron edge: Advantages in positron emission mammography. *RT Image Magazine.* 2009; February 2:32–33.

Scott B. An alternate takes center stage: Breast imaging on the molecular edge. *RT Image Magazine.* 2008; September 29:21–23.

TransScan Wins FDA Clearance FOR T-Scan 2000 Breast Imaging Device. http://www.cancernetwork.com/articles/transscan-wins-fda-clearance-t-scan-2000-breast-imaging-device. Accessed September 2016.

Wilcsek B, Aspelin P, Bone B, Pegerfalk A, Frisell J, Danielsson R. Complementary use of scintimammography with 99m-Tc-MIBI to triple diagnostic procedure in palpable and nonpalpable breast lesions. *Acta Radiol.* 2003; 44(3):288–293.

Interventional Procedures | 9

Keywords and Phrases
The Need for Interventional Techniques
Cytological Analysis
Histological Analysis
General Patient Preparation
Cyst Aspiration
Stereotactic Localization and Biopsy
   Add-on Units
   Dedicated Prone Units
   Blended Technology
   Stereotactic Procedure
   Limitations
Preoperative Needle Localization
   Localization Procedure Using Mammography
Breast Biopsy
   Fine Needle Aspiration Biopsy
   Core Biopsy Methods
Other Biopsy Methods
   Radiofrequency Biopsy
   The Mammotome Biopsy System
   The Automatic Tissue Excision and Collection
   Large Core Breast Biopsy
   Open Surgical Biopsy
   Modality-Based Core Biopsy
   Comparison of the Various Biopsy Methods
Specimen Radiography
   Specimen Radiography Units
Ductography or Galactography
Nipple Aspiration
Ductal Lavage
Pathology Review
Mammoplasty
   Cosmetic Intervention
   Reduction Mammoplasty

Mastopexy
Augmentation Mammoplasty
Summary
Review Questions
Chapter Quiz

Objectives

On completing this chapter, the reader will be able to:

1. Differentiate between cytological and histological sampling
2. State the purpose of cyst aspiration
3. Discuss the advantages and disadvantages of the different stereotactic breast biopsy units
4. Identify the purpose of breast needle localization
5. Discuss the advantages and disadvantages of the various core biopsy methods
6. Explain why magnetic resonance (MR)–guided biopsy is often necessary
7. Describe the process of imaging the specimen
8. Differentiate between ductography and ductal lavage

KEYWORDS AND PHRASES

- **Aspiration** is a method of removing fluid from a cyst using a needle. Aspiration is often done under ultrasound guidance.
- **BIRAD system—breast imaging reporting and data** is a standardized mammographic reporting system that can be used as a coding and assessment system.
- **Capsular contracture** or firm rigidity to the breast is a complication of implant augmentation. It is often the body's normal response to the presence of a foreign object. A fibrous capsule forms around the implant. Most often, the capsule remains pliable and slightly larger than the implant. Occasionally, however, the capsule contracts and becomes constrictive because the tendency is to revert to a sphere— the smallest possible area per volume. In cases with calcified capsule, ultrasound and magnetic resonance imaging (MRI) have proved invaluable as adjunctive evaluation tools.
- **Connective tissue disorders** can sometime be associated with silicone implants. These can include any collagen vascular disease with inflammation, such as rheumatoid arthritis and osteoarthritis.
- **Core biopsy** is a technique of obtaining core samples from a questionable area for histological analysis.
- **Ductography** is used to evaluate patients with abnormal nipple discharge. The lactiferous duct is cannulated, and a small amount of contrast is injected into the ducts. Craniocaudal (CC) and mediolateral (ML) radiographs are taken. The purpose of the examination is to determine the location and number of lesions. The examination will not determine whether the lesion is malignant or benign.
- **Radiofrequency (RF) waves** are part of the electromagnetic spectrum. The frequency of RF waves can range from 1605 kHz to 54 MHz, and these waves have multiple communication uses.

- **FNAB—fine needle aspiration biopsy**—is a technique used to obtain cellular material from a questionable area within the breast.
- **Informed consent** implies that the physician must tell the patient of all the potential benefits and risks of the procedure. The patient is also informed of any possible alternative procedures and must sign a written consent.
- **Mammoplasty** is the reshaping of the breast either by enlarging or reducing the breast size; breast size reduction; breast lift; breast lift with reduction; breast lift with augmentation.
- **Mastopexy** is cosmetic surgery to lift the breast.
- **Pneumocystography** was used to examine the inner walls of a cyst. The cyst was aspirated and air injected. This procedure was often done under ultrasound or mammography guidance; however, it is not a common procedure in today's imaging world.
- **Preoperative needle localization** is a technique usually performed on nonpalpable abnormalities, such as lesions or microcalcifications. Because the surgeon cannot feel the lesion, the lesion must be localized to a specific area—therefore minimizing the amount of tissue that will be removed.
- **Specimen** is the tissue that is removed during a biopsy. The specimen should be examined radiographically to confirm that the area in question was removed.
- **Stereotactic localization** is a system using computerized stereo equipment to localize a lesion before an open surgical, core, or fine needle biopsy. The computer calculates the proper coordinates of the lesion in the breast and enables placement of the needle in the exact location of the lesion.
- **Transducers** are devices that convert energy from one form into another. In ultrasound, the piezoelectric transducers convert electrical pulses to sound acoustic energy, and the returning acoustic energy is converted back into electrical energy.
- **Vasovagal reaction** is a condition in which the pulse rate drops. If left untreated, it can result in fainting.

> Only a histological or cytological analysis can definitely confirm a malignancy.

THE NEED FOR INTERVENTIONAL TECHNIQUES

Generally, the first step in breast cancer detection is the mammogram; however, if on mammography a lesion is discovered to be irregularly shaped, lobulated, obscured, or spiculated, or a definite area of architectural distortion is identified, the next step is to use supplemental projections to rule out malignancy. If malignancy cannot be ruled out, or even if the finding has a high probability of malignancy, only a histological or cytological analysis can be used to confirm malignancy. A finding can be palpable or nonpalpable, microcalcifications or masses, but only by removing a sample of the tissue in question and examining the removed breast tissue using microscopic analysis can a definitive diagnosis be made.

The type of intervention performed depends on several factors, such as how suspicious the abnormality is; its size, shape, and location; and the number of abnormalities present. Other factors that determine the type of biopsy performed include the patient's medical history, patient's preference, training of the radiologist or surgeon, and type of facility where the biopsy is being performed. Hormone markers are performed before surgery because chemotherapy, then lumpectomy or mastectomy, is recommended for certain tumor types.

Radiologists often use the Breast Imaging and Reporting and Data System (BI-RADS) to guide the choice of an interventional procedure. The BI-RADS categorizes lesions according to their suspiciousness and recommends fine needle aspiration biopsy (FNAB), fine needle biopsy, or core biopsy on category 4 lesions. Occasionally a category 3 lesion will be suggested for biopsy, although the routine procedure is short-term follow-up. Category 5 lesions are generally highly suggestive of malignancy, so a biopsy should be performed. The protocol is to perform a core biopsy before surgery if possible. A surgical biopsy also is recommended if the cytological or histological findings are inconsistent with the imaging.

CYTOLOGICAL ANALYSIS

If a cytological analysis is planned, the specimen is removed from the site with a 22- to 25-gauge needle. The needle can be washed out in an alcohol preservative solution or the aspirate placed in CytoPrep solution and sent to the laboratory for analysis. In the older conventional method, the aspirated is smeared on a glass slide, immediately fixed with 95% alcohol, dried, and sent to the laboratory. All smears are protected with coverslips. Alcohol-fixed smears are stained with a modified Papanicolaou (Pap) stain process. Air-dried stains are not common; however, if air dried, they are not stained with Pap stain. A cytopathologist will determine the benign or malignant nature of the aspirate from the cytomorphological appearance of the microscopic tissue fragments and individual cells. Routine cytological reports of an FNAB can be ready within 24 to 48 hours.

HISTOLOGICAL ANALYSIS

The histological process begins with macroscopic tissue blocks. The tissue specimen is surgically removed and preserved by fixing in a 10% formalin solution (formaldehyde in water). To process the specimen, it is dehydrated, defatted, and embedded in warm paraffin. After cooling, the embedded tissue can be sliced as thin as paper, collected on a slide, stained (using a hematoxylin and eosin technique), and protected with a coverslip. The pathologist can determine the morphology of the tissue fragment after microscopic analysis. Accurate results can be obtained in 3 to 5 days.

If a quick determination of the nature of a suspicious lesion is needed during an open biopsy, the surgeon can request a frozen section. The biopsy specimen is removed while the patient is under anesthesia and taken to the histology department, where a special machine called a cryostat is used to freeze-harden the tissue biopsy, allowing cutting into transparent centimeter slices. These slices are placed on a glass slide, stained with a water-soluble stain, coverslipped, and viewed under a microscope. Because the patient is "under anesthesia," this process has to be rapid, and results are usually given in approximately 10 minutes.

GENERAL PATIENT PREPARATION

For any interventional procedure, the patient must be informed and fully prepared. On the day of the procedure, the patient should wear comfortable clothing with no deodorant, powders, lotions, or creams in

> **Cytological analysis** is an examination of the cells; results in 24 to 48 hours.
>
> **Histological analysis** is a study of the tissues of a specimen. Preliminary frozen section results in 10 minutes; accurate results in 3 to 5 days.

the area of the breast. Patients are usually advised to have a light meal to minimize nausea, and to avoid all blood-thinning medication, such as aspirin. Some herbal remedies also have blood-thinning properties, and patients should inform their physician of all medications they are taking, including prescription drugs and herbal remedies. An informed consent is required before any interventional procedure.

The specific definition of informed consent may vary; however, in general, it implies that a physician (or other medical provider) must tell a patient all of the potential benefits, risks, and alternatives involved in any surgical or interventional procedure or treatment and must obtain the patient's written consent to proceed. This principle is based on the patient's rights, namely the right of the patient to refuse treatment or to prevent unauthorized contact. The informed consent therefore ensures that the patient can make a reasonable decision regarding the treatment or procedure, based on an understanding of that treatment or procedure. In many situations, the failure to obtain informed consent is a form of medical negligence and could lead to a charge of battery.

The physician or medical provider, and not the nurse or technologist, is the best authorized person to speak to patients about informed consent. The physician should also ensure that patients understand what they are hearing. The dialogue with the patient should include the following:

- The patient's current diagnosis
- The nature and purpose of the proposed intervention procedure or treatment and the procedure's probability of success
- The benefits and risks of the proposed treatment or procedure
- Any alternatives to the proposed treatment or procedure. Patients should be made aware of alternatives to the proposed treatment or procedure regardless of the cost of the alternatives and the patient health insurance status.
- The risks and benefits of any alternative treatment or procedure
- The risks and benefits of not receiving or undergoing the proposed treatment or procedure

All consent forms must be signed and dated by the patient or the patient's legally authorized representative. The patient should receive a copy of the informed consent, and a copy should be included in the patient's medical records.

If the patient is mentally disabled, an appointed guardian is usually authorized to make medical decisions and give informed consent.

Consent for minors, in most situations, can be given by the parents. However, some states in the United States allow adults younger than 18 years to play a more active role in their medical care and treatment, including the process of informed consent. A minor can give consent in a few other instances, such as if the minor is married, in active military service, is a mature minor (considered older than 14 or 15 in some states), or is emancipated.

CYST ASPIRATION

An aspiration can be indicated to relieve the pain associated with a cyst or to prove a cyst by aspiration of the contents. Cyst aspiration can also be performed if the echogenic contents suggest a solid mass. Aspiration

Cytological analysis:
- Results in 24–48 hours

Histological analysis:
- Preliminary results in 10 minutes
- Accurate results in 3–5 days

An informed consent must be obtained before any interventional procedure.

is often done under ultrasound guidance. For the ultrasound examination, in most cases the patient lies turned slightly to the contralateral side with the ipsilateral arm resting above the head if possible. For cyst aspiration, a large gauge needle, for example 18-gauge, should be used.

The transducer must be cleaned or covered, and lidocaine is administered along the proposed course of the needle tract. The transducer is then positioned over the lesion away from the site of entry of the needle. If the needle runs parallel to the transducer, it will be clearly visualized on the ultrasound image. When the needle is in position, aspiration is performed with multiple in-and-out movements of the needle. The needle should be manipulated to obtain material from different parts of the lesion.

If the cyst has both cystic and solid components or no fluids, it can be intracystic papilloma or intracystic carcinoma. If nonfluid is obtained and the cyst resolves, no further workup is necessary. Greenish-black or tan fluids can be discarded, but bloody or dark-red fluids should have a cytopathological review and may also require an excisional biopsy. Pneumocystography can also be used to assess the internal walls of the cyst. This technique injects a small amount of air into the cyst (0.5 to 2 mL, depending on the size of the cyst). A mammogram can be taken to assess the cyst. High-resolution, linear array ultrasounds (12–14 MHz) have also been used successfully in evaluating the inner walls of a cyst.

For this and all other invasive procedures, the patient must first sign a consent form, and preliminary scout images are always obtained.

STEREOTACTIC LOCALIZATION AND BIOPSY

If a lesion is not palpable and can be seen on mammogram, stereotactic localization is necessary because the location of the lesion must be identified before it can be biopsied (Fig. 9–1). Calcifications and deep lesions and lesions in large fatty breasts are ideal for stereotactic mammographic localization or biopsy. The procedure can be performed using a standard mammographic unit with an "add-On" attachment or a dedicated prone

Add-on unit:

- Units attach to a regular mammography unit
- Patient is seated during the examination

Prone unit:

- Patient lies prone; affected breast protrudes through table opening
- Physician and technologist can remain seated during examination
- Newer tables allow imaging 360 degrees around breast

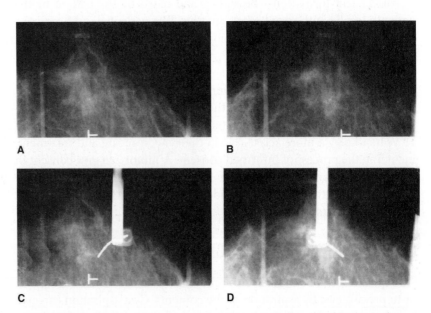

Figure 9–1. Stereo images showing the needle localized in the lesion.

table. The imaging can be done by analog or digital means. The advantage of digital imaging is the elimination of development time, therefore reducing the chance of poor positioning of the biopsy needle due to motion. Digital imaging also has the ability to postprocess the image for better visualization of fine details, especially microcalcifications.

The stereotactic equipment uses angled images to triangulate the depth of a lesion within the breast and to calculate its position in three dimensions. The horizontal (X axis), vertical (Y axis), and depth (Z axis) are calculated with the nonangulated radiograph. The tube is then angled 15 degrees to the left and right along the X axis for two scout images, used for calculations of the depth (Z axis). After the lesion is located, the stereotactic unit is then used to position a biopsy probe or localization needle within the breast at the calculated coordinates.

Add-On Units

With the add-on unit, the patient is imaged upright, usually seated (Fig. 9–2). These units have the advantage of being relatively less expensive compared with the prone units. The units also do not require a dedicated biopsy room. After using the equipment, the add-on attachment can be removed, and the unit can revert to a dedicated mammography unit. Add-on units can image the posterior breast and axilla area of the breast better than the older prone units, but there is less space to work in, and because the patient is upright there is a greater risk for vasovagal reactions by the patient. Motion of the patient can also be a problem with the add-on units, and patient motion will compromise the image.

> Add-on units have the patient imaged seated and have the advantage of being relatively less expensive compared with the prone units. However, these are not recommended for nervous or anxious patients. The prone unit is more expensive, but there is less chance for a vasovagal reaction. These units have essentially only one function.

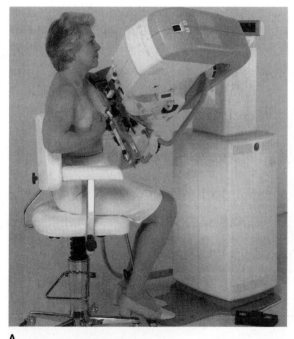

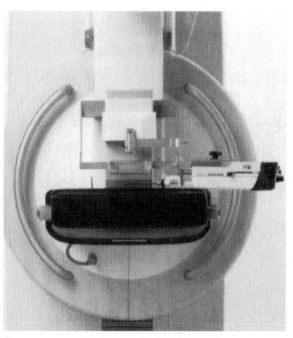

A **B**

Figure 9–2. Add-on units. **(A)** (Used with permission from GE Medical Systems.) **(B)** (Copyright © Siemens Healthineers 2017. Used with permission.)

Dedicated Prone Units

The prone units are dedicated procedure units, and the patient lies prone with the breast of interest extending through a hole near the middle of the table. Some units allow the affected arm and breast to extend through the hole, thus enabling access to far posterior lesions and 360-degree tube rotation around the breast (Figs. 9–3 and 9–4). The mammography unit and needle guidance device are under the table. This table can be raised or lowered to suit the radiologist's preference, and both the radiologist and the technologist can remain seated during the procedure. The prone unit is more expensive, but there is less chance for a vasovagal reaction. These units unfortunately have essentially only one function and tend to leave a room idle when not in use.

Blended Technology

The newer prone tables are a blended technology. They are add-on prone tables and include a "lateral arm," allowing biopsy along a path perpendicular to the plane of compression. Other improvements include the "target-on-scout" that allows the radiologist to choose the best of the two scout images coupled with a straight, 0-degree projection to target the lesion. The lesion's location is then calculated using images obtained 15 degrees apart rather than the traditional 30 degrees. This helps if the lesion is seen well in one of the stereo images but is not seen, or is poorly seen, in the other. The straight projection is also helpful because it shows the needle in a more straightforward manner.

Stereotactic Procedure

The lesion is optimally positioned in the center of the window where calculations of the depth are most accurate. When positioning is confirmed, the skin is marked to assess patient movement between

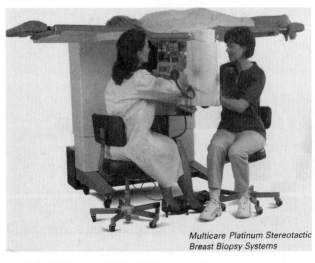

Multicare Platinum Stereotactic Breast Biopsy Systems

Figure 9–3. Prone units. (Used with permission from Hologic Inc.)

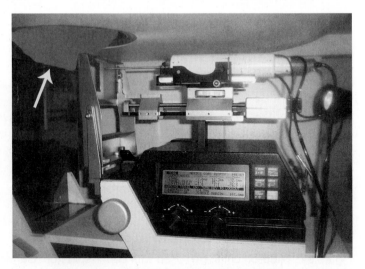

Figure 9–4. Breast placement in the prone unit.

imaging. The area within the window is cleaned, and stereo images are then obtained. The stereotactic unit will move automatically for imaging.

The biopsy needle is positioned and a prefire stereotactic pair of images obtained to confirm appropriate placement of the needle. Any fine adjustments can then be made, and if positioning is confirmed by another set of stereo images, the biopsy device is fired. Postfire stereotactic images are obtained to confirm that the biopsy needle traversed the area of interest.

If core tissue samples are needed, a postbiopsy stereo pair of images should be taken to confirm a decrease in the number of calcifications or small pockets of air throughout the target, verifying the accuracy of the core acquisitions. For microcalcifications, a specimen radiograph is often obtained to confirm adequate sampling before the biopsy device is removed. After removal of the core sample, a radiopaque marker is inserted to mark the site of the original microcalcifications.

Limitations

If the breast is compressed to less than 2.5 to 3 cm, a long-throw needle will hit the detector plate of the stereotactic unit or penetrate the skin on the other side of the breast. This problem should not occur for FNAB, or with short-throw needles, which have a 1.2 excursion when the spring-loaded gun is fired.

Patient movement after prefire images is often the most common cause of failure to obtain adequate samples of the lesion.

PREOPERATIVE NEEDLE LOCALIZATION

If stereotactic technology is not available, the preoperative needle localization technique can be used to locate any nonpalpable lesion or calcification within the breast. This procedure is done before a surgical incisional or excisional biopsy. The localization wire can be placed in the breast using either mammographic, sonographic, or magnetic resonance (MR) guidance (Fig. 9–5). Digital mammography units can be used. The use of digital imaging will reduce the time of the procedure because of the elimination of processing time.

Because this is a prelude to a surgical biopsy, the patient should be scheduled for needle localization only after review of images and coordination of plans with a surgeon. The needle here should be placed directly in the lesion with at least 5 mm of the wire within the lesion. The hook of the wire should be approximately 1 cm beyond the lesion. Any approach to the lesion should be parallel to the chest wall to minimize complications, and in general, the shortest approach should be used to reduce the risk of missing the lesion. The needle length can be calculated from the original image, and it is usually better to use a needle that is too long rather than run the risk of using too short a needle because needles that extend too deep can always be pulled back.

After the procedure, the relevant images with the lesion clearly marked should be sent with the patient to the surgeon. With digital technology, the images can be sent digitally. Some radiologists also

Comparison of Units

Prone Units
- Requires more space
- More expensive
- Most tables have a 300-lb weight limit
- Safe for most patients

Add-On Units
- Require less space
- Less expensive
- Not best option for elderly, anxious, or physically impaired patient
- Increased chance of vasovagal reaction

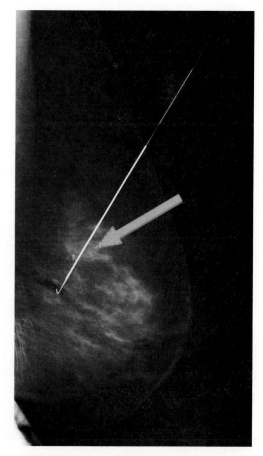

Figure 9–5. Preoperative localization with the needle and hook-wire in place in lesion. Arrow indicates lesion.

provide the surgeon with line diagrams or a brief description, showing or telling the approximate distance from the breast skin surface to the lesion.

Localization Procedure Using Mammography

Breast localization can be performed using mammography, ultrasound, or magnetic resonance imaging (MRI). In the mammography localization procedure, the technologist positions the breast with the patient seated or prone using a compression plate with an open window and alphanumeric grid (Fig. 9–6). The lesion is positioned within the open window.

The breast skin within the window is cleaned, and the localization needle tip is inserted into the breast directly above the lesion. Insertion should be as perpendicular as possible, and care should be taken not to penetrate to the opposite side of the breast. With the needle in place, the technologist takes another image. Compression is released without disturbing the needle, and the second image taken is at 90 degrees to the first. If the first image is in the CC projection, the second should be the ML, and vice versa. These two images are used to prove the location of the needle in the lesion.

If the needle is not within 5 mm of the lesion, it should be repositioned. Additional images can be taken as needed to confirm replacement of the needle. When the final position of the needle is

> Breast localization can be performed by using mammography, ultrasound, or MRI.

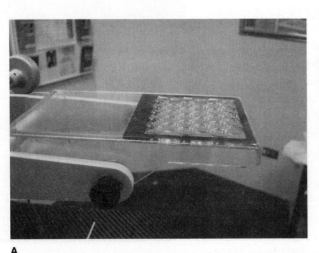

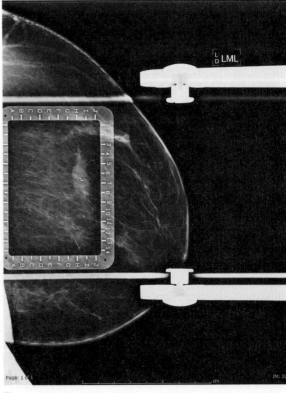

A **B**

Figure 9–6. **(A)** Breast localization compression plate. **(B)** Biopsy compression for use with digital units.

determined, the wire is inserted through the needle. Most localization wires have a hook or curved end, which should be placed just beyond the lesion. The needle is removed from over the wire, and a final image can confirm placement of the wire within the breast. Patients are sent for surgery with the wire covered and stabilized for safety. Commercially, wire covers are available that conform to the body and ensure that the localization wire is protected and visible at all times.

BREAST BIOPSY

A breast biopsy is the taking of a sample specimen for cytological or histological analysis. The type of biopsy performed depends on a number of factors, such as how suspicious is the abnormality; its size, shape, and location; and the number of abnormalities present. Other factors that should be checked include the patient's medical history, patient's preference, training of the radiologist or surgeon, and where the biopsy is being performed. The lesion should be properly evaluated, using multiple modalities if necessary, to rule out malignancy. If a nonpalpable mammographic mass is found to be oval, round, or slightly lobulated with relatively well-circumscribed margins, most radiologists will evaluate the mass with ultrasound. A simple cyst is generally recommended for routine follow-ups, but a solid or indeterminate lesion on ultrasound—not cystic or solid—or a complex mass with low-level internal echoes, and even a mass with fluid-debris or sponge-like clusters of cysts or thickened walls, will need a further study, such as aspiration. Masses that aspirate normal fluid with no blood present and masses that collapse during aspiration are generally determined to be benign and can be followed as part of a routine evaluation. Benign fluids can include secretory, inflammatory, benign epithelium, or apocrine cells. If, however, bloody fluid is aspirated, then the mass must be further evaluated by cytological examination or should be biopsied.

In the 1980s and even early 1990s, all women had to have open surgical biopsy to confirm a cancer. Today, however, minimally invasive biopsies, which require local anesthesia, are the norm. Nearly all minimally invasive biopsies can be performed as an outpatient versus an inpatient procedure, leaving less visible scarring. With smaller biopsies, there are fewer complications and less pain, bleeding, and scarring. Minimally invasive biopsies are also less costly because the patients are sent home the same day and normally will not require an overnight stay in the hospital. If a minimally invasive biopsy yields a benign concordant diagnosis, the woman is then spared the need for a surgical biopsy (Box 9–1).

Using the BI-RAD system, a minimally invasive biopsy is usually recommended on a category 4 or 5 lesion. Occasionally, a category 3 lesion will be biopsied, although the routine procedure is short-term follow-up. If short-term follow-up is not an option—either due to the remote location of the patient or because of patient noncompliance—a minimally invasive biopsy could be recommended. Examples could include

BIRAD™ categories

- BIRAD 0: Need additional imaging information and/or prior mammograms for comparison.
- BIRAD 1: Negative
- BIRAD 2: Benign finding
- BIRAD 3: Probably benign finding—short interval follow-up suggested
- BIRAD 4: Suspicious abnormality—biopsy should be considered
- BIRAD 5: Highly suggestive of malignancy—appropriate action should be taken
- BIRAD 6: Known biopsy proven malignancy—appropriate action should be taken

Box 9–1. Types of Breast Biopsy

- Fine needle aspiration biopsy (FNAB)
- Minimally invasive biopsy
  - Core needle biopsy (CNB)
  - Vacuum biopsy
- Large core biopsy
- Lymph node biopsy
- Open surgical biopsy

Cyst aspiration:

- Removal of the content of a cyst
- To relieve pain and assess contents

Preoperative localization:

- A prelude to the surgical biopsy
- Used if stereo localization is not available

FNAB analysis:

- Fastest and easiest biopsy method
- Less expensive and invasive than core biopsy
- Less accurate than core biopsy
- Rapid results
- No stitches or scarring
- Needs multiple needle sticks
- Cannot be used for chest wall lesions

the extremely anxious patient, or patients with plans for pregnancy, breast augmentation, or reduction. Category 5 lesions are generally highly suggestive for malignancy. However, a core biopsy is performed on almost all lesions before surgery.

One major controversy with any minimally invasive biopsy technique is the possibility of epithelial displacement. Even an anesthetic injection can lead to the displacement of benign or malignant epithelium into tissue away from the lesion. This displaced epithelium can also migrate to the skin and can lead to difficulties in histological interpretation or can mimic ductal carcinoma *in situ* (DCIS). Although there is no reported documentation addressing the implication of epithelial displacement, a vacuum-assisted biopsy reports a lower frequency of epithelial displacement. A possible reason is that with other techniques the needle is fired through the cancer. Firing through the cancer is not necessary using the vacuum-assisted method. Also, with vacuum-assisted biopsies, the needle is fired only once, and a larger volume of tissue is acquired. This suggests that the displaced cells are likely to be retrieved by the vacuum, which pulls cells into the probe rather than displacing them from the biopsy site.

If the minimally invasive biopsy leads to inconclusive results, a surgical biopsy is suggested. A surgical biopsy is also recommended if the histological findings and imaging findings are discordant. After any benign diagnosis at minimally invasive biopsy, a follow-up is essential. A short-interval follow-up mammogram of the affected breast is recommended at 6 months and a bilateral mammogram at 12 months.

Fine Needle Aspiration Biopsy

FNAB is used to obtain cellular material from the area in question for cytological analysis. This technique can be used to diagnose both cystic and solid lesions, such as fibroadenomas. If the lesion is not palpable, the needle can be guided to the lesion using ultrasound or mammographically using stereotactic breast localization (Fig. 9–7). FNAB can reduce the necessity for a surgical breast biopsy, but the accuracy of FNAB is dependent on the individual performing the procedure—the radiologist or surgeon—and the ability of the cytologist. Better

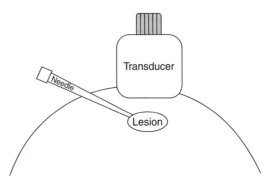

Figure 9–7. Ultrasound-guided cyst aspiration. Showing the needle position relative to the ultrasound transducer.

specimens are obtained for cytological assessment using gauges 22, 23, or 25. The needle length used will depend on the depth of the lesion in the breast, but in general, needles 3.8 cm long are adequate, although needles 7 or 9 cm are also available. Most FNAB needles are disposable. Other equipment needed for the examination includes a plastic syringe with Luer-Lok tip and a CytoPrep kit.

Patient Preparation

Candidates for an FNAB should not have bleeding disorders and should not be taking anticoagulants. All patients should avoid aspirin, fish oil, and nonsteroidal anti-inflammatory medications 5 days before and 5 days after the procedure. An FNAB can be performed on patients taking anticoagulants; however, this depends on the *concomitant aspirin* (ASA) therapy, and the International Normalized Ratio (INR). The INR is a blood test used to check the clotting time. The higher the INR, the longer it will take blood to clot and the higher the risk for bleeding. Because this is an invasive procedure, the patients must sign an informed consent. The risks and alternatives to the procedure should be explained to the patient, and the patient should obtain follow-up care instructions, including how to get results and when results will be available.

Procedure

FNAB requires aseptic but not sterile technique. The skin is cleaned with alcohol or povidone-iodine (Betadine), and a local anesthesia (e.g., lidocaine) is administered along the course of the proposed needle tract.

The needle for the FNAB ranges from 20 to 25 gauge. The technique involves the use of some degree of vacuum along with up, down, and rotational movements of the needle. FNAB collects cellular material, which can be placed in a CytoPrep tube and sent to the laboratory for analysis by a cytopathologist or cytotechnologist. For many FNAB procedures, the cytotechnologist or cytopathologist is present during the procedure to verify adequate specimen collection.

Care must be taken when sampling cells for FNABs. The aspirate material should not be filled with blood. Heavy blood or fluid is difficult to interpret. The preparation of aspirate should always be discussed with the cytopathologist.

If the result is a cancer diagnosis, the specimen is further scrutinized for antibodies, enzymes, or proteins. This will provide information on the prognosis and therapeutic options. One widely known marker is the HER2/neu protein. Cancers with this marker can be treated with the drug trastuzumab (Herceptin).

The most serious drawback to FNAB is the need for a highly trained breast cytopathologist or cytotechnologist. Also, because FNAB provides cytological rather than histological specimens, without a good cytotechnologist there is a higher probability of false-negative readings and a higher rate of insufficient specimen samples.

Postcare and Follow-Up

Postcare under FNAB is minimal. The skin is cleaned with alcohol, and bandages can be applied after the procedure. Postbiopsy imaging

Core biopsy analysis:

- Is more invasive than FNAB, but results are more definitive
- Requires several needle samples
- Poor for small or hard lumps

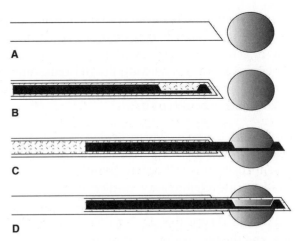

Figure 9–8. Core biopsy gun: prefire. **(A)** The trough without the needle. **(B)** The trough with the needle. **(C)** Postfire into the lesion. The needle moves forward, filling the trough with breast tissue. **(D)** Postfire. The outer sheath moves forward to cut the tissue and keep it in the trough.

of the site should be performed within 6 months to check for missed lesion.

Core Biopsy Methods

The core biopsy method is the most commonly performed minimally invasive technique. It is inexpensive, easy to perform, and highly accurate for many lesions. A core biopsy removes a larger sample of tissue than FNAB because tissue samples are obtained using an 11- to 14-gauge or larger needle (Fig. 9–8). Smaller needles will compromise the ability of the pathologist to make an accurate diagnosis. To yield adequate specimens, at least four core samples are needed for assessing a lesion and at least six for assessing calcifications.

The core samples are obtained using a gun-needle combination. Gun-needle combinations are available mainly as disposable equipment. They are also available as long throw, for which the distance the needle travels in the breast exceeds 2 cm, and short throw, for which the distance traveled by the needle is less than 2 cm. The length of the throw selection will be determined by the breast size and location of the lesion. Because of the larger needle size used, a core biopsy will require 1-mm skin nicks at the site.

The procedure uses either an automatic or mechanical core "gun." The gun-needle combinations have an inner needle with a trough extending within it. One end is covered by a sheath and attached to a spring-loaded mechanism. When the mechanism is activated, the needle moves forward, filling the trough with breast tissue. The outer sheath instantly moves forward to cut the tissue and keep it in the trough. The gun-needle combination is designed to move a cutting needle rapidly through the breast. It takes only a fraction of a second to obtain a sample, but for each sample it is necessary to withdraw the needle to collect the tissue. Because each reinsertion results in additional destruction of breast tissue and hemorrhage into the area of biopsy, there is a limit to the amount of insertion possible on a small lesion.

Radiofrequency analysis:

- Device is light and easy to hold and place
- Examination is quick
- Bleeding is the most common complication
- Not intended for compete removal of lesions

Tissue samples from the core biopsy are sent for histological analysis. Most radiologists recommend radiographing the specimen of core sample using magnification technique, especially after biopsy of calcifications. The specimen can be placed either on a microscope slide or Petri dish for the radiograph.

After the core biopsy, compression should be maintained at the site for 5 minutes to achieve hemostasis (termination of bleeding) and to minimize hematoma formation, and an ice pack can also be applied to minimize swelling. The skin at the incision site should be closed with closure strips to minimize scarring. After the core biopsy, patients should be instructed to keep the wound dry and to leave the dressing on for at least 3 days. They should also avoid strenuous activity for at least 1 to 2 days after the procedure.

OTHER BIOPSY METHODS

There are a number of commercial modifications of the basic core biopsy principle that will provide state-of-the-art alternatives to open surgical biopsies and are capable of giving accurate and consistent samples. These include the radiofrequency (RF) technique; automated core biopsy "guns" or automatic vacuum "guns" (Fig. 9–9), sometimes called minimally invasive breast biopsy; or vacuum-assisted core biopsy.

Radiofrequency Biopsy

A relatively new method of biopsy is performed using an RF introducer. An RF introducer is an electrosurgical device used in conjunction with a handheld biopsy system. The system includes an obturator and cannula with a stainless steel RF cutting tip on the obturator. To use, the introducer is attached to an electrosurgical hand piece or cord and is powered with a standard electrosurgical generator with a patient return pad.

The RF introducer is first used to penetrate the breast using RF cutting. Pure-cutting mode of 90 W is recommended. The introducer also serves as a means of introducing the biopsy device. When the introducer reaches the targeted lesion, the obturator on the device is

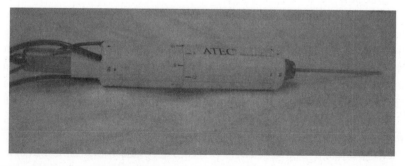

Figure 9–9. SiteSelect stereotactic breast biopsy guns allow for an intact specimen that is 2 cm in diameter.

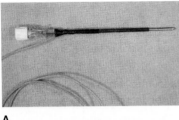
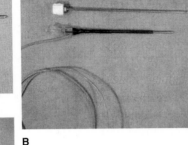
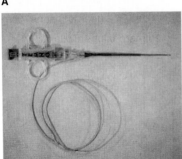

Figure 9–10. Radiofrequency biopsy tools. Biopsy introducer needle with the obturator in place **(A)**, with the obturator removed **(B)**, and with the biopsy needle in place **(C)**. All but the distal 2 cm of the needle shaft are insulated with shrink-wrapped plastic. (Reproduced with permission from Pritchard WF, et al. Radiofrequency cauterization with biopsy introducer needle. *J Vasc Interv Radiol.* 2004;15:183-187. © SIR 2004.)

removed. The cannula remains, and this serves as a passageway for the handheld biopsy system. The outer cannula of the RF introducer has a connector at its distal end, and this allows connection with the probe of the handheld biopsy system. When the system probe is in place in the outer cannula of the RF device, tissue samples can be collected. The procedure is often conducted under ultrasound guidance.

Advantages and Disadvantages of Radiofrequency Biopsy

The system is easy to use and easily penetrates dense connective tissue of the breast without the use of force. It also easily penetrates small dense lesion without pushing them away. In general, the total time of the procedure is comparable to mammography or ultrasound-guided biopsy techniques.

Excessive bleeding is the most common complication of this procedure, and it is not intended for the complete removal of lesions.

Vacuum biopsy analysis:

- Often provides definitive diagnosis
- Good for calcifications
- No stitches required
- Minimal scarring
- More accurate than FNAB and CNB
- Not recommended for hard-to-reach lesions

The Mammotome Biopsy System

The mammotome is attached to a standard prone stereotactic table. It is essentially a stereotactically guided core biopsy using a directional vacuum-assisted needle instead of a spring-loaded gun (Fig. 9–10). Instead of collecting tissue in a sample notch that must be withdrawn after each pass, as with the biopsy gun systems, the mammotome uses a vacuum to draw tissue from the lesion and pulls the tissue into a collecting chamber, which can be rotated through 360 degrees, thus enabling a volume of tissue to be sampled without removing the needle from the breast. The mammotome can be used with mammography, ultrasound, or MR. Small lesions can be removed entirely, and the mammotome canvasses a lesion very quickly. The mammotome offers a sutureless minimally invasive procedure that reduces the possibility of scarring or breast disfigurement. It is cost-effective and much more comfortable for the patient than the traditional core sampling techniques.

The mammotome consists of a probe with an outer, hollow sleeve and a tissue bowl at its end. Multiple small vacuum holes are at the bottom of the tissue bowl to pull the breast tissue into the bowl when the vacuum is activated. A hollow "cutter" with a sharpened end is positioned within the outer sleeve. After breast tissue is pulled into the tissue bowl, the cutter begins to rotate at high speed and is advanced through the tissue in the tissue bowl. A large breast specimen, equal in diameter to the probe but longer, is therefore collected. The probe can then be rotated in a clockwise position to place the sample bowl in another position for another acquisition. The cutter is withdrawn within the outer sleeve of the probe, and the tissue specimen is carried back to a pin that punches the captured specimen out into a specimen chamber. Suction is reapplied and the cutter readvanced into the new specimen. This process is repeated in a clockwise manner. One benefit of the suction is that most of the blood from the site is removed, making additional radiographs clearer and easier to assess for lesion removal. The larger volume of tissue and the directional abilities of the vacuum biopsy system will result in a more accurate acquisition of calcifications. Often, the radiologist will place a small stainless steel marker or a collagen plug with an embedded titanium marker to identify the biopsy site. This allows the radiologist to monitor the original site of the lesion in future mammograms.

The Automatic Tissue Excision and Collection

The automatic tissue excision and collection (ATEC) system, manufactured by Hologic, uses a vacuum-assisted biopsy system to aspirate tissue in a continuous basis. The system's operating console connects to a disposable biopsy needle, and a single foot pedal controls the hand piece. Tissue is acquired every 4.5 seconds in a single continuous cycle. The ATEC system can take up to 14 core biopsies per minute in a closed system that reduces fluid exposure and core handling. The automated vacuum capabilities shorten the procedure time, and the system can be operated under either ultrasound or MRI guidance. The hand piece tubing is tinted blue so that the patients will not be able to see their own blood during the procedure. A newer version of this biopsy method, the Eviva, offers updated features, including quieter firing and integrated pain management.

Large Core Breast Biopsy

The advanced breast biopsy instrumentation (ABBI) removes a cylinder of tissue from just under the skin of the breast. This is a stereotactically guided biopsy system, and the procedure is performed under sterile conditions with the patient prone on a stereotactic table. The procedure setup is similar to that for stereotactic imaging, but both local anesthesia and deep anesthesia are required. The procedure requires electrocautery and suturing, and often the surgeon, not the radiologist, will perform this procedure. After the procedure, patients may be instructed to take pain relievers for discomfort as needed.

The ABBI can remove lesions up to 2 cm in diameter. This has the advantage of being capable of removing the entire lesion. The ABBI, however, has a high complication rate because of the high volume of tissue removal and significant blood loss. It also tends to have a high abort rate, and there is a high risk for incompletely removing lesions at the margin of the specimen. Complete removal of the lesion does not ensure complete excision of the abnormality, resulting in residual cancer remaining in the breast. ABBI results in more scarring and cosmetic deformity, and the cost of the procedure is higher than core or even vacuum-assisted biopsy. Although this procedure is less invasive than an open surgical biopsy, many radiologists consider the disadvantages to outweigh the advantages, and the procedure is not very popular.

> **Large core analysis:**
> - Less invasive than open surgical biopsy
> - Mainly used for nonpalpable lesions and calcifications
> - Tissue margins may be inadequate for accurate diagnosis
> - Causes scarring, longer recovery

Open Surgical Biopsy

An open surgical breast biopsy is needed if the results of an FNAB or core biopsy are inconclusive. Sometimes, the lesions are so hard that the radiologist is unable to get a good sample or even penetrate the lesion. The open surgical biopsy is also the next step after the needle localization. Occasionally, the placement of lesion, very close to the chest wall or immediately behind the nipple, will make the FNAB or core biopsy contraindicated.

The surgical biopsy requires full sterile techniques. The procedure is generally performed as a same-day surgery, with the patient given a general anesthetic. There are very few complications, and most patients will have pain at the site only for 1 or 2 days. Generally, aftercare treatment includes taking pain medication and using a cold pack.

There are two types of surgical biopsies. In the incisional biopsy, a sample of the lesion is removed for histological testing while the patient is still under anesthesia. If the lesion proves malignant, an excisional biopsy can be performed in which the entire lesion with clean margins will be removed.

Modality-Based Core Biopsy

Ultrasound Guided Core Biopsy

Ultrasound can provide a highly accurate way to evaluate suspicious masses within the breast that are visible on ultrasound, whether or not they can be felt on the breast self-examination or the clinical breast examination. The procedure prevents the need to remove tissue surgically and is faster than stereotactic biopsies and uses no ionizing radiation. Tissue specimens are then taken using either an automatic spring-loaded or vacuum-assisted device (VAD).

Equipment Necessary

The equipment needed for the ultrasound biopsy includes a scalpel blade (#11), a jar of formalin, adhesive dressing, syringes, needles, local anesthetic, sponges, forceps, a specimen cup, and a linear array transducer of 12 to 14 MHz. The high transducer is necessary to focus the sound beam to the depth of the area of interest and to adjust gain and power settings on the unit to improve image quality. Most radiologists prefer to do freehand positioning of the needle within the breast, although there are needle guidance systems available. Whatever the method used, the same person must control both the transducer and the needle. An advantage of the freehand system is the ability to approach the lesion with the needle parallel to the chest wall, therefore avoiding puncturing the chest and also avoiding the expense of added equipment. Depending on the probe configuration, the geometry of the acoustic beam, and the route of the needle entry, either a small portion of the needle may be visible as an echogenic dot, or if the needle entry is aligned with the acoustic beam and is nearly perpendicular, the entire shaft, including the needle tip, may be visible.

Because of the throw of most core biopsy needles, the skin entrance site should be farther from the lesion and the transducer than in the case of FNAB, and the needle tip should always run parallel to the chest wall to avoid puncturing the pleura.

Procedural Considerations

The patient should be positioned either supine or turned slightly to the side. The ultrasound probe is then used to locate the lesion. The patient's ipsilateral arm should be raised, with the hand underneath the head. If the lesion is in the lateral aspect of the breast, the patient should be in the oblique position for the examination. If the lesion is in the medial aspect of the breast, the patient should be supine. Enough local anesthetic is injected to be sure that the patient feels no discomfort during the procedure. Ultrasound is also used to guide the injection of anesthetic along the route to and around the lesion.

A very small nick—4 mm or less—is made in the skin at the site where the biopsy needle is to be inserted. The radiologist, constantly monitoring the lesion site with the ultrasound probe, guides a hollow-core biopsy needle or the vacuum-assisted needle directly into the mass and obtains specimens. Usually, at least 5 to 10 samples are taken using the core biopsy method, and at least 12 when using the VAD. The needles are angled differently for each core to obtain core sampling from different parts of the lesion (Fig. 9–11). Frequently, the VAD will remove the entire mass, a process that can be continuously monitored with the ultrasound probe.

Generally, the biopsy is completed in less than an hour. Prefire and postfire images are always obtained to document the position of the needle relative to the lesion. It is not usually necessary to close the tiny skin incision with sutures, but a small compression dressing can be applied. Most patients are able to resume their usual activities later the same day, although any athletic activities should be avoided on the day of the biopsy.

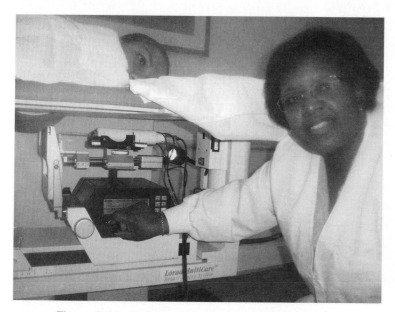

Figure 9–11. Patient positioned for a needle biopsy.

Limitations

- The ultrasound-guided method cannot be used unless the mass can be seen on an ultrasound examination.
- Calcifications within a cancerous nodule are not shown as clearly with this modality as when x-rays are used.
- Ultrasound cannot easily be used for deep lesions or large breast size because the 12- to 14-MHz transducers required for the necessary resolution in breast ultrasound do not penetrate well beyond 3 to 4 cm.
- Small lesions may be difficult to target accurately by an ultrasound-guided core biopsy.
- As with an x-ray–guided breast biopsy, ultrasound-guided biopsy occasionally will miss a lesion or underestimate the extent of disease. If the diagnosis remains uncertain after a technically successful procedure, surgical biopsy will be necessary.

MRI-Guided Core Biopsy

If a suspicious area is detected only on MRI or is not well seen on ultrasound or mammography, the radiologist may recommend an MR-guided core breast biopsy with clip placement to confirm a malignancy. If the pathology result comes back positive, the mammographic localization can be performed by localizing the clip. The localization procedure leaves a localization wire in place. Many also use titanium or stainless steel localizing markers to allow pinpoint accuracy in identification, localization, and confirmation of the lesion's position. The patient is generally taken to surgery, where the wire guide is used to identify the exact location of the lesion.

All the instruments used for the MR biopsy or localization must be MR compatible, and there are now new MRI pulse sequences that will significantly reduce artifacts from the needle (Fig. 9–12).

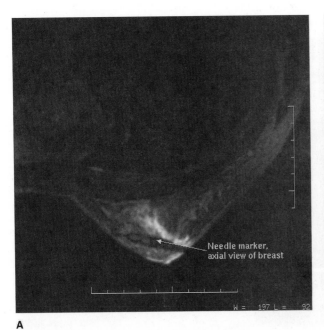

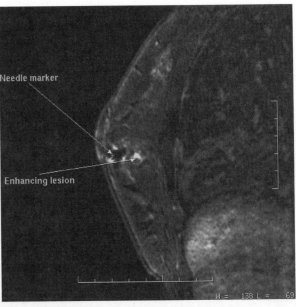

A

B

Figure 9–12. Magnetic resonance needle localization showing the needle placement in the axial **(A)** and sagittal planes **(B)**.

MR-Guided Biopsy Procedure

A MR-guided biopsy requires only a local anesthetic. As with ultrasound or mammographic minimally invasive biopsies, the MR biopsy is less costly than surgical biopsy, will leave little to no scarring, and can be performed in less than an hour. Many of the newer biopsy systems employ stereotactic guidance with a biopsy grid that precisely maps the location of the lesion.

The procedure can be performed using regular breast coils; however, there are newer dedicated types of biopsy coils that allow easy access to the area (Fig. 9–13). The biopsy coils setup allows compression and immobilization of the breast. When the area is localized from the MR scan, the radiologist cleans the area, applies local anesthetic, and then uses a scalpel blade to make a small incision at the area of interest. A 9-gauge stainless steel core needle is inserted into the introducer and advanced to the location of abnormality. The needle is then removed, and a titanium or stainless steel marker is inserted for imaging purposes. After verification, the radiologist uses a vacuum biopsy to take multiple samples of the area.

A final series of MR images are taken to confirm that the lesion was sampled. The patient can now be removed from the MRI unit. Generally, compression to the site is maintained to prevent excessive bleeding.

The biopsy samples are sent to the laboratory for microscopic analysis.

> MR biopsy is needed if the lesion can be identified only on MRI.

Comparison of the Various Biopsy Methods

Complications and severity of side effects of a biopsy will depend on the method used, and in the case of needle localization, inaccurate placement or displacement of the wire can occur. For any breast biopsy, patients can develop an unexpected allergic reaction to local anesthetic, and there is a risk for bleeding whenever the skin is penetrated. To avoid excessive bleeding, patients taking aspirin, fish oil, or blood thinners are advised to stop their medication 5 to 7 days before the procedure. Some herbal supplements also act as blood thinner; therefore, a through clinical history should be obtained from the patient. Other

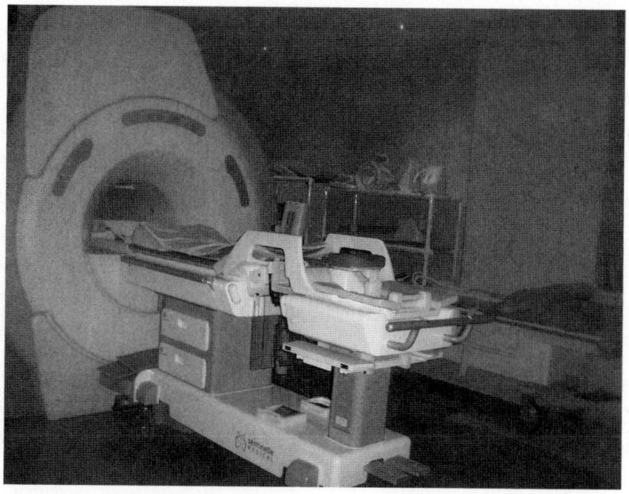

Figure 9–13. Magnetic resonance biopsy localization or biopsy setup.

complications can include a risk for hematoma or infection. Infection can occur whenever the skin is penetrated, but the chance of infection requiring antibiotic therapy is rare.

The open surgical biopsy has the lowest false-negative rates (<0.5%) but is the most invasive. The reported false-negative rate for malignancy with core biopsy is in the range of 2% to 6.7%, with a mean rate of 4.4%. False-negative findings are more likely to occur with microcalcifications. Stereotactic core-needle biopsy using a 14-gauge needle has a 90.5% sensitively rate and a 98.3% specificity in diagnosing breast masses, compared with 62.4% and 86.9%, respectively, for fine needle aspiration. The most common error in core biopsy is understaging of the cancer; for example, DCIS could be diagnosed on a core biopsy, but invasions may be found at excisional biopsy. The best core samples are obtained using a long-throw gun and are taken with a 14-gauge needle. Insufficient sampling is reduced if a minimum of five biopsy samples are taken. The 14-gauge automatic needle has two main limitations. If multiple specimens are obtained, the later samples are contaminated with blood rather than breast tissue, and retrieval of calcifications is often difficult if the calcifications are not along the line of fire of the needle or located in small lesions.

FNAB is less costly, less invasive, and faster than a core biopsy, although it is more likely to lead to errors because of insufficient sampling. The false-negative rate for the FNAB can range from 5% to 20% and is lower for experienced operators. The false-negative rate can be further reduced by using smaller gauges and by increasing the number of aspirates obtained. The most substantial limitation for FNAB is the high rate of insufficient sampling. The rate of insufficient samples for FNAB under stereotactic biopsy is higher and can reach 39.9% when compared with ultrasound guidance (8.5%). Rates are also higher for calcifications (46.1%) than for masses (26.6%). For all types of lesions (masses and calcifications), the rates are higher if there is no on-site cytopathologist (31.2%) compared with an on-site cytopathologist (14.5%).

Ultrasound-guided biopsy performed using any of the biopsy methods is faster and cheaper than stereotactic biopsy. It also takes less time than the surgical biopsy and causes less tissue damage. Ultrasound also offers real-time visualization of the needle, which means that the motion of the biopsy needle can be followed. It also gives no ionizing radiation and costs less than stereotactic mammography methods. Ultrasound is the better option when a biopsy of the axilla and chest wall areas is needed. These areas are difficult to biopsy using x-ray–guided methods with or without using a prone table or including stereotactic guidance. Most palpable lesions are best biopsied under ultrasound guidance, whereas calcifications are best biopsied mammographically. The palpable lesion, however, must be suspicious or indeterminate on ultrasound to warrant an ultrasound biopsy.

The prone table can be better for the patients than the add-on units because, in the dedicated prone unit, the patient does not see the procedure as it is being performed. The examination is faster, and the units allow the breast to be approached from a greater circumference of its surface. Also, because the patient is lying down, there is less chance of vasovagal reactions, and the breast is securely immobilized during the procedure, reducing the possibility of patient motion.

> Most palpable lesions are best biopsied under ultrasound guidance, whereas calcifications are best biopsied mammographically. Ultrasound-guided biopsy performed by using any of the biopsy methods is faster and cheaper than the stereotactic biopsy.

SPECIMEN RADIOGRAPHY

The specimen is the breast tissue removed during a surgical or core biopsy (Figs. 9–14A and B). Specimen radiographs are not typically requested after ultrasound or MRI needle biopsy or on a stereotactic biopsy that does not have calcification. The specimen radiography is normally requested to see calcifications on the needle biopsy or clips on the excisional biopsy. A specimen radiograph will confirm that the lesion was removed.

Specimen Radiography Units

There are a number of specimen radiography systems on the market. These standalone digital systems are small and offer immediate in-room specimen imaging. The images are available for viewing on the monitor within seconds, and most of these systems use a microfocused tube to produce high-quality, rapid specimen varication.

Mammography units that are used exclusively for special interventional purposes (localization, biopsy, or specimen radiography) do not need to be accredited.

When imaging the specimen after a surgical biopsy, speed and efficiency are important because the patient may be under anesthesia. The technologist should always use compression, and magnification is recommended for calcifications. If microcalcifications are present in a specimen, they should be counted and noted. The radiologist should

> **Specimen imaging** is done to confirm the removal of a lesion or calcifications.
>
> **Specimen radiograph** is always performed for microcalcifications.

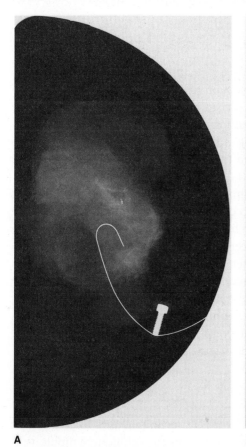 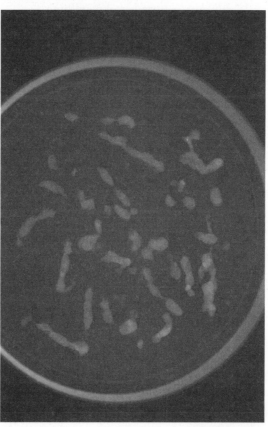

A   **B**

Figure 9–14. **(A)** Specimen from a surgical biopsy. **(B)** Specimen from a core biopsy.

> **Box 9–2.** Key Points When Imaging the Specimen
>
> - Always use compression
> - Magnification may help to visualize microcalcifications

indicate where the pathology is located on the specimen because this will help the pathologist in evaluating the lesion. This is especially important in patients with extensive microcalcifications.

If the tumor is close to the margins of the specimen or if the margins are positive for cancer, additional tissue must be exercised before the incision is closed. A check should also ensure that the total number of calcifications removed confirms the original number in the area of interest (Box 9–2).

DUCTOGRAPHY OR GALACTOGRAPHY

Ductography can be performed to evaluate a suspicious nipple discharge. The discharge can be spontaneous and unilateral and may be clear, serous, or bloody. The technique can only be done if nipple discharge is present. The examination is performed with the patient lying supine.

The affected nipple is cleaned, and the discharging duct is identified and annulated with a 30-gauge blunt-tip sialographic needle. The needle is secured to the breast with tape, and 0.2 to 0.3 mL of contrast agent is then injected slowly until reflux or the patient reports pain or fullness. Pain can indicate extravasation of the contrast. After the contrast injection, the breast is imaged in the CC or 90-degree ML projection.

Ductography can detect abnormalities, such as filling defects, wall irregularity, duct expansion, or abrupt cutoff of a duct; however, the procedure is rarely performed today because ultrasound can now be used to identify and localize abnormalities within the ducts, such as an intraductal papilloma.

NIPPLE ASPIRATION

Nipple aspiration uses gentle suction to collect fluid from the nipple to check for abnormal cells or precancerous abnormalities. The device used is similar to the breast pumps used by nursing women. The fluid-containing cells from the lining of the ducts can be obtained in approximately 75% of women. Only a tiny amount of fluid or a few cells are needed for testing.

The US Food and Drug Administration (FDA) has issued a warning about nipple aspirate tests. According to the FDA, the nipples aspirate test can lead to potential harm by issuing false-negative results, indicating the absence of breast cancer when cancer exists, and false-positive test results, indicating the presence of breast cancer when none exists.

The National Comprehensive Cancer Network 2013 guidelines state that the clinical utility of nipple aspiration is still being evaluated and that it should not be used as a breast cancer screening tool. The test should not be used to detect breast disease and cannot replace the mammography or other breast imaging tests or breast biopsy.

DUCTAL LAVAGE

Ductal lavage or "breast Pap smear" is a technique used to collect cells from the lining of the milk duct. In the original technique, a nipple aspiration is first used to drain a tiny amount of fluid to the surface of the nipple and to locate the milk ducts. A slender catheter is then inserted into the duct through the opening, and a small amount of anesthetic is injected, followed by a small amount of saline. The saline rinses through the ducts, collecting cells, and is then withdrawn. The resultant cells can be checked for atypical cells. The disadvantage of the single duct technique is that the sampling from a single duct cannot be used to assess the entire breast.

In another technique, a local anesthetic is applied to the entire breast, and a slender needle is randomly inserted into the inner quadrant of the breast and slightly withdrawn and reinserted 8 to 10 times. This process is called random fine needle aspiration and is repeated to ensure that all the quadrants of the breast are checked. Cells are checked by a pathologist for specific molecular changes that are common to many breast cancers.

In a third technique, an anesthetic cream is applied to the nipple area. Gentle suction is used to identify any duct producing fluid. In this method, only ducts producing fluids are tested. A thin catheter is inserted into the duct delivering first anesthetics and then normal saline. The saline rinse loosens cells, which are then withdrawn through the catheter.

The technique has received FDA approval in the United States, and clinical trials are under way to check its effectiveness. It is considered a means of giving women a preventive agent by eradicating abnormal cells, therefore preventing cancer from developing. In particular, the test will examine the cells for a specific gene called *RAR beta*. This gene regulates how breast cells use vitamin A to maintain their proper health. Studies have shown that *RAR beta* loses its ability to function in many women with breast cancer. Without *RAR beta,* vitamin A does not work, and breast epithelial cells eventually become cancerous.

Unfortunately, ductal lavage does not locate the abnormity or indicate the location of the cancerous cells. The test can give the general area of any abnormal cells, but other imaging tests would be needed to provide accurate localization.

PATHOLOGY REVIEW

The final stage of any interventional process is the pathology review. The purpose of the pathology review is to assess whether the initial diagnosis matches the confirmed findings. This is usually a measure of false-negative or false-positive rates and will provide a good overall evaluation of the entire imaging and diagnostic process. High false-positive rates could indicate poor imaging skill on the part of the technologists or, more often, poor interpretation skill of the radiologist. The review can be used for teaching, research, or as a statistical profile of the facility.

There should be good communication among the radiologist, the surgeon, and the pathologist. Most facilities now have a medical audit system in place to track all positive pathologic findings. A tracking system is required to comply with the US Mammography Quality Standards Act (MQSA), but regular reviews of medical audit will help the facility to identify areas of concern, including high false-negative or false-positive rates. By identifying possible problem areas, the facility can improve service and enhance breast cancer detection and screening.

Breast reduction:
- Removal of excess breast fat, glandular tissue, and skin
- Nipple and areola resized and shifted to a higher position

Cosmetic:
- At the request of the patient
- To achieve a breast size in proportion with body

Medical:
- To alleviate the discomfort or pain associated with overly large breasts

Mammoplasty procedures
- Cosmetic intervention
- Reduction mammoplasty
- Mastopexy
- Augmentation mammoplasty

MAMMOPLASTY

The number of mammoplasty procedures has increased significantly within the last few years. Reduction mammoplasty can be considered for enlarged breasts (macromastia); back, neck, or shoulder pain not relieved by supporting bras; deep grooves in the bra strap area of the shoulder; headache; sleeping problems from overlarge breasts; poor posture caused by large breasts; or arm or finger numbness caused by large breasts. If surgery is indicated for medical purposes, it is not considered cosmetic intervention. Also, reconstruction breast surgery after breast cancer does not fall under the heading of cosmetic intervention.

Any woman considering breast mammoplasty must first have an initial consultation with a plastic surgeon. The surgeon should be certified by the American Society of Plastic Surgeons (ASPS), and this information should be verified as part of the initial evaluation. In general, implants or reconstruction is not recommended for women before age 22 years because women's breasts are not fully developed before that age. Breast augmentations or reductions are definitely contraindicated for anyone younger than 18 years because, in addition to having immature breasts, they may not be mature enough to make an informed decision. Consultation and a mammogram to rule out breast diseases are recommended before the surgery. The surgeon will then determine which surgical technique is best for the woman.

Breast mammoplasty, whether for enlargement or reduction, can be performed in an outpatient surgical facility or in a hospital. General anesthesia is the preferred option.

Cosmetic Intervention

Cosmetic intervention can be requested by the patient for a variety of reasons. Over the years, breasts lose their shape and firmness from pregnancies, nursing, and the aging process. Some women will then consider mastopexy. Other surgeries available include breast augmentation and breast reduction to improve self-confidence or improve appearance.

Reduction Mammoplasty

In breast reduction, excess breast fat, glandular tissue, and skin are removed and the nipple and areola relocated higher on the breasts for better cosmetic results. A breast reduction surgical procedure can last up to 6 hours.

Reduction will achieve a breast size in proportion to the woman's body and can be used to alleviate the discomfort, such as pain in the chest, neck, back, and shoulders, associated with overly large breasts. Other problems associated with extremely large breasts can include chafing of chest and inframammary crease, indentations to the shoulder, and breathing difficulty. Reduction surgery can improve the physical health, mental health, and self-esteem of a woman. Tissue removal assessment is critical, and the amount of tissue removed will determine the functional sensitivity and lactational capability of the breasts.

Mastopexy

During a breast lift, incisions are made along the natural creases in the breast and around the areola. A keyhole-shaped incision above the areola is also made to define the new location for the nipple. Skin is removed from the lower section of the breast. The areola, nipple, and underlying breast tissue are moved up to a higher position. The nipple is moved, and incisions are closed with sutures. The procedure usually lasts approximately 2 hours, depending on the extent of the surgery. Mastopexy is most often done at the request of the patient and is considered a cosmetic interventional procedure.

Augmentation Mammoplasty

The technique of augmented mammoplasty has been around since the 1950s. Initially, liquid silicone was injected directly into the breast, but these procedures led to severe complications, and the procedure was soon discontinued (Fig. 9–15).

Silicone gel–filled implants were first used in 1962. The technique was to fill a silicone elastomer bag with silicone gel. In 1992, the FDA removed silicone from the public market because of complications, such as leakage and rupture, and its association with connective tissue disorder or immune disorders. The ban was finally lifted in 2006.

Implants today are often saline filled or silicone filled and come in various shapes and sizes. Both the saline-filled and the silicone-filled implants have an outer shell composed of silicone elastomer. Some implant shells are double lumen, giving an extra protective layer to reduce the risk for ruptures. The saline-filled implants use sterile saline and can be round or anatomically shaped. There are two types of shells: textured shells and smooth shells. There is also a thicker silicone implant called a form-stabilizing implant or "gummy bear" implant that is under investigation.

Implants can be placed in front of the pectoral muscle (subglandular or retromammary implants) or behind the pectoral muscle (subpectoral or retropectoral implants). Generally, during the implant surgery the empty saline sac is implanted. The surgeon later fills it to the desired size. The silicone-filled implant uses a silicone gel, which is less likely to leak in cases of rupture.

In general, silicone implants have a more natural feel. They are softer and smoother and less likely to wrinkle or ripple than the saline. The biggest disadvantage of the saline is the wrinkles that can sometimes be seen or felt, especially on thinner women. The smooth-textured saline implants are less likely to ripple. One advantage of textured implants was the lower risk for capsular contracture. However, the implants placed behind the pectoralis major muscle are less likely to experience capsular contracture.

In the case of rupture, the saline-filled implant will deflate, and the result will be obvious because the breast will become noticeably changed as the tissue fills with saline. Surgery will be needed to remove the silicone shell. The ruptured silicone is often harder to perceive and may not even be noticeable. The FDA recommends imaging of the breast every 2 to 3 years to monitor the silicone-filled implants. Saline and silicone implants have similar risks that can include poor reaction to anesthesia,

Figure 9–15. Mammogram of the MLO projection showing direct injection of silicone in the breast.

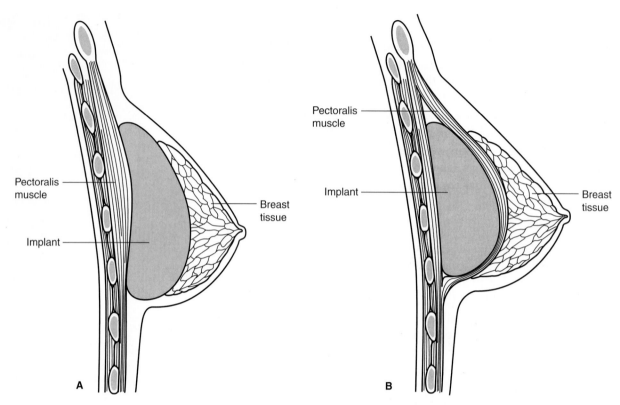

Figure 9–16. Implant placement **(A)** in front of and **(B)** behind the pectoralis muscle.

Augmented Mammoplasty

Saline Implants

- Saline solution contained in a silicone shell

Silicone Implants

- Silicone shell filled with silicone gel

Implant Placement

- Subglandular or retromammary placement
- Subpectoral or retropectoral placement

excessive bleeding, hematoma, breast pain, permanent changes in nipple or breast sensation, infection, scar tissue formation, capsular contracture, and implant leakage or rupture (Fig. 9–16). In 2011, the FDA noticed an extremely rare cancer, called anaplastic large cell lymphoma (ALCL), which affects cells in the immune system and can be found around the breast implant. ALCL is not a type of breast cancer. It can be found in the skin or lymph nodes and is mainly associated with textured surface implants. Symptoms of ALCL include fluid buildup, hardening or a mass around their implants, swelling, and redness around the breast implants. ALCL is described as a slow-growing cancer that is treatable when detected early.

Both ultrasound and MRI are better than mammography in imaging implants. Mammographically, the denser silicone will obscure more of the breast parenchyma than the more radiolucent saline, but regardless of the type of implant the presence of the implant and a calcified capsule will interfere with breast compression and therefore interpretation during a mammogram. Also, implants, whether saline or silicone, cannot be compressed for the mammogram. Ultrasound and MRI are better able to evaluate complications such as rupture, which can occur spontaneously or after a trauma, leaking from the implants, or capsular contracture.

MRI can easily distinguish between silicone and normal breast tissue and is the single best modality to image silicone implant ruptures because it has a sensitivity of 81% and a specificity of 92%. Ultrasound has a sensitivity of 70% and a specificity of 92%. Silicone gel implants are seen as echolucent on ultrasound. Mammography has a sensitivity of 11% and a specificity of 89%.

Summary

Only a histological or cytological analysis can definitely confirm malignancy.

- Cytological analysis is an examination of the cells, which gets results in 24 to 48 hours.
- The histological analysis is a study of the tissues of a specimen. Preliminary results can be obtained from a frozen section in 10 minutes; however, more accurate results can take 3 to 5 days.

Patient preparation includes the following:

- Wearing comfortable clothing and avoiding deodorant, powders, lotions, or creams in the area of the breast
- Eating a light meal to minimize nausea
- Avoiding blood-thinning medications such as aspirin or herbal blood-thinning remedies.

An informed consent is required before any interventional procedure.

Cyst aspiration: the removal of the content of a cyst. It is performed to relieve pain or to assess the cyst's contents and is often performed under ultrasound guidance.

Stereotactic localization: a method used to locate nonpalpable lesions. Images are taken at different angles to triangulate the exact coordinates of a lesion. A computer calculates the lesion's location. Lesions can be biopsied after localization, or the stereo localization can be a prelude to fine needle aspiration biopsy (FNAB) or core biopsy.

Types of Stereotactic Units

- **Add-on units** attach to a regular mammography unit, and the patient is seated during the examination.
- **Dedicated prone units** have the patient prone with the affected breast and arm protruding through the table opening. The radiologist and technologist can remain seated during the examination. Most prone tables allow imaging with 360-degree tube rotation around breast.

Comparison of Prone and Add-On Units

Prone Units
- Are more expensive and require more space
- Provide a better biopsy result
- Are safe for most patients
- Generally have a 300-lb weight limit

Add-On Units
- Are less expensive and require less space
- Are not the best option for elderly, anxious, or physically impaired patients
- Are associated with an increased risk for a vasovagal reaction

Preoperative localization: a prelude to the surgical biopsy that can be used if stereo localization is not available. It is performed under mammographic, ultrasound, or magnetic resonance (MR) guidance.

- The radiologist inserts a long needle with a hooked wire into the breast, to a point beyond the lesion.
- The needle is withdrawn, leaving the hooked wire in place.
- The hooked end of the wire holds the wire within breast tissue.

Types of Breast Biopsies

Fine needle aspiration biopsy (FNAB): removes cell samples. The procedure uses a 22- to 25-gauge needle, and it is generally recommended that the cytotechnologist evaluate the removed sample before the patient leaves the department.

Minimally invasive biopsy: removes tissue samples and can include the following:

- **Core needle biopsy (CNB)** uses large-core needles with a special cutting edge (11–16) to remove core samples.
- **Vacuum biopsy** sometimes uses a local anesthetic and can be performed with mammographic, ultrasound, or MR guidance.
- **Large core biopsy** removes a larger volume of tissue than the core biopsy and is not commonly used.
- **Radiofrequency biopsy** uses a radiofrequency (RF) device, an electrosurgical introducer with a stainless steel tip. The RF introducer is inserted into breast tissue using RF energy.
- **Open surgical biopsy** used with lesions that are difficult to approach or close to the breast surface. It is also used as confirmation after FNAB. Open surgical biopsy is a surgical procedure performed under general anesthesia.
  - Incisional: a sample of the lesion is removed
  - Excisional: the entire lesion is removed
- **Lymph node biopsy** used to determine cancer spread to the lymph nodes. General anesthesia is required, and the procedure is often performed with a lumpectomy or mastectomy.

Factors Determining Type of Biopsy Performed

- Preference of the radiologist
- Needs of the patient
- Availability and expertise of a cytopathologist
- Level of suspicion of the malignancy
- Location and type of lesion

Modality base biopsies: the modality used will depend on the modality used to visualize the lesion.

Ultrasound-Guided Biopsy

- Highly accurate, quick, and comfortable
- Lesion must be palpable or visualized on ultrasound
- Commonly used with FNAB or core biopsy
- Allows biopsy from any orientation and allows continuous imaging during the biopsy

MR-Guided Biopsy

- Needed if the lesion is visualized only on MRI
- All equipment used must be MRI compatible
- A slower and very expensive method of biopsy

Comparison of the Various Methods of Biopsy

FNAB
- Ideal for cyst evaluations and aspirations
- Fastest and easiest biopsy method
- Less expensive and less invasive than core
- No stitches or scarring after the procedure
- Offers rapid results
- Less accurate than core
- Cannot be used for chest wall lesions

Core Biopsy
- More invasive than FNAB, but results are more definitive
- Not recommended for small or hard lumps

Vacuum-Assisted Biopsy
- Often provides definitive diagnosis
- Very good method for calcifications
- No stitches and minimal scarring
- More accurate than FNAB or CNB
- Not recommended for hard-to-reach lesions

Radiofrequency Biopsy
- Easy to place under ultrasound guidance
- Light and easy-to-hold device
- Bleeding is a common complication

Large Core Biopsy
- Less invasive than open surgical biopsy
- Mainly used for nonpalpable lesions and calcifications
- Tissue margins may be inadequate for accurate diagnosis
- Causes scarring with a longer recovery

Lymph Node Biopsy

- To determine cancer spread to the lymph nodes
- Required axillary node dissection
- Needs general anesthesia
- Often performed at the time of lumpectomy or mastectomy

Open Surgical Biopsy

- Lowest false-negative rate but highest complication rate

Specimen imaging: used to verify the removal of the lesion or calcifications.

Ductography: used to evaluate nipple discharge. The procedure involves filling the ducts with contrast to check for duct-filling irregularities, duct expansions, or duct defects.

Nipple aspiration: removes fluid from the nipple to check the cells. A process called "ductal lavage" can be used to irrigate the ducts to assess for cancer cells.

Pathology review: involves a check of all biopsies to assess the false-negative or false-positive rates.

Mammoplasty: can involve breast reduction, mastopexy where surgery is used to lift and reshape the breast, or breast augmentation where the breast size is increased using implants.

- **Cosmetic intervention** includes breast augmentation, reduction, or any breast surgery performed at the request of the patient that is not medically needed.
- **Mastopexy** is a minimal surgical intervention to reshape the breast.
- **Augmentation mammoplasty** uses implants to increases the size of the breast.
  - Saline implant—solution contained in a silicone shell
  - Silicone implants—silicone shell filled with silicone gel

Implant Placement

- Subglandular or retromammary placement
- Subpectoral or retropectoral placement

REVIEW QUESTIONS

1. What is the difference between a cytological and a histological analysis?

2. Using the BI-RAD system, what category of lesion is usually referred for biopsy?

3. What is the name given to the procedure whereby the contents of a cyst are removed?

4. Name the technique used to triangulate a lesion within the breast and calculate its location in three dimensions.

5. Give three disadvantage of the add-on unit.

6. Give three disadvantages of the prone unit.

7. What is the needle gauge range commonly used for FNAB?

8. What is the difference between a long-throw and a short-throw biopsy gun?

9. The lowest false-negative rate is obtained using which biopsy technique: open, core, or FNAB?

10. Which modality will best detect biopsy breast calcifications?

11. Give two reasons for imaging the specimen.

12. Give one use of the ductogram.

CHAPTER QUIZ

1. Methods used to definitively classify a lesion include:

   (1) Cytological analysis

   (2) Mammography

   (3) Histological analysis

   (A) 1 and 2 only

   (B) 2 and 3 only

   (C) 1 and 3 only

   (D) 1, 2, and 3

2. A process by which the breast size is increased using implants is called:

   (A) Augmentation mammoplasty

   (B) Mastopexy

   (C) Ductography

   (D) Needle aspiration

3. A surgical procedure done at the request of the patient to lift the breast and improve firmness is:

   (A) Needle aspiration

   (B) Mastopexy

   (C) Augmentation

   (D) Postmastectomy breast reconstruction

4. A technique performed to evaluate suspicious nipple discharge is called:

   (A) Aspiration biopsy

   (B) Lumpectomy

   (C) Axillary dissection

   (D) Ductography

5. Which of the following biopsy techniques is the most accurate?

   (A) FNAB

   (B) Core biopsy

   (C) Stereotactic breast biopsy

   (D) Open surgical biopsy

6. Which biopsy method removes very small sections of tissue samples for analysis?

   (A) Ductography

   (B) Core biopsy

   (C) FNAB

   (D) Ductal lavage

7. An analysis of cell samples taken from the breast is referred to as:

   (A) Histological analysis

   (B) Cytological analysis

   (C) Open surgical biopsy

   (D) Core biopsy

8. Which procedure is considered a safer replacement for the lymph node biopsy?

   (A) Axillary dissection

   (B) Sentinel node biopsy

   (C) Lumpectomy

   (D) FNAB

9. If a lesion is seen only on MRI of the breast, which of the following can be used to biopsy the lesion?

   (A) Ultrasound-guided core biopsy

   (B) Stereotactic biopsy

   (C) Mammography-guided FNAB

   (D) MR-guided core biopsy

10. A minimally invasive procedure used to remove a nonpalpable lesion to determine its benign or malignant nature is:

    (A) Open surgical biopsy

    (B) Breast MRI

    (C) Ultrasound imaging

    (D) Stereotactic biopsy

BIBLIOGRAPHY

American Cancer Society. Breast Biopsy. Available at: http://www.cancer.org/treatment/understandingyourdiagnosis/examsandtestdescriptions/forwomenfacingabreastbiopsy/index. Accessed September 2016.

Andolina VF, Lille SL, Willison KM. *Mammographic Imaging: A Practical Guide.* 3rd ed. Philadelphia, PA: Lippincott Williams & Wilkins; 2010.

Assisted Breast Cancer Biopsy Treatment. Mammotome Breast Biopsy Devices. Available at: http://www.mammotome.com/mammotome-breast-biopsy-devices/ Accessed July 2017

Carlson GW, Wood WC. Management of axillary lymph node metastasis in breast cancer: Making progress. *JAMA.* 2011;305(6):606–607.

Interventional Breast Solutions. Available at http://www.hologic.com/products/intervention-and-treatment/breast-biopsy. Assessed September 2016.

Mammotome Biopsy System. Mammotome Fact. Available at http://www.mammotome.com/mammotome-breast-biopsy-devices/. Accessed September 2016.

Peart O. *Lange Q&A: Mammography Examination.* 2nd ed. New York, NY: McGraw-Hill; 2008.

Schwartzberg BS, Goates JJ, Keeler SA, et al. Use of advanced breast biopsy instrumentation while performing stereotactic breast biopsies: review of 150 consecutive biopsies. *J Am Coll Surg.* 2000;191(1):9–15.

Tortora GJ, Derrickson B. *Principles of Anatomy and Physiology.* 13th ed. New York, NY: John Wiley & Sons; 2011.

Venes D, Biderman A, Adler E. *Taber's Cyclopedic Medical Dictionary.* 22nd ed. Philadelphia, PA: F. A. Davis; 2013.

What Is Large Core Biopsy? Large Core Biopsy (brand name, ABBI). Available at http://www.imaginis.com/breast-health-biopsy/large-core-biopsy-brand-name-abbi. Accessed February 15, 2011.

Keywords and Phrases
Breast Cancer Treatment
   Staging Breast Cancer
   Mastectomy as an Option
   Preventative Surgical Options
   Lumpectomy as an Option
   Lymph Node Biopsy
   Comparisons of Surgical Treatment Options
   Other Options to Consider
Radiation Therapy
   Indications for Radiation Therapy
   Classifications of Radiation Therapy
   Side Effects of Radiation Therapy
   Intensity-Modulated Radiation Therapy
   Image-Guided Radiation Therapy
   Internal Radiation or Partial Radiation
   Intraoperative Electron Radiation Therapy
Chemotherapy
   The Cell Cycle
   Uses of Chemotherapy
   Drugs Used in Chemotherapy
   Chemotherapy Regimens
   Method of Giving Chemotherapy
   Side Effects of Chemotherapy
Targeted Treatment
   Molecular Treatment
   Hormone Therapy
Other Treatment Options
   Ablative Hormone Therapy
   Antiangiogenesis
   Circulating Breast Cancer Cell
Gene Therapy
   Methods of Gene Therapy

Hematopoietic Stem Cell Transplantation
Immunotherapy or Biotherapy
Photodynamic Therapy
Thermal Ablation
Cryotherapy
Prophylactic Surgery
Surgical Reconstruction Using Implants
   Nipple and Areola Reconstruction
   Implant Procedures
   Preparation for Surgery
   Immediate Reconstruction
   Delayed Reconstruction
   Other Reconstruction Options
   Autologous Tissue Reconstruction
   Pain Medication and Pain Management
Conclusion
Summary
Review Questions
Chapter Quiz

Objectives

On completing this chapter, the reader will be able to:

1. Describe the staging process and identify the five stages of breast cancer
2. Discuss the main difference between a modified radical mastectomy and a lumpectomy
3. Identify the risks and complications possible after a mastectomy
4. Identify the common contraindications to performing a lumpectomy
5. Identify the indications for radiation therapy
6. State the various types of radiation therapies
7. Describe the common side effects of radiation therapy
8. Identify indications for chemotherapy
9. Describe the common side effects of chemotherapy
10. Explain the use and benefits of tamoxifen
11. Describe the side effects of tamoxifen
12. Discuss the advantages and disadvantages of emerging treatment options
13. Identify reconstruction options

KEYWORDS AND PHRASES

- **Anesthesia** or loss of feeling or sensation can be a result of drugs or gases. General anesthesia causes loss of consciousness, whereas local anesthesia causes local or regional numbness.
- **Angiogenesis** is the process of forming blood vessels and is a common property of a malignant tumor.
- **Antigens** are agents that produce a reaction from the immune system that can lead to destruction of both the antigen and anything the

antigen is attached to. The word is derived from the Greek word *anti* meaning "against" and *gennan* meaning "to produce."

- **Axillary node dissection** is a surgical operation to remove lymph nodes from the axilla, generally as a check for breast cancer spread.
- **Brachytherapy** provides internal-beam radiation, which limits radiation treatment to the site of the cancer. It can start much earlier than external-beam radiation, and eligible patients are treated for 5 to 7 days, versus the 6 to 7 weeks of external radiation treatments.
- **Capsular contracture** refers to scar tissue formation around an implant that tightens and squeezes the implant. Contracture can be graded as soft to hard. It can be painful, leading to a distorted breast contour.
- A **dosimetrist** helps the physician to plan radiation treatments. The dosimetrist will calculate the doses of radiation at the various points in the body.
- **Free flap procedure** is a technique used for breast reconstruction in which tissue is moved entirely from one area in the body to another and the blood and nerve supplies are reattached under microsurgery.
- **Hematoma** or **microhematoma** is formed by the pooling of blood as a result of surgery or trauma. Over time, a hematoma may calcify, resulting in the formation of an oil cyst. Hematomas can show on a mammogram as mixed tissue composition, oval, or circular lesions.
- **HER2 (human epidermal growth factor receptor 2)** is a protein, which functions as a receptor on breast cells. The protein receptors control how the breast cells divide and repair themselves. It can be overexpressed in some women, resulting in a more aggressive form of breast cancer.
- **Image-guided radiation therapy** (IGRT) is the process of using two- and three-dimensional imaging during a course of radiation treatment. The imaging is used to direct, focus, and control the radiation therapy.
- **Infection** is the process by which microbes or microorganisms enter the body, multiply, and damage tissues, resulting in illness. There are four main types of microorganisms: bacteria, viruses, protozoa, and fungal organisms.
- **Intensity-modulated radiation therapy** (IMRT) uses a computer-controlled precision tool to delineate the three-dimensional volume of a tumor and uses a multileaf radiation beam to shape and focus the radiation only on the tumor during radiation treatment, thus sparing normal adjoining tissue and delivering an increased dose of radiation to the tumor.
- **Latissimus dorsi flap** is the removal of muscle, fat, and skin from the upper back; the flap is tunneled under the skin to create a breast mound and therefore a new breast.
- **Lymphedema** is the swelling of the arm or numbness in the arm and skin that can be caused by the removal of lymph nodes. It can lead to temporary or permanent limitation of arm and shoulder movement.
- **Lymphoma** is cancer of the lymphatic cells. The cancerous cells often originate in the lymph nodes and present as solid tumors.
- The **medical physicist** ensures that the imaging and therapeutic equipment are running properly per Mammography Quality

Standards Act (MQSA) standards. The physicist generally performs regular calibrations and checks as mandated, to ensure accurate exposure and quality imaging.

- **Microsurgery** uses microscopes and fine surgical instruments to reattach blood and nerve supplies to tissue.
- **Multicentric tumors** describe cancers located within multiple ducts of the breast.
- **Multifocal tumors** describe cancer, located at multiple sites within a single duct system.
- **Necrosis** is cell death due to lack of blood supply to the tissue.
- **One-stage** or **immediate breast reconstruction** describes the reconstructive surgery when it is performed at the same time as the mastectomy.
- **Phantom breast pain** is a sensation experienced by many mastectomy patients after the mastectomy. Symptoms can include unpleasant itching, pins and needles, pressure, or throbbing. The pain is often associated with small nerves between the breast tissue and skin, which were cut during surgery. The neural connection in the brain is undergoing a process of reorganization called neural plasticity. This results in the spontaneous firing of electrical signals from the ends of the cut or injured nerves causing the phantom sensations. Phantom pain can be relieved by massage and/or pain medication.
- **Preventive** or **prophylactic surgery** is done to remove body tissue that is not malignant but at risk for future malignancy.
- **Raloxifene** is a drug with properties similar to tamoxifen. Raloxifene was originally marketed to treat osteoporosis but has now been approved to treat invasive breast cancer in postmenopausal women.
- **Reconstructive** or **breast implant surgery** is performed by a plastic surgeon to restore the appearance of the breast after mastectomy.
- **Saline-filled implant** is an external silicone shell filled with saline solution used to create a new breast mound.
- **Sentinel lymph node biopsy** is a more conservative process than axillary node dissection. It involves the removal of only a few lymph nodes, including the sentinel node. Before a sentinel node biopsy, the sentinel node or first nodes in the lymphatic chain must be identified using any of the adjunctive detection options.
- **Seroma** describes the clear fluid trapped in the site after surgery.
- **Silicone gel–filled implant** is a silicone gel-filled sac used to create a breast mound.
- **Staging** is the process of determining how widespread the cancer is and whether it has spread to the lymph nodes in the axilla or to other parts of the body.
- **Transverse rectus abdominis muscle (TRAM) flap** uses tissue from the lower abdomen to reconstruct a breast mound. The mound can be attached by tunneling it under the skin to the breast area (pedicle technique), or by completely detaching tissue from the abdomen, then using a free flap technique (unattached) in which microsurgery would be required to attach blood vessels.
- **Two-stage reconstruction** describes the technique whereby the breast reconstruction is performed on a separate surgery date, after the mastectomy.

BREAST CANCER TREATMENT

Because breast cancer is not a medical emergency, most women are advised to consult more than one physician, thereby getting a second opinion, before proceeding with any treatment option. Breast cancer can be treated with surgery, radiation, or drugs (chemotherapy and hormonal therapy). Physicians can use one or more of these combinations, depending on the type and location of the cancer, whether the disease has spread, and the patient's overall health status.

Before beginning any treatment option, the stage of the cancer, its size, and its location must be known. It is essential to conduct tests for estrogen or progesterone receptors and tests to determine whether the cancer overexpresses any protein genes, such as *HER2,* before treatment starts. The medical professionals involved in a patient's care can include the radiologist, surgeon, oncologist, dosimetrist, radiation therapy technologist, radiographer, radiation therapy physicist, pathologist, reconstructive or plastic surgeon, gynecologist, and oncology social worker.

The need for breast cancer treatment will not start immediately. Many mammography patients are recommended for breast screening by their gynecologist. After the radiologist interprets the mammogram, if the readings are suspicious, the patient's primary care physician or gynecologist may recommend adjunctive imaging. Adjunctive testing will involve a cytotechnologist or pathologist who specializes in classifying breast cancer from histological or cytological analysis.

If adjunctive testing proves the lesion malignant, the patient is then referred to a breast surgeon or to a medical or radiation oncologist. The radiation oncologist specializes in the use of x-rays and other radiation methods to kill tumors, whereas the medical oncologist specializes in the use of chemotherapy or other drugs to treat breast cancer.

The breast surgeon will have to work closely with the oncologist and radiologist to coordinate the patient's treatment plan, which is determined by the stage of the breast cancer. If radiation treatment is an option, the radiation dose and distribution will be calculated by the radiation oncologist, working closely with the dosimetrist and the radiation therapy technologist (Fig. 10–1).

> Breast cancer can be treated with surgery, radiation, or drugs (chemotherapy and hormonal therapy).

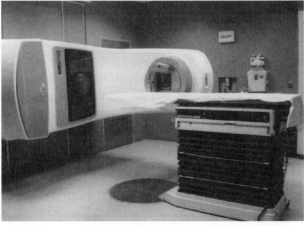
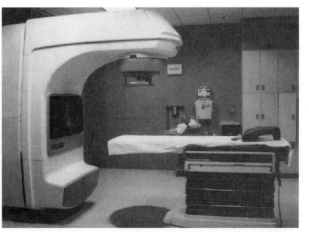

A **B**

Figure 10–1. **(A and B)** The radiation therapy unit can be turned in various directions depending on the radiation needs of the patient.

TABLE 10–1. Relative Survival Rates

Stage	5-Year Survival Rate (%)
0	100
I	98
IIA	88
IIB	76
IIIA	56
IIIB	49
IV	16

Patients, whether having a mastectomy or lumpectomy, may require breast reconstruction after surgery. The plastic surgeon and breast surgeon can work together during a single surgical procedure, or the patient can elect to have breast reconstruction after recovery from the mastectomy.

Most patients, after treatment, will continue follow-up visits every 3 to 4 months. After 5 years, patients might visit their physician every year. Patients on tamoxifen will have a pelvic examination every year. In the past, pregnancy after cancer was ruled out. However, with modern treatment, motherhood is no longer ruled out, and most studies have confirmed that pregnancy does not increase the risks for recurrence of breast cancer.

With better detection and treatment options, the relative survival rates, which refer to the percentage of patients who live at least 5 years after their cancer is diagnosed, have been steadily increasing. The relative survival rates will exclude patients dying of other diseases, such as heart disease (Table 10–1).

Research in all aspects of breast cancer is ongoing. Breast cancer covers numerous scientific disciplines, including biochemistry, cell biology, clinical research, and psychosocial and behavioral research. Researchers are constantly refining new therapies and techniques to both discover and treat breast cancer. The nature of complex interactions between cancer cells and the significant difference in how some women respond to the treatment are still not fully understood. Studies have shown that after radiation treatment begins, delayed or incomplete treatment often leads to worse outcomes and higher recurrence rates. As many as 20% of older women experience delayed or incomplete radiation treatment after breast-conserving surgery, and this can adversely affect disease-free survival. Delaying treatment by 8 weeks or more significantly increased the odds for recurrence or new breast malignancies in patients with early breast cancer. Delay in radiation therapy of 12 or more weeks after surgery, or 8 or more weeks after chemotherapy, was more than 4 times more likely to result in a subsequent breast cancer event, regardless of stage. For all patients, noncompletion of radiation therapy was not associated with an increased risk for recurrence or relapse, but among those with stage 1 disease, incomplete radiotherapy was associated with worse overall survival.

It is hoped that an understanding of how and why interactions take place in the body will be used to determine individual risk for breast cancer and add to the information about breast cancer. These conclusions will eventually affect screening, prevention strategies, and lifestyle choices and provide new perspectives on breast cancer.

Researchers have been especially excited about the development in recent years of various targeted therapies. Ideally, targeted therapies would be tailored to genetic mechanisms found to be responsible for the individual's tumor.

At present, all breast cancer detection and treatment tools have their advantages and disadvantages, and because of the wide variety, patients really need to explore their options wisely. Detecting cancers early still offers the best treatment options and the highest survival rate. This fact is driving the need to perfect a better breast cancer detection tool.

An important consideration in any cancer care is the physiological well-being of the patient. Cancer care cannot focus solely on eliminating the tumor. The psychological status of the patient can affect treatment options and even recovery time. Suffering can lead to depression, stress, and other mental and emotional conditions that should not be ignored. Good oncology care providers are often proactive in identifying patients' needs and helping them to find the resources and support that are often vital to achieving quality care and full recovery.

Staging Breast Cancer

Stage grouping classifies breast cancer patients in four stages (stages 0–IV). A cancer's stage depends on the extent of the primary tumor, the spread of the cancer to regional lymph nodes, and whether there is distant metastasis. In staging, lower numbers represent less spread of the cancer. The tumor, represented by the letter T, describes the size of the cancer. The lymph node involvement, represented by N, indicates the number of nodes involved. Men with breast cancer can be treated with the same criteria. Tumor margins, tumor size, and the number of positive axillary nodes are just as important in planning treatment for male breast cancer patients as for female patients (Table 10–2).

For example, stage 3C invasive breast cancer that may have spread to the skin is considered inflammatory cancer. Also, stage 3C cancer is

> Stage grouping classifies breast cancer patients in four stages (stages 0–IV). A cancer's stage depends on the extent of the primary tumor, the spread of the cancer to regional lymph nodes, and whether there is distant metastasis.

> **T: size or direct extent of the primary tumor**
> - Tx: tumor cannot be evaluated
> - Tis: carcinoma in situ
> - T0: no signs of tumor
> - T1, T2, T3, T4: size and/or extension of the primary tumor
>
> The higher the T number, the larger the tumor and/or the more it has grown into nearby tissues.
>
> **N: degree of spread to regional lymph nodes**
> - Nx: lymph nodes cannot be evaluated
> - N0: tumor cells absent from regional lymph nodes
> - N1, N2, N3: describes the size, location, and/or the number of nearby lymph nodes affected by cancer. The higher the N number, the greater the cancer spread to nearby lymph nodes.
>
> **M: presence of distant metastasis**
> - M0: no distant metastasis
> - M1: metastasis to distant organs (beyond regional lymph nodes)

TABLE 10–2. Staging of Breast Cancer

Stage	Tumor Size (T)	Lymph Node Involvement (N)	Metastasis (M)
0	Abnormal cell—no tumor, e.g., DCIS or LCIS	No	No
I	Less than 2 cm	No	No
IIA	No tumor or tumor less than 2 cm	Yes	No
	Between 2 and 5 cm	No	No
IIB	Between 2 and 5 cm	Yes	No
	Larger than 5 cm	No	
IIIA	No tumor	Yes. Spread to nodes in axilla and sternum	No
	Any size	Yes	No
IIIB	Any size	Yes. Spread to skin or chest wall or lymph nodes near sternum	No
IIIC	No tumor or any size tumor	Yes. Also to chest wall and/or skin of breast and nodes near clavicle and sternum	No
IV	Not applicable	Yes, on same side of breast	Yes—lungs, liver, bone or brain

> **Breast tumor classification**
>
> Luminal A (ER1 and/or PR1, ERBB2)
>
> Luminal B (ER+ and/or PR+, ERBB2+)
>
> HER-2-enriched (ER+, PR−. ERBB2+)
>
> Triple negative (ER−, PR−, ERBB2−, and/or CK 5/6+) − negative for all five markers

considered inoperable if the cancer has spread to the lymph nodes above the clavicle.

After a cancer is removed from the breast, the specimen is further scrutinized for antibodies, enzymes, or proteins. This provides information on the prognosis and guides the choice of therapeutic options. Breast tumors are currently classified using five immunohistochemical (IHC) tumor markers: estrogen receptor (ER+), progesterone receptor (PR+), human epidermal growth factor receptor 2 (HER-2), HER-1, and cytokeratin 5/6 (CK 5/6). Some breast cancers can be tracked because of the elevated levels of CK5/6 in the blood. Cytokeratins are proteins found in the lining of organs and glands; they help form the tissues of the hair, nails, and the outer layer of the skin. In the breast, CK5/6 protein is found on epithelial cells. If a blood test shows elevated levels of CK5/6, certain cytokeratins can be checked to provide a differential diagnosis of poorly differentiated metastatic carcinomas with unknown primary site. About two out of three breast cancers have at least either the ER+ or the PR+ receptors.

Some breast cancers overexpress the human epidermal growth factor receptor 2 (ERBB2, formerly HER2 or HER2/neu) protein, which functions as a receptor on breast cells. The protein receptors control how the breast cells divide and repair themselves. Cancers with the ERBB2 protein marker do not respond to treatment with tamoxifen or other anti-estrogen drugs. The ERBB2 gene makes HER2 protein and, normally, HER2 receptors help control how a healthy breast cell grows, divides, and repairs itself. When the gene malfunctions in a process known as gene amplification, HER2 receptors are overproduced, leading to uncontrolled growth of breast tissue cells. In about 25% of breast cancers, the *HER2* gene does not work correctly and makes too many copies of itself. Trastuzumab (Herceptin) has been found to be effective therapy for these aggressive cancers. Tumors with increased levels of ERBB2(Her2/neu) are referred to as HER2-positive. There is also triple negative cancer that does not contain receptors for estrogen, progesterone, or HER2.

Based on expression of these markers, breast tumors are classified into the following subtypes: luminal A (ER+ and/or PR+, ERBB2−); luminal B (ER+ and/or PR+, ERBB2+); HER-2-enriched (ER+, PR−. ERBB2+); and triple negative (ER−, PR−, ERBB2−, and/or CK 5/6+) — negative for all five markers

Mastectomy as an Option

Mastectomy is the surgical removal of the entire breast. Patients undergoing a mastectomy procedure will require general anesthesia.

In the past, the radical mastectomy, which removes the entire breast, lymph nodes, and chest wall muscles under the breast, was the standard of care. It is rarely performed today because the modified mastectomy is just as effective and is less debilitating and deforming.

The modified radical mastectomy removes the breast tissue and includes a removal of the nipple-areolar region. This is the most common mastectomy procedure performed today. The modified radical mastectomy generally requires a hospital stay of 0 to 1 days depending on the body's rate of healing.

The actual procedure usually lasts 2 to 3 hours. Mastectomy will include an axillary lymph node dissection or sentinel node dissection. The incision is closed with stitches or clips, which are removed within a week of the surgery. After the surgery, a draining tube is often placed in the breast to remove blood and lymph fluids, which accumulate during the healing process. Drainage tubes are removed in about 2 weeks.

Complications of Mastectomy

The mastectomy generally proceeds normally for most patients; however, it is a surgical procedure, and there can be complications. As with any surgery, the patient can develop an infection that will require treatment, or clear fluid (seroma) can become trapped in the surgical site. The patient can also develop a hematoma. However, none of these complications should cause major problems for the patient.

Other Mastectomy Options

Skin-Sparing Mastectomy

An increasingly popular choice for patients is a procedure known as skin-sparing mastectomy (SSM). This procedure includes resection of the nipple-areola complex, removal of any existing surgical or biopsy scar, and removal of all the breast parenchyma. Retaining much of the skin simplifies the reconstruction process and improves the cosmetic appearance of the breast. Preserving the areola is even possible in selected cases. This process, however, is contraindicated for patients with inflammatory carcinoma.

Another procedure, the quadrantectomy or partial mastectomy, is an option for some women. This surgery removes one quarter of the breast, including the tumor, with clear margins of up to 2 to 3 cm of breast tissue, skin, and some chest wall muscle. As in a mastectomy, the lymph nodes in the axilla will be removed to check for cancer spread.

Nipple-Sparing Mastectomy

In this procedure the breast tissue is removed, but the nipple-areola complex remains.

This procedure can be used by women with relatively small tumors or nonaggressive cancers, or as a prophylactic procedure. It provides improved cosmetic outcomes because the nipple is spared and, after implant placement, the breast looks intact.

A possible complication of the procedure is that all the tissue from the nipple core must be completely removed, yet a thin layer of fat with blood vessels must remain because if the blood supply to the nipple is disrupted, the patient can eventually lose the nipple. Risks of procedure are recurrent cancer at the nipple. The procedure is not recommend for patients with large breasts or patients with large or aggressive tumors.

> The psychological status of the patient can affect treatment options and recovery time.

Preventative Surgical Options

Preventive surgery is used to remove the entire breast when a woman has a very high risk factor for breast cancer. For example, women with the breast cancer gene *BRCA1* or *BRCA2* will sometimes consider a prophylactic mastectomy.

Lumpectomy as an Option

Lumpectomy is the most breast-conserving surgery. Lumpectomy is the removal of the breast cancer tumor and the surrounding margins of the normal breast tissue. During the procedure an incision is made along the perimeter of the breast close to the tumor, and the underlying tissue is cut free and removed. However, a number of factors will determine whether the patient should have a modified radical mastectomy or a lumpectomy. Some of these factors include tumor size, type, and cancer stage. Other patients may not have clear choices or may choose the modified radical mastectomy on the basis of anecdotal evidence rather than on survival statistics of the various options. Breast conservation should only be performed if the treatment provides a cure rate equal to that obtained from a mastectomy.

Indications and Contraindications for Lumpectomy

Lumpectomy is commonly performed for patients with ductal carcinoma *in situ* (DCIS) and those with stages I to III cancers.

Lumpectomy is often combined with other therapy options. Patients with a low-grade DCIS, elderly patients, or patients with other radiation contraindications will receive only the lumpectomy. However, for most other patients, the lumpectomy is followed by 6 weeks of radiation therapy to ensure that all cancer cells in the remaining breast have been destroyed. Radiation treatment usually begins 1 month after surgery, after giving the breast time to heal. In addition, chemotherapy to control the systemic spread of breast cancer or a 5-year treatment with the drug tamoxifen may be used with a lumpectomy.

Poor candidates for lumpectomy include the following:

1. Women with two or more areas of cancer in the same breast—multicentric disease—especially if the cancer areas are widely separated.
2. Women who have already undergone radiation in the breast or chest area. The patient cannot have radiation treatment after prior therapeutic irradiation to the breast or chest area. This also applies to patients treated for Hodgkin's disease in the past.
3. Women whose previous lumpectomy did not completely remove the cancer.
4. Women who are pregnant at the time of radiation therapy.
5. Women with large cancers in small breasts. The tumor size in relation to breast size is important because this may rule out breast-conserving treatment. The cosmetic result will not be good and will result in breast deformity.
6. Women with a tumor larger than 5 cm in diameter, regardless of breast size, or with extensive DCIS.

After the lumpectomy, the pathologist must check the specimen margins to make sure that the surgeon has removed the entire cancerous tumor. There should be no cancer present in the outermost edges of the specimen sample. Preliminary checks are usually made while the patient is still in the operating room, but the final result may not be available until days later. If the final results reveal positive margins, then additional surgery is necessary.

> Breast conservation should only be performed if the treatment provides a cure rate equal to that obtained from a mastectomy.

Lumpectomy can be performed using a local anesthetic, sedation, or general anesthesia depending on the extent of the surgery needed. The surgeon will make a small incision over or near the site of the lesion, then excise the entire lesion along with a clean margin of normal surrounding breast tissue. In addition to a lumpectomy, a sampling of the axillary nodes or an axillary lymph node dissection is necessary to determine whether the cancer has spread.

A seroma usually fills the surgical site after the operation and helps to naturally remold the breast shape. Gradually, the seroma is absorbed or drained, and the body replaces it with scar tissue over a period of months. Depending on the surgical technique used, the lumpectomy can be an outpatient procedure with no hospital stay or can require a 1- to 2-day hospital stay. Most patients can resume normal activity in 2 weeks. The extent of breast soreness will depend on the amount and location of tissue removed during surgery and the type of axillary dissection done. In rare instances, the seroma will reoccur after the lumpectomy, but these are easily aspirated on an outpatient basis. Compression or an injection of ethanol, an autologous fibrin clot, can also be used to fill and harden the space. These techniques can be painful. A newer technique is the injection of a fibrin sealant during the initial lumpectomy. The ingredient of fibrin sealant is fibrinogen, a protein from the blood that forms a clot when combined with thrombin—another blood protein that clots blood. The result is a reduction in the accumulation of serous fluid.

Lymph Node Biopsy

Lymph node biopsy or axillary lymph node dissection is a surgical procedure in which some or all of the lymph nodes are removed for testing. The procedure is also called a lymphadenectomy and usually follows a lumpectomy or mastectomy to determine whether the cancer has spread to areas outside the breast.

Complications of the lymph node biopsy include infection, abnormal sensations, fluid collection in the armpit, and lymphedema. A new study sponsored by the US National Cancer Institute suggests that a sentinel node biopsy plus radiation and/or chemotherapy is appropriate treatment for patients with early breast cancer identified as stage 1 or 2 with no palpable adenopathy and one to two sentinel lymph nodes containing metastases. The rationale is that chemotherapy and radiation are both designed to kill cancer cells leaving the breast. In the study, 27% of the patients had additional nodes removed after the sentinel node biopsy. These patients fared no better than those choosing not to have additional surgery. The 5-year survival rate was more than 90% for both groups.

Researchers concluded that the lymph node dissection is unnecessary and that the minimally invasive procedure, called the sentinel node biopsy, in which only one to four axillary nodes in the chain draining from the breast are removed, should be the treatment of choice. The sentinel node biopsy generally results in fewer complications than the lymph node biopsy. To identify the sentinel node, the oncologist gives the patient an injection of a radioisotope and/or a blue dye. The procedure, lymphoscintigraphy, is described in Chapter 8.

> Males with breast cancer can be treated with the same criteria that predicts a female's treatment option.

Comparisons of Surgical Treatment Options

Numerous studies have shown that there is statistically no significant difference in overall survival rates between women who undergo lumpectomy with radiation therapy and those who choose to undergo a mastectomy, when lumpectomy was a viable option.

Other Options to Consider

1. Generally, breast cancer in the first or second trimester of pregnancy can be treated with conservation surgery only if the pregnancy is terminated. In the third trimester, surgical treatment can proceed with radiation delayed until after delivery.
2. Usually, patients considering breast conservation surgery will also need radiation treatment, and traditional radiation treatments last several weeks. If the patient is unable to travel or undergo the full treatment, and if there are no other therapy treatment options available, the patient cannot have breast conservation surgery.
3. Males with breast cancer can be treated with the same criteria that predict a female's treatment option. Tumor margins, size, and the number of positive axillary nodes are just as important in planning the treatment in the male patient as in the female patient.
4. A quadrantectomy or partial mastectomy could be an option to consider for some women. The surgery removes one quarter of the breast, including the tumor, clear margins with up to 2 to 3 cm of breast tissue, skin, and some chest wall muscles. Like in a mastectomy, the lymph nodes in the axilla will be removed or sampled to determine metastasis.

Radiation therapy procedures kill cancer cells by exposing them to high-energy radiation while protecting healthy tissue from the radiation exposure.

RADIATION THERAPY

Radiation therapy was invented shortly after the discovery of x-ray in 1895. The technology was hampered by the inability to produce a consistently high-energy beam. This changed with the invention of the linear accelerator in the 1950s. The linear accelerator (LINAC) uses electromagnetic fields to propel charged particles and then allows the particles to collide with a heavy metal target.

The collision of the electrons in the target produces the high-energy x-rays needed. The technology was originally called "atom smashers" when it first appeared and was used by physicists to investigate the properties of subatomic particles. In the past, the beam exiting the accelerator was controlled by blocks, but in today's therapy world, multileaf collimators are incorporated into the units.

Radiation therapy is practiced by exposing a specific body area to various types of high-energy radiation to destroy cancer cells, while protecting other parts of the patient's body from radiation exposure (Fig. 10–2). Radiation therapy can be teletherapy (external), whereby the treatment is given once a day for 6 to 7 weeks, or brachytherapy (internal), whereby the treatment can last 5 to 9 days. Some oncologists suggest that even with a lumpectomy, a boost dose of radiation, especially for women younger than 40 years and those with early-stage breast cancer (stage 0), can be beneficial. Studies have also shown that

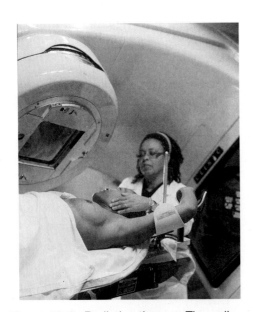

Figure 10–2. Radiation therapy. The radiation beam comes in from two directions and is directed to the breast (Reproduced from the National Cancer Institute. Photographer: Rhoda Baer.)

delaying or not completing postsurgical radiation treatment by 8 or more weeks will lead to worse outcomes—including increased recurrence rates.

External-beam radiation therapy (EBRT) can be delivered by means of a linear accelerator, which focuses and directs the beam to the area of the tumor. For breast cancer, EBRT is the most common type of radiation treatment. It involves the use of high-energy radiation to destroy cancer cells in the chest wall, axilla, or breast. The higher the energy of the x-ray beam, the deeper the x-rays will penetrate into the target tissue. The oncologist and the dosimetrist will determine the total dose of radiation and how that dose should be divided to give healthy tissue time to repair itself between treatments.

Therapeutic radiation uses photons (x-rays or gamma rays) or particulate radiation (electrons, protons, or neutrons). All are ionizing forms of radiation with different energy levels. High-energy photons are from radioactive sources, such as cobalt, cesium, or a linear accelerator. The electron beam will provide superficial treatment or treatment to a specific area but will avoid radiation to deeper tissues. Energy of the photons is expressed in electron volts (eV), and radiation therapy typically uses rays with energy values in the 120,000 to 18,000,000 eV range. For comparison, the energy of visible light, which is also a form of electromagnetic radiation, is approximately 2 to 3 eV.

Most of the biological effects of ionizing radiation are thought to be caused by the improper repair or damage to deoxyribonucleic acid (DNA) in the cells. The radiation will destroy cancer cells in the radiation field by damaging the DNA of the cells. DNA stores genetic information on cell growth, division, and function. Radiation, however, will also damage the DNA of normal cells, but normal cells are usually able to repair themselves and function properly after the treatment stops.

> The biological effects of ionizing radiation are caused by the improper repair or damage to deoxyribonucleic acid (DNA) in the cells.

Before the start of the treatment, the specific area of the body must be identified to avoid radiation to noncancerous areas. The patient starts radiation treatment only after the radiation oncologist has met with the patient for a full consultation and reviews the patient's medical history, physical examination, and all pertinent data. The oncologist could even request further testing. Generally, the radiation oncologist works with the dosimetrist to formulate a treatment plan and a simulation of the treatment. The oncologist and dosimetrist must first measure the correct angles for aiming the radiation beam at the specific area of the body and make ink marks on the patient's skin. Treatment planning will often include using a computer to obtain a virtual three-dimensional (3D) analysis of the breast. Computed tomography (CT) scans of the area can be taken, and digitally reconstructed radiographs (DRRs), sometimes called digital integrated radiographs (DIR), can be obtained. DRRs are used to document that the patient is being treated exactly as proposed. Use of DRRs will enable the oncology team to plan an individualized treatment plan or to design protective treatment shields to minimize radiation to the normal tissues.

EBRT usually begins 1 to 2 months after surgery, after giving the breast time to heal. However, brachytherapy, which limits radiation treatment to the site of the cancer, can start much earlier. If EBRT follows lumpectomy, then the radiation therapy can involve daily 15-minute sessions. The photon beam of radiation is actually on only for 1 or 2 minutes, but treatment is given 5 times per week for 5 weeks to the entire breast, plus 1½ weeks of booster (electron or photon beam) radiation directly to the tumor site. The electron-beam radiation falls off rapidly when directed at the tumor site and will not substantially affect tissues beyond the treatment sites.

The procedure itself is pain free, and the patients are generally outpatients. To ensure that the radiation beam is correctly aimed, special blocks are sometimes made to protect normal tissues and organs from the radiation. Because the therapist cannot stay in the treatment room during the procedure, the patient is monitored by a closed-circuit television. Another treatment option includes accelerated hypofractionated whole-breast irradiation, whereby the entire breast is treated with a more intensive 3-week course of radiation, generally given without a boost dose, and stereotactic body radiation therapy (SBRT) that completes the radiation treatment in 5 days rather than several weeks.

Indications for Radiation Therapy

Radiation therapy is often used in conjunction with other treatment options, such as after a mastectomy or lumpectomy to remove the missed cancer cells in the breast, chest wall, and axilla area. Postmastectomy radiation is also recommended for patients with tumors larger than 5 cm or who have cancer in four or more axillary lymph nodes.

Radiation therapy can also be used before surgery to shrink the size of the tumor, or it can be given immediately after surgery if no chemotherapy is planned.

If the patient is having chemotherapy, it is usually given before the radiation treatment starts because chemotherapy will ideally eliminate distant metastasis.

External radiation can damage not only any remaining tissue but also the breast skin, and because the outcome is unpredictable, some surgeons prefer women to delay reconstruction until after radiation therapy is completed. This minimizes the risks for unfavorable cosmetic outcomes, such as formation of scar tissue around the implant, a condition known as capsular contracture that can lead to fibrous (scar) tissue in the reconstructed breast and painful breasts. This practice, however, is controversial, and many surgeons maintain that if reconstruction is performed immediately, women fare better emotionally than if reconstruction is delayed.

> Radiation therapy is often used in conjunction with other treatment options.

Classifications of Radiation Therapy

Radiation therapy can be classified as primary, adjuvant, combined, or palliative:

1. *Primary radiation:* use of radiation alone to attempt to cure a cancer or as a substitute for more extensive surgery (breast-conserving therapy).

2. *Adjuvant radiation therapy:* use of radiation after surgery or chemotherapy for possible residual cancer
3. *Combined modality therapy:* use of radiation therapy with another treatment option, for example, chemotherapy, although this is rarely done with modern breast cancer chemotherapy regimens
4. *Palliative radiation therapy:* use of local radiation therapy to alleviate a distressing symptom or to prevent a problem. Common palliative situations can include treatment for pain, especially from bone metastases, or treatment for a tumor that is causing bleeding or obstruction.

Side Effects of Radiation Therapy

Side effects of radiation can occur early or late in the course of treatment. The side effects typically occur close to the time of treatment, and most will be completely gone within a few weeks of finishing treatment. Most late effects will take months or years to develop and are often permanent. Therapy is generally not interrupted because of a side effect because an interrupted treatment can alter the outcome of the treatment.

Side effects can include the following:

1. Fatigue—extreme tiredness
2. Swelling of the breast
3. Heaviness in the breast
4. Sunburn-type appearance of the breast skin

Most side effects can be controlled by medication. After radiation treatment, the breast will sometimes remain firm, and on rare occasions, it may become enlarged because of fluid buildup or smaller because of tissue changes. The breast skin can become more or less sensitive. During treatment, the patient is usually advised to do the following:

1. Avoid additional sun exposure to the area.
2. Wear loose-fitting clothing, preferably cotton or other material that breathes.
3. Use warm or tepid water when bathing instead of hot water and avoid using heating pads or warm compresses to relieve pain in any treated area.
4. Avoid constricting bras.
5. Use cool compresses, not cold or ice packs.
6. Avoid using lotions or powders on the treated area. The patient should check with the oncologist for specific creams that should be avoided or should consult on which creams can be used as soothing lotions or oils.

> Intensity-modulated radiation therapy (IMRT) is capable of giving radiation doses to 3D tumors while minimizing the dose to surrounding normal structures.

Intensity-Modulated Radiation Therapy

In traditional radiation therapy treatment, the x-ray beam is directed to the breast, but both normal and abnormal tissue will be irradiated. Therefore, portions of the pleura, lungs, and even the sternum can be irradiated. To deliver a more precise radiation dose to a malignant tumor, the technology of IMRT is employed.

IMRT uses a computer-controlled linear accelerator and is capable of giving radiation doses to 3D tumors while minimizing the dose to surrounding normal structures. Treatment is planned by using 3D CT images of the patient in conjunction with computerized dose calculations. This allows the oncologist and dosimetrist to determine precise dose intensity patterns.

The method of IMRT is to first define the tumor (size, shape, and location) and the radiation dose needed. The oncologist and dosimetrist will also determine which areas should be spared and which areas will need the maximum amount of radiation during the course of the treatment. A computer is then used to determine the optimal treatment fields.

Radiation treatment is always customized for each patient, but IMRT makes it possible to create radiation dose distribution patterns that are unique to the technology and are customized to each patient's clinical needs. During the course of a single treatment, the radiation beam, aimed at the tumor, is controlled and will change shape hundreds of times as it conforms to the 3D size of the tumor.

Advantages and Disadvantages of Intensity-Modulated Radiation Therapy

With the reduction in the amount of radiation going to normal tissue, IMRT may allow higher doses to cancerous areas, better sparing of normal tissues, or fewer side effects on normal tissues. However, the total dose of radiation and the number of treatments given will depend on the size, location, and type of cancer; the patient's general health; and other medical therapy the patient is receiving.

Women who received IMRT may experience significantly fewer skin color changes, less reddened or itchy skin, and less swelling of the breast compared with those who undergo traditional external radiation treatment. Research shows that 41% of women who received IMRT had reddened or itchy skin compared with 85% of women receiving conventional radiation. Also, only 1% of women receiving IMRT have breast swelling, compared with 28% of women receiving conventional treatment. Changes in skin color were reported in 5% of the women undergoing IMRT, compared with 50% of those who had conventional treatment.

Even the smallest movement, such as irregular breaths, can result in significant deviations from calculated doses. Immobilization is therefore vital during any IMRT procedure.

In addition, IMRT needs total integration of all the networks and software within a facility. The technology is complicated, and to avoid disastrous results it demands ongoing quality assurance to ensure that the information systems and software are compatible. IMRT plans cannot be fully optimized unless it is integrated with another modality image to manage organ motion and variations in tumor location due to weight loss.

Image-Guided Radiation Therapy

Image-guided radiation therapy (IGRT) is often used in the delivery of IMRT treatments. It involves using another diagnostic imaging modality to aid in the radiotherapy process and to optimize the accuracy of treatment delivery on a daily basis. The imaging modality is used only

> IMRT may allow higher doses to the cancerous areas, better sparing of the normal tissues, or fewer side effects on the normal tissues.

after the physician has determined that treatment is warranted, the extent of the treatment, and the methods of minimizing dose to normal tissue. IGRT for example could utilize general x-ray imaging using anteroposterior and lateral projections to check patient positioning or cone-beam CT or ultrasound to check tumor location.

X-ray tubes in multiple arrays or fluoroscopy images are used to create and reconstruct 3D radiographs for treatment planning. The cone-beam CT uses CT for image acquisition and 3D image reconstruction. The radiation travels in a coil-like trajectory around the area of interest, and a two-dimensional (2D) array of detectors measures the transmitted radiation. Ultrasound uses sound wave to scan the patient, thus integrating patient positioning with the radiation therapy delivery system. During the treatment, the imaging is capable of giving updated assessment of treatment, allowing modification as needed. Generally, a true IGRT system has the modality in the room, integrated with the linear accelerator used for the radiation therapy.

IGRT can also use a number of other modality imaging, such as CT, magnetic resonance imaging (MRI), positron emission tomography (PET), and tomotherapy, in the planning process. Tomotherapy uses discrete-angle, sliding beams to deliver precise radiation therapy. The unit uses a multileaf collimator and CT-style gantry technology. Newer units are expected with 3D imaging to perform image registration or image fusion, which allows planning and daily evaluation of position during treatment.

> IGRT involves another diagnostic imaging modality to aid in the radiation therapy process and to optimize the accuracy of treatment delivery on a daily basis.

Advantages and Disadvantage of Image-Guided Radiation Therapy

Unlike regular radiation treatment, IGRT relies directly on the imaging modality to coordinate the reference when localizing the tumor in the patient. The modalities may provide real-time tumor tracking, allowing adjustment for the patient's breathing movements, and even irregular breaths. Many of the newer IGRT imaging methods are experimental, and clinical trials will be needed to assess the impact of the technology on patient outcomes.

Internal Radiation or Partial Radiation

Internal radiation, also called brachytherapy, has been used in the past and is currently used in selected breast cancer patients. Instead of using radiation beams from outside the body, radioactive sources are temporarily placed directly into a balloon or catheter that has been inserted into the lumpectomy cavity. Brachytherapy is used to deliver a high dose of radiation to small areas. Eligible patients are treated for 5 to 7 days, versus the 6 to 7 weeks of external radiation treatments. This method of treatment is sometime referred to as accelerated partial breast irradiation (APBI). The shorter length of treatment is often more convenient for many patients, especially for those living long travel distance from treatment centers.

> Internal-beam radiation therapy places the radioactive source into a balloon or catheter inserted directly in the lumpectomy cavity.

Internal-beam radiation therapy therefore involves applying the radiation from inside out, and there are different types. The idea is to apply a more intense dose of radiation to the tumor while limiting the radiation dose to the adjacent healthy breast tissue. Brachytherapy is not the standard treatment for breast cancer. There are a number of methods.

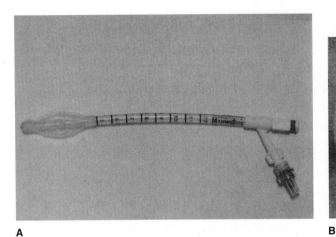

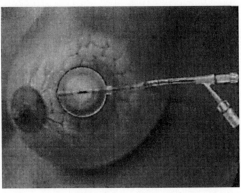

A   **B**

Figure 10–3A. A brachytherapy device before inflation **(A)**. Inflated in position in the breast **(B)**.

Figure 10–3B. The SAVI applicator, a multiple catheter breast brachytherapy device. (Used with permission from Cianna Medical.)

Balloon Catheter

This treatment is generally done after surgery but before chemotherapy. An inflatable balloon catheter device is placed in the breast at the time of the lumpectomy surgery. Bandages are placed over the site, and the patient is sent home. Treatment begins 2 days after surgery and generally takes 5 days, twice a day approximately 6 hours apart (Fig. 10–3A). The patient can come in the morning, go to work, and return in the afternoon. Each morning the patient's skin is checked for infection. Before each treatment, the balloon is inflated, and CT scans are taken each morning before treatment to ensure balloon placement and balloon volume. The CT scan also ensures that there is no potential saline leakage, there are no air pockets, and the balloon is no less than 7 mm from the skin. Corrections can be made to ensure dose accuracy and homogeneity.

The device remains in place for the duration of the treatment, but at each treatment session, a tiny radioactive bead attached to a wire is threaded into the inflated balloon. The wire device (radioactive bead plus wire) is moved to various places inside the inflated balloon delivering different levels of radiation to different areas. After completion of treatment, the balloon is deflated, and the system is removed. This method has no source of radiation remaining in the patient's body between treatments.

Single-Entry Multicatheter Device

The single-entry multicatheter device method uses a special mini-brachytherapy applicator. It is designed to be used in the smallest of lumpectomy cavities and is available in three sizes, allowing treatment to small, large, or irregularly shaped cavities (Fig. 10–3B). The device has six peripheral catheters surrounded by one central catheter. It is a single-entry interstitial-type brachytherapy applicator and similar to the balloon-based technology that allows a 5-day course of radiation. The device is inserted in the collapsed position and after insertion can be expanded to conform to the shape of the tumor cavity. Because each catheter has its own radiation source, the physician can customize the dose depending on the area to better focus the radiation where it is needed while sparing healthy tissue.

Multicatheter Method

The radiation can be given using unsealed radioactive sources with the implantation of radioactive seeds directly in the breast tissue next to the cancer. First, 10 to 20 plastic catheters are surgically placed into the breast tissues (Fig. 10–3C). Radioactive pellets—iridium-192—are inserted into the catheters. The catheters are connected to a high-dose rate brachytherapy machine for approximately 10 minutes 9 or more times in a week. The catheters are removed after approximately 1 week. In this method, the source radiation remains in the patient's body between treatments, and the patient will require a brief hospital stay.

Intraoperative Electron Radiation Therapy

Intraoperative electron radiation therapy (IOERT) involves the application of electron radiation directly to the tumor bed during cancer surgery. The radiation is applied to the tumor site and areas immediately surrounding the site while the site is still exposed during surgery. This direct treatment can be used to replace EBRT or brachytherapy and is considered effective in killing cancer cells when they are most vulnerable, before they have proliferated or migrated. Another advantage is that the beam is very precise, and because the dose falls off rapidly below the targeted site, healthy underlying tissues and organs are spared.

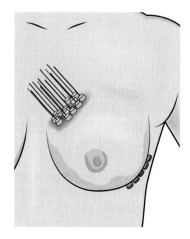

Figure 10–3C. Multicatheter system.

Advantages and Disadvantages of Brachytherapy

Brachytherapy will reduce the time of treatment from 6 weeks to 1 week. This means that there is less irritation of healthy breast tissue, and because the treatment can start immediately after a lumpectomy, there is less delay before the start of treatment. Other advantages of brachytherapy are that the treatment causes fewer skin reactions, such as redness, rashes, or irritation.

A limitation of brachytherapy is that although the technique aims to give a high, uniform dose to the site of the tumor, the dose falls off rapidly from the source, and surrounding tissues may not get the necessary high dose of radiation.

There is a general absence of randomized clinical trials of breast brachytherapy methods, which means that these methods have not been proved to be an alternative to whole-breast irradiation. This may change because there is an ongoing randomized trial comparing partial breast radiation therapy to conventional whole breast radiation.

Chemotherapy uses drugs to treat cancer that may have spread beyond the breast.

CHEMOTHERAPY

Chemotherapy is an adjuvant therapy and involves the use of drugs to treat cancer that may have spread beyond the breast. Chemotherapy is often a combination of drugs. There are more than 90 chemotherapy drugs available. Most physicians prefer a combination of lower doses of individual drugs instead of a high dose of one powerful drug. Low-dose drugs are also associated with fewer side effects. Chemotherapy is considered a systemic form of cancer treatment because the drug is distributed throughout the entire body through the bloodstream. Chemotherapy drugs will therefore affect all tissues and organs in the body.

Box 10–1. Phases of the Cell Cycle

- G_0 is the resting stage. The cells have not started to divide. This step can last for hours to years. When the cells are signaled to reproduce, the next stage begins.
- G_1 is the RNA and protein synthesis phase. During this phase, the cells start making more proteins to get ready to divide. This phase lasts approximately 18 to 30 hours.
- S is the DNA synthesis phase. During this period, the proteins containing the genetic code (DNA) are copied so that both of the new cells formed will have the right amount of DNA. This phase lasts approximately 18 to 20 hours.
- G_2 phase is the construction of mitotic apparatus. This phase starts just before the cells start splitting into two cells. It can last for 2 to 10 hours.
- Mitosis (M) refers to the phase in which the cell actually splits into two new cells. This phase lasts 30 to 60 minutes.

The drug tends to attack cells that are quickly dividing whether cancerous or not.

Generally, chemotherapy regimens are tailored for the individual patient and can vary tremendously. The type of treatment will depend on the patient's age, overall health, stage, grade of the cancer, past or future treatments, and other health problems. Some patients will receive chemotherapy as the only form of treatment because chemotherapy can be used to cure the cancer by totally destroying the cancerous cells in the body. However, chemotherapy can also be used to control the cancer and extend the patient's life by stopping an inoperable cancer from growing and spreading, or as a palliative treatment to relieve symptoms caused by the cancer. If chemotherapy is a part of the treatment option, it is given first before radiation or hormonal treatment, between or after surgery.

The Cell Cycle

The principle component of the human body is the cellular structure. Every human cell has a specific function. Cells will increase in number, grow, and reproduce to replace cells lost during injury or normal wear and tear. There are five phases in the cell cycle (Box 10–1). The phases are characterized by the structure of the chromosomes, which contain the genetic material DNA. The phases are M, G_0, G_1, S, and G_2. Chemotherapy drugs are designed to work only on actively reproducing cells (M or S phases), not on cells in the resting phase G_0. Unfortunately, however, the chemotherapy drugs cannot differentiate between reproducing cells of normal tissues and cancer cells.

Uses of Chemotherapy

1. Stop the spread of cancer to other parts of the body.
2. Slow the growth of cancer.
3. Kill cancer cells.
4. Relieve symptoms of cancer.

Drugs Used in Chemotherapy

1. Alkylating agents are chemotherapy drugs that work directly on the DNA to prevent the cancer cells from reproducing. These drugs work on all phases of the cell cycle.
2. Nitrosoureas act similarly to alkylating agents. They interfere with enzymes that help repair DNA.
3. Antimetabolites are drugs that work during the S phase of the cell cycle and interfere with DNA and ribonucleic acid (RNA) growth.
4. Antitumor antibiotics interfere with DNA by stopping enzymes and mitosis or by altering the membranes that surround cells. They are not the same as antibiotics used to treat infection and work in all phases of the cell cycle.
5. Mitotic inhibitors will inhibit or stop mitosis or inhibit enzymes from making proteins needed for reproduction of the cells. They work during the M phase of the cell cycle and are plant alkaloids or compounds derived from natural products.

6. Corticosteroid hormones are steroids and natural hormone-like drugs used to kill cancer cells or slow their growth. Many are used with other drugs to increase their effectiveness.

7. Sex hormones are drugs that alter the action or production of the female or male hormones and are used to slow the growth of cancer cells, if they are hormone receptor positive.

Chemotherapy Regimens

Chemotherapy can be given before or after cancer surgery with or without other treatments. Neoadjuvant chemotherapy is given before surgery to help shrink the size of the cancerous tumor. Adjuvant chemotherapy is chemotherapy given in addition to another breast cancer treatment, for example, mastectomy.

Method of Giving Chemotherapy

Some chemotherapy drugs are given orally as tablets or liquids or applied to the skin as a cream or lotion. Chemotherapy can also be given as an intravenous or intramuscular injection, topically or injected directly into the cancer area. Chemotherapy courses can be given daily, weekly, monthly, or by using other scheduling options depending on the patient's response to the drug. Generally, the treatment can last 3 to 6 months, but most chemotherapy sessions include built-in rest cycles to give the healthy cells recovery time.

The most used parenteral route is intravenous through a semipermanent catheter or vascular access device (VAD), implanted into a large vein in the arm or hand or the subclavian vein. VAD is useful to give several drugs at once, for long-term therapy and for continuous infusion chemotherapy. The peripherally inserted central catheter (PICC) and the implantable venous access port (e.g., port-a-cath) are examples of VAD. The PICC is placed in an arm vein and threaded through to the subclavian vein or another major vein in the thoracic cavity (Fig. 10–4A and B). No surgery is needed for this type of placement. The implantable venous access port requires surgery to implant a catheter and port access under the skin. This system provides continuous access to a large central vein.

Side Effects of Chemotherapy

Chemotherapy causes the most damage to bone marrow blood and cells of the hair follicles, the reproductive system, and the digestive tract. The side effects of chemotherapy reflect these damages. Side effects from chemotherapy will vary depending on the strength of drugs used, the dosage, and the duration of treatment. Some patients experience few side effects, whereas others experience many of the common side effects.

The closer the woman is to menopause when she undergoes chemotherapy, the more likely she is to experience premature menopause. Symptoms include hot flashes, vaginal dryness, and irregular menstrual cycles. Chemotherapy drugs can also cause birth defects; therefore, a woman should not be pregnant while receiving chemotherapy treatment.

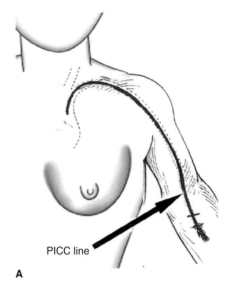

PICC line

A

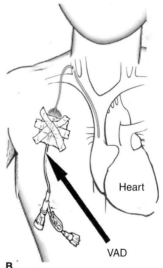

Heart

VAD

B

Figure 10–4. (A and **B)** Peripherally inserted central catheter (PICC) and vascular access device (VAD) parenteral routes for chemotherapy.

Side effects from chemotherapy will vary depending on the strength of drugs used, the dosage, and the duration of treatment.

Several drugs are available to counter the side effects of chemotherapy, and patients with low blood cell count during chemotherapy can be given medication to help raise the blood cell or platelet counts, or they can be given a transfusion.

Vasomotor symptoms can be managed with steroids or antidepressant drugs. A less invasive treatment for these side effects is acupuncture, which can be used to manage hot flashes, night sweats, and other vasomotor or related symptoms.

Main Side Effects From Chemotherapy

The main side effects from chemotherapy are nausea and vomiting, which are often caused by irritation to the lining of the stomach and duodenum, which in turn trigger the vomiting center in the brain. Other side effects can include the following:

1. Hair loss—alopecia
2. Fatigue caused by low blood count
3. Infection, often a result of low white cell count
4. Irritation of the lining of the stomach or intestine
5. Infertility or premature menopause

Minor side effects include the following:

1. Mouth sores
2. Taste changes
3. Decreased appetite
4. Diarrhea because of damage to the rapidly dividing cells in the digestive tract
5. Constipation due to loss of motility, poor diet, and certain medications
6. Tingling or burning sensations
7. Numbness in hands or feet
8. Skin irritations—redness, itching, peeling, or acne
9. Dark, brittle, or cracked fingernails or toenails

Hair loss is generally temporary and occurs because in some women the hair follicles are weakened by the chemotherapy drug, causing hair to fall out at a much faster rate than normal hair growth. Hair loss can occur 2 to 3 weeks after treatment begins, but hair grows back at the end of treatment—sometimes with a change in texture.

Low blood cell count (white and red cells and platelets) is the result of the chemotherapy's effect on the bone marrow.

A low white cell count will make the body more susceptible to infections. Patients undergoing chemotherapy need careful monitoring of their immune system because white cells are an essential component of the body's immune system. The normal white cell count can range from 4000 to 20,000 white blood cells per cubic millimeter. On average, white blood cells have a short life cycle and can live from a few days to a few weeks. A low white cell count is called leukopenia.

A reduction in red blood cells causes anemia, which is associated with fatigue, dizziness, headaches, irritability, and increased heart rate or breathing. Red blood cells bring oxygen to the tissues, and their life cycle is on average 120 days. The normal red blood cells per cubic millimeter are 4 to 6 million.

The main side effects of chemotherapy are nausea and vomiting.

Platelets help to prevent unnaturally long bleeding, and on average the life cycle of a platelet is 10 days. A low platelet count is referred to as thrombocytopenia. Symptoms include the tendency to bruise easily or to develop large and small bruises. Patients can also bleed longer than usual after cuts or have nosebleeds or bleeding gums. Severe cases can cause internal bleeding. The normal range for platelets is 150,000 and 450,000 platelets per cubic millimeter.

TARGETED TREATMENT

Targeted treatment involves addressing each individual's unique biology and disease structure but can result in a higher level of treatment efficiency plus more successful outcomes. There are two types of targeted treatments: molecular and hormonal.

Molecular Treatment

The earliest attempt at molecular treatment in cancer therapy was the drug tamoxifen. The idea of molecular treatment is to determine the exact genetic profile of the altered cancer cells and design a treatment plan based on the nature of these cells or subcells.

Tamoxifen

Tamoxifen has been in use since the 1970s to treat patients with ER-positive breast cancer. It is one of a number of antiestrogen drugs called selective estrogen receptor modulators (SERM). SERMS are drugs that block estrogen and can lower the risk for breast cancer recurrence in postmenopausal women. Some cancers have ERs on the surface of their cells. These cancers rely on a supply of estrogen to grow and are said to be ER positive. Tamoxifen prevents estrogen from latching onto tumor cell receptors and directing them to multiply. This slows or stops the growth of cancer cells in the body. Tamoxifen can be used after the initial mastectomy or lumpectomy. If the cancer recurs after treatment, tamoxifen can also be used to help prevent new cancers from developing in women who have already been treated for breast cancer. Tamoxifen is also used to shrink large tumors before surgery. Tamoxifen does not prevent ER-negative breast cancer; however, it could make these cancers more detectable. In a study, the ER-negative tumor was detected 77.4% of the time in a placebo group compared with 94.7% of the time with the tamoxifen group.

A study by the National Cancer Institute found that compared with the women on placebo, those taking tamoxifen had almost 50% fewer cases of invasive breast cancer. The benefits of tamoxifen were, however, found to be negligible after 5 years.

Tamoxifen is only recommended for women at high risk for breast cancer, including women older than 60 years and women between the ages of 35 and 59 years who have increased risk factors for breast cancer. Because of the side effects of tamoxifen, the breast cancer risks should be higher than the average woman's risk before a woman should consider taking tamoxifen.

Increased risk factors can be due to having the *BRCA* gene alteration, a previous history of breast cancer, family history of breast cancer, an

> Targeted treatment involves addressing each individual's unique biology and disease structure, but can result in a higher level of treatment efficiency plus more successful outcomes.

> SERMS drugs, such as Tamoxifen, block estrogen and can lower the risk for breast cancer recurrence in postmenopausal women.

atypical breast biopsy, not having any children, having a first child at age 30 years or later, starting menstrual periods before age 12 years, or going through menopause after age 50 years.

Side Effects of Tamoxifen
Minor Side Effects of Tamoxifen
1. Depression, tiredness, and dizziness
2. Vaginal dryness, itching, or bleeding
3. Menstrual irregularities
4. Loss of appetite, nausea, and/or vomiting
5. Weight gain
6. Mild allergic reactions, such as skin rashes
7. Temporary thinning of the hair
8. Headaches and changing patterns of headaches, especially in migraine sufferers
9. Visual problems, including blurred or reduced vision

Tamoxifen has some weak estrogen-like properties and will increase a woman's risk for certain cancers. The drug does not cause a woman to begin menopause, although it can cause some symptoms of menopause, such as hot flashes, night sweats, mood swings, and vaginal dryness. In most premenopausal women taking tamoxifen, the ovaries continue to act normally, and the drug does not reduce menopausal symptoms and may actually make them worse.

Serious effects of tamoxifen include increased risks for:

- Endometrial cancer
- Pulmonary embolism
- Stroke
- Deep vein thrombosis
- Blood clot in the lungs
- Uterine sarcoma

Major Side Effects of Tamoxifen
The more serious effects of tamoxifen have been linked to increased risks for endometrial cancer, pulmonary embolism, stroke, deep vein thrombosis, blood clot in the lungs, and uterine sarcoma, which is cancer of the connective tissue of the uterus. In general, the blood clots occurred more often in people with high blood pressure or diabetes, smokers, and those who are obese. Women with a hysterectomy do not have increased risk for endometrial cancer.

Tamoxifen is not recommended for women who have had blood clots or who develop blood clots; women taking blood thinners; women with a history of high blood pressure, smoking, obesity, or diabetes; women who are pregnant or are planning to become pregnant; women who are breastfeeding; women younger than 35 years or younger than 60 years who are not at increased risk for breast cancer; and women taking hormone replacement therapy or raloxifene.

Tamoxifen can cause defects if taken at the time of conception or during pregnancy, and it may affect fertility. Other organs, such as bone and the uterus lining, also have estrogen receptors, and tamoxifen reacts with these like estrogen, leading to increased bone density and higher risk for uterine cancer.

Physicians clearly have to weigh the benefits against the risks. For example, an older patient may benefit from tamoxifen, but because of a prior history of stroke, tamoxifen would not be recommended.

Benefits of Tamoxifen
The benefits of tamoxifen, however, are thought to outweigh some of the risks. The prophylactic effects of tamoxifen have been shown to last up to 5 years after treatment ends. Tamoxifen will slightly reduce the

risk for bone fracture of the hip, wrist, and spine in women who are past menopause, but it does not protect against heart attacks. Despite its benefits, studies have found that these benefits are only considered effective within a 5-year period, although recent studies suggest that tamoxifen can be beneficial up to 10 years. Some women have been known to develop a resistance to tamoxifen that may be associated with the function of another gene, *PAX2*. However, the exact way in which this works is still under investigation.

Other Selective Estrogen Receptor Modulators

Recently, a number of drugs with effects similar to tamoxifen have appeared on the market. Many of these drugs do not have the dangerous side effects of tamoxifen; however, many are not as effective as tamoxifen. Researchers are also measuring how patients are responding to SERMs by monitoring a molecule, Ki67. If Ki67 is diminishing or if it goes away, the treatment is effective. If Ki67 remains high during treatment, there is a greater chance of cancer recurrence.

Raloxifene

Raloxifene (Evista) is another SERM that is considered as a replacement for tamoxifen. Studies by the National Cancer Institute have confirmed raloxifene as a drug with fewer side effects while offering similar benefits as tamoxifen. The drug was originally marketed to treat osteoporosis in postmenopausal women. As a breast cancer treatment, raloxifene has been approved by the US Food and Drug Administration (FDA) since September 2007 for treatment of invasive breast cancer in postmenopausal women. Like tamoxifen, raloxifene decreases the risk for developing breast cancer by blocking the effects of estrogen and thereby stopping the growth of the cancer.

Side Effects

Raloxifene can result in side effects similar to tamoxifen, but not as severe. More common side effects of raloxifene are hot flashes, especially within the first 6 months of raloxifene therapy; swollen hands, feet, ankles, or lower legs; leg cramps; and joint pain.

Fulvestrant

Other antiestrogen drugs, such as fulvestrant (Faslodex), work in a somewhat different way to block estrogen's effects. Like SERMs, fulvestrant attaches to the estrogen receptor and functions as an estrogen antagonist. An antagonist is used to counteract the effects of certain drugs, whereas an agonist is a drug that readily combines with a receptor (organ) to enhances the body's natural response to stimulation (e.g., epinephrine given for bronchospasm because it enhances the natural bronchodilation capabilities of the body). Unlike SERMs, fulvestrant has no estrogen agonist effects. It is a pure antiestrogen. In addition, when fulvestrant binds to the estrogen receptor, the receptor is targeted for destruction.

Hormone Therapy

Hormone therapy (also called hormonal therapy, hormone treatment, or endocrine therapy) slows or stops the growth of hormone-sensitive

Side effects of raloxifene:
- Hot flashes
- Increased sweating
- Joint aches
- Leg cramps
- Blood clots
- Stroke

More common side effects of fulvestrant:

Bloating or swelling of the face, arms, hands, lower legs, or feet

Rapid weight gain

Tingling of the hands or feet

Unusual weight gain or loss

Wheezing

Less common side effects of fulvestrant:

Difficult or labored breathing

Shortness of breath

Tightness in the chest

Common side effect of anastrozole:

Blurred vision

Chest pain or discomfort

Dizziness and headache

Nervousness

Pounding in the ears

Shortness of breath

Slow or fast heartbeat

Swelling of the feet or lower legs

Less common side effects of anastrozole:

Pain in legs, feet, arm, back, jaw

Sore throat, cough, or hoarseness

Difficult or painful urination

Fever or chills and unusual tiredness or weakness

Increased blood pressure

Nausea and sweating

Vaginal bleeding (unexpected and heavy)

Side effects of exemestane:

• New or unusual bone pain

• Vision problems

• Swelling in your hands or feet

• Shortness of breath, even with mild exertion

• Chest pain

• Sudden numbness or weakness

• Sudden headache

• Confusion

• Problems with vision, speech, or balance.

tumors by blocking the body's ability to produce hormones or by interfering with hormone action. Tumors that are hormone insensitive do not respond to hormone therapy.

Drugs called aromatase inhibitors can be used to block the activity of an enzyme called aromatase, which the body uses to make estrogen in the ovaries and in other tissues. Aromatase inhibitors are used primarily in postmenopausal women because the ovaries in premenopausal women produce too much aromatase for the inhibitors to block effectively.

However, these drugs can be used in premenopausal women if they are given together with a drug that suppresses ovarian function.

Examples of aromatase inhibitors approved by the FDA are anastrozole (Arimidex) and letrozole (Femara), both of which temporarily inactivate aromatase, and exemestane (Aromasin), which permanently inactivates the enzyme.

Anastrozole

Anastrozole (Arimidex) can also be used to reduce the recurrence of breast cancer. Trials with anastrozole show that patients with hormone receptor–positive breast cancer were 65% less likely to have a relapse or a new tumor than women on tamoxifen.

In a study of more than 500 patients, approximately half were assigned tamoxifen, and half took anastrozole. Fourteen percent of the patients on tamoxifen had a recurrence of their breast cancer within 3 years versus 5.4% of the patients taking anastrozole.

Side effects of anastrozole include a higher rate of vaginal dryness, painful intercourse, and loss of interest in sex. It also caused a higher degree of bone demineralization, with possible risk for osteoporosis and osteoporotic fractures.

Exemestane

Exemestane (Aromasin) is an oral steroidal aromatase inhibitor that offers better protection against tumor development and carries fewer side effects than tamoxifen. Exemestane (Aromasin) lowers the blood levels of estrogen. It works by attaching to the aromatase enzyme and permanently deactivating it. Exemestane is used to treat early breast cancer in postmenopausal women.

Side effects of exemestane include hot flashes, hair loss, bone or joint pain, tiredness, unusual sweating, nausea, diarrhea, dizziness, and bone loss. Bone loss is of concern for patients with osteoporosis. Some studies suggest giving 2 years of tamoxifen followed by 2 to 3 years of exemestane.

Trastuzumab

More than 75% of breast cancers in the United States are ER-positive cancers; however, not all cancers are ER positive. There is another class of breast cancers that overexpress the *HER2* gene. Approximately 25% of the population have this overactive gene and will not respond to treatment with tamoxifen or other antiestrogen drugs. Trastuzumab (Herceptin) has been found to be very effective therapy for these aggressive cancers.

Side Effects

Minor side effects of trastuzumab are fever, chills, weakness, nausea, vomiting, cough, diarrhea, and headache. A major side effect of trastuzumab is possible damage to the heart muscle. HER2 is essential for early development and later growth of the muscles of the heart. Trastuzumab, by blocking HER2, can also increase risk for heart abnormalities. Heart abnormalities have been detected in 2% to 7% of patients taking the drug. This drug should not be given to patients with heart conditions.

Lapatinib

Another drug lapatinib (Tykerb) has also been found to be effective in the treatment of HER2-aggressive cancers and cancers that are both HER2 positive and ER positive. Lapatinib is in a class of medications called kinase inhibitors. It is effective in interrupting the HER2 growth receptor pathway. HER2 is essential for early development and later growth of the muscles of the heart. Trastuzumab, by blocking HER2, can increase risks for heart abnormalities. Heart abnormalities have been detected in 2% to 7% of patients taking trastuzumab, and this drug should not be given to patients with heart conditions.

Lapatinib may cause severe or life-threatening liver damage but less damage to the heart than trastuzumab. It can cause liver damage within several days or months after start of treatment.

Lapatinib is an effective treatment for patients who do not respond to trastuzumab.

Lapatinib and trastuzumab (Herceptin) both target HER2; however, Herceptin works on the surface of cancerous cells, whereas lapatinib is able to penetrate inside the cell to disable HER2.

Side Effects

Lapatinib may cause liver damage, which can be severe or life-threatening. Lapatinib can also cause liver damage as soon as several days or as late as several months after the start of treatment. Lapatinib is an effective treatment mainly for patients who do not respond to trastuzumab treatment.

OTHER TREATMENT OPTIONS

Ablative Hormone Therapy

This technique is used for men with advanced breast cancer. Surgery to remove the testes (orchiectomy) causes tumors to shrink, and often the metastases will clear up. Another technique removes the adrenal glands (adrenalectomy) and the pituitary gland (hypophysectomy). Orchiectomy or surgical castration is usually the initial ablative procedure because it has such a high response rate with minimal surgical complications. Men who fail to respond to the orchiectomy will often respond to either adrenalectomy or hypophysectomy.

Antiangiogenesis

Angiogenesis is the creation of tiny new blood vessels. The process takes place normally as new blood vessels develop in response to cuts or

---

Minor adverse effects of trastuzumab:

Fever, chills, weakness, nausea, vomiting, cough, diarrhea, and headache

A major adverse effect of trastuzumab:

Possible damage to the heart muscle

Severe side effect of lapatinib:

May cause liver damage

Other side effects:

Nausea, vomiting, and heartburn

Sores on the lips, mouth, or throat

Loss of appetite

Red, painful, numb, or tingling hands and feet

Dry skin

Pain in the arms, legs, or back

Difficulty falling asleep or staying asleep

Antiangiogenesis treatment is the use of drugs to stop tumors from developing new blood vessels.

other wounds. During cancer, the same process creates many new blood vessels to provide the tumor with its own blood supply.

Antiangiogenesis treatment is the use of drugs to stop tumors from developing new blood vessels. The idea is that without a blood supply the tumors cannot grow. Antiangiogenesis drugs do not kill normal cells; they act only at the place where new blood vessels are being formed. The drugs prevent the growth of the endothelial cells that form the inner lining of blood vessels. Without that first step, new blood vessels cannot form. The treatment, therefore, does not cause any side effects in the patient but can affect a developing fetus.

An advantage of antiangiogenesis drugs is that cancers do not develop a resistance to the drugs because the drugs do not attack cancer cells but rather affect the blood vessels serving the cancer. Cancer cells eventually become resistant to chemotherapy drugs by going through genetic changes that make the cancer less vulnerable to the drugs.

Antiangiogenesis worked well in reducing the size of tumors in laboratory animals, but in clinical trials, the expected results were not achieved. One theory is that, although the drugs blocked new blood formation, the old vessels were still able to supply the tumor. More clinical trials are in progress, with studies being done using antiangiogenesis drugs in combination with chemotherapy. Research is also being conducted on antiangiogenesis drugs that are monoclonal antibodies. The idea is to create antibodies that are normally made by the body's immune system and use them to fight diseases. These antibodies are designed to block the protein that speeds up endothelial growth, making the protein less effective. Other antiangiogenesis drugs stop the signal sent to the nucleus of the cells, in effect telling the cells not to grow.

Circulating Breast Cancer Cells

Circulating tumor cells are always present whenever there is a central tumor in the body because the primary tumor sheds cancer cells into the circulating system. A recently approved laboratory test will identify and count the circulating tumor cells from a patient's blood sample. The kit by Quest Diagnostic, Inc. is called the CellSearch Epithelial Cell Kit, and it was approved to count circulating tumor cells of epithelial origin in the whole blood. The number of circulating tumor cells can predict the effectiveness of treatment or the overall survival rate of patients with metastatic breast cancer.

GENE THERAPY

Genes are made up of the chemical DNA, and each gene is present in every cell of the body. Scientists believe that faulty genes are inherited and can become defective during one's lifetime, especially if the gene is exposed to dangerous chemicals or radiation. Gene therapy involves inserting specific genes into cells to restore a missing function or to give the cells a new function. The theory here is that missing or damaged genes cause certain diseases. Gene therapy is currently under clinical trial.

Examples of gene therapy could include replacing the tumor suppressor genes that could help prevent cancer from developing or

> Gene therapy involves inserting specific genes into cells to restore a missing function or to give the cells a new function.

stopping oncogenes or other genes important to cancer from working. Oncogenes are mutated forms of normal genes that cause cells to divide out of control, leading to cancer. Other genes are important in allowing cancer cells to metastasize. Stopping these genes or the proteins they make may prevent cancers from growing or spreading. Genes can be added to make cancer cells more vulnerable to chemotherapy or radiation; other genes can prevent the cancer cells from becoming resistant to chemotherapy drugs.

Scientists at the Johns Hopkins University School of Medicine in the United States have discovered that a protein that pushes calcium out of cells also works as a signal to get large quantities of calcium into cells. The focus of the study, supported by the National Institutes of Health and the National Health and Medical Research and Cancer Council Queensland, was the SPCA2 protein that is found in very high levels in breast cancer. Researchers suspect that the normal purpose if SPCA2 is to signal calcium channels to open so that a large volume of calcium comes into the cells of mammary tissue and gets pumped into human milk to make it extremely high in calcium. In some forms of breast cancer, the *SPCA2* gene is turned on when it should be off. This means that there is no regulation and that vast amounts of calcium gets pumped into the cells, stimulating the cell cycle and triggering proliferation.

Scientists are working on identifying women carrying certain biomarkers that are predictive of a higher risk for invasive cancer. When all three markers, *p16,* cyclooxygenase-2 (COX-2), and Ki-67, are present, a woman has a 20% risk for developing breast cancer over 8 years. If the markers are not present, the risk drops to 4%. COX-2 is the enzyme that converts arachidonic acid to prostaglandins. It is overexpressed in a variety of different tumors, including breast cancer. Cyclin-dependent kinase inhibitor 2A or the *p16* gene is a tumor suppressor protein that plays an important role in regulating cell cycle. Mutations of p16 increase the risk of developing a variety of cancers. Antigen Ki-67 is a nuclear protein that is associated with and may be necessary for cellular proliferation. Patients with atypia tissue that expressed COX-2 enzymes were more likely to develop breast cancer subsequently. Also, the more the enzyme expressed, the higher the risk. New studies are ongoing on perhaps using a COX-2 inhibiting agent, such as celecoxib or rofecoxib, to prevent breast cancer from developing.

> Oncogenes are mutated forms of normal genes that cause cells to divide out of control, leading to cancer.

Methods of Gene Therapy

1. Adding genes to tumor cells so that they are more easily detected and destroyed by the body's immune system
2. Adding genes to immune system cells (e.g., lymphocytes) to make them better able to detect cancer cells
3. Stopping genes that contribute to angiogenesis or adding angiogenesis inhibitor genes to cancer cells

HEMATOPOIETIC STEM CELL TRANSPLANTATION

Hematopoietic stem cell transplantation (HSCT) is used to describe bone marrow transplantation (BMT) and peripheral blood stem cell transplantation (PBSCT). These therapies use very high doses of

Immunotherapy aims to create an immune system response within each patient and to prime the body to kill cancer cells.

chemotherapy and/or total-body irradiation to kill cancer cells. In the process, normal cells, especially blood-forming cells or hematopoietic stem cells, are also killed. Stem cells removed from the patient before treatment or stem cells from a donor are used to replace these cells.

IMMUNOTHERAPY OR BIOTHERAPY

The idea behind a vaccine against breast cancer is to create an immune system response within each patient and to prime the body to kill cancer cells. Preliminary studies with such a vaccine show promise, but the conclusive results will only be obtained if the vaccine can shrink an existing cancer or lower the risk for breast cancer or cancer recurrence. Biotherapy or immunotherapy is intended to strengthen the immune system's ability to recognize and attack cancer cells.

There are two main types of immunotherapies: active immunotherapies, which stimulate the body's own immune system to fight the disease; and passive immunotherapies, such as antibody therapy. Passive immunotherapies use immune system components, such as antibodies, created outside the bodies. The intents of passive immunotherapies are to give a general boost to the immune system and improve the effectiveness of the vaccine.

Cancer vaccines are active specific immunotherapies. The hope is that by injecting cancer cells, parts of cells, or pure antigens, the substance will trigger the body's immune system to respond. The vaccines are considered specific because they do not bring about a generalized immune response. Cancer vaccines cause the immune system to produce antibodies to one or several specific antigens or to produce killer T cells to attack cancer cells that have those antigens. To date, cancer vaccines have only been used in clinical trials.

Tumor cell vaccines are another form of active immunotherapy. The tumor cells are killed, usually by radiation, so they cannot form more tumors, and then are injected into the patient. Antigens on the tumor cells' surface are still present and can stimulate a specific immune system response. As a result, the cancer cells carrying these antigens are recognized and attacked. Tumor cell vaccines have not yet been approved by the FDA and at this time are only used in clinical trials.

Monoclonal antibody therapy is a passive immunotherapy because the antibodies are produced in large quantities outside the body rather than by the immune system. This type of therapy can be effective even if the immune system is weakened. Clinical trial of monoclonal antibody therapy is already in progress, and the FDA has approved several of these drugs for the treatment of various types of cancers.

Nonspecific immunotherapy stimulates the immune system. The increase in activity can fight against cancer cells. Immunotherapies can be adjuvant—used in conjunction with other therapies or used alone.

Photodynamic therapy (PDT), or photo-chemotherapy, combines a light source and a photosensitizing agent to destroy cancer cells.

PHOTODYNAMIC THERAPY

PDT, or photochemotherapy, is a treatment that combines a light source and a photosensitizing agent (a drug activated by light) to destroy cancer cells. The technique has been approved by the FDA, but only for certain cancers (esophagus). Traditional PDT is based on the ability to kill cancer cells by releasing oxygen inside the malignant cells. The technique has been used successfully to treat deep body tumors. Oxygen released inside a tumor is effective in killing cells, but its destructive properties are limited because the oxygen depletes rapidly inside the tissues.

PDT works because the photosensitizing agent collects more readily in cancer cells than in normal cells. When the agent is exposed to light, it reacts with oxygen to create chemicals that can kill the cell. PDT is mainly used to treat cancerous areas just under the skin because the approved light sources can only penetrate to a limited depth through tissue. Recent advances have created supramolecular PDT drugs. These new drugs must remain inert as they pass through the healthy tissue. When they reach the cancerous tissue, specialized light waves trigger the activation of the drugs.

The ability to control when the lethal activity of the drug is activated will cut down on the toxicity of the drugs. Another advantage is that the drugs are nontoxic in normal tissue and only become destructive when activated by light; also, these new drugs do not need oxygen to operate.

Another advantage of PDT is that it can be repeated several times at the same site if necessary, and clinical trials have shown that it can be just as effective as surgery or radiation therapy in treating certain types of cancers and precancerous conditions.

PDT might be used in the treatment of other cancers and diseases in the future. Currently, newer photosensitizing agents—that will treat deeper tumors—are being developed, and researchers are looking at different types of laser and other light sources. Some agents may respond to small doses of radiation as well as to light, which would allow the use of smaller doses of radiation combined with PDT treatment.

Researchers are also looking at using PDT intraoperatively to prevent the recurrence of cancer, or using interstitial therapy whereby fiberoptics would be directed to the tumor sites using needles guided by computed tomography.

> Thermal ablation involves the use of thermal energy to kill cancer cells.

THERMAL ABLATION

Thermal ablation involves the use of thermal energy to kill cancer cells. The aim is to deliver the ultimate cosmetic results with no scars or deformity. Ablation therapy can use ultrasound or MRI to guide the thermal device to the cancer site. There are three types of ablation techniques: radiofrequency, laser, and cryotherapy. At present, there are ongoing clinical trials to test the effectiveness of ablation therapies.

Radiofrequency ablation involves the use of heat produced by an electric current to treat small breast cancers. Ultrasound imaging is used to guide a needle electrode directly to the site of the tumor. Temperatures between 140° and 200° F (60°–93°C) are applied for approximately 15 minutes, destroying the malignant tissue.

Laser ablation uses either ultrasound or MRI guidance to position an optic fiber to the site of the tumor. A concentrated beam of light is then used to kill the cancer cells.

CRYOTHERAPY

Cryotherapy uses liquid nitrogen delivered by a cryoprobe to freeze, kill, and destroy the tumor of interest. The radiologist first injects a local anesthetic into the breast tissue around the tumor. After an incision is made, the radiologist uses ultrasound to guide a cryoprobe to the site of the tumor. The cryoprobe is a thin probe shaped like a large needle. The cryoprobe produces a small ice-ball at temperatures of 20° to 40° C, which grows and encompasses the entire tumor, killing it. Surgeons generally try to provide at least a 1-cm treatment margin surrounding all aspects of the tumor. Multiple probes can be used to treat larger tumors. To keep the skin from freezing, a protective saline layer is injected between the skin and the tumor. Care should also be taken to keep the ice-ball from the chest wall. Ultrasound is often used to confirm cytotoxicity occurrence.

The process takes between 6 and 30 minutes, depending on the size of the tumor, and within 48 hours, the patient is back to normal activities. Minor side effects can include localized bruising and tenderness. Generally, because the patient leaves with the frozen tumor in place, the site of the tumor will feel larger for approximately a month after treatment until the swelling subsides. The technique is contraindicated for superficial tumors because freezing a tumor close to the skin surface can permanently damage the skin.

Ultrasound can be used as a follow-up study at 6 months to confirm shrinkage of the tumor. Cryoablative procedure was approved by the FDA in 2001 as a means of removing benign breast tumors, such as fibroadenomas, up to 4 cm in diameter. Long-term clinical trials will be needed to test its effectiveness and safety on malignant tumors.

> Cryotherapy uses liquid nitrogen to freeze, kill, and destroy the tumor.

PROPHYLACTIC SURGERY

Preventive surgery is used to remove the entire breast when a woman has a very high risk factor for breast cancer. For example, women with mutations in the breast cancer gene *BRCA1* or *BRCA2* will sometimes consider a prophylactic mastectomy. If surgery is indicated for medical purposes, it is not considered cosmetic intervention. The procedure can take place in an outpatient surgical facility or in a hospital. General anesthesia is the preferred option.

Any woman considering breast mammoplasty must first have an initial consultation with a plastic surgeon. The surgeon should be certified by the American Society of Plastic Surgeons, and this information should be verified as part of the initial evaluation. In general, implants and reconstruction are not recommended for women before age 22 years because women's breasts are not fully developed before that age. Breast augmentations or reductions are definitely contraindicated for anyone aged younger than 18 years because, in addition to having immature breasts, they may not be mature enough to make an informed decision. Consultation and mammography to rule out breast diseases are recommended before the surgery. The surgeon then determines which surgical technique is best for the woman.

SURGICAL RECONSTRUCTION USING IMPLANTS

Breast reconstruction is a surgical procedure to restore the appearance of the breast for women who have had a breast removed due to breast cancer. Breast reconstruction is generally done by a plastic surgeon after a mastectomy, but it is not usually necessary after a lumpectomy. The surgeon rebuilds the breast contour and can even include a nipple and areola. The goal of breast reconstruction is to provide symmetry to the breast and to permanently regain the breast contour to give the convenience of not needing an external prosthesis. Breast reconstruction could also involve reducting, enlarging, or reshaping the remaining breast to match the reconstructed breast.

Nipple and Areola Reconstruction

Some patients request a nipple and areola to make the new breast look more real. Tissue for the nipple and areola is taken either from the new breast, from the opposite nipple, or from the ear. Tissue for reforming the areola can be taken from the upper inner thigh. Tattooing is used to darken the areola to match the color with the opposite breast. Saving the nipple from the breast with cancer is often contraindicated because cancer cells can be present in the nipple.

Implant Procedures

Decisions about reconstruction can depend on the patient's overall health, stage of breast cancer, size of the natural breast, amount of tissue available (thin women may not have the excess body tissue to make a flap procedure possible), health insurance coverage, type of procedure, and size of the implant or reconstructed breast.

Many women elect to have breast reconstruction during the mastectomy to avoid further surgery. However, occasionally there are complications with reconstructive healing that interfere with chemotherapy or radiation treatment, or both. Some women choose to forgo surgical reconstruction and instead choose an external prosthesis.

Two common implants are the saline-filled and silicone implant. The saline implant sometimes has an external silicone shell and is filled with sterile saline. The silicone implant can be sized to the woman's preference. With implants, the differences between the two breasts will be minimal when wearing a bra, although the two breasts will never be identical.

> **Two common implants options:**
> - Saline
> - Silicone

Preparation for Surgery

Most surgeons recommend no smoking for 2 months before the surgery to aid healing after surgery. The use of tobacco can cause constriction of the blood vessels and reduce the supply of nutrients and oxygen to tissues. Smoking will therefore delay healing. Patients should also avoid anticoagulants, aspirin, and other nonsteroidal anti-inflammatory medications 5 days before and 5 days after the procedure. Patients with bleeding disorders need a careful evaluation. Certain herbal medications are also known to have anticoagulant properties, or to increase or decrease the anticoagulant effect of prescription medication. These herbals contain coumarin, salicylate, or peroxidase constituents or inhibit platelet aggregation.

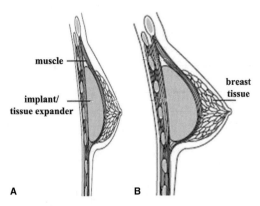

Figure 10–5. Implant or tissue expanders can be adjusted to required breast size. **(A)** small tissue expander; **(B)** larger implant (or tissue expander).

Autologous tissue reconstruction (ATR) or flap surgery uses skin, fat, and muscle from the abdomen, back, or buttocks to create a new breast.

Immediate Reconstruction

Breast reconstruction can be immediate or delayed. Reconstruction will not affect the recurrence of breast cancer.

Advantages of immediate reconstruction are that the chest tissues are undamaged by radiation therapy and there is one less surgery needed. After radiation, the first step in implant surgery is stretching the skin with a tissue expander, but this process is difficult when the skin is damaged by radiation. This leads to a higher rate of complications, such as poor healing, skin breaks, and implants that protrude or stiffen.

In the one-stage immediate breast reconstruction, after the mastectomy portion of the surgery is completed, the surgeon places the implant in the site where the breast tissue was removed.

Delayed Reconstruction

Although surgeons will have their preferences, so do women. Some patients may want their breast restored as quickly as possible; others prefer waiting until their cancer treatment is completed. Delayed reconstruction, done some time after surgery, may also be necessary if radiation immediately follows a mastectomy. Another reason to delay reconstruction would be if the skin is tight and flat. In such cases, an implant or tissue expander is placed under the skin and chest muscle. The tissue expander involves placing a balloon under the skin and then injecting a salt-water solution at regular intervals to fill the expander over time. When the skin has expanded to the required size, the expander is removed, and a permanent saline implant is put in place. Some expanders are left in place as the final implant.

If the implant is inserted before radiation, the scar tissue around the implant can contract because of the radiation, leading to distortion of the implant. Some surgeons will mold a slightly larger breast knowing that the radiation will tend to shrink the breast tissue. The other option for patients is the use of muscle, skin, and tissue from their own body to create a breast mound; this technique is called the autologous flap. This reconstruction can be performed at the time of the mastectomy because it will not be reshaped by radiation.

After a breast reconstruction, many physicians may not recommend a mammogram of the reconstructed breast. Studies show that 8% to 10% of women can have a recurrence in the scar after a mastectomy within the first 5 years. Also, before the mastectomy, the cancer could have already spread to the other areas of the body. Most research suggests either that patients should have a visual inspection of the site by an oncologist or that imaging should be done yearly for the first 5 years after the mastectomy.

Other Reconstruction Options

Other reconstruction options include autologous tissue reconstruction (ATR) or flap surgery, in which skin, fat, and muscle from the abdomen, back, or buttocks are taken to form a new breast. Flap surgery can be performed at the time of a mastectomy. It can complicate the surgical procedure and extend the length of surgery from 2 hours to 6 hours. Unfortunately, the reconstructed breast will not have nerve sensation

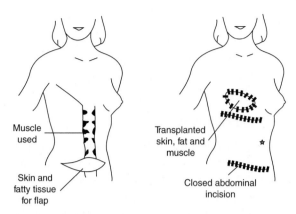

Figure 10–6. Transverse rectus abdominis muscle (TRAM) flap removes muscle from the abdomen to reconstruct the breast.

because the nerves cannot be transferred. There are, however, advantages for women because ATR tolerates radiation much better than implant reconstruction and is less likely to require follow-up surgical intervention. The process of ATR carries a low risk for weakening the muscles at the site from which the tissue is removed; however, there are newer techniques, such as muscle-sparing ATR. Recovery from ATR surgery can take up to 2 months. There are two methods of ATR surgery: the pedicle flap and the free flap.

The pedicle flap involves leaving the flap attached to its original blood supply and tunneling it under the skin to the breast area.

The free flap technique removes the flap—including the skin, fat, blood vessels, and muscle—from its original location and then attaches the flap to the blood vessels in the breast area. This procedure involves the use of microsurgery to reconnect the tiny blood vessels and will require longer surgery times. Some physicians recommend the technique because it gives a more natural result. However, if the microsurgery fails, the flap will have to be removed.

Two common types of ATRs are transverse rectus abdominis muscle flap (TRAM flap), which uses tissue from the abdominal area, and latissimus dorsi flap, which uses tissue from the upper back (Figs. 10–6 and 10–7). Newer flap techniques are the deep inferior epigastric artery perforator (DIEP) flap and the gluteal free flap. The flap techniques

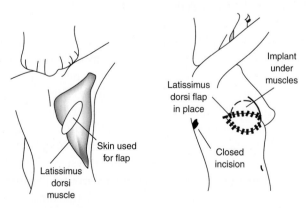

Figure 10–7. Latissimus dorsi flap removes muscle from the back to reconstruct the breast.

cannot be offered to diabetic patients, patients with connective tissue or vascular disease, or smokers. Also, thin women are not ideal candidates for tissue flaps because they do not have the area skin and tissue to donate to their breast.

Autologous Tissue Reconstruction

The transverse rectus abdominis myocutaneous (TRAM) flap is a technique used to create a breast mound by removing skin, fat, and blood vessels from the abdomen area plus at least one of the abdominal muscles. It is named by the muscle in the area, the rectus abdominis muscle. The procedure will result in a tightening of the lower abdomen or a "tummy tuck." The removed flap along with the superior epigastric arterial is tunneled under the skin to the beast area.

In a muscle-sparing TRAM flap technique, very little of the abdominal muscle is taken. The result is less possibility of loss of abdominal muscle function. This procedure is sometimes performed as a free flap technique, and microsurgery is needed to reconnect the blood supply.

The latissimus dorsi flap is the removal of muscle, fat, blood vessels, and skin from the back, moving them to the chest areas to create a breast mound. Generally, the flap is tunneled under the skin to the breast. This procedure can occasionally result in weakness in the back, shoulder, or arm.

The DIEP flap is named by the main blood vessel that runs through the area being removed. The technique creates a breast mound using fat and skin from the abdomen. Very little if any muscle is removed. The procedure is a "free" flap because the tissue is completely detached from the abdomen and moved to the breast area. This surgical technique is longer than both the TRAM flap techniques. It can take 5 to 8 hours of surgery if both breasts are reconstructed. The DIEP flap also requires microscope surgery to connect the tiny blood vessels. However, with no muscle removal, the recovery time is less with lower risk for losing abdominal muscle strength.

The gluteal free flap is a technique whereby the gluteal muscles from the buttocks are used to create a breast shape. The procedure is similar to the free TRAM flap and also requires the use of microsurgery to attach the tiny blood vessels.

Risks and Complications of Reconstruction

Regardless of the type of implant, they do not last a lifetime and must be eventually replaced. Implants can rupture, or scar tissue can form around the implant (capsular contracture). Capsular contracture occurs when the scar or capsule around the implant begins to tighten and squeeze the implant, making the breast feel hard. Capsular contracture can require surgery to remove the scar tissue, or the implant may be removed or replaced. Studies suggest that both saline and silicone-filled implants may cause a rare form of cancer, called anaplastic large cell lymphoma. The cancer involves the immune system. The lymphoma grows in the capsule of scar tissue that forms around the implants and can cause lumps, pain, and asymmetry from the fluid buildup and swelling. In some cases, removing the implant and scar tissue is the only cure needed. However, sometimes the patient will need

---

Two methods of ATR surgery:
- Pedicle flap
- Free flap

chemotherapy and radiation. The flap technique does not restore normal sensation to the breast, although some feeling may return, and women usually find that there are changes in nipple and breast sensation. Also, in any of the flap techniques, but especially in the free flap surgeries, tissue necrosis of all or part of the flap can occur. If this occurs, the flap must be removed.

Other risks and complications include the following:

- Anesthesia complications are a risk with any surgery.
- Infections can occur and can sometimes lead to unpleasant cosmetic results.
- Bleeding problems will occur in patients with poor clotting time. These patients should be premedicated before surgery or should not have this surgery.
- Fluid can collect around the implant, causing swelling and pain.
- Excessive scar tissue sometimes forms at the site. It could take 1 to 2 years for scar tissue to fade, but scars will never completely disappear.

Pain Medication and Pain Management

Pain receptors are located throughout our bodies in nerve endings in the skin and mucous membranes. When pain receptors are triggered by mechanical, chemical, or thermal stimuli, the pain signal is transmitted through the nerves to the spinal cord and then to the brain. Cancer pain can result from a number of factors, including blocked blood vessels causing poor circulation, bone fractures, metastasis to the bone, cancer invading the neural structures, tumors exerting pressure on a nerve, infection, inflammation, or adverse effects from treatments, such as chemotherapy or other drugs, radiation therapy, and surgery.

> Cancer pain can be acute or chronic.
> - Acute pain—less than 6 months.
> - Chronic pain continues for 6 months or more.

Although pain is commonly associated with advanced cancer, more than 30% of all cancer patients experience pain, and as many as 50% of patients are undertreated for cancer pain. Some studies estimate that 90% of cancer patients experience pain. Although analgesic use should be carefully monitored, there are still reports of reluctance to provide analgesics to cancer patients because of concerns about inappropriate use or dependence on opioids.

Cancer pain can be acute or chronic. Acute pain may last only a short time and can be the result of surgery or an immediate injury. Chronic pain continues for 6 months or more and, depending on the severity of the pain, can have life-altering implications for patients, such as diminished activities or dependence on aid for basic functions.

Pain management depends on the cause of the pain. Nevertheless, all cancer pain, whether acute or chronic, needs to be addressed. Patients also can experience breakthrough pain in which the medication they are taking no longer controls the pain, possibly because of changes in absorption, metabolism, or elimination of the drugs. In end-stage cancer, chemotherapy, radiation, or surgery can be used to reduce tumor size if the tumors are exerting pressure on a nerve. Newer pain medications are even more potent than morphine. Patients also can be given intrathecal anesthetics, which are pain-killing drugs injected directly into the cerebrospinal fluid. Nerve blocks also can be used to kill or deaden the nerve associated with the pain. Acupuncture, a technique using thin needles inserted into the skin to stimulate specific nerves, has also been found to be effective for some patients, although studies suggest that the benefits from the technique are subject to patients' expectations and beliefs.

CONCLUSION

Cancer has been with us for centuries, and it is becoming apparent that treatment will be individually based. The development of targeted therapies offers good news to patients. However, because the treatment is tailored to the individual's genetics, good oncology care will be needed to avoid generalized treatment that may not work for a specific cancer.

Treatment also cannot focus solely on eliminating the tumor. It must also assist with the patient's psychological well-being. The patient's psychological status can affect treatment options and even recovery time. Patient suffering can lead to depression, stress, and other mental and emotional conditions that should not be ignored. Good oncology care providers are often proactive in identifying patients' needs and helping patients find the resources and support they need.

At present, all breast cancer detection and treatment tools have advantages and disadvantages, and because of the wide variety, patients must explore their options wisely.

Detecting cancers early still offers the best treatment options and the highest survival rate. This fact is driving the need to perfect the breast cancer detection tools. However, despite the numerous adjunctive detection tools available, mammography is still the most comprehensive tool in the fight against breast cancer.

Summary

- Breast cancer is not an emergency, so in many cases the patients have the time to carefully weight their options. Treatment options include surgery, radiation, or drugs—or any combination of these—and should not begin until stage, size, and location of cancer are known.
- Surgical options are:
  - *Mastectomy:* surgical removal of the entire breast
  - *Radical mastectomy:* removal of the entire breast plus lymph nodes, chest wall, and muscle
  - *Modified radical mastectomy:* removal of the entire breast plus lymph nodes
  - *Lumpectomy:* removal of the tumor plus surround margins. Lumpectomy is often combined with radiation or drug treatment.
  - *Axillary node dissection:* removal of the lymph nodes in the axilla during the mastectomy. An alternative option is the sentinel lymph node biopsy, whereby only the first node on the chain and surrounding nodes are removed.
- Radiation treatment options are:
  - *External-beam radiation therapy:* generally begins 1 month after surgery. Treatment can be for 15 to 30 minutes 5 times per week

for 5 to 7 weeks. During treatment, patients are monitored by closed-circuit television. The side effects are often temporary and can include fatigue, swelling of breast, heaviness in the breast, or sunburn-type appearance of the breast skin.

- *Intensity-modulated radiation therapy (IMRT):* an advance method of external-beam treatment that uses a computer-controlled x-ray accelerator to deliver precise radiation doses to tumors. With IMRT, the radiation can be designed to conform to a three-dimensional tumor. Treatment will minimize radiation exposure to surrounding normal tissue while allowing higher dose to affected tissue.

- *Internal-beam radiation:* treatment involves less radiation to other healthy parts of the breast and body, such as the skin, lungs, heart, and ribs. There are fewer skin reactions with these treatments, but research is still ongoing. The available options are the multiple catheter technique, whereby 10 to 20 plastic catheters are surgically placed into the breast tissues, and the single catheter technique, which involves the surgical implantation of a single inflatable catheter. The catheter is inflated at each treatment.

- *Chemotherapy* is the use of drugs to treat cancers that may have spread beyond the breast. It is a systemic treatment because the drugs will affect all tissues and organs in the body. Treatment will depend on patient age, health, stage of cancer, past or future treatment, and other health problems. Chemotherapy drugs can be delivered through a catheter into a large vein, taken orally as tablets or liquids, given intramuscularly or topically, or injected directly into the cancer site. The regimen could be daily, weekly, or monthly, and depends on patient's response. Typically, the treatment can last 3 to 6 months. Side effects of chemotherapy are often temporary and can affect the cells in the digestive tract, hair follicles, and blood cells.

  - *Molecular treatments* can include the use of antiestrogen drugs. These drugs will prevent estrogen from latching onto tumor cell receptors. They are effective in shrinking or stopping the recurrence of breast cancer and also lowering the risks for breast cancer recurrence in postmenopausal women.

  - *Tamoxifen:* also sold as Nolvadex, Istubal, and Valodex, is a common antiestrogen drug. The major side effects of tamoxifen are increased risks for uterine cancer, endometrial cancer, pulmonary embolism, stroke, deep vein thrombosis and blood clots, and increased menopausal symptoms. Tamoxifen use is limited to 5 years or less for each patient.

  - *Raloxifene:* another antiestrogen drug. It was originally used to prevent osteoporosis. Raloxifene is not as effective as tamoxifen but has fewer side effects. It is sold as Evista.

  - *Hormone treatments* slow or stop the growth of hormone-sensitive tumors by blocking the body's ability to produce hormones or by interfering with hormone action. Examples of aromatase inhibitors approved by the FDA are anastrozole

(Arimidex) and letrozole (Femara), both of which temporarily inactivate aromatase, and exemestane (Aromasin), which permanently inactivates the enzyme.

- *Herceptin:* used for treatment of HER2-positive metastatic breast cancer.

- Surgical treatment options include the following:
  - *Reconstructive surgery:* after a mastectomy, the patient often undergoes a breast reconstruction that includes recreating the breast, nipple, and areola.
  - *Saline implants:* use a fixed or adjustable volume of a saline solution in a capsule. The capsule sac is inserted into the breast. Patients undergoing this type of implant surgery often have a temporary tissue expander to expand the breast tissue to the desired size.
  - *Silicone implants:* use a silicone shell filled with a silicone gel. The entire shell is implanted. The patient decides final breast size before the implant surgery.
  - *Autologous tissue technique:* can be "free flap," in which the tissue is completely removed and the breast mound created with microsurgery, or a "pedicle flap," in which the breast mound is tunneled to create a new breast and no microsurgery is needed to reattach blood vessels. The most common technique is called the TRAM flap, in which skin, fat, and muscle from the abdomen are removed to create a breast mound. In another procedure called the latissimus flap, skin, fat, and muscle from the upper back are removed. Other autologous tissue variations are named according to the artery preserved to supply the breast mound. These include DIEP (deep inferior epigastric perforator) flap—fat and skin from abdomen (no muscle), and SIEP (superficial inferior epigastric perforator) flap.

- *Pain management* is an important part of cancer treatment. Pain receptors are located throughout our bodies in nerve endings, in the skin and mucous membranes. Pain can be triggered by mechanical, chemical, or thermal stimuli. The pain signal is transmitted through the nerves to the spinal cord and then to the brain. Causes of pain include blocked blood vessels causing poor circulation; bone fractures; metastasis to the bone; cancer invading the neural structures; tumors exerting pressure on a nerve; infection; inflammation; and adverse effects from treatments, such as chemotherapy or other drugs, radiation therapy, and surgery.

REVIEW QUESTIONS

1. Give the three main breast cancer treatment options.

2. In a discussion of breast cancer, what is *stage*?

3. What is the difference between the modified radical mastectomy and the lumpectomy?

4. Lumpectomy is often combined with which two other treatment options?

5. How long does the average radiation therapy treatment last?

6. What are the two main methods used to deliver radiation therapy?

7. Give two advantages and two disadvantages of brachytherapy.

8. Why is chemotherapy considered a systemic form of treatment?

9. Identify four uses of chemotherapy.

10. Why does chemotherapy cause a low blood cell count?

11. What is the main use of tamoxifen?

12. Name the breast cancer treatment that combines a light source and a photosensitizing agent.

13. What is the main effect of antiangiogenesis drugs in cancer treatment?

14. Identify the technique that does not involve microsurgery, whereby skin, fat, and blood vessels, plus at least one abdominal muscle, are moved from the abdomen to the chest area as a breast mound.

CHAPTER QUIZ

1. A radiation treatment that typically lasts 5 to 9 days is termed:

   (A) External-beam radiation

   (B) Teletherapy

   (C) Brachytherapy

   (D) Proton-beam radiation

2. Of the following, which is the most breast-conserving surgery?

   (A) Chemotherapy

   (B) Mastectomy

   (C) Modified radical mastectomy

   (D) Lumpectomy

3. Side effects of chemotherapy include:

   (1) Nausea and or vomiting

   (2) Reduced white cell count

   (3) Increased red cell count

   (A) 1 and 2 only

   (B) 2 and 3 only

   (C) 1 and 3 only

   (D) 1, 2, and 3

4. If the margins of a specimen are positive, then:

   (A) All the cancer has been exercised

   (B) Additional tissue must be exercised

   (C) Lumpectomy was the wrong treatment option

   (D) The patient will have to undergo a mastectomy

5. Neoadjuvant chemotherapy refers to treatment:

   (A) Before surgery that uses antiestrogen drugs

   (B) After the mastectomy and is used to slow the growth of cancer

   (C) Before surgery to help shrink the size of cancerous tumors

   (D) After surgery that involves radiation

6. The benefits of tamoxifen are negligible after how many years?

   (A) 2

   (B) 3

   (C) 4

   (D) 5

7. What is tamoxifen?

   (A) An adjuvant therapy using drugs to shrink certain cancers

   (B) A drug used to block all hormones in the body

   (C) External radiation method of killing cancer cells

   (D) The removal of all cancer cells from the body

8. Advantages of brachytherapy include all of the following, *except:*

   (A) Reduced time of treatment

   (B) Less irritation of health breast tissue

   (C) Longer delays before start of treatment

   (D) Fewer skin reactions

9. Reconstructing the breast to remove tissue in a technique that involves relocating the nipple is called:

   (A) Augmentation mammoplasty

   (B) Reduction mammoplasty

   (C) Core biopsy

   (D) Ductography

10. The removal of tissue and muscle from the abdomen to reconstruct a breast mound using microsurgical techniques is called:

    (A) Latissimus dorsi flap

    (B) Pedicle flap

    (C) TRAM flap

    (D) Free flap

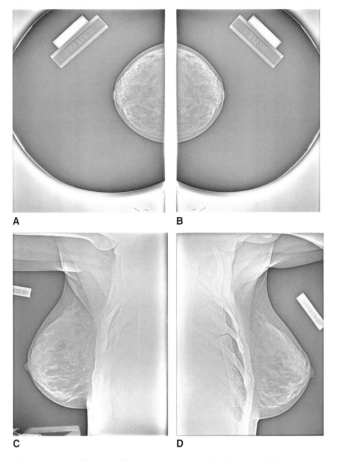

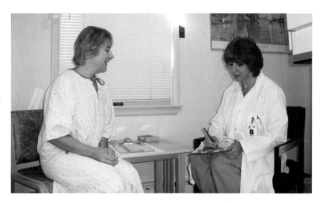

Figure 2–1. The technologist should sit and take a patient history before the mammogram.

Figure 1–9. Xeroradiograph series showing a wider latitude image (when compared with a mammogram) and a clear image of the patient's ribs. **(A)** L CC; **(B)** R CC; **(C)** Left lateral; and **(D)** Right lateral.

Figure 2–2. The patients' waiting room should provide videos, newspapers, or magazines.

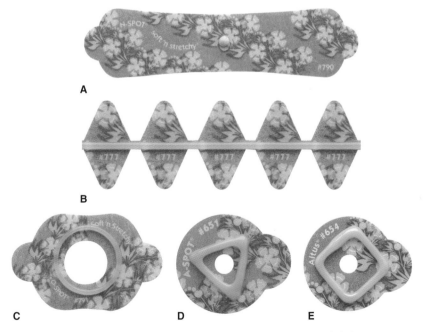

Figure 2–4. **(A)** Nipple marker (N-SPOT Designed for Digital Soft 'n Stretchy, 790). **(B)** Scar or surgical marker (S-SPOT Designed for Digital Soft 'n Stretchy, 777). **(C)** Mole or skin lesion marker (O-SPOT Designed for Digital Soft 'n Stretchy, 791). **(D)** Palpable lump marker (A-SPOT Light Image, 651). **(E)** Pain/nonpalpable area of concern marker (Altus Light Image, 654). (Used with permission from Beekley Medical.)

Half-wave, rectified DC voltage

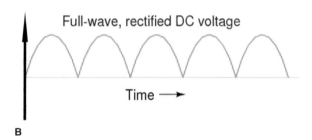

Full-wave, rectified DC voltage

Time ⟶

B

Figure 5–1. **(B)** DC waveform

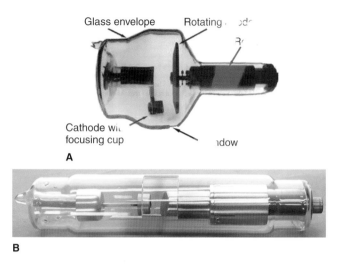

Glass envelope   Rotating anode

Cathode with
focusing cup   window

A

B

Figure 5–3. Actual x-ray tube: **(A)** with rotating anode; **(B)** with stationary anode.

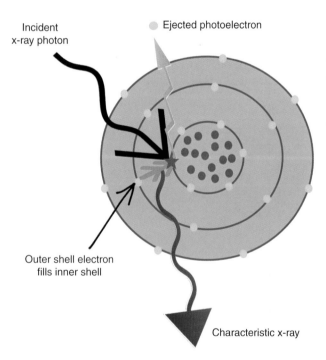

Incident
x-ray photon

Ejected photoelectron

Outer shell electron
fills inner shell

Characteristic x-ray

Figure 5–4. Characteristic radiation—an incident electron ionizes the atom by totally removing an inner shell electron. This creates a vacancy, which is filled by an outer shell electron falling into the void. This process is accompanied by a release of energy. The energy released is equal to the difference in the binding energies of the orbital electrons involved in the interaction.

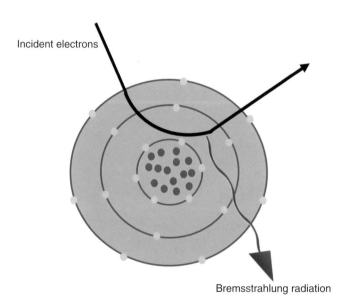

Incident electrons

Bremsstrahlung radiation

Figure 5–5. Bremsstrahlung radiation—an incident electron loses some of its kinetic energy and changes direction in an interaction with the nuclear field of an atom. This loss of energy is emitted as bremsstrahlung radiation. The electron can lose any amount of energy; therefore, the radiation emitted can have any energy up to the energy of the incident electron.

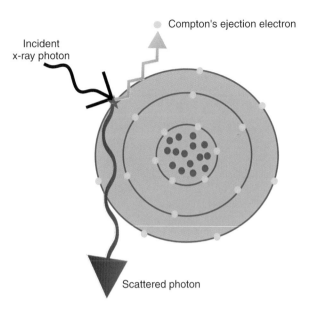

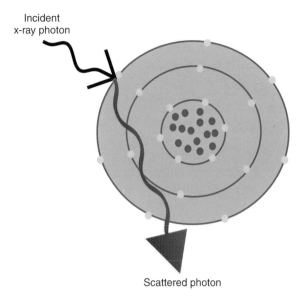

Figure 5–6. Compton's effect—a moderate-energy x-ray interacts with an outer shell electron. The x-ray is scattered and loses energy. It also causes an outer shell electron to be ejected—called *Compton's electron*.

Figure 5–7. Coherent scattering—low-energy x-ray interaction with an atom. The atom releases excess energy as a scattered x-ray. Both the incident x-ray and the scattered x-ray have the same energies, but the scattered x-ray will go off in a different direction.

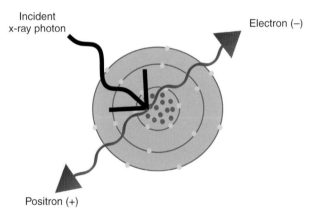

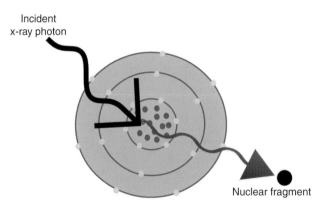

Figure 5–8. Pair production—high-energy x-ray interaction with the nucleus of the atom. The x-ray disappears, and two electrons appear: one positive and one negative.

Figure 5–9. Photodisintegration—very high-energy interaction with the nucleus. The nucleus becomes excited and emits a nucleon or other nuclear fragment.

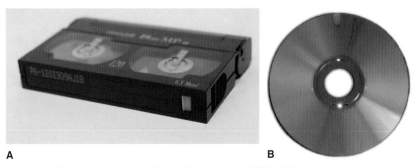

Figure 5–10. **(A)** Digital linear tape (DLT); **(B)** optical disk.

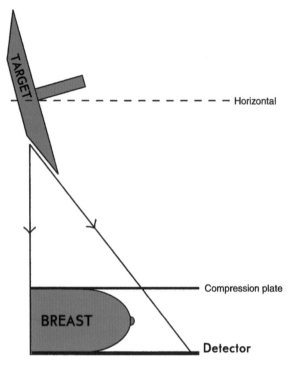

Figure 5–19. Mammography tube tilt. With the x-ray tube tilted approximately 6 degrees off the horizontal, the central rays run parallel to the chest wall so that no breast tissue is missed. Tilting the tube allows a smaller target angle and therefore a smaller effective focal spot size, while minimizing the heel effect. Tilting also allows greater anode heat capacity because the actual focal spot size is not further reduced.

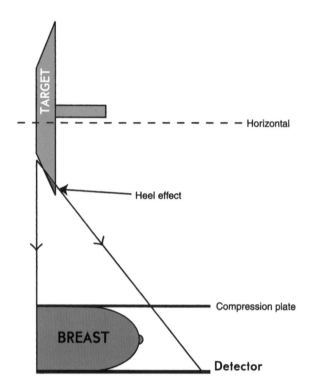

Figure 5–20. Anode heel effect. The *shaded area* indicates absorption of x-rays by the target. This results in the lower beam at the anode end of the tube. This effect is minimized in mammography by placing the anode toward the nipple (less thick) area of the breast.

Figure 5–21. Mammography compression plate.

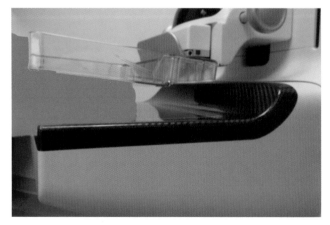

Figure 5–22. A compression plate showing how the flex paddle angles during compression.

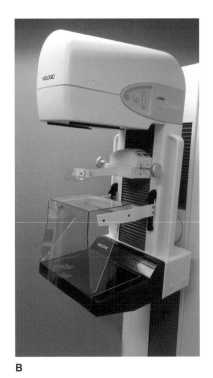

B

Figure 5–25. **(B)** mammography unit set up for spot magnification imaging.

Figure 5–26. Cathode ray tube (CRT) monitor.

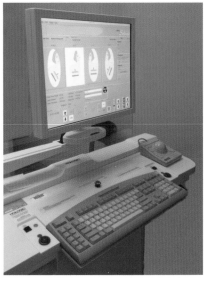

Figure 5–28. A digital unit showing the soft-copy display or acquisition workstation (AWS).

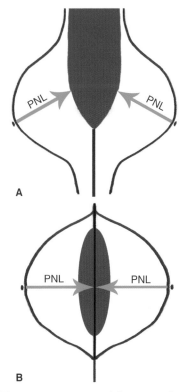

A

B

Figure 6–8. The measurement of the posterior nipple line (PNL) on the mediolateral oblique (MLO) **(A)** projection should be within 1 cm of the PNL measurement on the craniocaudal projection **(B)**.

Figure 5–29. Review workstation (RWS) monitor.

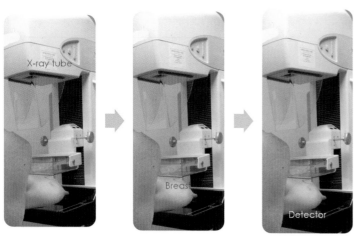

Figure 7–1. X-ray passes through the breast and strikes the detector.

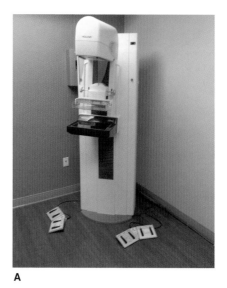

A

Figure 7–3. **(A)** Flat-panel Hologic unit.

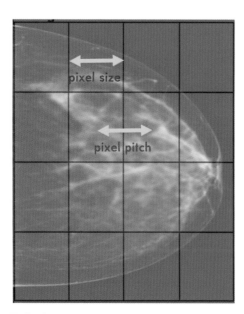

Figure 7–8. Schematic representation of the pixel pitch.

Figure 7–16. Computer reader.

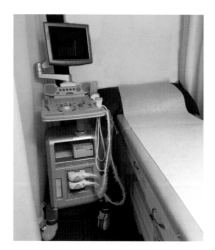

Figure 7–27. The ultrasound unit.

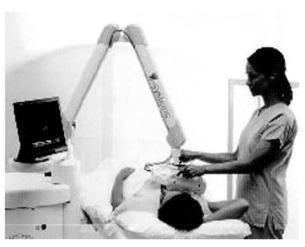

Figure 8–5. SomoVu Automated Breast Ultrasound System. Copyright © Siemens Healthineers 2017. Used with permission. https://www.siemens.com/press/en/pressrelease/?press=/en/pressrelease/2009/imaging_it/him200902016.htm

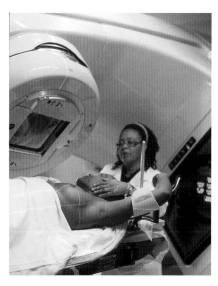

Figure 10–2. Radiation therapy. The radiation beam comes in from two directions and is directed to the breast (Reproduced from the National Cancer Institute. Photographer: Rhoda Baer.)

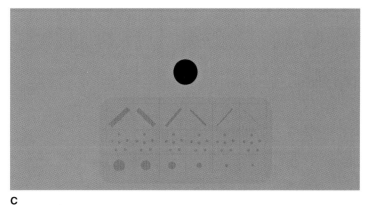

C

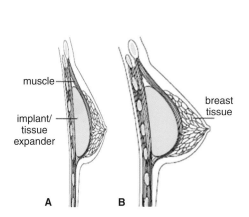

Figure 10–5. Implant or tissue expanders can be adjusted to required breast size. **(A)** small tissue expander; **(B)** larger implant (or tissue expander).

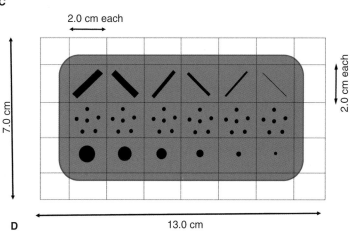

D

Figure 11–1. **(C)** Schematic diagram of the new ACR DM phantom. **(D)** Schematic diagram showing the relative position of the different objects embedded within the phantom.

Figure 11–2. Compression thickness indicator test showing a tape used as a phantom.

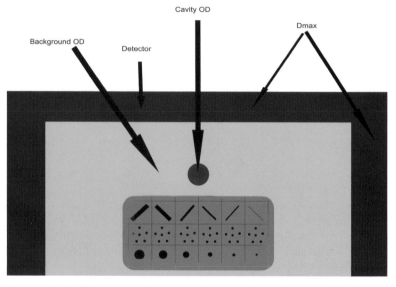

The phantom is placed on the detector and exposed. This diagram shows where readings are measured

Figure 11-6. Schematic diagram of the American College of Radiology digital mammography phantom showing where to measure background optical density (OD), cavity OD, and D$_{max}$.

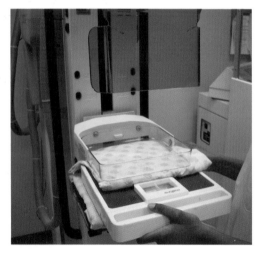

Figure 11-7. Demonstrating how to conduct the compression force test using a convention bathroom scale.

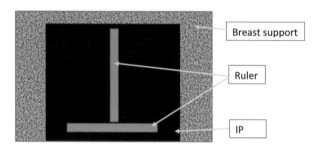

Figure 11-9. Computed mammography (CM) setup for the CM reader test for scanner performance.

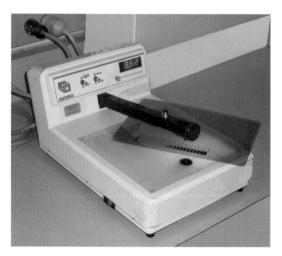

Figure 11–11. A densitometer.

BIBLIOGRAPHY

Acupuncture: What You Need To Know. National Center for Complementary and Alternative Medicine (NCCAM). http://nccam.nih.gov/health/acupuncture/introduction. Last updated July 2014. Accessed October 16, 2014.

American Cancer Society. How Is Breast Cancer Diagnosed? http://www.cancer.org/cancer/breastcancer/detailedguide/breast-cancer-diagnosis. Last medical review and revision: September 25, 2014. Accessed September 2016.

American Cancer Society. How Is Breast Cancer Staged? http://www.cancer.org/cancer/breastcancer/detailedguide/breast-cancer-staging. Last medical review and revision: September 25, 2014. Accessed October 2016.

American Cancer Society. Breast Reconstruction After Mastectomy. http://www.cancer.org/cancer/breastcancer/moreinformation/breastreconstructionaftermastectomy/index. Accessed October 2016.

American Cancer Society. Mammograms and Other Breast Imaging Procedures. http:http://www.cancer.org/Healthy/FindCancerEarly/ExamandTestDescriptions/MammogramsandOtherBreastImagingProcedures/mammograms-and-other-breast-imaging-procedures-toc. Accessed October 2016.

American Cancer Society. Tamoxifen. http://www.cancer.org/treatment/treatmentsandsideeffects/guidetocancerdrugs/tamoxifen. Accessed October 10, 2014.

Baron L, Baron P, Ackerman S, et al. *American Journal of Roentgenology* Sonographically Guided Clip Placement Facilitates Localization of Breast Cancer After Neoadjuvant Chemotherapy. http://www.ajronline.org/doi/full/10.2214/ajr.174.2.1740539. Accessed October 2016.

Beahm EK, Walton RL. Sensibility following innervated free TRAM flap breast reconstruction. *Plast Reconstr Surg.* 2006;117:2128–2126.

Breast Cancer. What Happens After Treatment for Breast Cancer? http://www.cancer.org/cancer/breastcancer/detailedguide/breast-cancer-after-follow-up. Last medical review and revision: September 25, 2014. Accessed April 9, 2016.

Burnstein HJ. Aromatase inhibitor-associated arthralgia syndrome. *Breast.* 2007;16(3):223–234.

Bushong SC. *Radiologic Science for Technologists—Physics, Biology and Protection.* 10th ed. St. Louis, MO: Mosby; 2012.

Dean L. Blood Groups and Red Cell Antigens. National Center for Biotechnology Information (US); 2005. http://www.ncbi.nlm.nih.gov/books/NBK2263/. Accessed April 2016.

Deandrea S, Montanari M, Apolone G. Prevalence of undertreatment in cancer pain: A review of published literature. *Ann Oncol.* 2008;19(12):1985–1991.

DeFrene B, Van Landuyt K, Hamdi M, et al. Free DIEAP and SGAP flap breast reconstruction after abdominal/luteal liposuction. *J Plast Reconstr Aesthet Surg.* 2006;59:1031–1036.

Feng M, Grice DM, Haddy HM, et al. Store-independent activation of Orai1 by SPCA2 in mammary tumors. *Cell.* 2010;143:84–98.

Grady D. Breast implants are linked to rare but treatable cancer, F.D.A. finds. *The New York Times.* January 27, 2011. A18.

Huntzinger C, Munro P, Johnson S et al. Dynamic targeting image-guided radiotherapy. DOI: http://dx.doi.org/10.1016/j.meddos.2005.12.014. Available at: http://www.meddos.org/article/S0958-3947(05)00200-1/fulltext?cc=y=Accessed May 2017.

Keller, A. The Deep Inferior Epigastric Perforator Free Flap for Breast Reconstruction. http://journals.lww.com/annalsplasticsurgery/Abstract/2001/05000/The_Deep_Inferior_Epigastric_Perforator_Free_Flap.3.aspx. Accessed October 2016.

Miotto A, Lins TA, Montero EF, et al. Immunohistochemical analysis of the COX-2 marker in acute pulmonary injury in rats. *Ital J Anat Embryol.* 2009;114(4):193–199.

Mukherjee S. *The Emperor of All Maladies.* New York, NY: Simon & Schuster; 2010.

Najaj AK, Chevray P.M, Chang DW. Comparison of donor-site complications and functional outcomes in free muscle-sparing TRAM slap and free DiEP flap breast reconstruction. *Plast Recontr Surg.* 2006;117:737–746.

National Institutes of Health. Initial Results of the Study of Tamoxifen and Raloxifene (STAR) Released: Osteoporosis Drug Raloxifene Shown to be as Effective as Tamoxifen in Preventing Invasive Breast Cancer. Available at: https://www.nih.gov/news-events/news-releases/initial-results-study-tamoxifen-raloxifene-star-released-osteoporosis-drug-raloxifene-shown-be-effective-tamoxifen-preventing-invasive-breast-cancer. Accessed May 2017.

Nelson R. Delay in Radiation Therapy Affects Outcomes in Breast Cancer. http://www.medscape.com/viewarticle/584851. *Cancer.* 2008;113:3108–3115. Accessed April 9, 2016.

Peart O. Breast intervention and breast cancer treatment options. *Radiol Technol.* 2015;86(5):535–563.

Peart O. *Lange Q & A Mammography Examination.* 2nd ed. New York, NY: McGraw-Hill; 2008.

Pennant M. Takwoingi Y, Pennant L, et al. A systematic review of positron emission tomography (PET) and positron emission tomography/computed tomography (PET/CT) for the diagnosis of breast cancer recurrence. *Health Technol Assess.* 2010;14(50):1–103.

Tartar M, Comstock C, Kipper M. Locally advanced breast cancer (CABC) and neoadjuvant chemotherapy. In: *Breast Cancer Imaging: A Multidisciplinary, Multimodality Approach.* 1st ed. Philadelphia, PA: Mosby Elsevier; 2008.224–381.

Tevaarwerk AJ, KolesarJM. Lapatinib: A Small-Molecule Inhibitor of Epidermal Growth Factor Receptor and Human Epidermal Growth Factor Receptor-2 Tyrosine Kinases Used in the Treatment of Breast Cancer. PubMed: 20110044. Available at: https://www.ncbi.nlm.nih.gov/labs/pubmed/20110044-lapatinib-a-small-molecule-inhibitor-of-epidermal-growth-factor-receptor-and-human-epidermal-growth-factor-receptor-2-tyrosine-kinases-used-in-the-treatment-of-breast-cancer/. accessed may 2017.

Tirona M, Breast cancer screening update. Edwards Comprehensive Cancer Center, Huntington, West Virginia. *Am Fam Physician.* 2013;87(4):274–278.

Varian Medical Systems. IMRT (Intensity Modulated Radiation Therapy) Available at: https://www.varian.com/oncology/treatment-techniques/external-beam-radiation/imrt. Accessed May 2017.

Venes D, Biderman A, Adler E. *Taber's Cyclopedic Medical Dictionary.* 22nd ed. Philadelphia, PA: F. A. Davis; 2013.

Quality Assurance and Quality Control | 11

Keywords and Phrases
Quality Assurance
  Benefits of Quality Assurance
  Agencies Responsible for Safety and Quality of Imaging
    Practices Within the United States
Responsibilities of the Various Quality Control Personnel
  The Radiologist
  Lead Interpreting Physician (Radiologist)
  Interpreting Physician (Radiologist)
  The Technologist
  The Medical Physicist
Quality Control Testing
General QC Tests
  Physicist Tests
Summary
Review Questions
Chapter Quiz

Objectives

On completion of the chapter, the reader will be able to:

1. Differentiate between quality assurance (QA) and quality control (QC)
2. Name the agencies at the federal, state, and local levels responsible for the safety and quality of imaging practices within the United States
3. Describe what is meant by the term "manufacturer-specific" quality control testing
4. List and describe the daily, weekly, monthly, quarterly, and semi-annual quality control tests of the technologist using digital imaging
5. Identify the duties of the radiologist in quality assurance
6. Identify the responsibilities of the medical physicist in quality assurance
7. List and describe the regular tests of the medical physicist

KEYWORDS AND PHRASES

- **Air kerma** is determined by multiplying the measured exposure rated by a conversion factor.
- **Corrective maintenance** indicates the actions taken to eliminate potential or actual problems.
- **Entrance skin dose** measures the exposure dose at the skin's surface. It is most often referred to as the *patient dose* but is not used as a measure of dose in mammography because the biological effects of radiation are more likely to be related to the total energy absorbed by the glandular tissues of the breast.
- **Environmental Protection Agency (EPA)** was established in July 1970 by the White House and US Congress in response to the growing public demand for cleaner water, air, and land. The EPA is led by the Administrator, who is appointed by the president of the United States. Its mission is to protect human health and to safeguard the natural environment.
- **Exposure linearity** is the ability of a radiographic unit to produce a constant radiation output for multiple combinations of mA and exposure time.
- **Feedback** is the process of implementing corrective action.
- **Glandular dose** in mammography refers to the average glandular dose or radiation dose delivered to the center of the breast during an exposure. The glandular dose will depend on the half-value layer (HVL) (the effective energy of the x-ray beam), the breast thickness, and tissue type.
- **Ghosting** occurs when an earlier exposure may change the sensitivity of the detector, leading to underresponse or overresponse to a later exposure. Ghosting will therefore negatively affect later exposures on the detector.
- **Half-value layer (HVL)** is the amount of filtration that will reduce the exposure rate to one-half of its initial value.
- **Illuminance** is the amount of light that falls on a surface. The unit for illuminance is the lux; 1 lux is equal to 1 lumen/m^2.
- **Image plate (IP)** is the name for the cassette-like device used in computed mammography (CM). The IP holds the photostimulable storage phosphor (PSP) that receives the x-ray photons. The PSP is sensitive to radiation but not to visible light.
- **Lag** occurs when residual signals (shadows) of prior images are retained. Lag can negatively affect later exposures on the detector.
- **Luminance** is the brightness of the illuminator and a measure of the amount of light either scattered or emitted by a surface. The unit for luminance is the candela per square meter (cd/m^2) or foot-lamberts; 1 cd /m^2 = 1 nit.
- **Occupational Safety and Health Administration (OSHA)** in the United States was created in 1971. The agency works to save lives, prevent injuries, and protect the health of America's workers. OSHA's job includes working in partnership with state governments to investigate workplace fatalities and occupational injuries.
- **Preventive maintenance (PM)** refers to action taken on a regular basis to prevent deterioration of image quality or any breakdown in the imaging chain.

- **Rejected** images are all images used or placed in the reject bin of the department, regardless of the reason.
- **Repeated** images are images that were exposed, resulting in a radiation dose to the patient. All images that have been repeated should be included in the repeat analysis, not just those rejected by the radiologist.
- **Standards** are reference objects or devices with known values used to check the accuracy of other measurement devices. Standards must be calibrated against references known to be certain, accurate, or consistent.
- **Tolerance** is the amount of measurement variation allowed under normal operating circumstances.
- **Trends** are serial measurements that follow a pattern of increasing or decreasing value that may or may not lead to readings that are out of the limits of tolerance.

QUALITY ASSURANCE

Quality assurance (QA) or *continuous quality improvement* (CQI) includes the total overall management of actions taken to consistently provide high image quality in the radiology department, with the primary objective being to enhance patient care. QA is very important in mammography, both in meeting the standards of the US Mammography Quality Standards Act (MQSA) and in ensuring quality patient care. Every QA program must include a quality control (QC) component, which includes the collection and evaluation of data with a systematic and structured mechanism to control variables such as repeat radiographs, image quality, processing variations, patient selection parameters, technical effectiveness, efficiency, and in-service education. A comprehensive QA plan would include efficacy studies, continuing education, QC tests, preventive maintenance, and calibration of the equipment. A QA program will not work without open communication and the active involvement of all personnel within the mammography department. In addition, a QC program should have clear assigned individual roles and assigned responsibilities.

QC involves actual testing to ensure that the standards of the QA program are met. All QC programs have a few basic elements. There should be a preventative maintenance program (PM), which includes actions taken on a regular basis to prevent deterioration of image quality or any breakdown in the imaging chain. There should be *corrective maintenance* whenever there is an actual problem or to eliminate potential problems. There should be well-defined *standards*, used to check the accuracy of the measurement devices against a calibrated or known standard. The person performing the QC testing should be familiar with the *tolerance* or amount of measurement variation allowed under normal operating circumstances and should be aware of *trends*, which could be a pattern of increasing or decreasing reading that might lead to readings that are outside of the limits of tolerance. Finally, all QC should have a *feedback* mechanism or a process of implementing corrective action.

> Quality assurance (QA) or *continuous quality improvement* (CQI) includes the total overall management of actions taken to consistently provide high image quality in the radiology department, with the primary objective being to enhance patient care.

Benefits of CQI

- Reduction of unnecessary radiation to the patient by reducing repeats
- Improve overall efficiency of service
- Improved patient satisfaction
- Consistency of image production
- Cost-effectiveness

Benefits of Quality Assurance

One of the benefits of having a good QA/QC program is the reduction of unnecessary radiation to the patient by reducing repeats. A QA program will also improve overall efficiency of the delivery of service, which will result in improved patient satisfaction. Overall, the facility will benefit in the achievement of improved image quality as well as an increased consistency of image production, reliability, efficiency, and cost-effectiveness of equipment use.

QA generally has a positive effect on personnel morale and always results in overall efficiency and improved customer service.

Agencies Responsible for Safety and Quality of Imaging Practices Within the United States

State Agencies

On the state level, state licensing agencies define rules to protect the health and safety of the patients or clients served by health care facilities.

Federal Agencies

Occupational Safety and Health Administration (*OSHA)* is a federal agency that establishes standards for safety in the workplace.

The *EPA* and the *Food and Drug Administration* (FDA) monitor matters ranging from blood-borne pathogens to disposal of processing chemicals. FDA certification of mammography units is required for reimbursement by Medicare. Many states have the authority to inspect on behalf of the FDA.

The *National Center for Devices and Radiological Health* (CDRH) (this was formerly called the Bureau of Radiation Health [BRH]) is a branch of the FDA that is responsible for ensuring the safety and effectiveness of medical devices and eliminating unnecessary human exposure to human-made radiation from medical, occupational, and consumer products. The CDRH collects and analyzes information on injuries from household and commercial, medical, and radiation-emitting devices and also provides technical assistance and ensures good manufacturing practices.

The *Nuclear Regulatory Commission* (NRC) serves to protect the public's health and safety, and also the environment, from the effects of radiation, nuclear reactors, materials, and waste facilities. The NRC is also involved in defense and security, plus policy and rule making.

The *Centers for Disease Control and Prevention* (CDC) is recognized as the main federal agency ensuring the protection, health, and safety of people—both at home and at work. The CDC develops and applies disease prevention and control and promotes many educational activities designed to improve the health of the people of the United States.

Professional Agencies and Societies in the United States

The *American College of Radiology* (ACR) provides certification of the administrative professional and technical aspects of mammography services. The ACR started the voluntary accreditation of mammography facilities, and many of its regulations continue today within the MQSA standards.

The *American Registry of Radiologic Technologists* (ARRT) is dedicated to promoting the standards of education of radiographers entering the profession. The organization supports high standards of patient care by certifying the professional behavior, ethics, responsibilities, and continuing education of practicing radiographers.

The *American Society of Radiologic Technologists* (ASRT) is the largest national professional association for technologists in radiologic science. It promotes quality patient care by providing career opportunities, continuing education credits, scholarships, and general knowledge and resources to its members.

The *American Registry of Magnetic Resonance Imaging Technologists* (ARMRIT) is a certifying organization that promotes formal magnetic resonance imaging (MRI) education with MRI clinical training.

The *American Registry of Diagnostic Medical Sonographers* (ARDMS) is an independent, not-for-profit organization that administers examinations and awards credentials in medial, cardiac, and vascular ultrasound technology.

The *Joint Commission (TJC),* formerly the *Joint Commission on Accreditation of Healthcare Organizations* (JCAHO), regulates the quality of care provided to patients and the way the facility is supervised and operated. Hospitals voluntarily subscribe to membership in the The Joint Commission, which conducts on-site visits to check hospitals' compliance with established guidelines. The Joint Commission's approval is linked to many reimbursements by the federal and state governments and many insurance companies.

The *Joint Review Committee on Education in Radiologic Technology* (JRCERT) is the only agency recognized by the US Department of Education to accredit educational programs in radiography and radiation therapy. The JRCERT also accredits educational programs in magnetic resonance and medical dosimetry. Accreditation of an educational program ensures that the students will acquire the knowledge, skills, and values needed to competently perform the range of professional responsibilities expected by potential employers nationwide. It also ensures that they will be eligible for licensure in each of the 50 states. Patients are also assured that students are appropriately supervised during the education process.

> The 2016 *Digital Mammography Quality Control Manual* identifies both a lead mammography radiologist responsible for general oversight of QA requirements and a mammography radiologist responsible for interpreting the images, ensuring quality imaging, and ensuring that corrective actions are taken in case of tests failures.

RESPONSIBILITIES OF THE VARIOUS QUALITY CONTROL PERSONNEL

The Radiologist

The FDA's MQSA regulation recommends that a lead physician should be responsible for general oversight of the MQSA QA program. It is the responsibility of the lead physician to ensure that all individuals' qualifications, performance, and assignments are adequate. The lead physician also monitors training and continuing education of staff; identifies a primary QC technologist; ensures the equipment, material, and time are available to perform QC tests; ensures that a medical physicist is available; and reviews the results of any physicist test. The radiologist should also oversee the activities of the technologist and the medical physicist. The level of commitment by the radiologist is often reflected in the quality of a mammography program.

The 2016 *Digital Mammography Quality Control Manual* identifies both a lead mammography radiologist responsible for general oversight of QA requirements and a mammography radiologist who is responsible for interpreting the images, ensuring quality imaging, and ensuring that corrective actions are taken in case of tests failures.

Lead Interpreting Physician (Radiologist)

In addition to interpreting the image, the lead radiologist is ultimately responsible for quality imaging and an effective QA program. Other responsibilities are the following:

- Ensuring that the facility has a QC program in place
- Providing an orientation for technologists
- Identifying a primary QC technologist
- Ensuring that the technologist has adequate training and is following the continuing education requirements
- Ensuring that the QC equipment, staffing, and time allotted for QC are adequate
- Reviewing the QC results at least quarterly
- Selecting a medical physicist and reviewing the physicist's testing results annually or more frequently as needed
- Providing regular feedback, both positive and negative, on clinical image quality on the new *Radiologist Image Quality Feedback* form
- Ensuring that all policy manuals are properly maintained and updated regularly

Interpreting Physician (Radiologist)

In addition to interpreting the image, the interpreting radiologist should do the following:

- Take corrective action in cases of poor image quality
- Participate in medical outcome audit
- Document current qualifications per MQSA or local states

The Technologist

The technologist is responsible for patient positioning, compression, and image production. The technologist plays a crucial role in QC and is usually solely responsible for the day-to-day testing details. The technologist should be knowledgeable enough to seek help, whether from the medical physicist, radiologist, or service personnel, at the first indication of potential problems with data from the QC tests. A primary technologist should be identified as the QC technologist responsible for conducting all QC tasks. If the QC technologist is not available, any other qualified technologist can perform the QC duties; however, it is the responsibility of the QC technologist to review and evaluate the data to ensure that all tests are performed correctly.

The Medical Physicist

The medical physicist performs annual evaluations of the mammographic unit and reviews all test procedures, including records, charts, and mammography policies and procedures on a yearly basis. The physicist is required to conduct a Mammography Equipment Evaluation (MEE) of new equipment and after any major repair and ensures that the

---

Main patient care responsibilities of the technologist:

- Patient positioning
- Compression
- Image production
- Image processing
- Infection control

equipment does not exceed the radiation dose limits. The physicist should address any issues of patient or operator safety. The medical physicist is also responsible for providing a written report, and any corrective recommendation should be reviewed with the radiologist and technologist in charge of QC. The medical physicist is also a good potential source to consult before contacting the equipment service personnel.

QUALITY CONTROL TESTING

QA and QC tests are absolutely essential in producing quality images. Guidelines as determined by the MQSA and the ACR provide the standard criteria. These criteria exceed the criteria for processing of radiographic studies done in a diagnostic radiology department. Under the MQSA standards, specific duties are required of the technologist, the radiologist, and the medical physicist who monitors the mammography imaging equipment. These duties include the imaging, processing, and viewing of the mammograms. Imaging is discussed in Chapter 6.

Starting in 2017, the FDA approved the use of the new *ACR Digital Mammography Quality Control Manual* for two-dimensional digital mammography (DM) systems. Facilities with tomosynthesis or contrast enhancement equipment must follow all manufacturers' QC guideless for both the DM application and the advance technology application of their system. Any facility using the ACR DM QC guideline must also use the ACR DM phantom with all the manual's procedures (Table 11-1 and 11-2).

> Starting in 2017, the FDA approved the use of the new *ACR Digital Mammography Quality Control Manual* for two-dimensional DM systems. Facilities with tomosynthesis or contrast enhancement equipment must follow all manufacturers' QC guideless for both the DM application and the advance technology application of their system.

TABLE 11–1. Digital Mammography Quality Control Tests for Technologists

Test	Minimum Frequency	Corrective Action
1. American College of Radiology (ACR) digital mammography (DM) phantom image quality	Weekly	Before clinical use
2. Computed mammography (CM) photostimulable storage phosphor (PSP) erasure *(if applicable)*	Weekly	Before clinical use
3. Compression thickness indicator	Monthly	Within 30 days
4. Visual checklist critical items	Monthly	Before clinical use; less critical items: within 30 days
5. Acquisition workstation (AWS) monitor quality control (QC)	Monthly	Within 30 days; before clinical use for severe defects
6. Radiologist workstation (RWS) monitor QC within 30 days	Monthly	Within 30 days; before clinical use for severe defects
7. Film printer QC *(if applicable)*	Monthly	Before clinical use
8. Viewbox cleanliness *(if applicable)*	Monthly	Before clinical use
9. Facility QC review	Quarterly	Not applicable
10. Compression force	Semiannually	Before clinical use
11. Manufacturer detector calibration *(if applicable)*	Per manufacturer recommendation	Before clinical use
Optional—repeat analysis	As needed	Within 30 days after analysis
Optional—system QC for radiologist	As needed	Within 30 days; before clinical use for severe artifacts
Optional—radiologist image quality feedback	As needed	Not applicable

TABLE 11–2. Digital Mammography Quality Control Tests for Physicists

Test	Minimum Frequency	Corrective Action
1. Mammography Equipment Evaluation (MEE), Mammography Quality Standards Act (MQSA) requirements	MEE	Before clinical use
2. American College of Radiology (ACR) digital mammography (DM) phantom image quality	MEE and annually	Before clinical use
3. Spatial resolution	MEE and annually	
4. Automatic exposure control system performance	MEE and annually	Within 30 days
5. Average glandular dose	MEE and annually	Before clinical use
6. Unit checklist	MEE and annually	Critical items: before clinical use; less critical items: within 30 days
7. Computed mammography (*if applicable*)	MEE and annually	Before clinical use
8. Acquisition workstation (AWS) monitor quality control (QC)	MEE and annually	Within 30 days; before clinical use for severe defects
9. Radiologist workstation (RWS) monitor QC	MEE and annually	Within 30 days; before clinical use for severe defects
10. Film printer QC (*if applicable*)	MEE and annually	Before clinical use
11. Evaluation of site's technologist QC program	Annually	Within 30 days
12. Evaluation of display device technologist QC program	Annually	Within 30 days
MEE or troubleshooting: beam quality (half-value layer) assessment	MEE or troubleshooting	Before clinical use
MEE or troubleshooting: kVp accuracy and reproducibility	MEE or troubleshooting	MEE: before clinical use; troubleshooting: within 30 days
MEE or troubleshooting: collimation assessment	MEE or troubleshooting	MEE: before clinical use; troubleshooting: within 30 days
Troubleshooting: ghost image evaluation	Troubleshooting	Before clinical use
Troubleshooting: viewbox luminance	Troubleshooting	NA

GENERAL QC TESTS

Sample forms with testing details are found in Appendix D.

Phantom Test

Purpose: To ensure consistent image quality and artifact-free imaging

Frequency: Weekly, after any service, and before clinical use or on installation of new equipment

Equipment Needed:

- ACR DM phantom. (Facilities may not use the old ACR phantom with these new tests or the new ACR DM phantom with the old test.) The new ACR phantom now has six fibers, six speck groups, and six masses. The phantom represents a 4.2-cm-thick compressed breast consisting of 50% glandular tissue and 50% adipose tissue. The phantom is designed to cover the majority of the digital detector area and provides the same attenuation as the previous screen/film ACR

phantom. The 2016 *ACR Digital Mammography Quality Control Manual* is now recommended for use with all digital units with standard features. Facilities with three-dimensional units or units with contrast enhancement features must follow the manufacturer's QC guidelines.
- ACR DM Phantom Image Quality Form

> Phantom test will ensure consistent image quality and artifact-free imaging.

Process:

- Access the exam input section on the acquisition workstation (AWS).
- Enter correct data to identify the phantom: last name (e.g., ACR DM Phantom), first name (e.g., Room 1), patient ID (e.g., Current Date), date of birth (e.g., current date).
- Use the largest paddle size and the type of paddle (flex or fixed) used with the majority of clinical cases. For CM, use the largest IP and corresponding paddle.
- Place the phantom on the detector. The same location should be used each time.
- The pink insert should be on top near the chest wall.
- The phantom should be centered on the detector.
- The edge of the phantom should aligned with the chest-wall edge of the detector.
- Compress to approximately 5 daN (12 lb). The same compression force should be used each time. The compression thickness indicator may not read 4.2 cm.
- Select the image mode for routine imaging of a 4.2-cm-thick compressed breast consisting of 50% glandular and 50% adipose tissue.
- Image the phantom.

Record the following. These should be consistent each week:

- Facility
- MAP ID number
- Room ID
- X-ray unit and manufacturer and model number
- Exposure mode
- Paddle size (IP size if using CM)
- Paddle type (flexed or fixed)
- Compression force
- AEC cell position, target, filter, kVp, and density setting if applicable

Results and Analysis:

- View the phantom on the AWS.
- The phantom should be scored and evaluated using lightening similar to that in the radiologist's reading room.
- The window width and level can be optimized for visualization of the test objects.
- Use the same window width and level adjustment for all scoring.
- Record the presence or absence of artifacts. This is a pass/fail option.

Clinically significant artifacts are:

- Gridlines
- Ghosting
- Collimator cutoff (often see on the chest-wall edge of the detector)

The method of scoring the new phantom is the same as with the old version:

- Count the number of visible objects (fibers, speck groups, and masses) until a score of 0 or ½ is reached. The objects must be in the correct location with the correct orientation.

Fibers:

- Each fiber is 10 mm in length. A score of 1 is given if the fiber length is 8 mm or more. A score of ½ is given if the fiber length is equal or greater than 5 mm and less than 8 mm.
- Small gaps in the fiber that are less than the width of the fiber do not count.

Speck Groups:

- Count each speck group as 1 point if four to six specks are visible.
- Count each speck group as ½ point if two or three specks are visible.

Masses:

- Count each mass as 1 point if it is in the correct location and the mass is circular with a continuous border that is ¾ of the total circumference.
- Count each mass as ½ point if it is in the correct location but greater than ½ but less than ¾ of the circumference is visible.

Record the final score of each test object (Fig. 11–1):

- The fiber score must be ≥2.
- The speck group score must be ≥3.
- The mass score must be ≥2.

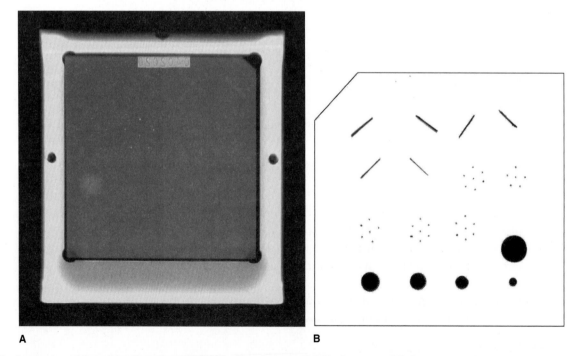

A   B

Figure 11–1. **(A)** Radiography of the older American College of Radiology (ACR) phantom. **(B)** Schematic diagram of the older ACR phantom showing the relative position of the different objects embedded within the phantom. Each phantom will have six fibers, five speck groups, and five masses. **(C)** Schematic diagram of the new ACR DM phantom. **(D)** Schematic diagram showing the relative position of the different objects embedded within the phantom.

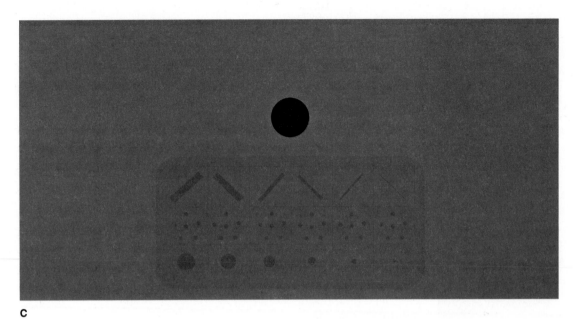

C

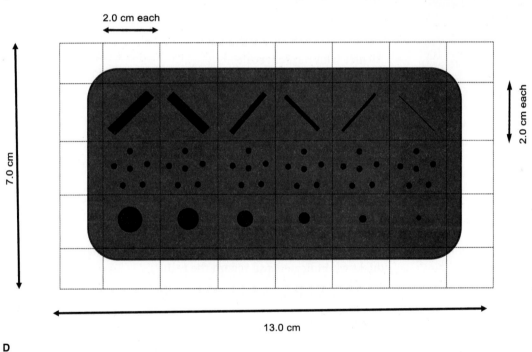

D

Figure 11–1. (*Continued*)

Artifacts must not be clinical significant. There is no subtraction of the score for artifacts; however, if the artifact is located in an area that could affect interpretation, then the test fails. The artifact must be resolved before imaging any patients.

If the test fails, it should be repeated, making sure that the exposure mode is correct. A repeat failure must be corrected before clinical use.

Computed Mammography (CM) PSP Erasure (If Applicable)

Purpose: To ensure the IP is free from information that would affect image quality

Frequency: Weekly

> The final score of each test object in the new ACE DM phantom:
> - The fiber score must be ≥2.
> - The speck group score must be ≥3.
> - The mass score must be ≥2.

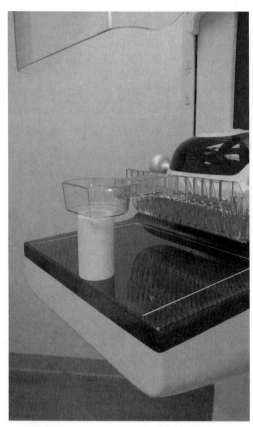

Figure 11–2. Compression thickness indicator test showing a tape used as a phantom.

The thickness indicator should be accurate to within ±0.5 cm (± 5 mm).

Equipment Needed: All CM IPs

Process: Select CM erasure on the CM reader. Insert one IP at a time. Repeat the process if the erasure test fails. With a second failure, record date, IP number, reason for failure, and action taken. Initial report and remove the IP from use.

Results and Analysis: IPs that cannot pass the test must be pulled from service and the problem investigated. The IP must be repaired before it can be used again. When putting an IP back into service, document date and the type of service performed.

Compression Thickness Indicator

Purpose: To ensure that the compression thickness indicates the correct degree of compression

Frequency: Monthly, on installation of new equipment and whenever the indicator is suspected to be wrong

Equipment Needed:

- An object to use as a phantom. It must have a known thickness, e.g., a 2-inch roll of medical tape
- A ruler with mm/cm scale
- Compression thickness indicator form

Process:

- Record a description of the type of phantom.
- Measure and record the thickness of the phantom.
- Center the phantom on the detector, aligned with the chest-wall edge (Fig. 11–2).
- Compress the phantom using the spot compression paddle and record the thickness. Do not use the flex feature. The compression should be 10 to 15 lb (4.4–6.7 daN).
- Subtract the actual measured thickness of the phantom from the thickness indicated and record the results.

Results and Analysis: If the test fails, it should be repeated. A service representative must be contacted is there is a repeat failure. The source of the problem must be identified and corrective action taken within 30 days.

The same phantom should be used whenever possible.

Visual Checklist

Purpose: To ensure that all indicator lights detents and mechanical locks are in proper working order

Frequency: Weekly, after any service and before clinical use or on installation of new equipment

Equipment Needed: Visual checklist form

Process:

- Review all items on the checklist.
- Rotate the C-arm of the unit and test all locks.
- Initial and date the checklist.

Results and Analysis: Check and report any defect. If unsure, consult with the medical physicist or radiologist.

The following items on the checklist are considered critical and must be repaired or replaced before clinical use if they are defective or missing:

- Cleaning solution
- All locks must work correctly.
- No cracked paddles/face shields
- No cracked or defective breast support
- The image plate holder for the CM units must hold and lock the IP properly (small and large).
- The CM IP must be free of cracks and dents; latches should function properly.
- CM IPs should load into the CM reader smoothly.

The following items on the checklist are considered noncritical and must be repaired or replaced within 30 days of a report of defect or damage.

- All equipment in the room, including the magnification stands and extra paddles, must be free from dust.
- The mammography room and countertops must be free from dust.
- Indicators and collimator lights must be working.
- Cables must be safely positioned without cracks.
- The C-arm motion must be smooth.
- The compression paddle motion must be smooth.

Acquisition Workstation (AWS) Monitor

Purpose: To ensure that monitors are free of dirt and dust or other marks that can interfere with clinical information. This test also checks the manufacturer's specification and calibrations, including the brightness and contrast settings.

Frequency: Monthly, after any service and before clinical use or on installation of new equipment

Equipment Needed:

- A soft, dry, lint-free cloth or cleaning material recommended by the manufacturer
- AWS monitor QC form
- AQCAPMTG18 test pattern or SMPTE test pattern (Fig. 11–3A and B)

Process: Visually inspect monitor and wipe away any foreign material using the cloth or cleaning material recommended by the manufacturer. After the monitor dries, it should be rechecked and recleaned if needed (Fig. 11–4).

Results and Analysis: If using any of the previous test patterns, reduce the lighting to that of the radiologist reading room and evaluate the following areas on the test pattern:

- 0% to 5% contrast boxes should be visible.
- 95% to 100% contrast boxes should be visible.
- Line-pair images at the center and four corners should be visible and distinguishable.

> **Acquisition workstation (AWS) monitor** is the technologist monitor. Tests will ensure that monitors are free of dirt and dust or other marks that can interfere with clinical information. This test also checks the manufacturer's specification and calibrations, including the brightness and contrast settings.

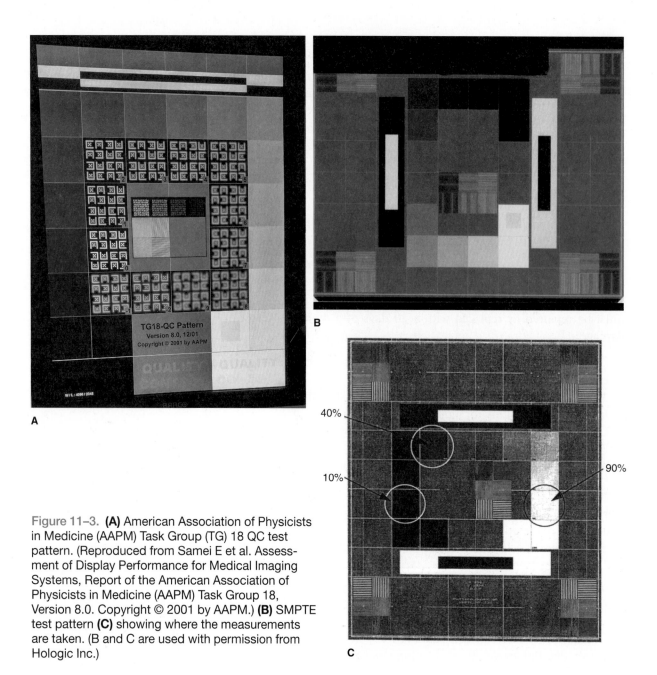

Figure 11–3. **(A)** American Association of Physicists in Medicine (AAPM) Task Group (TG) 18 QC test pattern. (Reproduced from Samei E et al. Assessment of Display Performance for Medical Imaging Systems, Report of the American Association of Physicists in Medicine (AAPM) Task Group 18, Version 8.0. Copyright © 2001 by AAPM.) **(B)** SMPTE test pattern **(C)** showing where the measurements are taken. (B and C are used with permission from Hologic Inc.)

If using the monitor automated test:

- Open the monitor's QC program.
- Review results and verify that all tests have passes.
- Record pass/fail.

All failures must be corrected within 30 days. Blemishes, anything interfering with visualization of the image, must be corrected before clinical use.

Radiologist Workstation (RWS) Monitor

Purpose: To ensure that monitors are free of dirt and dust or other marks that can interfere with clinical information. This test also checks the manufacturer's specification and calibrations, including the brightness and contrast settings.

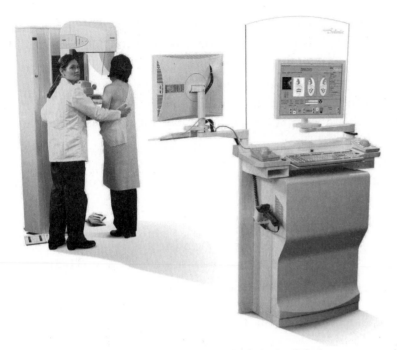

Figure 11–4. Acquisition workstation (AWS): Selenia digital mammography machine. (Images used with permission from Hologic Inc.)

Frequency: Monthly, after any service and before clinical use or on installation of new equipment. This monitor should also be cleaned if significant dirt/dust is noted during daily use.

Equipment Needed:

- A soft, dry, lint-free cloth or cleaning material recommended by the manufacturer
- RWS monitor QC form
- ACR DM phantom
- AQCAPMTG18 test pattern or SMPTE test pattern

Process:

- Visually inspect monitor and wipe away any foreign material using the cloth or cleaning material recommended by the manufacturer.
- After the monitor dries, it should be rechecked and recleaned if needed.
- Acquire and display a phantom image. The phantom test objects must be evaluated according to the ACR DM Phantom Image Quality guidelines.
- There should be no clinically significant artifacts on the phantom.
- The criteria for the number of objects on the phantom necessary to pass the new ACR DM phantom are:
  - The fiber score must be ≥2.
  - The speck group score must be ≥3.
  - The mass score must be ≥2.

If using the monitor automated test (This is the typical scenario.)

- Open the monitor's QC program.
- Review results and verify that all tests have passes.
- Record pass/fail.

Results and Analysis: If using any of the previous test patterns, reduce the lighting to that of the radiologist reading room and evaluate the following areas on the test pattern:

- 0% to 5% contrast boxes must be visible.
- 95% to 100% contrast boxes must be visible.
- Line-pair images at the center and four corners must be visible and distinguishable.

All failures must be corrected within 30 days. Blemishes and anything interfering with visualization of the image must be corrected before clinical use.

Film Printer (If Applicable)

Purpose: To ensure adequate and consistent image quality on printed images provided to physicians, patients, and others (Fig. 11–5)

Frequency: Monthly after any service and before images are printed. This test is not necessary if printed images are not used. The facility should document in the QC logs if printed images are not used clinically.

Equipment Needed:

- ACR DM phantom
- Densitometer
- Film printer QC form

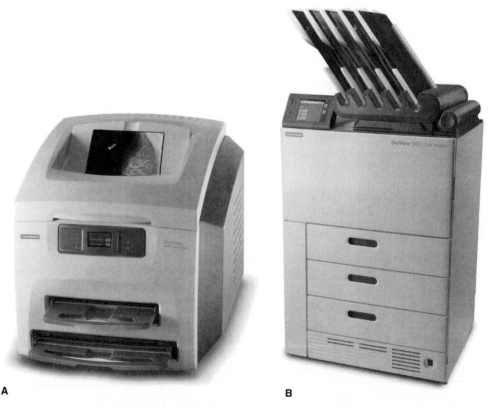

A   **B**

Figure 11–5. **(A)** Kodak DryView 6850 laser. **(B)** Kodak DryView 6850 laser. (Images used with permission from the Eastman Kodak Company.)

Process:

- Print the ACR DM phantom without adjusting any parameters, such as window level, window width, or sizing. The size used should be similar to that used for the majority of clinical printing. The printed image should be without magnification or minification and as close to "true size" as possible. The recommendation is within 25% of actual phantom size.
- Record the workstation used to print the image.
- Evaluate the image for artifacts.
- Score the phantom.
- Measure and record the background optical density (OD) outside of the cavity on the printed phantom image (Fig. 11–6).
- Measure and record the OD inside the cavity.
- Subtract the background OD from outside the cavity from the OD inside the cavity. This is recorded as the *contrast*.
- Measure the OD near the outside edge of the film at a location where the phantom image is not located. This is the D_{max}. If the phantom image covers the entire film and there is insufficient room, print a separate breast image and measure the nonbreast area.
- Score the phantom.

Contrast = cavity OD − background OD

Results and Analysis:

- The phantom should not have any clinically significant artifacts. There should be no artifacts larger than the objects in the phantom or

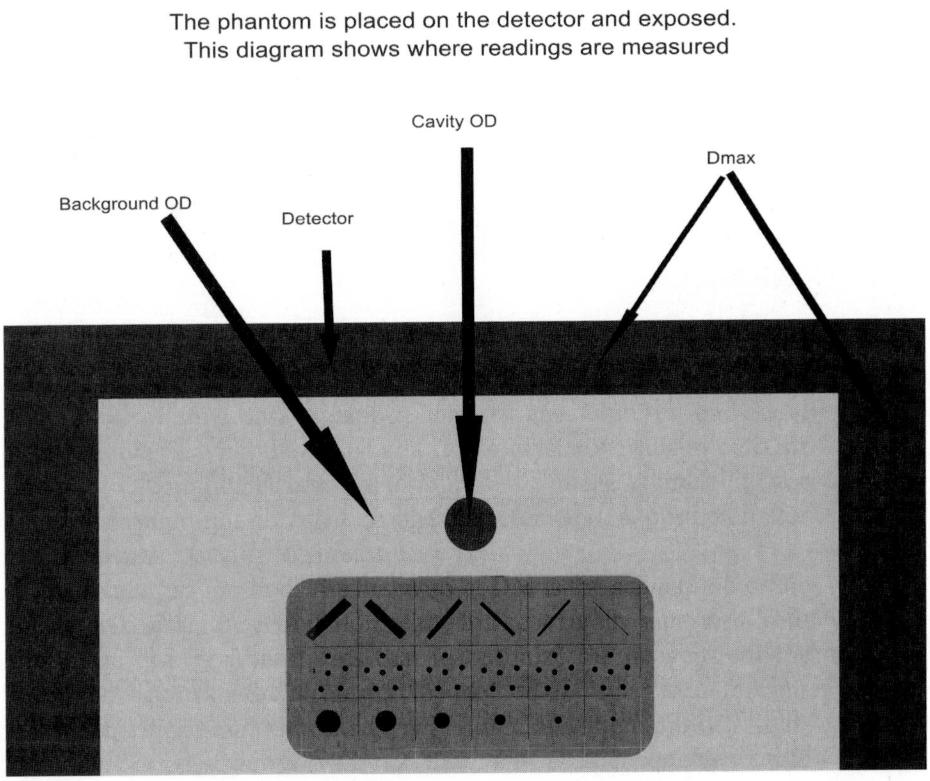

The phantom is placed on the detector and exposed.
This diagram shows where readings are measured

Figure 11–6. Schematic diagram of the American College of Radiology digital mammography phantom showing where to measure background optical density (OD), cavity OD, and D_{max}.

more prominent. Artifacts should not obscure the test objects in the phantom.

- Clinically significant artifacts must be correct before printing patient images.
- If the artifact evaluation fails, determine whether the artifact also appears on the workstation. If there is no artifact on the workstation, retest and check the printer.

Phantom Scoring:

- The fiber score must be ≥ 2.
- The speck group score must be ≥ 3.
- The mass score must be ≥ 2.

Background Optical Density:

- The background optical density must be ≥ 1.6 (background optical densities between 1.7 and 2.2 are recommended; 2.0 is optimum).
- Contrast *must be* ≥ 0.1.

D_{max}:

- D_{max} must be ≥ 3.1 (≥ 3.5 is recommended).

All failures must be corrected before clinical use.

Viewbox Cleanliness (If Applicable)

Purpose: To achieve optimal viewbox and viewing conditions. Even with digital imaging, comparison with a film may be needed, and the viewing conditions can affect interpretations.

Frequency: Monthly, after any service and before clinical use or on installation of new equipment

Equipment Needed:

- Cleaner recommended by the viewbox manufacturer
- Soft towels
- Viewbox cleanliness form

Process:

- Clean the viewbox surface using window cleaner and soft paper towel.
- Check and remove all marks.
- Visually inspect the viewboxes for uniformity of luminance.
- Ensure that the masking equipment is functioning.

Results and Analysis: Any marks not easily removed must be cleaned with a safe or manufacturer-recommended cleaner. The diagnostic potential of the mammogram is affected by the luminance of the viewbox and the ambient room illumination or amount of light falling on the viewbox surface. Viewboxes should not face bright lights and should avoid light from windows. The viewboxes provide a relatively high level of luminance, and areas around the film should be masked to exclude extraneous light. In addition, any nonuniform fluorescent lamps must be replaced as soon as possible.

Any failures must be corrected before viewing clinical images on the viewbox.

Facility QC Review

Purpose: This test will ensure that the lead interpreting radiologist and facility manager are aware of all QC tests and their frequencies. The data from QC tests must be collected at the appropriate intervals, the results reported, and corrective action taken. This test also ensures that clinical exams are conducted only when the imaging system is optimized.

Frequency: Quarterly or more frequently if problems are noted

Equipment Needed:

- Facility QC forms and data
- Facility QC review form

Process: All QC data must be reviewed by both the lead interpreting radiologist and the facility manager. This review can be conducted by video conference or teleconference.

- All results must be entered before the review.
- Each QC test result should be reviewed.
- If a test fails, the reason for the failure should be discussed, and the corrective action should be documented.
- The medical physicist's DM QC test summary for each mammography unit should be reviewed.
- Documentation from this review should be recorded on the facility QC review form.

Results and Analysis:

- The FDA Final Rules specify that the responsibility for ensuring an effective and efficient QA program cannot be delegated to the medical physicist or the QC technologist.
- If tests are not being done appropriately or if corrective actions are not implemented or documented as required, the reason must be determined and appropriate action taken.
- The medical physicist is an important resource for questions on the conduct of a test or data interpretation.

> **Facility QC review** will ensure that the lead interpreting radiologist and facility manager are aware of all the QC tests and their frequencies.

Compression Force

Purpose: This is a check to ensure that sufficient force is applied by the hands-free initial power-drive mode and that adequate compression can be maintained throughout the image acquisition. The test also ensures that the system provides adequate manual fine adjustment.

Frequency: Every 6 months, if unstable compression is suspected, after any service and before clinical use or on installation of new equipment

Equipment Needed:

- Several towels. The towels are folded and used to protect the detector.
- Calibrated analog bathroom scale. Digital scales often do not respond properly when added pressure is applied and are not recommended. Digital gauges or compression force tools may also be used.
- Compression force form

> **Compression force**
>
> Test will check to ensure that sufficient force is applied by the hands-free initial power-drive mode and that adequate compression can be maintained throughout the image acquisition.

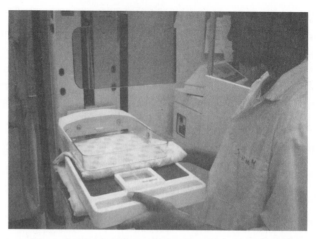

Figure 11–7. Demonstrating how to conduct the compression force test using a conventional bathroom scale.

Process:

- Place one folded towel on top of the scale and another on the detector (Fig. 11–7).
- The scale is placed on the detector, and initial power drive force is applied. Checks should be made that the force is within 25 to 45 pounds (111–200 N). This compression should hold for the average length of time for one projection exposure.
- Release the compression and repeat using the manual mode to compress to at least 25 pounds. Again, the compression force should be maintained for an average exposure.

Results and Analysis:

- A compression force of at least 25 lb (111 N) must be provided for the initial power drive and for the manual mode. The force must be maintained for the time it takes to make an average exposure.
- For the initial power drive, the maximum compression force should not exceed 45 lb (200 N).

If the test fails, the problem must be corrected before any examinations are performed.

Units such as the Siemens Mammomat are designed to terminate the compression when the breast is taut. With this unit, care should be taken to press the foot pedal more than once to accurately measure the maximum force. Failure to do so may report artificially low maximum compression force.

Repeat Analysis

Purpose: The repeat analysis will determine the number and cause of repeats. The test is performed to help identify problems with the equipment, improve efficiency, and reduce patient exposures.

Frequency: For meaningful repeat rate, this test should be performed with a minimum of 250 patients.

Equipment Needed:

- Data on all repeats
- A method of counting patient exposures

The minimum and maximum forces on the initial power drive are the forces available for use on the unit. These are not the minimum or maximum force that must be applied to the patient.

- A compression force of at least 25 lb (111 N) must be provided for the initial powerdrive and for the manual mode. The force must be maintained for the time it takes to make an average exposure.
- For the initial power drive, the maximum compression force should not exceed 45 lb (200 N).

- A repeat analysis form. Some equipment types have a built-in automated system to collect, record, and analyze repeated clinical images. For acceptable use, these systems must give a count of the total number of exposures during the evaluation period and the percentage of repeats during the period.

Process: Document the total number of patient exposures and repeated images made during the test dates. Document the reason for each repeat.

Results and Analysis: Total repeat rate is equal to the number of repeated images divided by the total number of exposures made times 100. The casual repeat rate is the percentage of repeats from a specific cause. This is equal to the number of repeats due to a specific cause divided by the total number of repeats times 100.

Rejects are all images thrown out, whereas repeats are images that resulted in a dose to the patient. Exposure taken for wire or seed localizations and added images taken to image a large breast or to image the entire breast (e.g., XCCL) do not count as repeats. Repeats can be related to poor positioning, patient motion, patient or equipment artifacts, incorrect patient ID, excessive noise image, overexposures (at image saturation), software or equipment failure, or aborted exposure. Repeats due to any other miscellaneous reason (such as good image with no apparent reason) will count as a repeat.

The repeat rate must be calculated for each mammography unit, but the rates can be combined or reported separately. Any reporting must be consistent. (See Repeat Charts in Appendix D.)

Performance Criteria: The overall repeat rate ideally should not exceed 2% but a rate of 5% is acceptable. Low repeat percentages can indicate that poor-quality images are tolerated. High repeat rate can indicate the need for improvement. Rate changes of more than 2% should be investigated.

If a problem is identified, it must be corrected within 30 days.

$$\% \text{ Repeats} = \frac{\text{Total \# repeat exposures}}{\text{Total \# exposures}} \times 100$$

$$\% \text{ Causal repeats} = \frac{\text{\# Repeat exposures in a specific category}}{\text{Total \# repeat exposures}} \times 100$$

System QC for Radiologist

Purpose: To allow the radiologist to perform a quick check of relevant QC tests, in particular the detector and monitor test

Frequency: This test is optional and performed as needed.

Equipment Needed: System QC form for radiologist

Process: The technologist pulls up a case (mediolateral oblique image) and fills out the top of the form based on information on the image. The radiologist pull up the same image and completes the rest of the form.

Results and Analysis: If any image quality problem is identified, corrective action must be taken.

Radiologist Image Quality Feedback

Purpose: This test ensures that the radiologist provides feedback on image quality to the technologists.

Frequency: This is an optional test that is performed as needed.

Equipment Needed: Radiologist image quality feedback form

Process: The radiologist documents image quality assessment and follows the protocols of the department that will ensure that the technologist receives all pertinent feedback—both positive and negative.

Results and Analysis: If there is severe deficiency in image quality, the patient should be called back for additional images. The radiologist may also need to know if the problem is due to equipment or technologist error or if it is related to the patient's body habitus.

Physicist Tests

The medical physicist is required to conduct an MEE of new equipment and after major repairs and conduction of an annual test. The medical physicist should also perform an annual survey on each unit (14 months between surveys is acceptable). During the survey the physicist must review all the technologist QC test results and provide written recommendation if there are problems or suggestions for improvement. For all tests conducted, the medical physicist must provide a written report and make recommendations for corrective actions according to the test results. The report must also include a summary of the pass/fail test results, along with all data related to the test (Tables 11–3 and 11–4).

The MQSA inspectors will check the physics report to ensure that the facility is in compliance with MQSA regulations and that the physicist recommendations have been considered by the facility.

> The medical physicist is required to conduct an MEE of new equipment and after major repairs and conduction of an annual test.

Note:

- MEEs and annual surveys of the radiologist workstation must be conducted on site by the medical physicist since the quality of the image displayed on the monitor is being evaluated.
- MEEs and annual surveys of the film printer (if applicable) may be conducted remotely by the medical physicist because the quality of the image displayed can be evaluated from the resultant film shipped to medical physicist for review.

Compression Paddle Deflection

- The compression paddle must be flat and parallel to the breast support table and should not deflect from the parallel by more than 1 cm at any point on the surface of the compression paddle when compression is applied. (This does not apply to compression paddles not designed to be "flat and parallel.")

All failures must be corrected before clinical use.

X-Ray Beam Limiting Devise Illumination

- The light must provide an average illumination of not less than 160 lux (15 foot-candles) at 100 cm or the maximum source–image receptor distance (SID), whichever is less.

Signal-to-Noise Ratio (SNR) and Contrast-to-Noise Ratio (CNR)

Some systems do not permit CNR measurements, whereas others do. The SNR and CNR can be calculated from values obtained from the phantom image. In some units this is an automatic calculation.

- The SNR must be ≥40.
- The CNR must be ≥2.

TABLE 11–3. Adjustment or Repairs That Require or Do Not Require Medical Physicist Involvement

Item	Component	Major Repair	Medical Physicist Involvement
Automatic exposure control (AEC)	AEC replacement	Yes	On-site
	AEC recalibration that affects dose	Yes	On-site
	AEC sensor replacement	Yes	On-site
	AEC circuit board replacement	Yes	On-site
	Density control—internal adjustment*	No	Oversight
	Thickness compensation internal adjustment*	No	Oversight
Bucky replacement	AEC sensor also replaced	Yes	On-site
	AEC sensor not replaced	No	Oversight
	Digital mammography (DM) detector also replaced	Yes	On-site
	DM detector not replaced	No	Oversight
Collimator	Replacement	Yes	On-site
	Reassembly with blade replacement	Yes	On-site
	Adjustment	No	Oversight
Compression device	Pressure adjustment	No	Optional
	Thickness scale accuracy adjustment but only if it affects AEC performance	No	Oversight
	Repair of auto decompression	No	Optional
Compression paddle	Paddle (new to facility)	No	Oversight
	Deflection adjustment	No	Oversight
	Adjustment due to extension beyond allowable limits, or visible on images	No	Oversight
X-ray unit	Installation	Yes	On-site
	Reassembly	Yes	On-site
	X-ray tube replacement	Yes	On-site
	High-voltage generator replacement	Yes	On-site
	Filter replacement	Yes	On-site
	Manufacturer's software upgrade or modifications	Yes	On-site
	DM detector replacement or repair	Yes	On-site
	kVp, mA, or time interval adjustments*	No	Oversight
Display devices	New installation or replacement	Yes	On-site
	New video card or software upgrade	Yes	On-site
	Relocation	No	Oversight
Computed mammography (CM) and photostimulable storage phosphor (PSP) plates	New installation or replacement of CM reader	Yes	On-site
	Replacement of all PSP plates	Yes	On-site
	Replacement of 1 or 2 PSP plates	No	Oversight

*Internal adjustments refer to equipment adjustments that typically cannot be made by the operator.

TABLE 11–4. Summary Table of Annual Quality Control Tests

Required documentation

Test	Final Regulation Citation	Regulatory Action Levels	Scope	Required Documentation	Timing of Required Corrective Action*
Automatic exposure control (AEC)	900.12(e)(5)(i)	Optical density (OD) exceeds the mean by more than ±0.15 (>2- to 6-cm thickness range), or the phantom image density at the center is less than 1.20.	All x-ray units, 2- to 6-cm thickness range; using appropriate kVp—required in contact mode only. Magnification mode testing is required only for equipment.	The two most recent survey reports	Within 30 days of the date of the test
kVp	900.12(e)(5)(ii)	Exceeds ±5% of indicated or selected kVp Coefficient of variation (C.O.V.) exceeds 0.02	All x-ray units at 3 kVp—lowest measurable clinically, most frequently used clinically, and highest clinically obtainable	"	"
Focal spot condition	900.12(e)(5)(iii)	System resolution	All x-ray units, all clinically used target materials and focal spots Resolution measurement Contact mode: at most used source–image receptor distance (SID) Magnification mode (if clinically used): at SID w/magnification value closest to 1.5 Must be done for all clinically used screen/film combinations	"	"
Half-value layer (HVL)	900.12(e)(5)(iv)	See table in regulations	All x-ray units, all clinically used target-filter combinations	"	"
Air kerma and AEC reproducibility	900.12(e)(5)(iv)	Reproducibility C.O.V. exceeds 0.05	All x-ray units	"	"
Dose	900.12(e)(5)(vi)	Exceeds 3.0 mGy (0.3 rad) per exposure	All x-ray units, all clinically used screen/film combinations, targets, and filters used for the standard breast	"	Before any further examinations are performed using the x-ray unit
X-ray field, light field, compression device alignment	900.12(e)(5)(vii)	Exceeds 2% SID at chest wall Paddle visible on image	All x-ray units, all combinations of collimators, image receptor sizes, targets, and focal spots clinically used for full-field imaging in the contact mode	"	Within 30 days of the date of the test

TABLE 11–4. Summary Table of Annual Quality Control Tests (*Continued*)

Required documentation

Test	Final Regulation Citation	Regulatory Action Levels	Scope	Required Documentation	Timing of Required Corrective Action*
Screen speed uniformity	900.12(e)(5)(viii)	OD variation exceeds 0.30 from the maximum to the minimum	All image plates (IPs) May be grouped by size and speed Limit holds within groups Groups must be identifiable to the technologist	"	"
System artifacts	900.12(e)(5)(ix)	Determined by physicist	All x-ray units and processors, all clinically used IP sizes, focal spots and target-filter combinations. Also see approved alternative standard substituting "all targets and filters" for all "target-filter combinations"	"	"
Radiation output	900.12(e)(5)(x)	Less than 7.0 mGy/sec (800 mR/sec)	All x-ray units	"	"
Automatic decompression control	900.12(e)(5)(x)	Failure of override or manual release	All x-ray units (if auto is provided)	"	"
Any applicable annual new mammographic modality tests	900.12(e)(6)	As specified by the equipment manufacturer	All x-ray units	"	Before any further examinations are performed
Phantom image	900.12(e)(9) see (e)(2)	As specified by the facility's accreditation body	All x-ray units, all target-filter and screen/film combinations used clinically for the standard breast	"	"

*Refer to 900.12(e)(8)(ii)(A) or (B) as applicable (http://www.fda.gov/radiation-emitting-products/mammographyqualitystandardsactandprogram/guidance/policyguidancehelpsystem/ucm052678.htm).

- The CNR change must be ≥85% of the previous year's CNR.
- The current CNR should be ≥85% of the MEE CNR.

All failures should be corrected before clinical use.

Spatial Resolution

Purpose: To measure limiting spatial resolution, which will indicate the detector performance

Frequency: Annually, as part of the MEE and after any relevant

Equipment Needed:

- ACR DM phantom
- Line-pair pattern with frequencies up to 10 lp/mm
- Spatial resolution form

Process:

- Place the ACR DM phantom on the breast support. It should be rotated 180 degrees from the normal orientation of the phantom.
- Place the line-pair pattern on the phantom at a 45-degree angle (Fig. 11–8).
- Make an exposure using manual technique.
- Repeat using the magnification setup.

Results and Analysis: The spatial resolution must be ≥4 lp/mm for contact mode and 6 lp/mm for magnification mode. All failures must be corrected within 30 days.

Automatic Exposure Control (AEC)

Purpose: To assess the performance of the AEC function and verify consistency in detector SNR level for a range of breast thickness

Frequency: As part of the MEE, after significant service and before clinical use or on installation of new equipment

Equipment Needed: Compression paddles and AEC system performance form

> Spatial resolution is the minimum separation between two objects at which they can be distinguished as two separate objects in the image.

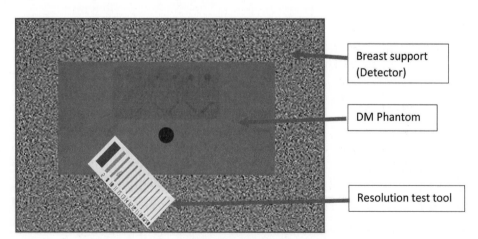

Breast support (Detector)

DM Phantom

Resolution test tool

Figure 11–8. Spatial resolution testing.

Process:

- An exposure is made with the 2-cm tissue-equivalent attenuator on the detector.
- The compression should be 2 cm or 5 daN.
- Select a large focal spot with the AEC sensor set to the center of the phantom.
- The test is repeated for magnification.
- Record the AEC mode, exposure compensation or density control step, target, filter, kVp, mAs, and indicated AGD.

Results and Analysis:

- The SNR must be ≥40 for the 4-cm phantom in contact mode.
- SNR must be ≥−15% of the previous year's SNF (not lower than 85% of the previous year's SNR, for each thickness and mode tested).
- Failures must be corrected within 30 days.

Average Glandular Dose (AGD)

Purpose: To measure the entrance exposure for an average patient (approximately a 4.2-cm-thick compressed breast consisting of 50% glandular and 50% adipose tissue). Ensure that the unit-indicated average glandular dose is reasonable accurate.

Frequency: As part of the MEE after relevant service and before clinical use or on installation of new equipment

Equipment Needed:

- An ionization chamber and electrometer or other appropriate dosimetry device that is calibrated at mammographic x-ray beam energies
- An integrated, solid-state instrument (one that automatically measures kVp, half-value layer (HVL). It must be calibrated for the target/filter combination.
- Average glandular dose form

Process: A lead sheet is placed on the detector to protect it from repeated exposures. Place the dosimeter at a height of 4.2 cm above the breast support just under the paddle. The dosimeter should be centered approximately 4 cm from the chest wall edge of the detector. The detector is not moved during the test. Set the kVp, target material, and filtration at the same value for the ACR DM phantom image. Use manual exposure to make three exposures. Record the exposure, HVL, and average glandular dose.

Results and Analysis: The physicist will also do a manual calculation of the AGD.

1. The AGD for an average 4.2 cm compressed breast *must* not exceed 30 mGy (0.3 rad) per projection.
2. The AGD should be within ±25% of the calculated average glandular dose.

Dose in excess of 3 mGy must be corrected before clinical use. Failure of the unit-indicated AGD *must* be corrected within 30 days.

Radiation dose in mammography

- Entrance skin dose = dose to the patient's skin
- Glandular dose = dose to the glandular tissue of the breast
- ESE for a typical single exposure ≈ 800–1200 mrad (812 mGy)
- Glandular dose ≈ 100 mrad (1.0 mGy)

ACR recommends

- 0.3 rad (300 mrad or 3m Gy) with a grid
- 0.1 rad (100 mrad or 1m Gy) without a grid

Unit Checklist

Purpose: To ensure that all locks, detents, angulation indicators, and mechanical support devices for the x-ray tube and breast support assembly are operating properly, and that the DICOM (Digital Imaging and Communications in Medicine) image file headers are correctly populated.

Frequency: As part of the MEE after relevant service and before clinical use or on installation of new equipment

Equipment Needed: Checklist form

Process: The physicist will review the following critical test:

1. Verify that the mammography unit is mechanically stable under normal operating conditions.
2. Set and test each lock and detent independently to ensure that mechanical motion is prevented when the lock or detent is set.
3. Verify that the detector or image receptor holder assembly is free from wobble or vibration during normal operation.
4. On CM systems, verify that the IP is held securely for any orientation of the holder assembly.
5. Verify that in normal operation the patient and operator are not exposed to sharp or rough edges or other hazards, including electrical hazards.
6. Verify that the compression paddles are all intact with no cracks or sharp edges.
7. Verify that the operator is protected by adequate radiation shielding during exposure.
8. Verify that automatic decompression can be overridden to maintain compression (for procedures such as needle localizations) and its status displayed continuously (if automatic decompression is available).
9. Verify that compression can be manually released in the event of a power or automatic release failure by turning power off to the equipment.

The physicist will review the following less critical tests:

1. Verify that the mammography area is clean and free from significant dust and debris that may cause artifacts.
2. Verify that all indicators work properly.
3. On CM systems, verify that the IP slides smoothly into the proper position in the holder assembly.
4. Verify that all moving parts move smoothly, that cushions or bumpers limit the range of available motions, and that no obstructions hinder the full range of motions within these limits.
5. Verify that the audible exposure indicator is at an appropriate volume level.
6. Verify that current and accurate technique charts are posted, confirmed by consulting with the mammography technologist if possible.
7. Add other unit-specific checks as necessary.

All test items are given a pass or fail on the form.

Results and Analysis: Failure of a critical test must be corrected before clinical use. Less-critical test must be corrected within 30 days.

Example of when to conduct the MEE

- After installation of new (or used) x-ray unit or processor
- After x-ray unit or processor disassembled and reassembled at the same or new location
- After x-ray tube replacement
- After collimator replacement
- After filter replacement
- After AEC unit or sensor replacement

Test for the CM Systems

Purpose: To check the uniformity of the PSPs in the IPs

Frequency: As part of the MEE after relevant service and before clinical use or if new IPs are acquired. The technologist can perform this test for new IPs.

Equipment Needed: ACR DM phantom and CM forms

Process:

- **Plate-to-Plate Uniformity:** Expose the DM phantom using a force of 5 daN. Record the mAs. Without making any postprocessing adjustments, calculate the SNR for each image. (Measured by recording the Region of Interest [ROI] 5 cm² in the center of each image.)
- **Plate-Specific Artifact Analysis:** Window and level the image to optimize viewing and record any clinical significant artifacts.
- **CM Reader Scanner Performance:** Place a pair of thin steel rulers on the top of the IP and make an exposure using 25 kVp at 4 mAs. Wait 15 minutes before processing the plates. The same time delay between exposure and processing should be used for each plate (Fig. 11–9).

Results and Analysis:

- In the plate-to-plate uniformity test, the mAs of any plate must not differ by more than ±10% from the mean mAs of all plates of the same size.
- The SNR of any plate must not differ by more than ±15% from the mean SNR of all plates of the same size.
- In the Plate-Specific Artifact test, any artifacts must be removed by cleaning the plate. Artifacts appearing on all plates could be caused by the x-ray unit or CM reader.
- In the CM reader scanner performance, there should be no jagged or nonlinear edges visible on the edges of the steel ruler.

  Failure of any of the test items must be corrected before clinical use.

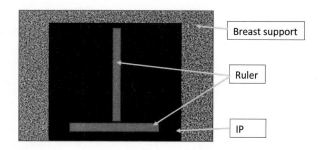

Figure 11–9. Computed mammography (CM) setup for the CM reader test for scanner performance.

AWS and RWS Monitor QC

Purpose: To check that the monitors are clean and free of any artifacts that could interfere with clinical information or interpretation. This test also ensures correct calibration of the monitors and the brightness and contrast setting are correct.

Frequency: As part of the MEE after relevant service and before clinical use or on installation of new equipment

Equipment Needed: The SMPTE test pattern or the preferred American Association of Physicists in Medicine (AAPM) Task Group (TG) 18 QC test pattern, a luminance meter, and an AAPM TG18 UNL80 test pattern (or something similar) for the luminance check

Process: The monitor should be visually inspected for dirt and defects. The test pattern should be examined in room lighting similar to that in the radiologist reading room. (Some systems allow the test pattern display on the monitor.) The luminance meter is used to measure the

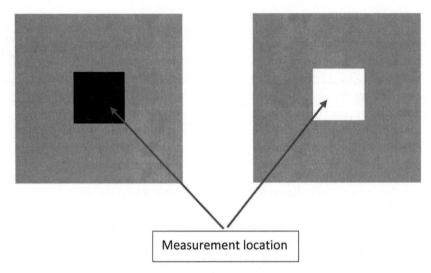

Figure 11–10. Schematic diagram of the luminance test tool (AAPM TG18UUNL80) showing where the measurements are taken.

minimum and maximum luminance (Lmin and Lmax) on the luminance test pattern (Fig. 11–10).

Results and Analysis: The monitor should be free of dust and dirt. The reading from the SMPTE test pattern or the preferred AAPM TG18 QC test pattern is recorded. The luminance is calculated from the luminance test pattern. Many units have automatic features that will calculate and display an overall pass/fail.

All failures must be corrected before clinical use.

Film Printer QC

Purpose: To ensure that the printed image provides adequate and consistent image quality

Frequency: As part of the MEE after relevant service and before clinical use or on installation of new equipment

Equipment Needed: An ACR DM phantom, densitometer (Fig. 11–11), ruler, and appropriate form

Process: The process is similar to the technologist Film Printer QC test.

Results and Analysis: All failures must be corrected before clinical use.

Evaluation of the Site's Technologist QC Program

Purpose: To ensure that all technologist QC tests are being performed at the correct intervals and to identify areas where image quality or QC testing can be improved

Frequency: Annually

Equipment Needed: The appropriate form listing all QC tests

Process: The physicist should review all the technologist QC tests. Checks should be made for correct analysis, correct results, records of deficiencies, missing data, incorrect scoring, incorrect calculations, missing corrective action documentation, or any other problems. The

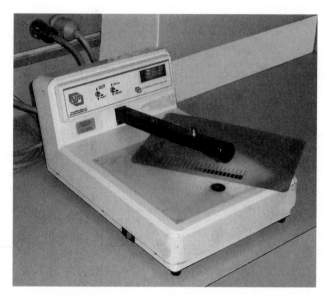

Figure 11–11. A densitometer.

FDA level 2 noncompliance items can be used as guide to determine the effectiveness of the technologist QC documentation.

Results and Analysis: All failures must be corrected within 30 days.

Beam Quality (Half-Value Layer)

Purpose: To ensure that the HVL of the x-ray beam is adequate to minimize radiation dose to the patient's breast dose without excessive loss of contrast

Frequency: Annually as part of the MEE, if the technical factors are changed or after relevant service

Equipment Needed:

- A ionization chamber and electrometer or other appropriate dosimetry device calibrated for mammography units
- An 0.1-mm thick sheet of 99.9% pure aluminum—large enough to cover the dosimeter
- An instrument to measure the kVp, HVL, and dose

Process: During this test the medical physicist must determine the HVL for all target and filter combination and for a range of clinical kVp

Results and Analysis: All failures must be corrected before clinical use.

kVp Accuracy and Reproducibility

Purpose: To ensure that the kVp is accurate and reproducible

Frequency: Annually as part of the MEE on new equipment, after any relevant service and after component replacement

Equipment Needed: kVp meter, lead sheet, and an instrument to measure the kVp, HVL, and dose

Process: With the test tools in place, exposures are made using the same manual setting mode.

> The kVp should be accurate to within 65% of the indicated kVp.

Results and Analysis: The kVp should be accurate to within ±5% of the indicated kVp. The kVp should have a coefficient of variation equal to or less than 0.02.

When the test is performed for an MEE, all failures must be corrected before clinical use. When the test is performed for troubleshooting, all failures must be corrected within 30 days.

Collimation Assessment

Purpose: To ensure that the x-ray field aligns with the light field. The x-ray field should cover the entire detector but should not allow significant radiation beyond the edges. The compression paddle should align with the chest-wall edge of the detector.

> The collimation assessment ensures that the x-ray field aligns with the light field.

Frequency: As part of the MEE, before clinical use, after any service, and on installation of new equipment

Equipment Needed:

- Five coins, four of one size (e.g., pennies) and one larger size (e.g., nickel). Another radiopaque object may be used in place of coins.
- Collimator detector test tool (IP, electronic radiation "rulers")
- Collimation assessment form

Process: Select the largest compression paddle. Place the external radiation detector on top of the breast support. (If using CM, use the largest compression paddle with the largest IP size.)

Align the edge of each of the four smaller coins with the edge of the light field. Position the center of the coin within the light field. Take the larger coin to the underside of the compression paddle so that the outer edge of the coin is aligned with the inner chest wall edge of the paddle. The coin should not interfere with the smaller coin indicating the chest-wall edge light-field border (Fig. 11–12).

Position the paddle 4 to 6 cm from the breast support and make an exposure. Repeat for all fields of view and all anode tracks. If the unit shifts for the mediolateral oblique projection, the exposures should be repeated for all three positions (left, right, and center).

Results and Analysis:

- **Computer mammography:** If the radiation field extends beyond the edge of the light field, record the value as positive. If the radiation field does not extend fully to the edge of the light field, record the value as negative.
- **FFDM (full-field digital mammography):** If the coin is fully visible, the difference will be positive. If the coin is only partially visible, the difference will be negative.
- For each edge, total the value and record this as the deviation between the radiation field and the visible image.
- For the larger coin, measure the distance between the diameter of the coin in the DM image and the dimension of the coin perpendicular to the chest-wall edge as the distance from the paddle edge to the detector edge. If the coin is fully visible, record a positive value. If the coin is not fully visible, record a negative value.
- The chest-wall edge of the compression paddle should be aligned just beyond the chest-wall edge of the detector such that the chest-wall edge of compression paddle does not appear in the image.

Figure 11–12. Coins are used to outline the light field and provide a visual check in a quick field light test.

- The chest-wall edge of the compression paddle should not extend beyond the chest-wall edge of the detector by more than 1% of the SID.
- The congruence of the light field with the radiation field should be such that the total misalignment is within 2% of the SID. The x-ray field should not extend beyond any edge of the detector by more than +2% of the SID. A tolerance of –2% of the SID on the right and left sides is acceptable. A tolerance of –4% of the SID is acceptable on the side away from the chest-wall edge of the detector (to allow for appropriate image information).

Note:

- When this test is performed as a part of the MEE, all failures must be corrected before clinical use.
- When the test is performed for troubleshooting, all failures must be corrected with 30 days.

- The chest-wall edge of the compression paddle should not extend beyond the chest-wall edge of the detector by more than 1% of the SID.
- The x-ray field should not extend beyond any edge of the detector by more than +2% of the SID.

Manufacturer Detector Calibration (If Applicable)

Purpose: To ensure that the detector is calibrated according to the manufacturer's recommendations

Frequency: The frequency is generally specified by the manufacturer. This test should also be performed on installation of new equipment.

Equipment Needed: Manufacturer detector calibration form

Process: Many units have an automatic calibration feature in their software. Access the QC mode and follow the step-by-step guidelines.

Results and Analysis: These tests can include the detector flat-field calibration or detector calibration. The manufacturer will specify the precautions and caveats.

All failures must be corrected before any clinical use.

Ghost Image Evaluation

Purpose: To evaluate the extent to which previous exposures can affect later exposures on the detector. Earlier exposure may change the sensitivity of the detector, leading to underresponse or overresponse to a later exposure—a phenomenon called ghosting. Lag occurs when residual signals (shadows) of prior images are retained.

Frequency: As part of the MEE, before clinical use after any service, and on installation of new equipment

Equipment Needed: ACR DM phantom, 0.1-mm-thick aluminum, timer, and evaluation forms

Process: The physicist takes a series of images to calculate the ghosting index.

Results and Analysis: The ghosting index should not exceed ± 0.3. Failures must be corrected before clinical use.

Viewbox Luminance

Purpose: To ensure that the luminance of the viewboxes used for interpretation exceeds minimum levels

Frequency: As part of the MEE, before clinical use, after any service, and on installation of new equipment

Equipment Needed: A luminance meter and appropriate forms

Process: The luminance meter is used to measure the luminance. The measured luminance is then compared with the limits to determine pass/fail.

Results and Analysis: If the luminance level of the viewbox is less than 3,000 cd/m^2 or if the colors of the light from individual bulbs appear different, then all the viewbox bulbs should be replaced.

> Ghost image evaluation test is used to evaluate the extent to which previous exposures can affect the later exposures on the detector.

Summary

Quality Assurance

Quality Assurance (QA) or continuous quality improvement (CQI) includes the total overall management of actions taken to consistently provide high image quality in the radiology department, with the primary objective being to enhance patient care. QA is very important in mammography, both in meeting the standards of the US Mammography Quality Standards Act (MQSA) and in ensuring quality patient care. Every QA program must include a quality control (QC) component, which includes the collection and evaluation of data with a systematic and structured mechanism to control variables such as repeat radiographs, image quality, processing variations, patient selection parameters, technical effectiveness, efficiency, and in-service education.

Benefits of Quality Assurance

One of the benefits of having a good quality assurance (QA)/quality control (QC) program is the reduction of unnecessary radiation to the patient by reducing repeats. A quality assurance QA program will also improve overall efficiency of the delivery of service, which will result in improved patient satisfaction. Overall, the facility will benefit in the achievement of improved image quality as well as an increased consistency of image production, reliability, efficiency, and cost-effectiveness of equipment use. QA generally has a positive effect on personnel morale and always results in overall efficiency and improved customer service.

Agencies Responsible for Safety and Quality of Imaging Practices Within the United States

Federal Agencies
OSHA
EPA and the *Food and Drug Administration* (FDA)
The *National Center for Devices and Radiological Health* (CDRH)
The *Nuclear Regulatory Commission* (NRC)
The *Centers for Disease Control and Prevention* (CDC)

Professional Agencies and Societies in the United States
The *American College of Radiology* (ACR)
The *American Registry of Radiologic Technologists* (ARRT)
The *American Society of Radiologic Technologists* (ASRT)
The *American Registry of Magnetic Resonance Imaging Technologists* (ARMRIT)
The *American Registry of Diagnostic Medical Sonographers* (ARDMS)
The *Joint Commission (TJC)*, formerly the *Joint Commission on Accreditation of Healthcare Organizations* (JCAHO)
The *Joint Review Committee on Education in Radiologic Technology* (JRCERT)

Responsibilities of the Various Quality Control Personnel
Radiologist

The FDA's MQSA regulation recommends that a lead physician should be responsible for general oversight of the MQSA QA program

Lead Interpreting Physician (Radiologist)
The mammography radiologist who is responsible for interpreting the images, ensuring quality imaging and ensuring that corrective actions are taken in case of test failures. The lead radiologist is ultimately responsible for quality imaging and an effective QA program.

Interpreting Physician (Radiologist)

In addition to interpreting the image, the interpreting radiologist will take corrective action in cases of poor image quality and participate in medical outcome audit.

The Technologist

The technologist is responsible for patient positioning, compression, and image production. The technologist plays a crucial role in quality control (QC) and is usually solely responsible for the day-to-day testing details.

The Medical Physicist

The medical physicist performs annual evaluations of the mammographic unit and reviews all test procedures, including records, charts, and mammography policies and procedures on a yearly basis.

Quality Control Testing

General QC Tests

Phantom Test: To ensure consistent image quality and artifact-free imaging

Computed Mammography (CM) PSP Erasure (If Applicable): To ensure the IP is free from information that would affect image quality

Compression Thickness Indicator: To ensure that the compression thickness indicates the correct degree of compression

Visual Checklist: To ensure that all indicator lights detents and mechanical locks are in proper working order

Acquisition Workstation (AWS) Monitor: To ensure that monitors are free of dirt and dust or other marks that can interfere with clinical information. This test also checks the manufacturer's specification and calibrations, including the brightness and contrast settings.

Radiologist Workstation (RWS) Monitor: To ensure that monitors are free of dirt and dust or other marks that can interfere with clinical information. This test also checks the manufacturer's specification and calibrations, including the brightness and contrast settings.

Film Printer (If Applicable): To ensure adequate and consistent image quality on printed images provided to physicians, patients, and others

Viewbox Cleanliness (If Applicable): To achieve optimal viewbox and viewing conditions. Even with digital imaging, comparison with a film may be needed, and the viewing conditions can affect interpretations.

Facility QC Review: This test will ensure that the lead interpreting radiologist and facility manager are aware of all QC tests and their frequencies. The data from QC tests must be collected at the appropriate intervals, the results reported, and corrective action taken. This test also ensures that clinical exams are conducted only when the imaging system is optimized.

Compression Force: This is a check to ensure that sufficient force is applied by the hands-free initial power-drive mode and that adequate compression can be maintained throughout the image acquisition. The test also ensures that the system provides adequate manual fine adjustment.

Repeat Analysis: The repeat analysis will determine the number and cause of repeats. The test is performed to help identify problems with the equipment, improve efficiency, and reduce patient exposures.

System QC for Radiologist: To allow the radiologist to perform a quick check of relevant QC tests, in particular the detector and monitor test

Radiologist Image Quality Feedback: This test ensures that the radiologist provides feedback on image quality to the technologists.

Specific QC Tests

Physicist Tests

Compression Paddle Deflection

X-Ray Beam Limiting Devise Illumination

Signal-to-Noise Ratio (SNR) and Contrast-to-Noise Ratio (CNR)

Spatial Resolution

Automatic Exposure Control (AEC)

Average Glandular Dose (AGD)

Unit Checklist

Test for the CM Systems

Acquisition Workstation (AWS) and Radiologist Workstation (RWS) Monitor QC

Results and Analysis

Film Printer QC

Evaluation of the Site's Technologist QC Program

Beam Quality (Half-Value Layer)

kVp Accuracy and Reproducibility

Collimation Assessment

Manufacturer Detector Calibration (if applicable)

Ghost Image Evaluation

Viewbox Luminance

REVIEW QUESTIONS

1. What type of maintenance includes actions taken on a regular basis to prevent deterioration of image quality or any breakdown in the imaging chain?

2. Give an example of how a good quality assurance program will benefit the patient.

3. Name one federal agency responsible for ensuring the safety and quality of imaging practices within the United States.

4. Identify some of the causes of repeats.

5. How often is the compression test performed?

6. What are the minimum and maximum allowed compressions on the initial power drive?

7. What federal agency is responsible for protection of the health and safety of people?

8. What are the main patient care responsibilities of the technologist?

9. What mammography imaging systems can use the new ACR digital mammography QC manual?

10. Collimation must cover the entire image receptor but should not extend beyond any edge of the image receptor by more than what percentage of the SID?

11. Who is responsible for medical auditing and patient communication?

12. Which assessment is used to ensure that the x-ray field aligns with the light field and the collimation covers the entire image receptor?

13. How is the effect of subjective judgment reduced when viewing the phantom image test?

14. For a statistically meaningful reject/repeat analysis, what is the recommended patient volume?

15. Ideally, the overall repeat rate should not exceed what percentage?

16. What is the MQSA recommended next step to take if the compression test fails?

CHAPTER QUIZ

1. Test on beam quality assessment (half-value layer measurement) is performed by the:
   - (A) Mammographer
   - (B) Medical physicist
   - (C) Technologist
   - (D) Radiologist

2. In quality control, the standards refers to:
   - (A) Action taken on a regular basis
   - (B) Action taken to eliminate potential problems
   - (C) Reference calibrations or measurements to check accuracy
   - (D) Measurement variations allowed

3. The person responsible for overseeing all the activities of the mammographer and the medical physicist is:
   - (A) Quality control radiographer
   - (B) State inspector
   - (C) MQSA inspector
   - (D) Radiologist

4. Which of the following QC duties is the responsibility of the medial physicists?
   - (A) Phantom image
   - (B) Visual checklist
   - (C) Average glandular dose
   - (D) Compression force

5. The ability of an x-ray tube to always produce the same intensity of radiation when the same set of technical factors is used to make an exposure is called:
   - (A) Radiation output
   - (B) Exposure time
   - (C) Exposure reproducibility
   - (D) kV accuracy

6. The MQSA standards were enacted:
   - (A) Because mammography was overregulation
   - (B) To address the poor quality of mammograms that were available
   - (C) To enforce continuing education for the radiologic technologist
   - (D) To enforce continuing education for the radiologist

7. The repeated image is:
   - (A) Any image that is rejected
   - (B) An image that is repeated and results in radiation dose to the patient
   - (C) The magnified image requested by the radiologist
   - (D) An image taken during quality control tests

8. The person responsible for performing evaluations of and generating reports on the mammographic unit is the:
   - (A) Radiologic technologist
   - (B) Mammographer
   - (C) Medical physicist
   - (D) Radiologist

9. A test used to determine whether the time of exposure indicated by the control panel is the same as that delivered by the x-ray tube is performed by the:
   - (A) Technologist
   - (B) Physicist
   - (C) Quality assurance technologist
   - (D) Radiologist

10. A quality control test performed to check the random disturbance in digital signals that obscures or reduces image clarity is:
    - (A) Detector calibration
    - (B) SNR
    - (C) Flat-field evaluation
    - (D) Monitor calibration

BIBLIOGRAPHY

Berns EA, Baker JA, Barke LD, et al. *Digital Mammography Quality Control Manual.* Reston, VA: American College of Radiology; 2016.

Bushberg JT, Seibert JA, Meidholdt EM, Boone JM. *The Essential Physics of Medical Imaging.* 3rd ed. Philadelphia, PA: Lippincott Williams & Wilkins; 2012.

Bushong SC. *Radiologic Science for Technologists—Physics, Biology and Protection.* 10th ed. St. Louis, MO: Mosby; 2012.

Centers for Disease Control and Prevention. Breast Cancer. http://www.cdc.gov/cancer/breast/index.htm. Accessed October 11, 2016.

Joint Review Committee on Education in Radiologic Technology (JRCERT). http://www.jrcert. org. Accessed October 11, 2016.

Nuclear Regulatory Commission. What We Do. http://www.nrc.gov/. Accessed October 11, 2016.

U.S. Environmental Protection Agency. http://www.epa.gov. Accessed October 11, 2016.

U.S. Food And Drug Administration. Policy Guidance Help System. http://www.fda.gov/Radiation-EmittingProducts/MammographyQualityStandardsActandProgram/Guidance/PolicyGuidanceHelpSystem/default.htm#Quality. Accessed July 2016.

Venes D, Biderman A, Adler E. *Taber's Cyclopedic Medical Dictionary.* 22nd ed. Philadelphia, PA: F. A. Davis; 2013.

The US Mammography Quality Standards Act (MQSA) | 12

Keywords and Phrases
Purpose of the Mammography Quality Standards
Interim Regulation
Final Regulations
Accreditation and Certification
Lawful Mammography Services in the United States
Quality Assurance Regulations of MQSA
    Clinical Images
    Levels of Findings or Noncompliance
    Withdrawal of Accreditation
    Suspension or Revocation
    Adverse Reporting
    Digital Mammography
    Quality Control
    Printers and Monitors
    Monitors
    Mobile Units
    Infection Control
    Xeroradiography
    Certificate Placement
    Consumer Complaints Mechanism
    Self-Referrals
    Record Keeping
    Content of Records and Reports
    Assessment Categories
    MQSA Assessment Categories
    Communication of Results to the Patient
    Medical Audit
    MQSA and Health Insurance Portability and
        Accountability Act
Final Regulation Documentation
    Requirements for the Radiologic Technologists
    Requirements for the Interpreting Physician
    Requirements for the Medical Physicist
Breast Cancer Screening Today

Enhancing Quality Using the Inspection Program (EQUIP) Initiative
  Details on EQUIP Questions
Summary
Review Questions
Chapter Quiz

Objectives

On completion of the chapter, the reader will be able to:

1. Describe the purpose of the Mammography Quality Standards Act (MQSA)
2. Explain the differences between accreditation and certification
3. List the names of accreditation bodies in the United States
4. State the purpose of the provisional certificate
5. State the time period for which certification is valid
6. Describe the levels of findings possible during an MQSA inspection
7. Discuss what a facility must do before the site can stop performing mammograms
8. Identify how a facility certifies the digital breast tomosynthesis (DBT), full-field digital mammogram (FFDM), and the mobile units
9. Explain where a facility would place an MQSA certificate
10. Explain how a facility could legally satisfy the consumer complaints mechanism regulation
11. Explain the meaning of "self-referral"
12. State how long the facility must keep a patient's records
13. List what must be included in the mammographic report
14. List the items in the assessment categories
15. State how a facility can communicate results to the patient
16. Explain the purpose of a medical audit
17. Discuss what effect, if any, the Health Insurance Portability and Accountability Act (HIPAA) regulations will have on MQSA regulations
18. State what requirements the radiologic technologist must meet before performing mammograms
19. State what requirements the interpreting physician must meet before independently interpreting mammograms
20. State what requirements the medical physicist must meet before legally performing surveys of mammography facilities or overseeing a facility's quality assurance program

KEYWORDS AND PHRASES

- **Accreditation** is the initial approval step required by all facilities in the United States before they can legally perform mammograms.
- **ARRT (American Registry of Radiologic Technologists)** is dedicated to promoting the standards of education of radiographers entering the profession. The organization supports high standards of patient care by certifying the professional behavior, ethics, responsibilities, and continuing education of practicing radiographers.
- **Certification** means that a mammography facility has been certified either by the US Food and Drug Administration (FDA) or an approved

state certification agency as capable of providing quality mammography. Certified facilities have either completed the accreditation process or are in the process of completions.

- **Continuing Education Unit (CEU)** refers to 1 hour of direct instructional training or 1 contact hour of training. CEU, CE (continuing education), and CME (continuing medical education) are acronyms for various continuing education processes sponsored by various accrediting agencies.

- **Digital mammography** is a technological advancement in the field of early detection of breast cancer. The equipment allows fast imaging and electronic storage, eliminating lost radiographs and enabling the digital transfer of images by the Internet or telephone lines.

- **Direct supervision** occurs when supervision is provided by a certified radiographer/clinical instructor. With direct supervision, the radiographer or clinical instructor will review the request to determine the examination in relation to the trainee's level of achievement, evaluate the condition of the patient in relation to the trainee's knowledge, be physically present in the radiography room while the trainee performs the radiographic examination, and then reviews and approves the final radiographs.

- **FDA (US Food and Drug Administration)** is responsible for protecting the public health of Americans. The agency oversees food, cosmetics, medicines, medical devices, and radiation-emitting products within the United States.

- **HIPAA** (Health Insurance Portability and Accountability Act) of 1996 was designed to protect the privacy of patients by limiting the ways that health plans, pharmacies, hospitals, and other covered facilities can use patients' personal medical information. The final regulations, which covered health care institutions, became effective on April 14, 2003. The regulations protect medical records and other individually identifiable health information, whether it is on paper, in computers, or communicated orally.

- **Interpreting physician** is the physician qualified to read mammograms.

- **Mammography Equipment Evaluation (MEE)** is a complete inspection of the mammography unit and all its component parts by a medical physicist.

- **Medical audit** is a method of tracking positive mammography results in order to show a relationship between the pathology results and the radiologist's findings.

- **Medical physicist** is the specialist in radiologic science who is qualified to inspect mammography facilities and oversee the quality assurance program of mammography units.

- **MQSA** (Mammography Quality Standards Act) was passed by the US Congress in 1992 to assure high-quality mammography for early breast cancer detection. Under the standards, all mammography facilities must be accredited, certified, undergo annual MQSA inspections, and prominently display their certificate.

- **Physical science** refers to physics, chemistry, radiation science (including medical physics and health physics), and engineering.

- **Self-referrals** are patients who come for diagnostic testing without a known health care provider.

- **Semester hours** or an academic semester hour should be considered equal to 10 class hours and thus 10 continuing education credits. This 10:1 ratio is useful when combining academic credit and continuing education units to determine whether a requirement is met; however, if the facility can document that it took a different number of class hours than 10 (e.g., 12, 14, and so on) to earn 1 semester hour of credit, then the actual number of class hours should be used in making the conversion.

PURPOSE OF THE MAMMOGRAPHY QUALITY STANDARDS

The US Mammography Quality Standards Act (MQSA) was enacted on October 27, 1992 to establish minimal national quality standards for mammography. After the act was passed, the US Congress authorized the US Food and Drug Administration (FDA) to develop and implement the MQSA regulations. Initially, the interim regulations were passed in December 1993 and became effective in February 1994. Then, in 1995, the FDA began enforcing MQSA with an inspection program. The FDA website publishes a standardized guidance system, which was developed to provide the public with clarifications or comments on the final regulations. The guidance system provides multiple paths that a facility can follow to achieve compliance with the standards. This allows greater flexibility, based on the individual facility, in complying with the law. Included in the guidance are words such as "could," "may," "can," and "recommend," indicating that the facility cannot be cited for failure to follow the guidance; however, sometimes the regulation states "must," "required to," or "shall," here indicating that failure to follow the guidance can lead to fines or citations.

The MQSA established quality standards for mammography equipment and practice, including quality assurance and quality control protocols. The quality control ensures that equipment performance and testing are conducted daily, weekly, monthly, semiannually, and annually by both the radiologic technologist and the physicist. The inspection program also enforced initial and continuing education qualifications for the radiologic technologist, the interpreting physician, and the medical physicist. Other components of the standards are the medical audit and outcome analysis records; the regulation governing medical records, including mammography reports and radiographs, which mandates how long records should be kept; specific requirements for mammography reporting; and notification of results to the patients.

By law, the MQSA inspection program is supported by inspection fees. Inspection fees are required to pay for the cost associated with the scientific, administrative, and data system support for the mammography inspection program. Inspection fees are determined by the FDA or the state certifying agencies.

INTERIM REGULATION

Interim regulations refer to the regulations entitled "Requirements for Accrediting Bodies of Mammography Facilities," published by the FDA on December 21, 1993, and amended on September 30, 1994.

> The US Mammography Quality Standards Act (MQSA) was enacted on October 27, 1992 to establish minimal national quality standards for mammography.

These regulations established the standards that had to be met by mammography facilities to lawfully operate between October 1, 1994 and April 28, 1999.

FINAL REGULATIONS

The final regulations were issued on October 28, 1997 and became effective on April 28, 1999. The final MQSA regulations are national quality standards for mammography services. They were written by the FDA and are based on the MQSA of 1992 and the MQSA reauthorization of 1998. Although there have been amendments to the final regulations since then, the basic contents have remained the same.

ACCREDITATION AND CERTIFICATION

Before a facility can legally perform mammograms in the United States, it must first be accredited, then certified. Under the MQSA, to be accredited, the facility must contact one of the four accreditation bodies (ABs) authorized under the FDA to accredit mammography facilities, conduct facility inspections, and enforce the MQSA standards. These are the American College of Radiology (ACR), the State of Arkansas, the State of Iowa, and the State of Texas (Box 12–1).

State accreditation bodies (SABs) can accredit only those facilities that are located in their respective states. State laws may also require facilities to have state accreditation or state certification. After accepting a facility's application for review, the AB will notify the FDA or the certifying state. The certifying agency will then issue an MQSA provisional certificate or a 45-day interim notice, which would then allow the facility to begin performing mammography as soon as possible.

Certification is a process that is separate from accreditation. The FDA has established procedures for the application, approval, evaluation, and withdrawal of certification. Certification is administered by a certifying agency (FDA or an FDA-approved certifying state). The FDA and the states as certifiers (SACs) are the only organizations authorized to issue MQSA certification. The FDA will not certify facilities in approved certifying states, and certifying states can only certify facilities within their state borders. The FDA-approved certifying states are the State of Illinois, the State of Iowa (August 2004), the State of Illinois (August 2004), the State of South Carolina (April 22, 2005), and the State of Texas (May 27, 2008). The certifying agency will not give facilities an original MQSA certificate until they have successfully completed the accreditation process (Box 12–2).

A provisional certificate is generally valued for 6 months, although under certain instances the facility can qualify for a one-time 90-day extension of the provisional MQSA certificate. During these 6 months, the facility must collect clinical images and other data needed to complete the accreditation process. According to the regulation, the AB must review the facility's equipment, personnel (interpreting physician, radiologic technologists, and medical physicists), and practices. After the review, if the AB can establish that the facility meets the quality standards under MQSA, it will be accredited. If the facility does not

---

Box 12–1. Table of State Accreditation Bodies and Contact

- American College of Radiology (ACR) Mammography Accreditation Program (800-227-644)
- Arkansas Department of Health Division of Radiation Control and Emergency Management (501-661-2301)
- Iowa Department of Health Bureau of Radiological Health (515-281-3478)
- Texas Department of Health Bureau of Radiation Controls (512 834-6688, ext. 2246)

Source: http://www.fda.gov/Radiation-EmittingProducts/MammographyQualityStandardsActandProgram/Regulations/ucm110823.htm.

Box 12–2. Table of Certifying States

- State of Illinois, Office of Radiation Safety, 1035 Outer Park Drive Springfield, IL 62704 (217-785-9974)
- State of South Carolina, Department of Health and Environmental Control, Bureau of Radiological Health, 2600 Bull Street Columbia, SC 29201 (803-545-4400)
- State of Iowa, Bureau of Radiological Health Iowa, Department of Public Health 401 SW 7th Street, Suite D Des Moines, IA 50309 (515-281-3478)
- State of Texas, Texas Department of State Health Services, Mammography Certification Program, P.O. Box 149347, Mail Code 2835 Austin, Texas 78714-9347 (512-834-6688, ext. 2247)

Source: http://www.fda.gov/Radiation-EmittingProducts/MammographyQualityStandardsActandProgram/Regulations/ucm110823.htm.

complete the accreditation process within the 6-month time period, the facility must cease performing mammograms.

Facilities can also get a 45-day interim notice and interim accreditation. This is issued when a facility's MQSA certificate has expired or is about to expire. These 45-day interim notices are only given after a facility has been granted interim accreditation by the AB but there is a delay in completing the reaccreditation because a final decision has not been made, or if the accreditation agency has already given the facility a positive accreditation decision but the facility has not yet received the original MQSA certificate.

LAWFUL MAMMOGRAPHY SERVICES IN THE UNITED STATES

A facility that has completed the accreditation process and certification review process will be considered fully certified and will be issued an MQSA certificate that is valid for 3 years. These facilities can legally perform mammograms in the United States. To maintain certification the facility must be inspected annually by FDA or state inspectors; have specific mammography equipment that is periodically surveyed by a medical physicist; employ specially trained personnel to perform mammograms and conduct regular quality control tests; have a quality assurance program; have a system in place to track mammograms that require follow-up studies or biopsies; and have a tracking system in place to obtain biopsy results.

QUALITY ASSURANCE REGULATIONS OF MQSA

Under the MQSA, each facility must establish and maintain a quality assurance (QA) program and quality control (QC) program to ensure the safety, reliability, clarity, and accuracy of mammography services performed at the facility. The regulations for analog mammography units specify testing details, including the frequency and specific equipment that must be tested. The regulations and testing details for digital mammography units were manufacturer specific. However, in February 2016, the FDA approved the new *ACR Digital Mammography Quality Control Manual* and digital mammography QC phantom as an alternative for all digital mammography units. These QC standards cannot be used for mammography systems with advanced capabilities, such as tomosynthesis or contrast enhancement. The MQSA quality assurance requirements are discussed in detail in Chapter 12.

Clinical Images

The clinical images in digital mammography can be submitted as hard copy or digital images and must be close the true image size, without magnification or minimization. The assessment will be on the following:

- Positioning
- Compression
- Exposure
- Contrast
- Sharpness

> Under the MQSA, each facility must establish and maintain a quality assurance (QA) program and quality control (QC) program to ensure the safety, reliability, clarity, and accuracy of mammography services performed at the facility.

- Noise
- Artifacts
- Labeling

The images should be the facility's best work and must include two mediolateral oblique (MLO) and two craniocaudal (CC) projections. *The entire breast must be imaged in a single exposure on each projection and all must be "negative" images:*

BI-RADS 1 ("nothing to comment on . . . breasts are symmetrical . . . no masses, architectural disturbances, or suspicious calcifications")
BI-RADS 2 ("benign") with prior approval and report

All images must be reviewed and approved by supervising radiologist. The FDA does not allow images from models or volunteers.

Images must be one dense and one fatty (Fig. 12–1A–H). The clinical and phantom images *from each unit must be taken within 30 days of each other and must be within the time period shown on the laser film printer QC chart. All clinical images should be clearly dated.*

Fatty image criteria:

BI-RADS Composition Category 1: composed almost entirely of fat

BI-RADS Composition Category 2: scattered fibroglandular densities

Dense image criteria:

BI-RADS Composition Category 3: heterogeneously dense

BI-RADS Composition Category 4: extremely dense

Fatty Categories

BI-RADS Composition Cat 1
Composed Almost Entirely of Fat

A

BI-RADS Composition Cat 2
Scattered Fibroglandular Densities

B

Dense Categories

BI-RADS Composition Cat 3
Heterogeneously Dense

C

BI-RADS Composition Cat 4
Extremely Dense

D

Figure 12–1. **(A to D)** Schematic diagram of fatty and dense categories. **(A)** Composed almost entirely of fat. **(B)** Scattered fibroglandular densities. **(C)** Heterogeneously dense. **(D)** Extremely dense. **(E to H)** Mammograms showing fatty and dense categories. *Continued*

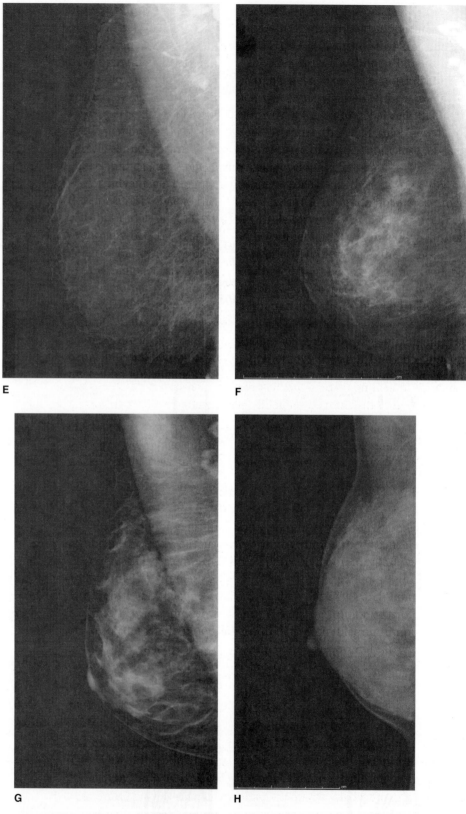

Figure 12–1. (*Continued*) **(E)** BI-RADS Composition Category 1—composed almost entirely of fat. **(F)** BI-RADS Composition Category 2—scattered fibroglandular densities. **(G)** BI-RADS Composition Category 3—heterogeneously dense. **(H)** BI-RADS Composition Category 4—extremely dense.

Levels of Findings or Noncompliance

Every year a facility must renew the FDA or SAC certification through an inspection process. There are a number of possible levels of findings, also called levels of noncompliance or levels of observations resulting from an MQSA inspection (Box 12–3 and Table 12–1).

Level 1 observation represents the most serious noncompliance with the MQSA standards. A level 1 observation indicates that the inspector found one or more deviations from the MQSA standards that may seriously compromise the quality of mammography services offered by the facility.

For a level 1, repeat level 1, or repeat level 2 observation, the facility must correct the problem(s) as soon as possible. Facilities must not wait to receive a warning letter; the facility should send a written response to the FDA within 15 days after the inspection.

A level 2 observation indicates that the facility's performance is generally acceptable; however, the inspector did find one or more deviations from the MQSA standards that may compromise the quality of mammography services offered by the facility.

For a level 2 or repeat level 3 observation, the facility must correct the problem(s) as soon as possible. The facility should respond to the FDA within 30 days after the inspection.

A level 3 observation represents minor deviations from the MQSA standards indicating that the facility's performance is generally satisfactory. Although satisfactory, the facility should adopt policies to correct these minor deviations.

For a level 3 observation, the facility should correct the problem(s) as soon as possible, with a recheck at the next annual inspection.

If, after receiving a level 2 or more severe observation, the facility's response to the inspector's observation appears adequate, the FDA may decide that no further action is needed and will notify the facility that the matter is closed.

If the facility's written response to the inspector's observation is not sufficient or the facility fails to respond, the FDA may need to reinspect the facility to verify that the problems have been corrected.

The FDA will usually reinspect facilities after sending warning letters to see if serious problems continue. Continuing violations at mammography facilities could result in the FDA taking regulatory action, such as directed plans of correction, civil money penalties, or suspension or revocation of the MQSA certificate.

Withdrawal of Accreditation

After a facility is accredited and certified, and before that facility can stop performing mammograms, it should inform its AB and its state radiation control program. The facility must also transfer each patient's medical record (both the original radiographs and the reports) to one of the following: the facility where the patient will be receiving future care, another physician or health care provider, or the patient. The facility must also make reasonable attempts to contact and notify each patient that the facility is no longer performing mammograms. If the facility is not performing mammograms, but it is still in existence at the site, it can choose to keep the patient's records on site. Patients, however, can request a transfer of their personal records. Facilities can also store the

---

Box 12–3. Levels of Findings

- Level 1: Serious deviations from MQSA standards that could severely compromise mammography service
- Level 2: One or more deviations that may compromise the quality of mammography services
- Level 3: Acceptable findings with minor deviations
- No findings: Facility is in full compliance with the MQSA standards
- Repeat findings: Uncorrected findings or deviations that have recurred since the facility's last MQSA inspection

TABLE 12–1. Sample Level of Findings

CATEGORY: COMMUNICATION OF RESULTS

LEVEL 1

No system in place to provide medical reports (to referring health care providers or self-referred patients) within 30 days

No system in place to provide lay summaries (to all patients) within 30 days

No system to communicate serious (suspicious or highly suggestive of malignancy) cases to health care providers and patients as soon as possible (typically interpreted to be 3-5 days)

CATEGORY: EQUIPMENT EVALUATION

LEVEL 1	LEVEL 2	LEVEL 3
The facility does not have an annual medical physicist survey for two successive years	Time interval between the current and previous survey exceeds 14 months Time interval between inspection date and the most recent survey exceeds 14 months Unsigned report or report without the ID of the person who conducted or supervised the survey Incomplete survey, e.g., any of the following tests are missing or incomplete: Spatial resolution AEC performance Phantom image Average glandular dose System artifacts New modality QC tests	Incomplete survey (e.g., any of the following tests/tasks are missing or incomplete): Pass/fail list in the report Recommendations for failed items Physicist's evaluation of the technologist's QC tests Any of the tests not listed under Level 2 e.g., collimation assessment/kVp accuracy/kVp reproducibility/beam quality (HVL) measurement/uniformity of screen speed/ radiation output/decompression (compression release) Not taking timely corrective actions for items that failed in the survey

CATEGORY: QUALIFICATION

LEVEL 1	LEVEL 2
For interpreting physicians missing Current State license Board certificate from an FDA-approved board or specific training in mammography interpretation	**For interpreting physicians missing** 40 CME credits in mammography if the person qualified under the interim regulations (60 category I CME credits under the final regulations) Initial experience (240 examinations in 6 months) New modality training (8 hr, if applicable) Continuing experience (960 examinations in 24 months) Continuing education (15 category I CMEs in 36 months)
For radiologic technologists missing Current State license or board certificate from an FDA-approved board	**For radiologic technologists missing** 40 contact hours (final regulations) or specific training (interim regulations) in mammography New modality training (8 hr, if applicable) Continuing experience (performing 200 mammograms in 24 months) Continuing education (15 CEUs/CMEs in 36 months)
For medical physicists missing Current State license or State approval letter or board certificate from an FDA-approved board Masters or higher in a physical science or Bachelors in a physical science if obtained prior to 4/28/99	**For medical physicists missing** Training in conducting surveys (20 contact hours for Masters or higher degree, and 40 contact hours for Bachelors degree) Experience in conducting surveys (1 facility and a total of 10 units for Masters or higher degree, and 1 facility and a total of 20 units for Bachelors degree) New modality training (8 hr, if applicable) Continuing experience (surveying 2 facilities and a total of 6 units in 24 months) Continuing education (15 CEUs/CMEs in 36 months)

TABLE 12–1. Sample Level of Findings (*Continued*)

CATEGORY: MEDICAL AUDIT AND OUTCOME ANALYSIS
LEVEL 2

Not ALL positive mammograms are entered in system

Biopsy results are not present (or no attempt is made to get them)

No audit (reviewing) interpreting physician has been designated

Analysis not done annually

Analysis not done separately for each individual

Analysis not done for the facility as a whole

CATEGORY: DAILY PROCESSOR QC

LEVEL 1	LEVEL 2
Not conducting the test and recording data and/or corrective actions for	**Not conducting the test and recording data and/or corrective actions for**
30% or more of clinical processing days in any month	10% or more but under 30% of clinical processing days in any month
5 or more consecutive days of clinical processing in a year	2–4 consecutive days of clinical processing in a year
Processing mammograms when either of MD, DD, or B + F is outside its action limits for 5 or more working days in a year	Processing mammograms when either of MD, DD, or B + F is outside its action limits for 2–4 working days in a year
	Not taking timely corrective action for failed items

CATEGORY: WEEKLY PHANTOM TESTING

LEVEL 1	LEVEL 2
Not conducting a phantom image test for 4 or more weeks in a consecutive 12-week working period	Not conducting a phantom image test for 2–3 weeks in a consecutive 12-week working period
A score of less than three fibers, less than two speck groups and /or less than two masses.	Failure to conduct a phantom image test at clinical settings
	Having a background optical density less than 1.2 at phantom center
	Not taking timely corrective action for failed items
	Not conducting a performance verification test after a mobile unit is moved

The most serious violation is a level 1 violation. Level 3 violations are not considered as serious as levels 2 and 1 and will not require a facility response but the FDA expects facilities to take corrective action as soon as possible on all violations. Missing any required documentation will result in a citation.

- No findings indicate that the facility met all the MQSA requirements and is in compliance with the MQSA standards.
- A facility can also receive a "repeat finding" if the MQSA inspection notes findings have not been corrected or have recurred since a facility's last MQSA inspection.[1]

medical records of patients in a hospital or make appropriate warehouse storage as long as former patients are notified and are aware of the means of obtaining their radiographs and records. If the facility is unable to store the records elsewhere, then the facility will remain responsible for the records. Under MQSA, facilities will not be held responsible for records (radiographs and reports) before October 1, 1994; however, state and local regulations may impose other restrictions. After closing, the facility must remove ACR and MQSA certificates from any display.

Suspension or Revocation

Generally, the facility's certificate is revoked if the facility is unwilling or unable to correct violations that were the basis for suspension or if the facility used or engaged in fraudulent documentation to obtain or continue certification. The FDA can also suspend or revoke the facility's certificate if it has failed to comply with any of the standards; has failed to comply with a reasonable request of the agency or the AB for records, information, reports, or material that the FDA believes is necessary; or has violated a reasonable request for permission to inspect the facility by a designated FDA inspector, a state inspector, or the AB representative. If a facility receives a suspension, the suspension will remain in effect until all violations or allegations are proved false, are not substantiated, or have been satisfactorily corrected.

Before the suspension or revocation of the certificate can take effect, the FDA must notify the owner or operator of the facility, providing them with the opportunity for an informal hearing not later than 60 days from the effective date of the suspension; however, the FDA may suspend the certificate of a facility before holding a hearing if the FDA determines that the violations present a serious risk to human health or if there is proof that the violations are intentional or fraudulent.

Adverse Reporting

As part of the MQSA, Congress mandated that there be annual reporting of adverse actions taken against mammography facilities. Congress stipulated that the report be made available to physicians and the general public and that it should include information that is useful in evaluating the performance of mammography facilities nationwide. This information allows patients and providers to make informed decisions about choosing a mammography facility. The FDA.gov website now posts information on facilities to provide this notification to the public in a timely manner. The information will include the following information on actions taken against mammography facilities against which there was an adverse action:

- Facility name and address
- A description of the adverse event
- The action taken
- The corrective action
- The current status of the facility

Digital Mammography

If the facility has an approved analog unit, clinical use of a full-field digital mammography (FFDM) can begin under current MQSA certification under the following circumstances:

- There is an approved AB for the unit.
- The medical physicist reports a pass on the mammography equipment evaluation (MEE) for the unit.
- The MEE will include an evaluation of the acquisition workstation (AWS) and review workstation (RWS) display systems and the laser printer (if applicable).
- The facility sends a completed application for the new unit (with the MEE results) to the AB.

When approved, the AB will notify the FDA or SAC of the application within 2 business days. The Centers for Medicare and Medicaid Services (CMS) will not reimburse for examinations performed on an

FFDM unit until the FDA has received notification that the facility has applied for accreditation of the new FFDM unit (Box 12–4).

If the facility does not have an approved analog unit, and there is an approved AB for the FFDM unit, clinical use cannot begin until satisfactory completion of all the AB's procedures for accrediting that unit.

Facilities with full field digital mammography-only (FFDM-only) units do not need to apply to the FDA for an extension of their MQSA certification to include the use of an FFDM-only unit. Currently all accrediting bodies (ABs) are approved by the FDA to accredit all FFDM-only units and all FFDM-only units must be accredited by one of these ABs.

If the facility is not already certified, patients cannot be imaged until receipt of the MQSA certificate or interim notice from the FDA or the state certifying agency.

Within 9 months of imaging, the AB needs results of tests obtained during the first 6 months of clinical imaging after beginning FFDM. All QAs and QCs must follow manufacturer's guidelines. The AB also needs a list of all personnel working with digital images (i.e., interpreting, performing, and surveying).

Box 12–4. Approved Full-Field Digital Mammography Units

The US Food and Drug Administration (FDA) approved, cleared, or accepted the following full-field digital mammography (FFDM) and digital breast tomosynthesis (DBT) units for use in mammography facilities as indicated by date:

- GE Senographe Pristina with Digital Breast Tomosynthesis (DBT) Option on 03/03/17
- Fujifilm ASPIRE Cristalle with Digital Breast Tomosynthesis (DBT) Option on 1/10/17
- Siemens Mammoth Fusion on 09/14/15
- Siemens Mammomat Inspiration With Tomosynthesis Option (DBT) System on 4/21/15
- GE SenoClaire Digital Breast Tomosynthesis (DBT) System on 8/26/14
- Fuji Aspire Cristalle Full-Field Digital Mammography (FFDM) System on 03/25/14
- Siemens Mammomat Inspiration Prime Full-Field Digital Mammography (FFDM) System on 06/11/13
- iCRco 3600M Mammography Computed Radiography (CR) System on 04/26/13
- Philips MicroDose SI Model L50 Full-Field Digital Mammography (FFDM) System on 02/01/13
- Fuji Aspire HD Plus Full-Field Digital Mammography (FFDM) System on 09/21/12
- Fuji Aspire HD-s Full-Field Digital Mammography (FFDM) System on 09/21/12
- Konica Minolta Xpress Digital Mammography Computed Radiography (CR) System on 12/23/11

(continued)

Box 12–4. *(Continued)*

- Agfa Computed Radiography (CR) Mammography System on 12/22/11
- Fuji Aspire Computed Radiography for Mammography (CRM) System on 12/8/11
- Giotto Image 3D-3DL Full-Field Digital Mammography (FFDM) System on 10/27/11
- Fuji Aspire HD Full-Field Digital Mammography (FFDM) System on 9/1/11
- GE Senographe Care Full-Field Digital Mammography (FFDM) System on 10/7/11
- Planmed Nuance Excel Full-Field Digital Mammography (FFDM) System on 9/23/11
- Planmed Nuance Full-Field Digital Mammography (FFDM) System on 9/23/11
- Siemens Mammomat Inspiration Pure Full-Field Digital Mammography (FFDM) System on 8/16/11
- Hologic Selenia Encore Full-Field Digital Mammography (FFDM) System on 6/15/11
- Philips (Sectra) MicroDose L30 Full-Field Digital Mammography (FFDM) System on 4/28/11
- Hologic Selenia Dimensions Digital Breast Tomosynthesis (DBT) System on 2/11/11
- Siemens Mammomat Inspiration Full Field Digital Mammography (FFDM) System on 2/11/11
- Carestream Directview Computed Radiography (CR) Mammography System on 11/3/10
- Hologic Selenia Dimensions 2D Full Field Digital Mammography (FFDM) System on 2/11/09
- Hologic Selenia S Full Field Digital Mammography (FFDM) System on 2/11/09
- Siemens Mammomat Novation S Full Field Digital Mammography (FFDM) System on 2/11/09
- Hologic Selenia Full Field Digital Mammography (FFDM) System with a tungsten target in 11/2007
- Fuji Computed Radiography Mammography Suite (FCRMS) on 07/10/06
- GE Senographe Essential Full Field Digital Mammography (FFDM) System on 04/11/06
- Siemens Mammomat Novation DR Full Field Digital Mammography (FFDM) System on 08/20/04
- GE Senographe DS Full Field Digital Mammography (FFDM) System on 02/19/04
- Lorad/Hologic Selenia Full Field Digital Mammography (FFDM) System on 10/2/02
- Lorad Digital Breast Imager Full Field Digital Mammography (FFDM) System on 03/15/02
- Fischer Imaging SenoScan Full Field Digital Mammography (FFDM) System on 09/25/01
- GE Senographe 2000D Full Field Digital Mammography (FFDM) System on 01/28/00

An MEE must be performed after installation of a new (or used) x-ray unit or processor, after an x-ray unit or processor is disassembled and reassembled at the same or new location, after x-ray tube replacement, after collimator replacement, after filter replacement, after AEC unit or sensor replacement, and after any software or hardware upgrade.

Currently, the FDA-approved ABs are not prepared to accredit three-dimensional (3D) tomography because standards have not been developed for this imaging modality. However, the FDA developed the certificate extension program to cover situations in which a mammography device has been approved for marketing but accreditation standards have not yet been established. The certificate extension program that allows facilities to legally use 3D tomography requires that an FDA-approved AB accredit the two-dimensional (2D) imaging component of the 3D tomography unit first before applying for an extension of their certification to include the use of the DBT portion of the unit.

The exception is the State of Arkansas. The State of Arkansas is now approved by the FDA to accredit both the FFDM and DBT portions of certain DBT units.

Quality Control

An alternative standard for FFDM systems and for systems without advanced imaging capabilities was approved and became effective on February 17, 2016. It has no time limit. The alternative standard allows facilities with digital mammography units to use the new *ACR Digital Mammography Quality Control Manual* as an alternative to the QA program recommended by the individual digital manufacturers. The FDA has determined that the ACR's QC manual is, as required in §900.18(a)(1): Alternative Requirements, "at least as effective in assuring quality mammography" as following the manufacturers' QC manuals.

Printers and Monitors

The US FDA recommends that facilities use only the printers cleared by the FDA Office of Device Evaluation (ODE) for FFDM. However, a facility may use other printers and monitors. Facilities need to ensure that all printers and monitors used by the facility with its FFDM unit comply with a QA program that is substantially the same as that recommended by the FFDM manufacturer and pass the facility's AB phantom and clinical image review process (see Appendix C for more details) (Fig. 12–2).

Monitors

If a monitor/workstation has been approved by the FDA ODE for FFDM, the monitor's QC manual should be followed.

If the monitor has not been approved by the FDA ODE, the facility must follow the quality control guidelines of the FFDM manufacturer.

FDA MQSA regulations state that facilities must check with their monitor or workstation manufacturer for information on the devices' FDA clearance (see Appendix C for more details).

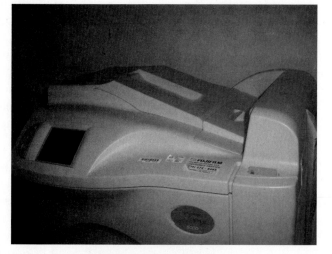

Figure 12–2. Laser Printer

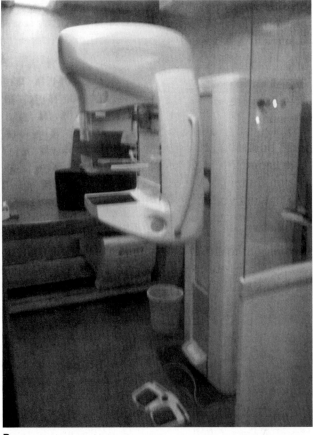

Figure 12–3. **(A)** Interior of a mobile mammography van. **(B)** Philips Micro Dose Mobile Mammography Unit.

Mobile Units

A mobile unit is recognized as a unit on wheels or a unit located in a coach, van, or truck that can provide mammography services to various locations (Fig. 12–3A and B). Under the ACR's policy, each mobile unit must be accredited as a separate facility with unique FDA MQSA identification numbers, and each unit will be issued its own MQSA certificate. Other ABs may accredit mobile or multiple units as belonging to a stationary facility as additional units. In this case, the units have the same identification number as the facility with the request of an additional MQSA certificate. The MQSA certificate on mobile units is set by the state agencies involved.

Accreditation and certification for mobile units are similar to the requirements for stationary units, with a few exceptions. Whenever the unit is moved, the following test must be completed. These tests must be completed after each move even if the unit moves to two locations within the same day:

1. ACR (DM) phantom image quality—after each move and before examining patients
2. Printer QC (for mobile film printers only)—after each move and before printing patient images
3. RWS monitor QC (for mobile RWS only)—after each move and before interpretation
4. Compression thickness indicator—after each move and before examining patients

These tests for the mobile unit must be performed each time the unit is moved:

1. ACR (DM) phantom image quality—after each move and before examining patients
2. Printer QC (for mobile film printers only)—after each move and before printing patient images
3. RWS monitor QC (for mobile RWS only)—after each move and before interpretation
4. Compression thickness indicator—after each move and before examining patients

A

B

C

Figure 12–4. All the patient contact surfaces of the mammography unit must be cleaned between patients. This includes **(A)** face-plate, **(B)** compression plate, and **(C)** detector.

If a problem is detected, corrective action must be taken before clinical radiographs are taken. In addition, the mobile unit is subject to all the QC requirements and frequencies that apply to the stationary mammography systems.

Infection Control

The FDA requires that all facilities establish procedures for cleaning and disinfecting the mammography equipment after contact with blood, other body fluids, or potentially infectious materials (Fig. 12–4). Each facility must document infection control procedures. The mammography unit should also be cleaned and/or disinfected after each patient.

All surfaces in contact with the patient must be wiped clean with a facility-approved disinfectant at the end of each examination.

If the mammography equipment comes into contact with blood or other potentially infectious agents, the mammography personnel must document on a log that infection control procedures were performed.

Xeroradiography

Xeroradiography, described in Chapter 1, has been banned in most of the United States, but ABs will accredit facilities with xeroradiography units if the unit meets the dose limit and other accreditation requirements.

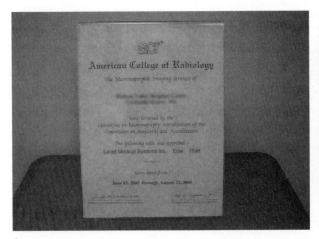

A
B

Figure 12–5. MQSA accreditation and certification proof. **(A)** Accreditation for a specific unit. **(B)** Certification certificate.

Certificate Placement

According to the MQSA regulation, the certificate must be prominently displayed where the mammography patients can easily see it. One suggested place is the patient waiting area. If a facility has more than one waiting area, each area should have an MQSA certificate (Fig. 12–5). Additional certificates can easily be obtained by contacting the FDA. If the facility moves or relocates, the MQSA certificate is still valid, but the facility must notify its AB. Also, any mammography unit or processor that is disassembled and reassembled at the same or a different location must have a mammography equipment evaluation.

Consumer Complaints Mechanism

Each mammogram facility must have a written policy for collecting and resolving patients' complaints. If the facility is unable to resolve a complaint, the patient must be given instructions on how to file complaints with the facility's AB. Complaints can include poor image quality, missed cancers, personnel not meeting applicable training requirements, or failure to send medical or layperson reports. The facility must maintain a record of each complaint for 3 years, and all serious complaints must be forwarded to the FDA, SAC, or a third party.

The complete consumer complaint policy need not be posted. Facilities can satisfy the consumer complaints mechanism by posting the facility's address with the name of a contact person, plus the appropriate email, website, or telephone number, along with the following notice: "If there are any questions or comment about your mammogram, please contact . . . . " Claims submitted by patient or facility must include the patient's name, address, and telephone number; the name and location of the accredited facility where the mammogram was performed; a description of the complaint; and copies of supporting documentation. Complaints must be signed and dated by the patient or facility.

Examples of serious complaints:

- Poor image quality
- Missed cancers
- Personnel not meeting applicable training requirements
- Failure to send medical or layperson reports

Self-Referrals

The FDA recognizes self-referred patients. Self-referred patients are those who come for a mammogram but have no health care provider, whether or not by choice. The MQSA does not require that a facility accept the self-referred patient. If a facility decides to accept self-referral, it must send both a medical mammography report and a report written in lay terms to the patient within 30 days. The facility should also have a system in place whereby if the results are positive or are of concern, the patient can be referred to a health care provider.

Patients who come to a facility without a health care provider but who are willing to accept a health care provider recommended by the facility and understand that their mammogram reports will be sent to that provider are no longer considered self-referred. Instead, they should be treated like referred patients with respect to communication of results. In some cases, the radiologist can also act as the referring physician, as long as he or she accepts responsibility for the patient's medical care and this arrangement is acceptable to the patient.

Record Keeping

Mammography radiographs and the medical records of patients must be kept for a period of not less than 5 years. If the patient has no additional mammograms, the radiographs and medical records must be kept for not less than 10 years. Some state or local laws may require longer storage times. A facility is not required to store copies of the patients' records if the patient requests a permanent or temporary transfer of her records to another medical institution, physician, or herself.

FFDM records can be stored in digital format, the original or lossless compressed full-field digital data, or in hard copy format. All records must retain their interpretation quality.

> **Content of reports:**
> - Patient name plus an additional patient identifier, e.g., date of birth
> - Date of the examination
> - Name of interpreting radiologist
> - An overall final assessment of findings

Content of Records and Reports

All mammographic reports should include the name of the patient. Because patients can have the same or similar names, to avoid mistakes the records should have an additional patient identifier. This could take the form of a hospital medical record number, admission number, or number tracking system unique to the facility. The reports should also include the date of the examination, the name of the radiologist interpreting the mammogram, and an overall final assessment of findings. The assessment findings must follow the assessment guidelines as established by the MQSA and are listed in the next paragraph. Records must be transferred at the request of the patient. Digital records can be stored in hard copy or digital format (e.g., film or CD), but they must be interpretation quality—and must be originals or lossless compressed.

Facilities can charge a fee for the transfer of mammography records, but the charge cannot exceed the total cost for the service. The charges could include the administrative costs associated with logging in the request, retrieving the appropriate images and reports, and packaging and mailing the images; photocopying costs incurred in making copies of reports; or costs associated with having the patient sign a release.

Assessment Categories

In an effort to guide the referring physician and radiologist in the breast cancer decision-making process, the ACR came up with the Breast Imaging Reporting and Data System (BI-RADS), which is a standardized mammographic reporting system that can be used as a coding and assessment system. To ensure recognition of the system, the ACR collaborated with the National Cancer Institute (NCI), the Centers for Disease Control and Prevention (CDC), the Food and Drug Administration (FDA), the American Medical Association (AMA), the American College of Surgeons, and the American College of Pathologists. The BI-RAD system is now well recognized and used by radiologists, physicians, and surgeons across the United States (Box 12–5).

The MQSA also has an assessment category; however, the ACR assessment categories were already in existence before the publication of the final regulations of MQSA. The assessment categories developed by the ACR are so widely recognized, the FDA will accept the BI-RADs categories as long as the category includes a verbal identifier (e.g., BI-RADS 1: negative). All mammographic reports should include at least one assessment, whether from the BI-RAD or the MQSA category (Box 12–6). Standardized assessment categories have reduced the confusion in mammography interpretation.

MQSA Assessment Categories

A "negative" assessment means that there is nothing to comment on. Although a mammogram can be assessed as negative if the interpreting physician is aware of clinical findings or symptoms, such findings should be noted and explained in the mammography report. A "benign" assessment is also negative. Benign assessment can be used to explain findings on the mammograms that are 100% normal. An example could be a wart or mole. A finding that is "probably benign" has a high (≥98%) probability of being benign but the interpreting physician is not absolutely sure. With this finding, the interpreting physician could suggest a short-interval follow-up mammogram. A probable benign finding could be a cyst. A "suspicious" finding is one without the characteristic size, shape, or imaging of breast cancer yet has some probability of being malignant. Malignant-looking calcifications often fall into this category. A finding that is "highly suggestive of malignancy" suggests that the interpreting physician is almost sure that the findings are malignant. Although mammography can show a lesion with a high probability of being malignant, mammography testing will not provide a definitive diagnosis. A patient with this finding would need a biopsy with cytology or histology analysis. A category of "known biopsy-proven malignant" is used for patients with a prior history of positive biopsy; the assessment of "incomplete" is assigned when a comparison mammogram is needed, or it could be assigned if additional workup, such as spot imaging, ultrasound, or other adjunctive imaging, is necessary before the interpreting physician can come to a conclusion. Whenever this category is used, the interpreting physician should state why a category was not assigned and should make recommendations to the patient's physician or health care provider on what additional actions, if any, should be taken.

Regardless of the assessment category, the interpreting physician should address all clinical questions raised by the health care provider even if the assessment is negative or benign.

Communication of Results to the Patient

All facilities must communicate results not only to the physician but also to the patient. Communication to the patient must be in writing as long as the results are written or stated in lay terms that the patient can readily understand. If assessments are "suspicious" or "highly suggestive of malignancy," the facility will make reasonable attempts to ensure that the results are communicated to the patient as soon as possible. "Incomplete, need additional imaging evaluation" should be communicated to the patient as soon as possible. Generally, "as soon as possible" is interpreted to mean 3 to 5 days from the date of interpretation, not from the date of performance of the mammogram. Other results should be sent within 30 days after the mammogram is performed. Even if the facility verbally transmits the results to the patient, the facility must provide, within *30 days of the examination*, a written lay summary. There must be documented evidence of verbal reports, for example, as a log entry in the patient's report. A computerized telephone system cannot be used in place of a written system for communicating mammographic results to patients.

Although the mammography report must be signed by the interpreting physician, the lay summary does not have to be signed and does not have to include a final assessment category

- The results should be sent to the patient within *30 days of the mammographic examination*.
- Concern findings should be communicated to the patient within *5 days from the date of interpretation*.
- Incomplete findings should be communicated as soon as possible.

Computerized reporting systems are accepted by the FDA. The computerized reporting system must be capable of generating printouts for FDA inspection.

Facilities using fax or email should have policies in place to ensure that delivery failures are recognized.

Medical Audit

All mammography facilities must keep medical outcome audits to follow positive mammography results and to correlate pathology results with the interpreting physician's findings. A positive mammogram is identified as one given as final assessment category of "suspicious" or "highly suggestive of malignancy." The medical audit either can be done on a computerized tracking system or using a paper logging system, and a facility may or may not designate a specific staff person to conduct the follow-up of positive mammograms.

To meet the FDA's criteria, the facility must show that there is a system in place to track or obtain pathology results for all their positive cases and to correlate pathology results with the interpreting physician's findings. The facility must also document that the medical audit is reviewed by the interpreting physician. The audit must begin no later than 12 months after date of certification.

The facility should, therefore, collect biopsy data on whether the tissue sample is benign or malignant. A facility does not need to track mammograms recommended for short-term follow-up, ultrasound, or cases assessed "incomplete," or "need additional imaging evaluation," but if an "addendum" or comparison report results in an upgrade to "suspicious" or "highly suggestive of malignancy," the facility should then include this information in the medical audit with the original case.

The medical audit must be reviewed annually, and this review should assess the data for all the interpreting physicians as well as each individual interpreting physician. Statistics from a medical audit outcome can be useful to the facility in tracking their false-positive or false-negative rates. The false-positive rate is an assessment of a result stated positive but later (perhaps after a cytological or histological analysis) proved to have a negative pathology report. The false-negative rate gives the ratio of mammograms read as negative, which are actually positive on the pathology report. Medical audits can be shared with other facilities or be part of a statewide or private system. Facilities can also track the staging and size of tumors, although this information is not required by the FDA.

MQSA and Health Insurance Portability and Accountability Act

Implementation of the Health Insurance Portability and Accountability Act (HIPAA) will not affect the keeping of medical audits. HIPAA regulations allow a mammography facility to release patient information to a MQSA inspector without patient authorization because the MQSA inspectors are performing health oversight activities required by law. Also, HIPAA regulations allow referring physicians, pathology departments, or surgeons to release patient biopsy information to a mammography facility for the purposes of the MQSA medical outcomes audit without patient authorization because such a disclosure is to "a person subject to FDA jurisdiction" and concerns an FDA-regulated activity.

Elements of a mammography medical audit system:

1. A definition of positive mammograms requiring follow-up

2. A method to follow-up positive mammograms

3. A system to attempt to collect pathology results for all biopsies performed

4. A method to correlate pathology results with the final assessment category indicated by the interpreting physicians

5. A method to include any cases of breast cancer among patients imaged at the facility that subsequently became known to the facility

6. A review of medical outcomes audit data for the aggregate of interpreting physicians as well as each individual interpreting physician at least once every 12 months

Section 164.506 of the HIPAA privacy regulations permits release of information for treatment, payment, or health care operation purposes without a specific patient authorization. Consequently, the regulation allows a mammography facility to transfer medical records to another covered entity in most situations without a specific patient authorization. The Office of Civil Rights, the Department of Health and Human Services (DHHS) office with primary responsibility for HIPAA implementation, has also stated that "a covered health care provider may share protected health information with another health care provider for treatment purposes without a business associate contract."

> HIPAA regulations allow a mammography facility to release patient information to a MQSA inspector without patient authorization.

FINAL REGULATION DOCUMENTATION

There are three distinct documentation paths that radiologic technologists, radiologists, and medical physicists can use to satisfy the current MQSA requirements. Specific documentations were accepted before October 1, 1994. Individuals meeting these initial requirements will always be qualified to work independently. If the individual fails to meet any of the continuing education or experience requirements, their original qualification date remains the same.

There are regulations for the period October 1, 1994 to April 28, 1999, and the final regulations cover the period from April 28, 1999 to present.

Currently, the MQSA does not have personnel requirements for individuals performing interventional mammographic procedures.

The term *mammographic modality* currently refers to a technology for radiography of the breast. Examples are screen/film mammography, digital mammography, and digital breast tomosynthesis (DBT) mammography. Since April 28, 1999, technologists, radiologists, and physicists who were trained solely with screen/film mammography are required to obtain 8 hours of training in digital mammography or DBT before they can independently perform mammograms on these units. This rule does not apply to technologists using digital mammography before April 28, 1999. The 8-hour training requirement can include special courses, continuing medical education, or training provided by the manufacturer.

Under MQSA, each manufacturer's DBT system is currently considered a separate new mammographic modality. MQSA defines a mammographic modality as "a technology for radiography of the breast." Because of the technological differences between the various DBT systems, and differences in their FDA-approved Indications for Use (IFU), each manufacturer's DBT system is currently treated as a separate mammographic modality under the MQSA definition. The personnel requirements for new modality training will therefore apply to the following:

- Interpreting physicians qualified to interpret DBT mammograms
- Radiological technologists qualified to perform DBT and the manufacturer-recommended QA tests
- Medical physicists qualified to perform equipment evaluations and/or surveys of DBT mammography units

All the above personnel will need 8 hours of initial training before independently using any new DBT units. However, the FDA Division of Mammography Quality Standards (DMQS) recently changed the training requirements. Under the new MQSA requirements, 8 hours of training obtained on any DBT system or general DBT training is considered sufficient to meet the MQSA new modality requirement of DBT. Training documentation can be as follows:

1. Course certificate or letter clearly indicates that the training included the unique features of a particular system, *or* general DBT training.
2. Signed attestation, using the DMQS recommended form or a form with similar elements, to document training in the unique features of any DBT system or general DBT training.

Requirements for the Radiologic Technologists

Under the final regulations, to perform a mammographic examination independently, the radiologic technologist must:

1. Have a state license to perform general radiographic procedures or have a general certification from an FDA-approved body to perform radiologic examinations.
2. Have completed at least 40 hours of documented training specific to mammography under the supervision of a qualified instructor, including:
   a. Training in breast anatomy and physiology, positioning and compression, QA/QC techniques, and imaging of patients with breast implants
   b. The performance of a minimum of 25 examinations under the direct supervision of a qualified individual.
3. Complete 15 CEUs every 36 months. To be valid, CEU topics must be mammography related.
4. Maintain experience by completing 200 examinations every 24 months.
5. Have at least 8 hours of training in each mammography modality to be used by the technologist in performing mammography examinations.

Technologists failing to meet any of the continuing requirements must requalify before independently performing mammographic examination by:

1. Completing the CEU requirements by bringing the total up to 15 CEUs per 36 months.
2. Meeting the continuing education experience by performing a minimum of 25 mammography examinations under direct supervision.

Mammographic training will not expire. If a technologist earned the ARRT(M) in the past but allowed it to expire, the expired certificates can be accepted as documentation and will still count as 24 hours toward meeting the initial training requirement. However, maintenance of either a state license or the ARRT(R) is necessary to show that general qualifications are being maintained.

Copies of certificates earned or other documentation from the training provider will suffice for initial mammography-specific training. If documentation is not available, proper attestation will be acceptable for records dated up to October 1, 1994. The FDA will continue to accept a limited form of attestation for CME or CEU credits received after October 1, 1994, in certain cases (Table 12–2).

TABLE 12–2. Summary of the Types of Documentation That Radiologic Technologists (RT) May Use to Document Their Initial Qualifications, as Well as Their Continuing Requirements and Requalification Requirements, Prior to MQSA and Under the Interim and Final Regulations

Requirement	Obtained Prior to 10/1/94	Obtained 10/1/94–4/28/99	Obtained After 4/28/99
State licensure	State license/copy with expiration date Confirming letter from State licensing board Pocket card/copy of license	State license/copy with expiration date Confirming letter from State licensing board Pocket card/copy of license	State license/copy with expiration date Confirming letter from State licensing board Pocket card/copy of license
Board certification	Original/copy of current certificate	Original/copy of current certificate	Original/copy of current certificate
(ARRT or ARCRT)	Confirming letter from certifying board Pocket card/copy of certificate	Confirming letter from certifying board Pocket card/copy of certificate	Confirming letter from certifying board Pocket card/copy of certificate
Initial training (~40–h—interim regulations) (40 h–25 supervised examinations—final regulations)	Attestation Letter or other document from training program CEU certificates Letter or other document confirming in-house or formal training ARRT (M) mammography certificate California mammography certificate Arizona mammography certificate Nevada mammography certificate	Letter or other document from training program CEU certificates Letter or other document confirming in-house or formal training Approved RT training courses ARRT (M) mammography certificate California mammography certificate Arizona mammography certificate Nevada mammography certificate	Letter or other document from training program CEU certificates Letter or other document confirming in-house or formal training
Initial mammographic modality-specific training–8 h—final regulations	Attestation for training or experience with investigational units Mammographic modality-specific CEU certificates CEU certificates plus agenda, course outline, or syllabus Confirming letters from CEU granting organizations Letters, certificates, or other documents from manufacturers' or other formal training courses Letter from facility where experience was obtained documenting experience in the new mammographic modality	Attestation for experience with investigational units Mammographic modality-specific CEU certificates CEU certificates plus agenda, course outline or syllabus Confirming letters from CEU granting organizations Letters, certificates, or other documents from manufacturers' or other formal training courses Letter from facility where experience was obtained documenting experience in the new mammographic modality	Mammographic modality-specific CEU certificates CEU certificates plus agenda, course outline, or syllabus Confirming letters from CEU granting organizations Letters, certificates, or other documents from manufacturers' or other formal training courses
Continuing experience (200/24 months—final regulations)	N/A	N/A	Letter, table, facility logs, or other documentation from training program or mammography facility

(continued)

TABLE 12–2. Summary of the Types of Documentation That Radiologic Technologists (RT) May Use to Document Their Initial Qualifications, as Well as Their Continuing Requirements and Requalification Requirements, Prior to MQSA and Under the Interim and Final Regulations (*Continued*)

Requirement	Obtained Prior to 10/1/94	Obtained 10/1/94–4/28/99	Obtained After 4/28/99
Continuing education (15 CMEs/36 months)	N/A	CEU certificates Confirming letters from CEU granting organizations Formal training courses Letters, certificates, or other documents from manufacturers' or other formal training courses	CEU certificates Confirming letters from CEU granting organizations Formal training courses Letters, certificates, or other documents from manufacturers' or other formal training courses
Continuing mammographic modality-specific education—final regulations	N/A	Mammographic modality-specific CEU certificates CEU certificates (plus agenda, course outline, or syllabus) Confirming letters from CEU granting organizations Letters, certificates, or other documents from manufacturers' or other formal training courses	Mammographic modality-specific CEU certificates CEU certificates (plus agenda, course outline, or syllabus) Confirming letters from CEU granting organizations Letters, certificates, or other documents from manufacturers' or other formal training courses
Requalification-experience—final regulations—done under direct supervision	N/A	N/A	Letter, table, facility logs, or other documentation from training program or mammography facility (done under direct supervision)
Requalification-education	N/A	CEU certificates Confirming letters from CEU granting organizations Letter or other document confirming in-house or formal training Letters, certificates, or other documents from manufacturers' or other formal training courses	CEU certificates Confirming letters from CEU granting organizations Letter or other document confirming in-house or formal training Letters, certificates, or other documents from manufacturers' or other formal training courses

Source: Mammography Quality Standard Act (MQSA) website. Available at www.fda.gov/chrh/mammography/rohelp.

Requirements for the Interpreting Physician

To independently interpret mammograms, the physician must either qualify as an interpreting physician under the interim regulations before April 28, 1999 or after April 28, 1999, or have documented all of the following requirements:

1. A valid state license to practice medicine
2. Board Certification in Diagnostic Radiology by an FDA-approved body, or 3 months of formal training in mammography

3. Sixty category I CME credits in mammography, with at least 15 obtained in the 3 years immediately before qualifying as an interpreting physician
4. Interpreted 240 mammographic examinations under direct supervision in the 6 months immediately before qualifying as an interpreting physician. If the physician passes the certifying board in diagnostic radiology, the first allowable time in the 6-month period can be anytime in the last 2 years of the residency program.
5. After April 28, 1999, if the physician wants to begin reading and interpreting mammograms produced by a new mammographic modality (e.g., digital mammography or DBT) in which he or she has not been trained, the physician will need to get at least 8 hours of training in the new mammographic modality.

After meeting all the initial requirements, all interpreting physicians must:

1. Maintain a valid state license to practice medicine.
2. Maintain continuing education of 15 category I CME credits each 36 months.
3. Maintain continuing experience by reading 960 examinations each 24 months.

Physicians failing to maintain the continuing requirements must requalify before performing independent mammographic interpretation by:

1. Obtaining continuing education to bring the total up to 15 CME credits every 3 years.
2. Obtaining continuing experience by interpreting 240 examinations under direct supervision or by interpreting a sufficient number under direct supervision, to bring the total to 960 every 24 months.

Each mammographic facility must select one interpreting physician to be the lead interpreting physician. The lead interpreting physician is responsible for ensuring that the facility's QA program meets the requirements of the FDA.

Each mammographic facility must also have a designated interpreting physician as the reviewing interpreting physician. This physician is responsible for analyzing the medical outcome audit. The reviewing interpreting physician may or may not be the lead interpreting physician.

The FDA guidelines allow physicians in training to work at facilities as long as they are under the direct supervision of a qualified interpreting physician (Table 12–3).

Requirements for the Medical Physicist

The medical physicist will independently conduct surveys of mammography facilities and provide oversight of a facility's quality assurance program. Many facilities call on the services of the medical physicist throughout the year to assess a problem or conduct a routine check of the mammographic unit. Under the MQSA, the facility must identify their qualified medical physicist at the time of the MQSA inspection, or they will be subject to citation.

TABLE 12–3. Summary of the Types of Documentation That Interpreting Physicians May Use to Document Their Initial Qualifications, as well as Their Continuing Requirements and Requalification Requirements, Prior to MQSA and Under the Interim and Final Regulations

Requirement	Obtained Prior to 10/1/94	Obtained 10/1/94–4/28/99	Obtained After 4/28/99
State license	State license/copy with expiration date Confirming letter from State licensing board Pocket card/copy of license	State license/copy with expiration date Confirming letter from State licensing board Pocket card/copy of license	State license/copy with expiration date Confirming letter from State licensing board Pocket card/copy of license
Board certification	Original/copy of certificate	Original/copy of certificate	Original/copy of certificate
(ABR, AOBR, or RCPSC)	Confirming letter from certifying board Confirming letter from ACR Listing in ABMS directory	Confirming letter from certifying board Confirming letter from ACR Listing in ABMS directory	Confirming letter from certifying board Confirming letter from ACR Listing in ABMS directory
Formal Training (2 months—interim regulations)	Letters or other documents from US or Canadian residency programs	Letters or other documents from US or Canadian residency programs	Letters or other documents from US or Canadian residency programs
(3 months—final regulations)	Documentation of formal mammography training courses Category I CME certificates	Documentation of formal mammography training courses Category I CME certificates	Documentation of formal mammography training courses Category I CME certificates
Initial medical education	Attestation	Letter from residency program	Letter from residency program
(40 h—interim regulations)	Letter from residency program	CME certificates	CME certificates
(60 h/15 in last 3 yr—final regulations)	CME certificates Letter or other document confirming in-house or formal training	Letter or other document confirming in-house or formal training	Letter or other document confirming in-house or formal training (category I)
Initial experience (240) (any 6-month period—interim regulations)	Attestation	Letter or other document from residency or training program or mammography facility—done under direct supervision	Letter or other document from residency or training program or mammography facility—done under direct supervision
(Last 6 months vs. 6 months in last 2 yr of residency—final regulations)	Letter or other document from residency or training program or mammography facility		

TABLE 12–3. Summary of the Types of Documentation That Interpreting Physicians May Use to Document Their Initial Qualifications, as well as Their Continuing Requirements and Requalification Requirements, Prior to MQSA and Under the Interim and Final Regulations (*Continued*)

Requirement	Obtained Prior to 10/1/94	Obtained 10/1/94–4/28/99	Obtained After 4/28/99
Initial mammographic modality-specific training–8 h—final regulations	Attestation for training or experience with investigational units Mammographic modality-specific CME certificates (category I or II) CME certificates (category I or II) plus agenda, course outline or syllabus Confirming letters from CME granting organizations Letters, certificates, or other documents from manufacturers' or other formal training courses Letter from facility where experience was obtained documenting experience in the new mammographic modality	Attestation for experience with investigational units Mammographic modality-specific CME certificates (category I or II) CME certificates (category I or II) plus agenda, course outline or syllabus Confirming letters from CME granting organizations Letters, certificates, or other documents from manufacturers' or other formal training courses Letter from facility where experience was obtained documenting experience in the new mammographic modality	Mammographic modality-specific CME certificates (category I or II) CME certificates (category I or II) plus agenda, course outline, or syllabus Confirming letters from CME granting organizations Letters, certificates, or other documents from manufacturers' or other formal training courses
Continuing experience (960/24 months)	N/A	Letter, table, facility logs, or other documentation from residency or training program or mammography facility CME certificates (category I or II)	Letter, table, facility logs, or other documentation from residency or training program or mammography facility CME certificates (category I)
Continuing education (15 CMEs/36 months—interim regulations) (15 category I CMEs/36 months—final regulations)	N/A	Confirming letters from CME granting organizations Letters, certificates, or other documents from manufacturers' training courses	Confirming letters from CME granting organizations

(continued)

TABLE 12–3. Summary of the Types of Documentation That Interpreting Physicians May Use to Document Their Initial Qualifications, as well as Their Continuing Requirements and Requalification Requirements, Prior to MQSA and Under the Interim and Final Regulations (*Continued*)

Requirement	Obtained Prior to 10/1/94	Obtained 10/1/94–4/28/99	Obtained After 4/28/99
Continuing Mammographic modality-specific education—final regulations	N/A	Mammographic modality-specific CME certificates (category I or II) CME certificates (category I or II) plus agenda, course outline, or syllabus Confirming letters from CME granting organizations Letters, certificates, or other documents from manufacturers' or other formal training courses	Mammographic modality-specific CME certificates (category I) CME certificates (category I) plus agenda, course outline, or syllabus Confirming letters from CME granting organizations
Requalification-experience—done under direct supervision	N/A	Letter, table, facility logs, or other documentation from residency or training program or mammography facility	Letter, table, facility logs, or other documentation from residency or training program or mammography facility
Requalification-education	N/A	CME certificates (category I or II) Confirming letters from CME granting organizations	CME certificates (category I) Confirming letters from CME granting organization

Source: Mammography Quality Standard Act (MQSA) website. Available at www.fda.gov/chrh/mammography/rohelp.

The FDA recognizes medical physicists certified with specialties in diagnostic radiologic physics or radiologic physics by the American Board of Radiology (ABR) or with a specialty in diagnostic imaging physics from the American Board of Medical Physics (ABMP).

A medical physicist who is state licensed or state approved in one state is qualified to conduct surveys in any other state under MQSA; however, the MQSA permits states to have more stringent requirements than the MQSA standards. If the second state has regulations, policies, guidelines, or some other method that allows it to regulate medical physicists, then, under state law, the medical physicist must meet their requirements in order to practice. States can require the physicist to have state approval to practice within its borders in addition to meeting one of the options under MQSA. The physicist would still be qualified under MQSA, but physicists lacking such approval would be cited under the state regulations.

Attestation typically is not acceptable for licenses or certification, and physicists will need to provide copies of certificates earned or other documentation from the training provider as proof of the initial training qualification. If documentation is not available, proper attestation will be acceptable for records dated up to October 1, 1994. The FDA will continue to accept a limited form of attestation for CME credits received after October 1, 1994, in certain cases.

To qualify under the final regulations, a medical physicist must document all of the following requirements:

1. A State license or approval or have certification
2. A master's degree or higher in a physical science with no less than 20 semester hours in physics
3. A minimum of 20 contact hours of specialized training in conducting mammography facility surveys
4. The experience of conducting surveys of at least one mammography facility and a total of at least 10 mammography units

MQSA allows various types of training to meet the requirement of having specialized training in conducting surveys. These can include CME credits related to the diagnosis or treatment of breast disease, formal academic training, or other types of training programs related to mammography facilities.

In accordance with the MQSA guidelines, physical science means physics, chemistry, radiation science (including medical physics and health physics), and engineering. A master's degree in radiologic technology will also satisfy these guidelines.

An alternative path is available for medical physicists who qualified under the interim regulations and maintained active status of any licensure, approval, or certification required under the interim regulation. Before April 28, 1999, the physicist was required to have the following:

1. A bachelor's degree or higher in a physical science from an accredited institution with no less than 10 semester hours or equivalent of college undergraduate or graduate level physics
2. Forty contact hours of documented specialized training in conducting surveys of mammography facilities
3. The experience of conducting surveys of at least one mammography facility and a total of at least 20 mammography units. Only one survey of a specific unit will be counted within a 60-day period.

Medical physicists cannot begin the training and experience requirements until they have completed their degree requirements. After meeting the initial qualifications, all medical physicists (whether they initially qualified under the interim or final regulations) must meet the continuing qualifications:

1. Continuing education of 15 CEUs every 36 months. Continuing education of the medical physicist must include hours of training appropriate to each mammographic modality evaluated by the medical physicist during his or her surveys.
2. Continuing experience by inspecting two facilities and six units every 24 months
3. Maintaining a valid state license, state approval, or certification

Medical physicists failing to maintain the continuing requirements must requalify before independently conducting surveys of mammography facilities by:

1. Obtaining continuing education to bring the total up to 15 CEUs per 3 years
2. Obtaining continuing experience to bring the total up to two facilities and six units per 24 months under direct supervision (Table 12–4)

TABLE 12–4. Summary of the Types of Documentation That Medical Physicists May Use to Document Their Initial Qualifications, as Well as Their Continuing Requirements and Requalification Requirements, Prior to MQSA and Under the Interim and Final Regulations

Requirement	Obtained Prior to 10/1/94	Obtained 10/1/94–4/28/99	Obtained After 4/28/99
State licensure or approval	State license or approval/copy with expiration date Confirming letter from State licensing board FDA approval letter	State license or approval/copy with expiration date Confirming letter from State licensing board FDA approval letter	State license or approval/copy with expiration date Confirming letter from State licensing board FDA approval letter
Board certification	Original/copy of certificate	Original/copy of certificate	Original/copy of certificate
(ABR or ABMP)	Confirming letter from certifying board Pocket card/copy of certificate Confirming letter from ACR FDA approval letter	Confirming letter from certifying board Pocket card/copy of certificate Confirming letter from ACR FDA approval letter	Confirming letter from certifying board Pocket card/copy of certificate Confirming letter from ACR FDA approval letter
Degree in a physical science—final regulations	Original/copy of diploma Confirming letter from college or university	Original/copy of diploma Confirming letter from college or university	Original/copy of diploma Confirming letter from college or university
(Master's pathway)			
(Bachelor's pathway)	FDA approval letter	FDA approval letter	FDA approval letter
Initial physics education—final regulations	College or university transcripts	College or university transcripts	College or university transcripts
(20 semester hours—Master)	Confirming letter from college or university	Confirming letter from college or university	Confirming letter from college or university
(10 semester hours—Bachelor)	Master or Bachelor degree specifically in physics FDA approval letter	Master or Bachelor degree specifically in physics FDA approval letter	Master degree specifically in physics FDA approval letter
Survey training—final regulations	Attestation		
(20 contact hours—Master)	Letter or other document from training program CME/CEU certificates	Letter or other document from training program CME/CEU certificates	Letter or other document from training program CME/CEU certificates
(40 contact hours—Bachelor)	Letter or other document confirming in-house or formal training Training gained performing surveys FDA approval letter	Letter or other document confirming in-house or formal training Training gained performing surveys FDA approval letter	Letter or other document confirming in-house or formal training Training gained performing supervised surveys FDA approval letter

TABLE 12–4. Summary of the Types of Documentation That Medical Physicists May Use to Document Their Initial Qualifications, as Well as Their Continuing Requirements and Requalification Requirements, Prior to MQSA and Under the Interim and Final Regulations (*Continued*)

Requirement	Obtained Prior to 10/1/94	Obtained 10/1/94–4/28/99	Obtained After 4/28/99
Initial experience— final regulations (one facility—10 units—Master) (one facility—20 units—Bachelor)	Attestation Copy or coversheet of survey Letter from facility or listing from company providing the physics survey services documenting performance of survey done FDA approval letter	Copy or coversheet of survey Letter from facility or listing from company providing the physics survey services documenting performance of survey done FDA approval letter	Copy or coversheet of survey done under direct supervision Letter from facility or listing from company providing the physics survey services documenting performance of survey done under direct supervision FDA approval letter
Initial mammographic modality-specific training—8 h—final regulations	Attestation for training or experience with investigational units Mammographic modality-specific CME/CEU certificates CME/CEU certificates plus agenda, course outline, or syllabus Confirming letters from CME/CEU granting organizations Letters, certificates, or other documents from manufacturers' or other formal training courses Letter from facility where experience was obtained documenting experience in the new mammographic modality	Attestation for experience with investigational units Mammographic modality-specific CME/CEU certificates CME/CEU certificates plus agenda, course outline, or syllabus Confirming letters from CME/CEU granting organizations Letters, certificates, or other documents from manufacturers' or other formal training courses Letter from facility where experience was obtained documenting experience in the new mammographic modality	Mammographic modality-specific CME/CEU certificates CME/CEU certificates plus agenda, course outline or syllabus Confirming letters from CME/CEU granting organizations Letters, certificates, or other documents from manufacturers' or other formal training courses
Continuing experience 2 facilities—6 units/24 months—final regulations	N/A	N/A	Copy or coversheet of survey Letter from facility or listing from company providing the physics survey services documenting performance of survey done
Continuing education CMEs/36 months	N/A	CME/CEU certificates Confirming letters from CME/CEU granting organizations Letters, certificates, or other documents from manufacturers' or other formal training courses	CME/CEU certificates Confirming letters from CME/CEU granting organizations Letters, certificates, or other documents from manufacturers' or other formal training courses

(continued)

TABLE 12–4. Summary of the Types of Documentation That Medical Physicists May Use to Document Their Initial Qualifications, as Well as Their Continuing Requirements and Requalification Requirements, Prior to MQSA and Under the Interim and Final Regulations (*Continued*)

Requirement	Obtained Prior to 10/1/94	Obtained 10/1/94–4/28/99	Obtained After 4/28/99
Continuing mammographic modality-specific education—final regulations	N/A	Mammographic modality-specific CME/CEU certificates CME/CEU certificates (plus agenda, course outline, or syllabus) Confirming letters from CME/CEU granting organizations Letters, certificates, or other documents from manufacturers' or other formal training courses	Mammographic modality-specific CME/CEU certificates CME/CEU certificates (plus agenda, course outline, or syllabus) Confirming letters from CME/CEU granting organizations Letters, certificates, or other documents from manufacturers' or other formal training courses
Requalification-experience—final regulations—done under direct supervision	N/A	N/A	Copy or coversheet of survey done under direct supervision Letter from facility or listing from company providing the physics survey services documenting performance of survey done under direct supervision
Requalification-education	N/A	CME/CEU certificates Confirming letters from CME/CEU granting organizations Letters, certificates, or other documents from manufacturers' or other formal training courses	CME/CEU certificates Confirming letters from CME/CEU granting organizations Letters, certificates, or other documents from manufacturers' or other formal training courses

Source: Mammography Quality Standard Act (MQSA) website. Available at www.fda.gov/chrh/mammography/rohelp.

BREAST CANCER SCREENING TODAY

As it now stands, screening mammography has the potential of not only saving lives but also dramatically increasing the already high health care cost in America; however, there is still a lack of experience on the part of some radiologists and technologists, and women may not be receiving the high-quality mammograms they deserve and are paying for. Mammography is also at a crossroad. Increased malpractice lawsuits because of false-negative readings and increased surgical biopsies for minimally suspicious lesions with their associated costs (namely, physician and surgical consultations, preoperative localization fees, hospital fees, and pathologic evaluation fees) are factors that are adversely affecting the cost of screening mammography.

But despite the numerous challenges to mammography, it remains the number one breast cancer–screening tool. Expert interpreting physicians have shown that mammography can have a sensitivity of 86% and a specificity of 80%. Nevertheless, the radiologist cannot diagnose a cancer that is not present on the radiograph because of improper positioning. Lesions in the tail of the breast and lesions close to the chest wall will be missed if care is not taken to image the entire breast. The quality of the mammography image is the responsibility of the technologist, and improving the image quality is the first step toward improving the true-positive rates of breast cancer detection.

Placing an undue emphasis on the mammographic equipment and processing will not guarantee that a patient is receiving good quality mammograms. The MQSA recommendation for accreditation involves not only the equipment and processing methods but also the technologist, the radiologist, and the medical physicist. Such an accreditation could improve the quality of screening mammography or quality mammograms and could also lead to the reduction in the health care cost because the number of advanced, costly breast carcinomas will be detected earlier and treated at less cost to society.

By using a dedicated mammography machine, by practicing consistent quality control, and with proper positioning, technologists will do their part in improving the overall mammography image quality.

ENHANCING QUALITY USING THE INSPECTION PROGRAM (EQUIP) INITIATIVE

At a September 15, 2017 meeting of the National Mammography Quality Assurance Advisory Committee (NMQAAC) the FDA introduced the concept of Enhancing Quality Using the Inspection Program (EQUIP). On October 27, 2016, the Division of Mammography Quality Standards launched this new initiative to coincide with the 25th anniversary of the passage of the MQSA. This initiative includes three inspection questions to emphasize the importance of daily image quality review, regular facility-wide quality reviews, and the personnel responsibilities for image quality. The main purpose of EQUIP is to address the MQSA requirements for the lead interpreting physicians' (LIPs') oversight of quality, the Interpreting Physicians' (IPs') responsibility to follow facility corrective action procedures if asked to interpret poor quality images, and a facility's responsibility to continuously comply with the clinical image quality standards established by its accrediting body. The first year of EQUIP (2017) is considered an educational year and facilities will not receive citations related to the new EQUIP questions. However, in the second year after implementation if inspection questions are not in place this will be considered a Level 2 citation and will be subject to corrective action. In year 3 and beyond, noncompliant facilities will receive a Level 1 citation and will have to undergo clinical image review by their AB, corrective action, and possibly a warning letter.

EQUIP questions are designed to check the processes that are already in place for image quality feedback and review, and allows a facility to propose and implement individualize processes that will satisfy the EQUIP inspection questions. The aim is to identify image quality issues early and rapidly correct possible errors before they can negatively impact the patient. As a continuous quality improvement process,

EQUIP should benefit both the patients and facilities by keeping mammography image quality high.

MQSA also hopes that the process could reduce the need for facilities to undergo Additional Mammography Reviews by ABs and that the need for mandatory Patient and Provider Notifications ordered by the FDA.

The first question is aimed at making sure the facility has a process in place for giving feedback on any poor-quality images presented for interpretation, and for documenting any corrective action taken, such as repeating the image. Documentation should include regular reviews of sample images from each technologist and each IP. The goal of this daily review is to produce consistent high-quality mammograms with a focus on image quality attributes such as positioning. Inspectors inquire about the process but are not required to review corrective actions. The method of documentation will be at the discretion of the facility.

The second question is an annual review of all images performed by all technologists as well as the quality of images accepted for interpretation by all the interpretation physicians. This question looks at the facility as a whole and considers image quality trends across the facility to identify any necessary improvement. The facility can decide size of the image samples and how the sampling is conducted. The inspection will request documentation of meetings or signed statements indicating that the facility performed the minimum of an annual review.

The third new inspection question assesses the procedure the facility has in place for LIP oversight of QA/QC records, including a review of the frequency of performance of all required tests and determining whether corrective actions were taken when needed. This requirement should also meet Practice Quality Improvement (PQI) requirements for the American Board of Radiology (ABR).

Details on EQUIP Questions

Question1: Quality Assurance – Clinical Image Corrective Action
Does the facility have procedures for corrective action (CA) when clinical images are of poor quality?

The facility must have a mechanism for the IP to provide feedback to RTs or other designated facility personnel when images are of poor quality. The facility must have a mechanism to document corrective action taken and the effectiveness of the corrective action.

Question 2: Clinical Image Quality
Does the facility have procedures to ensure that clinical images continue to comply with the clinical image quality standards established by its accreditation body?

The facility must perform at minimum an annual review of sample mammograms performed by each active RT and sample interpretations by each active IP. Inspectors will ask for documentation that the facility performed a clinical image review at least once since the last inspection. Documentation could include a summary report, signed statement by the LIP, clinical image review records, and memos to RTs and IPs.

Question 3: Quality Control
Does the facility have a procedure for LIP oversight of QA/QC records and corrective actions?

The LIP is responsible for providing oversight of all QA and QC records, including review of the frequency of the QC test and review of any corrective action. The inspector can request verbal answers or the LIP can provide a written attestation.

Information on EQUIP, including videos is available at the FDA website at https://www.fda.gov/Radiation-EmittingProducts/Mammography QualityStandardsActandProgram/FacilityCertificationandInspection/ ucm114134.htm

Summary

- The mammography quality standards were implemented to guarantee that all facilities had an overall management plan to ensure the highest quality images and to enhance patient care. All programs must include a quality control component with the collection and evaluation of data. The Mammography Quality Standards Act (MQSA) was enacted October 27, 1992, and enforcement by FDA began in 1995. All mammography units must be accredited, certified, and inspected. The MQSA has implemented equipment testing and enforced continuing education qualification for technologists, radiologists, and physicists, a medical outcome audit, and regulations governing medical records.

- *Certifying agencies* in the United States are the FDA or FDA-approved states as certifiers (SACs). Certifying agencies will grant a provision certificate valued for 6 months, allowing the facility to collect clinical images and data to complete the accreditation process.

- The FDA or SAC certificate is valued for 3 years. Medicare and Medicaid will only reimburse for mammograms performed at an MQSA-certified facility. The FDA will not certify facilities in approved certifying states, and SACs can only certify facilities within their state borders. The FDA-approved certifying states are Illinois, Iowa, and South Carolina.

- *State accreditation bodies* or the FDA can only accredit specific full-field digital mammography (FFDM) models.

- Accreditation bodies in the United States are the American College of Radiology (ACR) and the states of Arkansas, Iowa, and Texas. The state accreditation bodies can accredit only those facilities that are located in their respective states

- The *clinical images* should be the facility's best work and must include two mediolateral oblique (MLO) and two craniocaudal (CC) projections.

- *The entire breast must be imaged in a single exposure on each projection, and all must be "negative" images.* All images must be reviewed and approved by the supervising radiologist. The FDA does not allow images from models or volunteers. There must be one dense image and one fatty image. The clinical and phantom images from each unit must be taken within 30 days of each other and must be within the time period shown on the laser film printer quality control (QC) chart. All clinical images should be clearly dated.

- *Level of finding or noncompliance:* A facility must renew the FDA or SAC certification through an annual inspection process. There are a number of possible levels of findings/observances resulting from an MQSA inspection:

- Level 1: most serious—will compromise image quality
- Level 2: performance is generally acceptable
- Level 3: minor deviations from MQSA standards
- No findings: facility meets all requirements, "all items in compliance"
- Repeat finding: findings not corrected or reoccurred

- Initial qualification for the *technologist* after April 28, 1999:
  - Complete 40 hours of a mammography course, including the completion of 25 supervised examinations
  - Complete 8 hours of initial mammography modality-specific training
  - State-specific qualifications include state license and board certification (ARRT or ARCRT)
  - The mammography course must include training in breast anatomy and physiology, positioning and compression, quality assurance/quality control techniques, and imaging of patients with breast implants

- Initial qualification for the *radiologist* after April 28, 1999:
  - State license
  - Board certification (ABR, AOBR, or RCPSC)
  - Complete 3 months of formal mammography training
  - Initial medical education of 60 hours, with 15 hours in last 3 years
  - Complete 240 supervised interpretations within the last 6 months or 6 months in last 2 years of residency
  - Complete 8 hours of initial mammography modality-specific training

- Initial qualifications for the *medical physicist* after April 28, 1999:
  - State license or approval
  - Board certification (ABR or ABMP)
  - Degree in physical science
  - Initial physics education: 20 semester hours in the physical sciences
  - Survey training: 20 contact hours—master's degree/40 contact hours—bachelor's degree
  - Initial experience: one facility, 10 units—master's degree/1 facility, 20 units—bachelor's degree
  - Complete 8 hours of initial mammography modality-specific training

- *Continuing education* for the technologist, radiologist, and medical physicist must include 15 continuing education units in mammography during the 36 months immediately preceding the date of the facility's annual MQSA inspection
  - Technologist must perform 200 cases in 24 months
  - Radiologist must interpret 960 mammograms in 24 months
  - Physicist must inspect two facilities or six units within 24 months

- *Key components of MQSA*
  - Personnel requirements—initial training records, new modality-training records, continuing education records, and experience records
  - Certificate placement—the certificate must be prominently displayed where all patients can see it.

- Consumer complaints mechanism—all facilities must have a system in place to collect and resolve serious consumer complaints. The written policy for collecting and resolving the patient's complaints need not be posted; however, instructions on how to proceed with serious unresolved complaints must be posted. The facility must maintain a record of each complaint for 3 years.
- Infection control—facilities should have a policy in place and show evidence of infection control practices.
- Self-referrals—refer to a patient who comes for a mammogram but has no health care provider. A facility can decide to accept or not to accept self-referrals.
- Record keeping and transferring of records—the patient's medical records must be kept for not less than 5 years in either digital or hard copy format. If the patient has not had additional mammograms at the facility, records must be kept for not less than 10 years; however, state or local laws may require longer times. In any record transfer, the patient is allowed to request an original.
- *Content of records*
  - Name of the patient plus an additional patient identifier, such as medical record number
  - Date of examination
  - Name of radiologist interpreting the mammogram
  - Final assessment findings. The assessment categories are necessary for standardization of the mammographic reporting system.
- *Communication of results to the patient*
  - The results must be sent to the physician, and if it is a concern finding (BI-RAD 0, 4 or 5), the health care provider must be notified within 3 to 5 days. Results must also be sent to the patient within 30 days of the mammogram. The patient's results must be written and must be in lay terms.
- *Medical audits*
  - Audits are used to follow positive interpretations and to correlate them with pathology results. The medical audits must track BI-RAD 4 and 5. They must be reviewed annually by the lead interpreting physician. The Health Insurance Portability and Accountability Act (HIPAA) does not affect the medical audit.

EQUIP Initiative – Enhancing Quality Using the Inspection Program

- EQUIP is designed to spot image quality issues early so that they can be rapidly corrected, which is beneficial not only for patients but also for facilities. The initiative includes three component questions.
  - Does the facility have procedures for corrective action (CA) when clinical images are of poor quality?
  - Does the facility have procedures to ensure that clinical images continue to comply with the clinical image quality standards established by the facility's accreditation body?
  - Does the facility have a procedure for lead interpreting physician (LIP) oversight of QA/QC records and corrective actions?

REVIEW QUESTIONS

1. What was the main purpose of the MQSA?

2. What is the basic difference between accreditation and certification?

3. For how long is the provisional certificate valid?

4. Apart from the ACR, name the three other accreditation bodies in the United States.

5. For how long is the certification valid?

6. If a facility receives a "level 2 finding or observation," what does this mean?

7. Before a site can stop performing mammograms, how does the site handle the patient's medical records?

8. If a facility has a certified digital mammography unit, can it legally operate a 3D unit?

9. What test must be performed after moving a mobile?

10. How could a site satisfy the consumer complaints mechanism policy under MQSA?

11. How long must a facility keep the patient's medical records?

12. Name the four basic content categories of a mammography report.

13. Describe the meaning of a "suspicious" assessment category.

14. If the mammogram reports suspicious findings, within what time frame should the physician be notified of the results?

15. What is a medical audit?

16. How have the HIPAA regulations affected the keeping of medical audits under MQSA?

17. Before independently performing mammograms, what is the required minimum number of examinations the technologist must perform under direct supervision of a qualified technologist?

18. To continue interpreting mammograms, what is the minimum number of examinations the interpreting physicians must read each 24 months?

19. To meet continuing qualifications, the medical physicist must complete how many continuing education credits every 36 months?

20. What would be the first step in improving the true-positive rates at any facility?

CHAPTER QUIZ

1. The pathology review is a:

   (A) Method of tracking and reviewing all positive findings

   (B) Process of eliminating all positive findings from the record

   (C) Method of tracking cosmetic intervention

   (D) Process of identifying all mammoplasty

2. Elements considered essential to implementing a viable quality assurance program include:

   (1) Evaluation

   (2) Review

   (3) Tolerance

   (A) 1 and 2 only

   (B) 1 and 3 only

   (C) 2 and 3 only

   (D) 1, 2, and 3

3. The advantages of the MQSA include all of the following *except:*

   (A) Increases efficiency

   (B) Is cost-effective

   (C) Allows manipulation of the final image

   (D) Improves patient satisfaction

4. The patient will not have to authorize the release of biopsy results as required under HIPAA regulation because:

   (A) MQSA inspectors are allowed to authorize the transfer of any patients' records.

   (B) HIPAA regulations allow a person subject to FDA jurisdiction to collect patients' biopsy information for MQSA medical outcome audits.

   (C) Generally, on admission to a hospital or medical facility, the patient will sign a release of records agreement authorizing their medical results transfer.

   (D) Patients' medical results can be shared with other facilities.

5. If a mammography report is highly suspicious, the facility is required to communicate the results to the patient's physician within:

   (A) 30 days

   (B) 1 day of the mammogram

   (C) 3–5 days of the mammogram

   (D) 5 days of the interpretation

6. The BI-RAD category for a negative mammogram is:

   (A) BI-RAD 6

   (B) BI-RAD 5

   (C) BI-RAD 1

   (D) BI-RAD 0

7. Under MQSA guidelines, if the patient does not return to a facility for additional mammograms, the radiographs and medical records must be kept:

   (A) Forever

   (B) Not less than 10 years

   (C) Not less than 5 years

   (D) At least 1 year

8. If there are two or more mammography units at a facility, the MQSA certificate must be placed in the:

   (A) Main mammography room

   (B) Office of the chief technologist

   (C) Waiting area visible to all patients

   (D) Office of the radiologist

9. All technologists are required to complete:

   (A) 15 CEUs in mammography every 2 years

   (B) 15 CEUs in mammography during the 36 months immediately preceding the last inspection

   (C) 5 CEUs in mammography during the 36 months immediately preceding the last inspection

   (D) 15 CEUs in mammography every year

10. The technologist's initial mammography qualifications required by MQSA include:

    (1) Completion of a 40-hour mammography course

    (2) Completion of 25 supervised examinations

    (3) Completion of 8 hours of training specific to digital

    (A) 1 and 2 only

    (B) 2 and 3 only

    (C) 1 and 3 only

    (D) 1, 2, and 3

BIBLIOGRAPHY

Berns EA, Baker JA, Barke LD, et al. Digital Mammography Quality Control Manual. Reston, VA: American College of Radiology; 2016.

US Food and Drug Administration. Policy Guidance Help System. http://www.fda.gov/Radiation-EmittingProducts/MammographyQualityStandardsActandProgram/Guidance/PolicyGuidanceHelpSystem/default.htm#Quality. Accessed July 2016.

US Food and Drug Administration. EQUIP: Enhancing Quality Using the Inspection Program Avialable at: https://www.fda.gov/Radiation-EmittingProducts/MammographyQualityStandardsActandProgram/FacilityScorecard/ucm526238.htm. Accessed July 2017

Venes D, Biderman A, Adler E. *Taber's Cyclopedic Medical Dictionary*. 22nd ed. Philadelphia, PA: F. A. Davis; 2013.

Quality Control Tests—Digital Breast Tomosynthesis

DIGITAL BREAST TOMOSYNTHESIS

Digital breast tomosynthesis (DBT) is a new mammographic modality. Before imaging with DBT, faculties must meet all Mammography Quality Standards Act (MQSA) applicable requirements: (1) personnel must obtain at least 8 hours of DBT training; (2) the unit must undergo a mammography equipment evaluation before use; and (3) the facility must follow the manufacturer's recommended quality control procedures.

All the radiologists, medical physicists, and technologist will need 8 hours of initial training before independently using any new DBT units. Under the new MQSA requirements, 8 hours of training obtained on any DBT system or general DBT training is considered sufficient to meet the MQSA new modality requirement of any DBT unit. Training documentation can be:

1. Course certificate or letter that clearly indicates that the training included the unique features of a particular system, *or*
2. Signed attestation, using the DMQS recommended form or a form with similar elements, to document training in the unique features

Currently, the US Food and Drug Administration (FDA)–approved accreditation bodies are not prepared to accredit three-dimensional (3D) tomography because standards have not been developed for this imaging modality. This situation is similar to when FDA approved full-field digital mammography (FFDM) for marketing, but the accreditation bodies had not yet developed standards and could not accredit FFDM when it first came on the market. FDA developed the certificate extension program to cover situations in which a mammography device has been approved for marketing but in which accreditation standards have not yet been established. The certificate extension program that is currently in effect so that facilities can legally use 3D tomography requires that an FDA-approved accreditation body accredit the two-dimensional (2D) imaging component of the 3D tomography unit. The exception is the State of Arkansas. The State of Arkansas is now FDA approved to accredit both the FFDM and DBT portions of certain DBT units.

Accreditation and Certification Options for Facilities Using a 3D System With Either 2D FFDM Images or 2D Images Generated From the 3D Image Set (i.e., 2D Synthesized Images)

There are two accreditation/certification options:

1. If your facility's practice routinely uses 3D imaging with acquired 2D FFDM images, then you may submit those 2D FFDM images to your accreditation body for accreditation of the 2D component of your unit.
2. If your facility's practice routinely uses 3D imaging with 2D images generated from the 3D image set (i.e., synthesized 2D images), then you may submit those synthesized 2D images to your accreditation body for accreditation of the 2D component of your unit. Your accreditation body may have specific information on the submission of images.

Requests for DBT certification extension need to include all the information listed in the document MQSA Facility Certification Extension Requirements (similar to the documentation of Appendix B) and should be forwarded to:

FFDM and DBT Certification Extension Program
Division of Mammography Quality and Radiation Programs
FDA/CDRH/OCER
10903 New Hampshire Ave., WO66-4528
Silver Spring, MD 20993-0002
Phone: 301-796-5919
Fax: 301-847-8502

Source: US Food and Drug Administration. Available at http://www.fda.gov/Radiation-EmittingProducts/MammographyQuality Standards-ActandProgram.

REIMBURSEMENT FOR DBT IMAGING

Medicare's reimbursement for DBT beginning in 2015 is $30.39 for the professional component and $56.13 when billed globally (using the national Medicare fee schedule in effect January through June 2015) for code 77063 (screening) or G0279 (diagnostic). These payments are made in addition to the regular reimbursement for the associated digital mammogram. This means that documentation of DBT must be included in the report.

The patient must complete either a screening or diagnostic digital mammogram for reimbursement under the new Medicare rules because the codes that the Centers for Medicare and Medicaid Services (CMS) has defined are to be used as add-on codes. There is no way to receive reimbursement from Medicare for a standalone DBT exam when it is performed separately.

MAMMOGRAPHY QUALITY STANDARDS ACT INSPECTION QUESTIONS UNDER THE FINAL REGULATIONS

Facility
Inspection information
Inspector ID # and name
Date (of inspection) (mm/dd/yyyy)
Accomplishing district
Annual inspection type (select one): basic, joint audit, or mentored
Accompanying inspector (if joint audit or mentored inspection type is
  selected)
Inspection time (hours)
On-site (time spent at the facility)
Other (pre- and post-activities)
Total auto-calculated value from on-site and other time entered
Travel time (hours)
Software version

Facility Information
Facility ID
Facility name
(Facility) EIN
(Facility) FEI
Facility type
Facility category (nonfederal or federal)

Certificate
Continuously operating with a valid certificate (y/n)
Displayed (y/n)
Expiration date (mm/dd/yyyy)

Additional Sites (if applicable)
Additional site name
Mailing address

Contacts
Facility Accreditation Contact (full name)
Title
Contact methods: phone, ext/fax/email
Mailing address

Facility Inspection Contact
First name, middle initial, last name
Title
Contact methods: phone, ext/fax/email
Mailing address

Most Responsible Individual
First name, middle initial, last name
Title

Contact methods: phone, ext/fax/email
Mailing address

Billing Contact
First name, middle initial, last name
Title
Contact methods: phone, ext/fax/email
Mailing address

Inspection Report Contact
First name, middle initial, last name
Title
Contact methods: phone, ext/fax/email
Mailing address

Image Output Quality Control (QC)
Processor Performance QC
Processor QC records:
- Done on all days films processed (y/n)
- C/A (before further exams) documented (y/n/NA)

Laser Printer QC
Laser printer QC records:
- Done at least weekly when hard copy printed (NA)
- C/A (before further images) documented (NA)

RWS Monitor QC
RWS Monitor QC Records
- Done at frequency specified by output device manufacturer when clinical images are interpreted (y/n/NA)
- C/A (before further images) documented (y/n/NA)

Medical Records
Site information
Evaluate (y/n)
Evaluation
System (to communicate results) adequate (y/n)
- System to provide medical reports within 30 days (y/n)
- System to provide lay summaries within 30 days (y/n)
- System to communicate serious cases ASAP (y/n)
Random written reports:
- Number of random written reports reviewed
- Number with assessment categories
- Number with qualified interpreting physician identification

Medical Audit and Outcome Analysis
Site information
Evaluate (y/n)
Evaluation
All positive mammograms entered in system (y/n/NA)
Biopsy results present (or attempt to get) (y/n/NA)
An audit (reviewing) interpreting physician designated (y/n/NA)
Analysis done annually (y/n/NA)

Done separately for each individual (y/n/NA)
Done for the facility as a whole (y/n/NA)

Quality Assurance (QA)

Site information
Evaluate (y/n)
Evaluation
- QA personnel assigned (y/n)
- Written standard operating protocols (SOPs) for QC tests (y/n)

S.O.P. for infection control (y/n)
S.O.P. for handling consumer complaints (y/n)

Repeat Analysis QC

Site information
Site name (selected site)
Evaluate (y/n)
Evaluation
Repeat analysis QC is adequate (y/n)
- Done at least quarterly (y/n)
- Evaluation done (y/n)
- C/A documented (y/n)

Units

Unit Evaluation

Information
Unit number prefilled: accreditation body (AB) data
Mobile check box
Room name or number
Serial number
X-ray unit still in use
(No/evaluate records only temporarily out of service/yes)
Removed from service date (mm/dd/yyyy)
Unit type (screen/film, FFDM, computed mammography, DBT)
Manufacturer
Model
AB model
Manufacture date (mm/dd/yyyy)

Evaluation

The x-ray system includes the following:
- Appropriately sized compression paddle(s) (y/n)
- Postexposure display in automatic exposure control (AEC) mode for focal spot (y/n/NA)
- Postexposure display in AEC mode for target material (y/n/NA)

This unit is accredited (y/n/pending/NA)
This unit is new (y/n/NA)
Mammography equipment evaluation (by medical physicist) done (y/n/NA)

Phantom Image Quality Evaluation

Phantom image display method (inspector) (acquisition workstation [AWS]/review workstation [RWS]/hard copy)
Phantom image display method (facility) (AWS/RWS/hard copy)
Phantom used "RMI156" displayed

Evaluation Image 1
- Number of fibers
- Number of fiber artifacts
- Number of speck groups
- Number of specks in last group
- Number of specks artifacts
- Number of masses
- Number of mass artifacts

Image 2
- Number of fibers
- Number of fiber artifacts
- Number of speck groups
- Number of specks in last group
- Number of specks artifacts
- Number of masses
- Number of mass artifacts

Scores Image 1 Calculations
- Fibers score
- Fibers pass/fail
- Specks score
- Specks pass/fail
- Masses score
- Masses pass/fail

Scores Image 2 Calculations
- Fibers score
- Fibers pass/fail
- Specks score
- Specks pass/fail
- Masses score
- Masses pass/fail

Quality Control
Phantom Image QC
Number of operating weeks missing in which test not done at least once
Image taken at clinical (+/-1 kVp) or manufacturer recommended setting (y/n)
C/A (before further exams) documented (y/n/NA)
For mobile units (e.g., van, truck)
- Performance verification after each move (y/n/NA)

Compression Force QC
Compression QC adequate (y/n)
- Done at least semiannually (y/n)
- C/A (before further exams) documented (y/n/NA)

Signal-to-Noise Ratio (SNR)/Contrast-to-Noise Ratio (CNR) QC
CNR QC
- Done at frequency specified by unit manufacturer (y/n)
- C/A (before further exams) documented (y/n/NA)

SNR QC
- Done at frequency specified by unit manufacturer (y/n)
- C/A (before further exams) documented (y/n/NA)

Survey Report
Information
Survey report available (y/n/NA)
Date of previous survey (mm/dd/yyyy)
Date of current survey (mm/dd/yyyy)
Dose value measured by physicist (y/n)
- Dose value (mGy) reported (x.xx)
- C/A taken before resuming clinical use (y/n)
Survey conducted or supervised by:
Action taken (y/n/NA)

Survey Report Part 1
Resolution measurement (y/n)
AEC performance—reproducibility (mAs) (y/n/NA)
AEC performance capability (y/n/NA)
Phantom image (y/n)
CNR (y/n/NA)
SNR (y/n/NA)
Artifact evaluation (y/n)

Survey Report Part 2
Pass/fail list (y/n)
Recommendations for failed items (y/n/NA)
Physicist's evaluation of technologist's QC tests (y/n)
- Processor QC (y/n/NA)
- Laser printer QC (NA)
- RWS QC (y/n/NA)
- Phantom image (y/n)
- CNR (y/n/NA)
- SNR (y/n/NA)
- Repeat analysis (y/n)
- Analysis of fixer retention (y/n/NA)
- Darkroom fog (y/n/NA)
- Screen/film contact (y/n/NA)
- Compression (y/n)
Collimation (y/n)
- X-ray field—light field (y/n/NA)
- X-ray field—image receptor alignment (y/n/NA)
- Compression device edge alignment (y/n/NA)
kVp accuracy (y/n)
kVp reproducibility (y/n)
Beam quality (half-value layer [HVL]) measurement (y/n)
Uniformity of screen speed (y/n/NA)
Radiation output (y/n)
Decompression (y/n/NA)

Personnel
Interpreting Physicians
Information
Status (evaluate/hold)
First name, middle initial, last name
Lead interpreting physician (check box)

Evaluation
Rules qualifying under (final/interim)
(If the inspector selected the "interim" rules):
Initial qualifications under interim rules met (y/n)
- Licensed (y/n)
- Certified or 2 months of training (y/n)
- 40 continuing medical education (CME) hours (y/n)
- Initial experience adequate (y/n)

(If the inspector selected the "final" rules):
Initial qualifications met (y/n)
- Licensed (y/n)
- Certified or 3 months training (y/n)
- 60 category I CME hours (y/n)
- Initial experience adequate (y/n)

Date completed initial requirements (mm/dd/yyyy)
Currently licensed (y/n)
Trained in all applicable mammographic modalities
Trained mammographic modalities (check all that apply): [_]S/F [_] FFDM [_]DBT (y/n)
Continuing experience
- Continuing experience adequate (y/n/NA)
- Number of exams in 24 months
Continuing education
- CME credits adequate (y/n/NA)
- Number of CME credits in 36 months

Technologists
Information
Status (evaluate/hold)
First name, middle initial, last name

Evaluation
Rules qualifying under (final/interim)
If the inspector selected the "interim" rules):
Initial qualifications under interim rules met (y/n)
- Licensed or certified (y/n)
- Training specific to mammography (y/n)

(If the inspector selected the "final" rules):
Initial qualifications met (y/n)
- Licensed or certified (y/n)
- 40 supervised hours of training adequate (y/n)

Date completed initial requirements (mm/dd/yyyy)
Currently licensed or certified (y/n)
Trained in all applicable mammographic modalities
Trained mammographic modalities (check all that apply): [_]S/F [_] FFDM [_]DBT (y/n)

Continuing experience adequate (y/n/NA)
Continuing education
* CEU credits adequate (y/n/NA)
* Number of CEUs in 36 months

Medical Physicists
Information
Status (evaluate/hold)
First name, middle initial, last name

Evaluation
Degree qualifying under (bachelor's/master's (or higher)/none)
(If the inspector selected the "master's (or higher)" degree):
Initial qualifications met (y/n)
* Certified or state licensed/approved (y/n)
* Master's (or higher) degree in a physical science (y/n)
* 20 contact hours of training in surveys (y/n)
* Experience in conducting surveys (y/n)
(If the inspector selected the "bachelor's" degree):
Alternate initial qualifications met before 04/28/1999 (y/n)
* Certified or state licensed/approved (y/n)
* Bachelor's degree in a physical science (y/n)
* 40 contact hours of training in surveys (y/n)
* Experience in conducting surveys (y/n)
(If the inspector selected the "none" degree):
* Currently certified or state licensed approved
Date completed initial requirements (mm/dd/yyyy)
Currently certified or state licensed/approved (y/n/certified)
Trained in all applicable mammographic modalities
Trained in mammographic modalities (check all that apply): [_]S/F [_]
    FFDM [_]DBT (y/n)
Continuing experience adequate (y/n/x)
Continuing education
* CME credits adequate (y/n/NA)
* Number of CME credits in 36 months

Summary
Evaluation
Required personnel documents available (y/n/NA)

Inspection Information
Facility ID
Facility name
Facility address
Inspection ID
Inspection date
Inspection type
Inspector ID and name
Annual inspection type (basic, joint audit, or mentored)
Accompanying inspector ID and name

Noncompliances

Noncompliance statement, level, repeat indicator list of noncompliances

Report Delivery

Delivery method

Date delivered/sent (mm/dd/yyyy)

Policy guidance help system

US MQSA Facility Certification Extension Requirements for Digital Breast Tomosynthesis System

1. **Manufacturer of full-field digital mammography (FFDM) digital breast tomosynthesis (DBT) unit:** _____
2. **Facility status information**
   a. Facility name and US Food and Drug Administration (FDA) facility ID number
   b. FDA certificate expiration date
   c. Current accreditation body for the 2D unit
   d. Accreditation expiration date
   e. Facility contact person for DBT unit
   f. Contact person's title
   g. Contact person's telephone, fax, email
   h. Facility address
   i. Facility owner
3. **DBT Unit Identification**
   a. Machine manufacturer
   b. Machine model
   c. Year of manufacture
   d. Serial number
   e. Accreditation body unit number
4. **DBT Digital Image Receptor Identification** (if interchangeable)
   a. Receptor manufacturer
   b. Receptor model
   c. Year of manufacture
   d. Serial number (if applicable)
5. **Final Interpretation Review Monitor Identification** (if soft copy display is available)
   a. Monitor manufacturer
   b. Monitor model
   c. Year of manufacture
   d. Serial number
6. **Phantom Identification**
   a. Phantom manufacturer
   b. Phantom model

7. **Submit either a hard copy or soft copy three-dimensional (3D) phantom image. Soft copy CD or DVD must be in DICOM (Digital Imaging and Communications in Medicine) format and verified that the image opens properly before forwarding the 3D phantom image to the FDA. (Failure to include a 3D phantom image will delay review of the application.)**

8. **Personnel Qualifications**
   a. Interpreting physicians who are qualified to interpret DBT mammograms
   b. Radiology technologists who are qualified to perform DBT mammography examinations and the manufacturer recommended quality assurance tests
   c. Medical physicists who are qualified to perform equipment evaluations and/or surveys of DBT mammography units

9. ***Complete detailed* report of Mammography Equipment Evaluation (MEE) (must have been conducted in accordance with 900.12(e)(10) within the 6 months before the request for use approval) must be included when submitting application.**
   a. Statement that equipment performance, as required under the following sections of the MQSA final regulation 21 CFR 900.12(b), is met:
      (1) Prohibited equipment
      (2) Specifically designed for mammography
      (3) Motion of tube-image receptor assembly
      (4)(iii) Removable grid (if applicable to the DBT system used)
      (5) Beam limitation and light fields
      (6) Magnification
      (7) Focal spot selection
      (8) Compression
      (9) Technique factor selection and display
      (10) Automatic exposure control
   b. The results of quality control tests as required under the following sections of the MQSA final regulations 21CFR 900.12(e):
      (4)(iii) Compression device performance
      (5)(i) Automatic exposure control performance (if applicable to the DBT system used)
      (5)(ii) Kilovoltage peak accuracy and reproducibility
      (5)(iii) Focal spot condition (resolution)
      (5)(iv) Beam quality and half-value layer
      (5)(v) Breast entrance air kerma and AEC reproducibility (if applicable to the DBT system used)
      (5)(vi) Dosimetry
      (5)(vii) X-ray field/light field/image receptor/compression paddle alignment
      (5)(ix) System artifacts
      (5)(x) Radiation output
      (5)(xi) Decompression (or alternative standards allowed for these requirements)
      (6) Quality control tests—other modalities (facilities must perform all DBT manufacturer-recommended quality control tests, including the medical physicist's tests for soft copy display system)

    c. The results of the phantom image quality tests, including a sample image

    d. If any of the requirements in 8 a, b, or c are not met, submit documentation of successful corrective action

    e. If any of the requirements in 8 a or b are not performed, explain why the requirement is not applicable

    f. Date of the MEE

    g. Name and address of the physicist(s) who performed the MEE

10. DBT Manufacturer's Quality Control Program

    a. Name of the quality control manual

    b. Year published

    c. Revision number, if not the original

    d. Printing number, if not the original

11. Signature of facility contact person for the DBT unit

Qualified Personnel

Interpreting Physicians

PERSONNEL QUALIFICATIONS: INTERPRETING PHYSICIANS WHO ARE QUALIFIED TO INTERPRET DBT MAMMOGRAMS

List the current interpreting physicians who:

(1) Meet all the requirements of 21 CFR 900.12(a)(1) "Mammography Quality Standards; Final Rule" that became effective on April 28, 1999; and

(2) Have 8 hours of initial new-modality training in DBT, either including or supplemented by training in the unique features of the specific manufacturer's DBT system.*

_____

_____

_____

_____

_____

_____

_____

* Supporting documentation for these requirements will be checked during annual MQSA inspections.

Radiology Technologists

PERSONNEL QUALIFICATIONS: RADIOLOGY TECHNOLOGISTS WHO ARE QUALIFIED TO PERFORM DBT MAMMOGRAMS

List the current radiology technologists who:

(1) Meet all the requirements of 21 CFR 900.12(a)(3) "Mammography Quality Standards; Final Rule" that became effective on April 28, 1999; and

(2) Have 8 hours of initial new-modality training in DBT, either including or supplemented by training in the unique features of the specific manufacturer's DBT system.*

_____

_____

_____

_____

_____

_____

_____

* Supporting documentation for these requirements will be checked during annual MQSA inspections.

Medical Physicists

PERSONNEL QUALIFICATIONS: MEDICAL PHYSICISTS WHO ARE QUALIFIED TO PERFORM DBT SURVEYS

List the current medical physicists who:

(1) Meet all the requirements of 21 CFR 900.12(a)(3) "Mammography Quality Standards; Final Rule" that became effective on April 28, 1999; and

(2) Have 8 hours of initial new-modality training in DBT, either including or supplemented by training in the unique features of the specific manufacturer's DBT system.*

_____

_____

_____

_____

_____

_____

_____

* Supporting documentation for these requirements will be checked during annual MQSA inspections.

Lead Interpreting Physician Attestation to Staff Personnel Qualifications

To the best of my knowledge and my belief, the information provided in this document is true and correct. I understand FDA may request additional information to substantiate the statements made in the document. I understand that knowingly providing false information in a matter within the jurisdiction of an agency of the United States could result in criminal liability, punishable by up to $10,000 fine and imprisonment of up to 5 years, or civil liability under MQSA, or both.

Signature (Lead Interpreting Physician)

_____

Print Name _____

Date _____

Source: http://www.fda.gov/radiation-emittingproducts/mammographyqualitystandardsactandprogram/facilitycertificationandinspection/ucm413117.htm

Monitor and Laser Printer Test Procedures

PRINTER AND MONITOR QUALITY CONTROL—IF NO MANUAL IS PROVIDED

The manufacturers of some full-field digital mammography (FFDM) units do not provide instructions on quality control for the printer and the monitor. Instead, the manufacturer will instruct the user to test monitors and printers according to the component's quality control (QC) manual.

In such cases, it becomes the responsibility of the facility to obtain and follow the component's QC manual for all monitors and printers.

Facilities can use the same printer or monitor with FFDM units from different manufacturers. However, if each FFDM manufacturer QC manual requires that the same or equivalent test be done, but at different times or frequencies, the facility will have to perform the test at the more stringent frequency.

If each FFDM manufacturer QC manual requires different but equivalent tests, the facilities may perform only one of the tests at the more stringent frequency.

In both cases, the facility should get input from the medical physicist in the form of a written statement for the facility's QC records, verifying that tests are equivalent.

If each FFDM manufacturer QC manual requires tests that are different and not equivalent, the facility will need to perform each test at the frequency required in the respective FFDM manufacturer QC manual.

FFDM System	QC Manual	Weekly/Daily	Annual	MEE	QC Procedures—Comments
GE–All systems	All	Yes*	No	No	*Per printer mfr. QC manual
Fischer	Rev. 10–10/07	Daily check	No	No	Follow printer mfr. QC manual
Selenia	Rev. 7–8/07	Yes	No	Yes	Follow the Selenia QC manual
Siemens	Rev. 5–4/07	Before clinical use	No	Yes	Follow printer mfr. QC manual
Fuji	3rd Ed.–4/07	Yes	Yes	Yes	Follow applicable printer QC manual

Source: http://www.fda.gov/Radiation-EmittingProducts/MammographyQualityStandardsActandProgram/Guidance/ucm219622.htm

Quality Control Forms

CONTENTS

1. American College of Radiology Digital Mammography Phantom Image Quality
2. Computed Mammography Image Plate Erasure Quality Control
3. Compression Force Quality Control
4. Compression Thickness Indicator Quality Control
5. Facility Quality Control Review
6. Film Printer Quality Control
7. Manufacturer Detector Calibration Quality Control
8. Repeat Analysis (A)—Summary Form Quality Control
9. Repeat Analysis (B)—Tally Sheet
10. Repeat Analysis (C)—Daily Counting Form
11. Acquisition Workstation Monitor Quality Control
12. Review Workstation Monitor Quality Control
13. System Quality Control for Radiologists
14. Viewbox Cleanliness Quality Control
15. Visual Checklist Quality Control

American College of Radiology Digital Mammography Phantom Image Quality (Weekly)

Facility _____ Room identification _____

MAP ID—Unit # _____ Unit manufacturer & model # _____

Year		largest	largest	largest	largest	largest
Date (month & day)						
Tech initials						
Resulting techniques	Image receptor size	largest	largest	largest	largest	largest
	AEC mode					
	Target/filter					
	kVp					
	mAs					
ACR DM Phantom	Artifact					
	Fiber score					
	Speck group score					
	Mass score					
	Overall pass/fail					

P = Pass   F = Fail

Analysis

Scoring		Full Point	Half Point
	Fiber	≥8 mm long	≥5 and <8 mm long
	Specks	4–6 specks	2–3 specks
	Masses	≥¾ border	≥½ and <¾ border

Action Limits	**Required:** ACR DM Phantom Image must be free of clinically significant artifacts Fiber score must be ≥2.0: speck group score must be ≥3.0; mass score must be ≥2.0
	Time frame: Failures must be corrected before clinical use

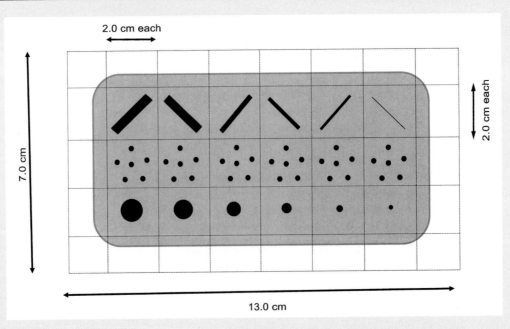

Computed Mammography Cassette Image Plate Erasure Quality Control
(Weekly) (if applicable)

Facility _____ Room identification _____
MAP ID–Unit # _____ Unit manufacturer & model # _____

Month	Jan			Feb		Mar			April		May		Jun	
Date Tech initials														

Month	July		Aug		Sept		Oct		Nov		Dec	
Tech initials												

Action taken (include date and ID or # of IP)

_____

_____

_____

_____

_____

Action Limits	**Required:** Each IP must pass the CM IP erasure process
	Time frame: Failures must be corrected before clinical use

Compression Force Quality Control

(Semiannually)

Facility _____ Room identification _____

MAP ID—Unit # _____ Unit manufacturer & model # _____

Year				
Date (month & day)				
Tech initials				
	Compression Force	Units (circle)	Compression Force	Units (circle)
		lb or daN		lb or daN
Manual fine-adjustment compression force				
Force at least 25lb (11.1 daN) P/F				
Initial power-drive compression force		lb or daN		lb or daN
Force at least 25 lb (11.1 daN) but no more than 45 lb (20.0daN) P/F				
Compression remains at least 25 lb (11.1 daN) throughout typical exposure P/F				
Overall Pass/Fail				

P = Pass   F = Fail   lb =pounds   daN = decanewton

Action Limits	**Required:** Manual fine-adjustment compression force must be at least 25 lb (11.1 daN) Initial power-drive compression force must be at least 25 lb (11.1 daN) but no more than 45 lb (20.0daN) Compression must remain at least 25 lb (11.1 daN) throughout a typical exposure
	Time Frame: Failures must be corrected before clinical images are performed

Compression Thickness Indicator Quality Control
(Monthly)
Facility _____ Room identification _____
MAP ID—Unit # _____ Unit manufacturer & model # _____

Year												
Month	Jan	Feb	Mar	Apr	May	Jun	Jul	Aug	Sept	Oct	Nov	Dec
Date												
Tech initials												
Description of compression thickness indicator problem												
Actual thickness of phantom	_____cm _____mm											
Indicated thickness												
Difference between indicated and actual thickness (indicated – actual)												
Pass/fail												

P = Pass   F = Fail

Action Limits	**Required:** Compression thickness indicator must be accurate to within ±0.5 cm (±5 m) of the actual thickness
	Time frame: Failures must be corrected within 30 days

Facility Quality Control Review

(Quarterly)

Facility _____ Date of QC _____ Overall Pass/Fail _____

1. Review Medical Physics Surveys and Results   Reviewed ____

	Room 1	Room 2	Room 3	Room 4	Room 5
Room ID					
Date of last Medical Physicist (MP) survey					
MP DM QC Test Summary reviewed by radiologist?					
All MP corrective actions completed?					
ACR DM Phantom Average Glandular Dose (mGy)					
Fiber score					
Speck Score					
Mass Score					

2. Review Technologist QC   Reviewed ____

Test	**Frequency**	**Summary comments from last quarter**
1. ACR DM Phantom Image Quality	Weekly	_____ ____

Scores of most recent phantom image		Room 1	Room 2	Room 3	Room 4	Room 5
	Date					
	Fiber score					
	Speck group score					
	Mass score					

Test	Frequency		
2. CR IP Erasure (If applicable)	Weekly	_____	____
3. Compression Thickness Indicator	Monthly	_____	____
4. Visual Checklist	Monthly	_____	____
5. AWS Monitor QC	Monthly	_____	____
6. RWS Monitor QC	Monthly	_____	____
7. Film Printer QC	Monthly	_____	____
8. Viewbox Cleanliness (if applicable)	Monthly	_____	____
9. Facility QC Review	Monthly	_____	____
10. Compression Force	Semiannual	_____	____
11. Manufacturer Detection Calibration (if applicable)		_____	____
12. Repeat Analysis (optional)	As Needed	% repeat ____	

3. Review and verify completion of all "corrective action"   ____
4. Technique chart review for each room   ____
5. Infection control procedures followed   ____
6. Off site RWS and Film Printer(s) QC reviewed   ____
7. Past and future service or service upgrades discussed (if applicable)   ____
8. Past and future state and/or MQSA inspections discussed (if applicable)   ____
9. Past and future ACR accreditation issues discussed (if applicable)   ____

                                                                                          Follow-up confirmed

10. Notable findings during QC meeting _____   ____

11. Items for quality improvement from QC meeting _____   ____

12. Other QC notes: _____   ____

_____   _____   _____

Lead Interpreting Radiologist   Facility Manager   QC Technologist

Action Limits	**Required:** Supervising radiologist and facility manager must review QC quarterly. Technologist and supervision radiologist should review techniques chars at least annually
	Time Frame: Not applicable

Film Printer Quality Control

(Monthly) (if applicable)

Facility _____ Film Printer Location and Identification _____

Workstation for printing _____ Film size _____

Note: Do not change window level setting from acquisition to printing

Year													
Month		Jan	Feb	Mar	Apr	May	Jun	Jul	Aug	Sept	Oct	Nov	Dec
Date													
Tech initials													
ACR DM phantom	Artifact P/F												
	Fiber score												
	Speck group score												
	Mass score												
	Phantom score P/F												
Background	Background OD (outside cavity)												
	Background OD ≥1.6 (P/F)												
Contrast	Cavity OD												
	Background OD (value from above)												
	Contrast												
	Cavity OD— Background OD												
	Contrast ≥0.1(P/F)												
D_{max}	D_{max} OD												
	D_{max} OD ≥3.1 (P/F)												
Monthly Check—MRF Automated test P/F (if available)													
Overall Pass/Fail													

P= Pass   F = Fail   N/A = Not applicable

Action Limits	**Required:** The ACR phantom must be free of significant artifacts Fiber score must be ≥2.0; speck group score must be ≥3.0; mass score must be ≥2.0 Background OD must be ≥1.6 (1.7–2.2 recommended; approximately 2.0 optimal) Contrast (Cavity OD – Background OD) must be ≥0.1 D_{max} must be ≥3.1 (≥3.5 recommended)
	Time Frame: Failures must be corrected before printing clinical images

Manufacturer Detector Calibration Quality Control
(if applicable)

Facility _____ Room identification _____

MAP ID—Unit # _____ Unit manufacturer & model # _____

Year				
Date (month & day)				
Tech initials				
Unit description				

P = Pass   F = Fail

Action Limits	**Required:** Unit must pass the manufacturer's prescribed periodic calibration test
	Time Frame: Failures must be corrected before clinical images are performed

Acquisition Workstation Monitor Quality Control
(Monthly)

Facility _____   Room identification _____

MAP ID—Unit # _____   Unit manufacturer & model # _____

Year		Jan	Feb	Mar	Apr	May	Jun	Jul	Aug	Sept	Oct	Nov	Dec
Month		Jan	Feb	Mar	Apr	May	Jun	Jul	Aug	Sept	Oct	Nov	Dec
Date													
Tech initials													
Monitor Condition P/F (significant findings)													
Test Pattern Image Quality (if applicable)	0%-5% contrast boxes visible?												
	95%–100% contrast boxes visible?												
	Line-pair images distinct (center)?												
	Line-pair images distinct (corners)?												
	Test pattern P/F												
Monthly Check—MRF Automated test P/F (if available)													
Overall Pass/Fail													

P = Pass   F = Fail   N/A = Not applicable

Action Limits	**Required:** Any blemish that interferes with clinical information must be removed Test pattern image quality must pass all visual tests Manufacturer's automated test (if available) must pass specifications
	Time frame: Significant items must be corrected before clinical use All other items must be corrected within 30 days

Review Workstation Monitor Quality Control (Monthly)

Facility _____ RW Location & Identification _____ MAP ID—Unit# _____

Unit manufacturer & model # _____ SN: Right _____ Left _____

Year		Jan		Feb		Mar		Apr		May		Jun		Jul		Aug		Sept		Oct		Nov		Dec	
Month		Jan		Feb		Mar		Apr		May		Jun		Jul		Aug		Sept		Oct		Nov		Dec	
Date																								
Tech initials																								
Monitor		R*	L*	R	L	R	L	R	L	R	L	R	L	R	L	R	L	R	L	R	L	R	L	R	L
Monitor Condition P/F (significant findings)																								
ACR DM phantom	Artifact P/F																							
	Fiber score																							
	Speck group score																							
	Mass score																							
	Phantom score P/F																							
Test Pattern Image Quality (if applicable)	0%-5% contrast boxes visible?																							
	95%–100% contrast boxes visible?																							
	Line-pair images distinct (center)?																							
	Line-pair images distinct (corners)?																							
	Test pattern P/F																							
Monthly Check—MRF Automated test P/F (if available)																								
Overall Pass/Fail																								

P = Pass   F= Fail   N/A = Not applicable   * Right & left monitors; if only one monitor, use R column

Action Limits	**Required:** Any blemish that interferes with clinical information must be removed Test pattern image quality must pass all visual tests Fiber score must be ≥2.0; speck group score must be ≥3.0; mass score must be ≥2.0 Manufacturer's automated test (if available) must pass specifications
	Time frame: Phantom must pass and significant items must be corrected before clinical use All other items must be corrected within 30 days

Repeat Analysis (B)—Tally Sheet

Facility _____ Room identification _____

MAP ID—Unit # _____ Unit manufacturer & model # _____

Total number of repeats for time period _____

Reason	Comments/Notes	Total # of Repeat Exposures	% Repeat
Patient-Related Repeats			
Poor Positioning			
Patient Motion			
Patient-Caused Artifacts			
Incorrect Patient ID			
Technical Repeats			
Exposure too low (excessive noise)			
Exposure too high (image saturation)			
Equipment-caused artifacts			
X-ray Equipment failure			
Software failure			
Aborted AEC exposure			
Miscellaneous Repeats			
Blank Images			
Good Image (No apparent reason)			
Other—miscellaneous			
Do Not Count as Repeats			
Wire Localization images			Not Included in repeat analysis
1–125 Seed localization images			
Additional projections of entire breast			
Quality Control			
	Total		

% Repeats = (# of Repeat Exposures/Total # of Exposures) * 100

Notes	Some equipment have an automated system to collect, record, and analyze repeated clinical images These system are acceptable as long as they include the following: 1. Method to count the total # of exposures made during the evaluation period 2. Method to calculate the repeats percentage during the period (# repeats exposures / total # of exposures) * 100

Repeat Analysis (C)—Daily Counting Form

Facility _____ Room identification _____

MAP ID—Unit # _____ Unit manufacturer & model # _____

Patient Name and/or ID	Total number of images	Total number of repeat exposures	Positioning	Patient motion	Patient-caused artifacts	Incorrect patient ID	Exposure too low	Exposure too high	X-ray equipment failure	Software failure	Aborted AEC exposure	Blank image	Good image (no reason)	Other miscellaneous	Localization images	Added projections	QC images
Total																	

Notes	Some equipment have an automated system to collect, record, and analyze repeated clinical images These systems are acceptable as long as they include the following: 1. Method to count the total # of exposures made during the evaluation period 2. Method to calculate the repeats percentage during the period (# repeats exposures/total # of exposures) * 100

Repeat Analysis (A)—Summary Form

(As needed)

Facility _____ Room identification _____

MAP ID—Unit # _____ Unit manufacturer & model # _____

	Monthly Analysis				Quarterly Analysis			
	Total # of exposures	# of repeat exposures	% repeat	Pass/Fail	Total # of exposures	# of repeat exposures	% repeat	Pass/Fail
January								
February								
March								
April								
May								
June								
July								
August								
September								
October								
November								
December								

% Repeats = (# of Repeat Exposures/Total # of Exposures) * 100

P = Pass   F = Fail

Action Limits	Required: If the repeat rate changes from the previously determined rate by more than 2% of total images included in the analysis, the reason(s) for the change must be determined
	Time Frame: Failures must be corrected within 30 days after analysis

System Quality Control for Radiologists

(As needed) (optional test for quality improvement)

Facility _____ Room identification _____

MAP ID—Unit # _____ Unit manufacturer & model # _____

Procedure for Radiologist	Room ID
Step 1. Complete the patient demographics:	DM or CM unit mfr & model
	Monitor ID
Step 2. Pull up the recent mammographic study from the above-listed DM unit and record ID and study date	Radiologist name
	Date of evaluation
	Image ID
Step 3. Place the same MLO image on each monitor	Study date

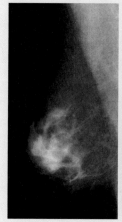 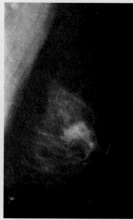

Left Monitor   Right Monitor

Step 4. Evaluate the image for artifacts and check appropriate boxes

Step 5. If any failures noted document them on the "Corrective Action Log" form and ensure items are resolved

	Yes	No
Comparing monitors side by side; do background areas (outside the breast) appear different (e.g., darker or lighter)?		
Is there a difference in contrast between monitors/sides?		
Does the image contain excessive nose (not patient motion)?		
Do you see ghosting?		
Do you see "bad pixels" (singular or clusters) (white or black)?		
Do you see white dots—could be from excessive dust?		
Do you see image distortion (not architectural distortion)?		
Do you see gridlines?		
Do you see "line artifacts" (single or multiple pixels that form lines extending across image—horizontally or vertically)?		
Are there any or significant artifacts present—impeding interpretation?		

Action Limits	**Required:** If any box checked "YES," seek service
	Time Frame: Image quality problems and/or artifacts that impede clinical interpretation must be corrected before further imaging or interpretations
	If artifact does not impede interpretation, it must be corrected within 30 days

Viewbox Cleanliness Quality Control
(Monthly) (if applicable)

Facility _____ Viewbox Location & Identification _____

After cleaning, the viewbox must be inspected for uniformity of luminance and to ensure masking equipment is functioning

Year												
Month	Jan	Feb	Mar	Apr	May	Jun	Jul	Aug	Sept	Oct	Nov	Dec
Date												
Tech initials												
Viewbox destination												

P = Pass   F = Fail

Action Limits	**Required:** Viewbox must be free of marks and have uniform brightness
	Time Frame: Failures must be corrected before viewing clinical images

Visual Checklist Quality Control (Monthly)

Facility _____ Room identification _____

MAP ID—Unit # _____ Unit manufacturer & model # _____

Inspect the unit and record any defects according to the check list below:

Year		Jan	Feb	Mar	Apr	May	Jun	Jul	Aug	Sept	Oct	Nov	Dec
Month		Jan	Feb	Mar	Apr	May	Jun	Jul	Aug	Sept	Oct	Nov	Dec
Date													
Tech initials													
Room Cleanliness	Mag stand & paddles free from dust												
	Room & countertops free from dust												
	Cleaning solution available *												
X-ray Unit	Indicators working												
	Locks (all)*												
	Collimator light working												
	Cables safely positioned												
	Smoothness of C-arm movement												
	Smoothness of compression paddle motion												
	Paddles/face shields not cracked*												
	Breast support not cracked*												
CM only (if applicable)	IP holder and lock												
	Condition of IPs												
Scanning detector system (if applicable													
Other													

P = Pass   F = Fail   N/A = Not applicable

Action Limits	**Required:** All items both critical (*) and noncritical must pass
	Time frame: Failures of a critical item must be corrected before clinical use Less critical items must be corrected within 30 days

Analog Imaging, Quality Assurance, and Quality Control

Keywords and Phrases
Analog Imaging
   Double Versus Single Emulsion
   Mammography Films
   Mammography Screens
   The Characteristic Curve
   The Cassette
   Image Quality: The Screen/Film System
   Kilovoltage Range
   Automatic Exposure Control
Quality Assurance
   Benefits of Quality Assurance
Quality Control Testing
   Technologists
   General Technologist Quality Control Tests
Testing Details—General
   Phantom Images—Analog Imaging
   Visual Checklist—Analog
   Repeat/Reject Analysis—Analog
   Viewbox and Viewing Conditions—Analog
   Compression—Analog
   Light Field and X-Ray Field Congruence—Analog
   Darkroom Cleanliness
   Processor Quality Control
   Screen Cleaning
   Analysis of Fixer Retention in Film
   Darkroom Fog and Safelight Testing
   Screen/Film Contact and Identification
Summary
Review Questions
Chapter Quiz

Objectives

On completing this chapter, the reader will be able to:

1. Describe the factors influencing image quality in analog imaging
2. List and describe the daily, weekly, monthly, quarterly, and semiannual quality control test of the mammographer using analog imaging
3. Identify the duties of the radiologist in quality assurance
4. Identify the responsibilities of the medical physicist in quality assurance
5. List and describe the regular tests of the medical physicist

KEYWORDS AND PHRASES

- The **air kerma** is determined by multiplying the measured exposure rated by a conversion factor.
- **Contrast** in analog imaging refers to the difference in optical density (OD) between two adjacent structures or the variation in optical density on the radiograph.
- **Control charts** are used to plot data for monitoring the measurement of quality control (QC) testing. The date and the individual performing the test should be noted on each control chart. Data outside the control limits should be circled, the cause of the problem noted, corrective action taken, and an in-control data point plotted.
- **Corrective maintenance** indicates the actions taken to eliminate potential or actual problems.
- **Control crossover** is a process of comparing the old films to the new batch of QC films in order to establish new operating levels whenever a new box of film is opened.
- **Entrance skin dose** measures the exposure dose at the skin's surface. It is most often referred to as the patient dose but is not used as a measure of dose in mammography because the biologic effects of radiation are more likely to be related to the total energy absorbed by the glandular tissues of the breast.
- The **Environmental Protection Agency (EPA)** was established in July 1970 by the White House and US Congress in response to the growing public demand for cleaner water, air, and land. The EPA is led by the administrator, who is appointed by the president of the United States. Its mission is to protect human health and to safeguard the natural environment.
- **Exposure linearity** is the ability of a radiographic unit to produce a constant radiation output for multiple combinations of mA and exposure time.
- **Feedback** is the process of implementing corrective action.
- **Flooded replenishment** is a method of replenishment that provides a consistent quantity of replenishment solution to the developer and fixer tanks, independent of the number of films being processed.
- **Fog** is defined as noninformational density that occurs because silver grains are exposed in areas that do not represent any of the anatomic structures within the patient.

- **Glandular dose** in mammography refers to the average glandular dose or radiation dose delivered to the center of the breast during an exposure. The glandular dose will depend on the half-value layer (HVL) (the effective energy of the x-ray beam), the breast thickness, and tissue type.
- **Half-value layer (HVL)** is the amount of filtration that will reduce the exposure rate to one-half of its initial value.
- **OSHA** (the Occupational Safety and Health Administration in the United States) was created in 1971. The agency works to save lives, prevent injuries, and protect the health of America's workers. OSHA's job includes working in partnership with state governments to investigate workplace fatalities and occupational injuries.
- **Preventive maintenance (PM)** refers to action taken on a regular basis to prevent deterioration of image quality or any breakdown in the imaging chain.
- **Rejected** images are all images used or placed in the reject bin of the department, regardless of the reason.
- **Repeated** images are images that were exposed, resulting in a radiation dose to the patient. All images that have been repeated should be included in the repeat analysis, not just those rejected by the radiologist.
- **Safelight** is a light source that emits wavelengths to which particular types of film are not sensitive.
- **Standards** are reference objects or devices with known values used to check the accuracy of other measurement devices. Standards must be calibrated against references known to be certain, accurate, or consistent.
- **Tolerance** is the amount of measurement variation allowed under normal operating circumstances.
- **Trends** are serial measurements that follow a pattern of increasing or decreasing value that may or may not lead to readings that are out of the limits of tolerance.

ANALOG IMAGING

Double Versus Single Emulsion

In analog imaging, the image formed on the film is due not to the action of x-rays, but to the action of light striking the film emulsion. The light comes from the intensifying screen placed in the cassette. When exposed to radiation, the intensifying screen will emit light, which then exposes the film emulsion and forms a latent image. If both sides of the film base are coated with emulsion, as in a double-emulsion system, the x-ray photons interact with the front screen to emit light. But photons also pass through the film, interacting with the back screen and also emitting light. In effect, a mirror image forms on both sides of the film base because light from both screens exposes both emulsions, significantly reducing the radiation necessary to image the part.

However, the light exposing the front emulsion can cross over to the other side of the base to expose the back emulsion of the film. This phenomenon is referred to as *crossover* or *crosstalk* (Fig. A1). Because the light diverges as it travels through the base, the image formed at the back

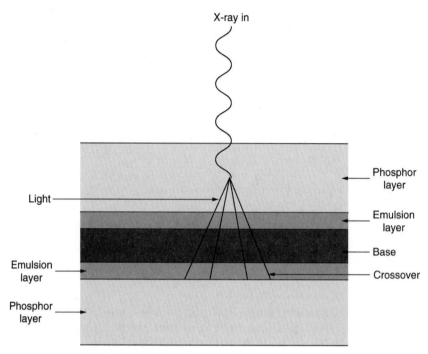

Figure A1. Crossover. Light from the intensifying screen crosses the base to expose the opposite emulsion.

is not an exact replica of the one formed in front—effectively decreasing the recorded detail of the image. In general radiography, crossover is reduced by using screens that emit short-wavelength light (blue or ultraviolet). Such light is more strongly absorbed by the silver halide crystals. Also, the polyester base is not transparent to ultraviolet light, so there is no crossover with ultraviolet-emitting screens. Crossover is also reduced by making the bases thinner, thus reducing the distance between the emulsion layers and reducing the divergence of the light.

In screen/film mammography, this is not an option. Recorded detail is extremely critical in imaging lesions in the breasts; therefore screen/film mammography systems use a single-emulsion film designed to be exposed with a single radiographic intensifying screen. This eliminates the effects of crossover. In addition, the surface base opposite to the screen is coated with a special light-absorbing dye, called the *antihalation coating*, to reduce reflection of screen light. The antihalation coating is removed during processing for better viewing.

Mammography Films

The film in analog mammography is used to record the image, display the image, and provide archival storage. All radiographic films consist of two main parts: the base and the emulsion. Other components of the film include the adhesive layer—between the base and the emulsion to ensure proper contact and integrity, especially during processing. The emulsion also has a protective covering of gelatin called the *overcoat* or *antiabrasion layer* to protect the emulsion from scratches, pressure, and storing, handling, and processing artifacts. During processing, as the emulsion swells and shrinks, the clear gelatin will balance the emulsion and prevent curling.

> The film is used to record, display, and archive the image.

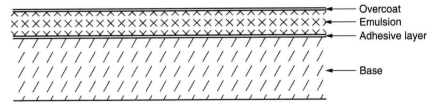

Figure A2. Cross-section diagram of the single-emulsion mammographic film showing the main parts. The base is actually the largest layer of the film.

The emulsion of the film contains silver halide crystals (Fig. A2). The latent image forms by the action of x-ray photons and light photons on the silver halide crystal. With processing, the latent image becomes the manifest image. General radiography uses emulsion on both sides of the base and is called double-emulsion film. In mammography, the emulsion is coated on one side of the base only.

Mammography Screens

In general radiography, the double-emulsion film is matched with double intensifying screens. In analog mammography, the single-emulsion film is matched with a single intensifying screen. In the direct exposure systems used in the past, the image was formed only by the action of x-ray on the film emulsion. These systems offered increased radiographic contrast but are no longer used because of the high radiation dose to the patient.

Mammography single-emulsion screens are slower than general radiography screens but provide improved contrast and resolution. The resolution of the screen will depend on the phosphor size, the phosphor layer thickness, and the phosphor concentration. As the phosphor size and layer thickness decrease, resolution increases, but smaller phosphor size or layer thickness would necessitate higher factors and would therefore result in increased patient dose. The converse is also true for increasing phosphor size and layer thickness. If the screen speed is decreased, there is also a gain in resolution, but the patient dose must be increased. In analog mammography imaging, the trade-off is to maximize the image detail while keeping patient dose to the minimum. This involves a careful balance of phosphor size and phosphor layer thickness.

> The screen resolution will depend on the phosphor size, layer thickness, and concentration.

The Characteristic Curve

In analog imaging, a graph of the optical density of a film and the log of the relative exposure units will form a curve called the *characteristic curve* or *H and D curve* (after Hurter and Driffield, who first described the relationship) (Fig. A3). The curve can be used to study the relationship between the exposure of a film to light and the degree of blackness produced in a process called *sensitometry*.

The film is first exposed to various light intensities using a sensitometer. After processing, a densitometer is used to measure the density steps obtained on the film. The curve consists of three distinct parts:

1. The *toe* represents the underexposed region of the radiograph. The toe will never begin at zero because the blue dye added to the base, plus chemical fog, will always result in a density reading.

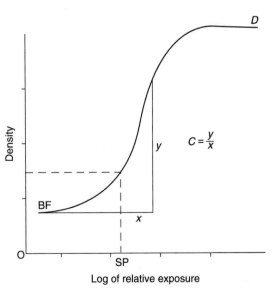

Figure A3. Characteristic curve (H and D curve) showing base-plus fog (BF), speed (SP), contrast (C), and maximum density (D) for a single exposure.

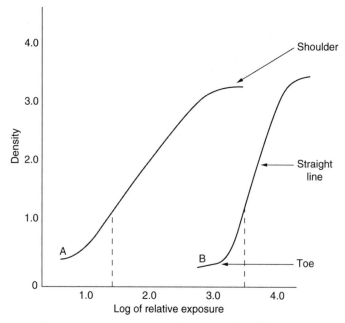

Figure A4. Characteristic curve (H and D curves) for two different types of radiographic films. Film A has a faster speed than film B because its speed point is to the left of Film B. Film B has a higher contrast than Film A because the slope of its curve is steeper than that of Film A.

2. The *straight-line portion* represents the area of useful densities on the radiograph.
3. The *shoulder* represents areas of overexposure where increased exposure to light no longer results in increasing density.

The characteristic curve is used in screen/film imaging as a means of quality control (QC) to monitor processing conditions by recording changes in density values. The curve can interpret characteristics of the film such as contrast: the steeper the slope, the higher the contrast (Fig. A4). Also, by plotting the characteristic curves of two films, the speed or sensitivity of the different films can be compared: the curve of the faster film will be positioned to the left of the curve of the slower speed film because the faster film will require less exposure than the slower film to produce any specific optical density. The characteristic curve can give important information about the film's exposure latitude.

The Cassette

The cassette or image plate holds and stores the film during exposure. There are newer, thinner cassettes designed for analog daylight and darkroom imaging systems. These have an identification slot, capable of recording patient information on the radiograph.

All cassettes or image plates are easy to open, are durable, and have low absorption characteristics relative to the kV. Analog mammography cassettes are designed for use with a single-emulsion film and are, therefore, matched with a single intensifying screen (Fig. A5). This special arrangement places the film between the x-ray tube and the screen with the emulsion side to the screen for better spatial resolution. The emulsion surface of the film must be in contact with the intensifying

> Screen/film image quality is affected by speed, resolution, contrast, and latitude.

Figure A5. The mammogram cassette has a single screen at the base.

screen, and the film must be closer to the x-ray tube than the intensifying screen. This position arrangement is important because x-rays interact primarily with the entrance surface of the screen (Fig. A5). If the screen is between the x-ray tube and the film, excess screen blur will occur (Fig. A6).

Image Quality: The Screen/Film System

Many components together play important roles in achieving the highest possible resolution. These include the use of a small focal spot, compression, and low kV and, in the past, the use of a slow speed screen/film combination coupled with extended processing. In any evaluation, generally the qualities of both the screen and the films are discussed in combination because image quality is affected by the properties of both the film and intensifying screen. The most important characteristic properties are the speed, resolution, contrast, and latitude. However, although film resolution can be a factor in assessing the overall resolution of a screen/film system, because the phosphor crystals in the intensifying screen are much larger than the silver halide crystal in the film emulsion, the resolution generally means an assessment of the screen resolution.

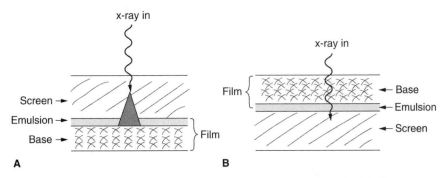

Figure A6. Position of cassette and screen is important in analog mammography imaging. **(A)** If the intensifying screen is between the primary beam and the film, excessive screen blur will result. **(B)** Spatial resolution is improved if the film is placed between the primary beam and the intensifying screen.

> The AEC detector must be positioned under the densest area of the breast.

In screen/film mammography systems, lower quality mammograms can be due to poor screen/film contact, dirty screens, or improper exposure settings, which will result in underexposed radiographs. Improperly maintained processors will also affect the resultant image by reducing the already low inherent contrast of the film. The effect of all these is that real lesions will be obscured or poorly visualized.

The kV selection or the penetrating power of the beam will influence the *subject contrast* and *exposure latitude* and, therefore, the *radiographic contrast*. Consideration should also be given to the patient's breast tissue structure. Excessive kV will result in loss of contrast (image too dark or gray). Insufficient kV will result in a high contrast in the medium-density range but loss of visibility of detail in the dense breast tissue. The ideal exposure on an analog mammography image will require a "bright light" to visualize the skin line and, generally, if the skin line is visualized, the image is underexposed or a check should be made of the processing conditions.

Most modern mammography units provide an automatic exposure control (AEC) system that automatically selects optimal exposure factors for each breast tissue type. For AEC to work, however, the position of the detector must vary depending on the different breast sizes and breast tissue composition. The AEC detector should be positioned under the most dense area of the breast.

The *speed* or *sensitivity* of the screen is determined by the same factors that control resolution. High-speed films will respond faster to exposure than lower speed films. Increasing phosphor size, layer thickness, and phosphor concentration will increase film speed. Double-emulsion films will have a higher speed than single-emulsion systems. As the kV increases, the density will also increase because higher kV increases the probability of light producing interaction within the phosphors. Temperature increases over 100° F or 38° C will decrease screen speed, but this should not be a factor in automatic processing.

Commercially, screen/film combinations are quoted as relative speed (RS) values. Higher RS numbers generally reduce patient dose, decrease latitude, and decrease resolution.

Latitude of the screen/film combination is primarily dependent on the latitude of the film. A narrow-latitude screen/film combination has high contrast. High-contrast film produces very black and white images, whereas a low-contrast film has more areas of gray. Films with lower contrast have larger grains with a wider range of sizes. Films with high contrast have smaller silver halide grains with relatively uniform grain size.

Image receptor (film) contrast is determined by the steepness of the slope of the characteristic curve—the steeper the slope, the higher the contrast. Films with steeper slopes or higher contrast will have a narrower exposure range/latitude. Films with lower contrast will have a wider exposure range/latitude. Although films with lower contrast will be more tolerant to exposure errors, the loss of contrast will have a significant effect on the visibility of the recorded detail.

Older mammography films required an extended processing cycle to improve contrast. In the extended cycle, the films spend a longer time in the developer solution, the developer temperature is increased, or both.

This allows more time to process the silver halide crystals. The advantage of extended processing is increased radiographic contrast plus reduced radiation to the patient; the disadvantage is increased radiographic noise and decreased resolution. Most newer mammography films have a higher contrast and can be processed under standard processing conditions.

Resolution is basically the ability to image two separate objects and visibly detect them as separate entities. Resolution is categorized as *spatial resolution*—which is the ability to detect small objects with high subject contrast—and *contrast resolution*, which is the ability to distinguish anatomic structures with similar contrast.

Resolution is directly related to focal spot size. As the focal spot size increases, resolution decreases. The aim, then, is always to minimize the effective focal spot while maximizing the actual focal spot to absorb heat. A large focal spot size will also result in focal spot blur (penumbra or geometric unsharpness). Focal spot blur is caused because the focal spot represents an area on the target, not a point. Focal spot blur can be reduced by small focal spot size, large source-to-image receptor distance (SID), and short object-to-image receptor distance (OID).

Sharpness refers to visibility of the image detail and sharpness of the image detail or the recorded detail. Sharpness is sometimes used interchangeably with resolution and is directly related to the focal spot size. It is often the degree of geometric sharpness of the structural lines or borders of tissue actually recorded on the radiographic image and the clarity or blur of the image. Increased SID will increase sharpness of the image detail, but increased OID will decrease the sharpness of the image detail.

Visibility of detail refers to the ability to see the detail on the radiograph. The loss of visibility can be due to any factor obscuring the image detail. Visibility of detail is, therefore, affected by fog, beam-restriction devices, the use of grids, and other factors used to prevent scattered radiation reaching the film; however, visibility of detail should be controlled so that it does not affect image quality. When a fine detail is lacking, the image often appears blurred to the naked eye.

Resolution is affected by focal spot size, SID, OID, and the intensifying screen (also by poor screen/film contact).

The other factor affecting resolution and sharpness is motion. Motion can be voluntary or involuntary. Voluntary motion is under the direct control of the patient and is best reduced through patient communication. Communication should be given to the patient before the start of the mammographic examination, to ensure patient cooperation in reducing motion. Involuntary motion is not under the conscious control of the patient and is best reduced by decreasing exposure time and immobilization. Involuntary motion is also controlled in mammography imaging by good compression, which reduces breast thickness, allowing for shorter exposure times.

All images will have less resolution and sharpness than the object itself. Although changes in the SID, OID, and size of the collimated field will not affect focal spot size, in mammography, in order to produce the sharpest possible image detail, the smallest possible focal spot size is coupled with the longest SID and the smallest possible OID (Box A1).

Box A1. Factors Affecting Resolution and Sharpness in Analog Imaging

- Motion due to long exposure times
- Poor screen/film contact
- Increase in the focal spot size
- Increase in the object-to-image receptor distance (OID)
- Decrease in the source-to-image receptor distance (SID)
- The relationship between the OID and the SID
- Characteristics of the screen (faster screens exhibit greater unsharpness)

Radiographic noise is the undesirable changes in the optical density on the image and will decrease the ability of the interpreting physician to identify microcalcifications in the breast. The primary sources of radiographic noise are quantum mottle and scattered radiation. Other factors that will affect noise are the film graininess—the distribution in size and space of the silver halide grains in the emulsion; and structural mottle, which is a characteristic of the phosphor of the radiographic intensifying screen. Radiographic noise is greater in faster screen/film systems.

Quantum mottle occurs when there are not sufficient incident photons reaching the intensifying screen. With insufficient photons, not enough phosphors are activated, and the light emitted is not enough to cover the entire surface of the film. The result is a mottled appearance on the radiography. Quantum mottle is more likely to occur on high-contrast films simply because the mottle will be more visible. Quantum mottle can only be corrected by increasing the mAs; however, in mammography imaging, the technology is now advanced enough and the screens are so sensitive that it is possible to achieve adequate radiographic density at low mAs levels.

Kilovoltage Range

A high-contrast radiograph has a lot of black and white areas compared with a low-contrast image with a lot of gray areas. Methods of controlling contrast include the use of beam-restriction devices or grids to prevent scattered radiation from reaching the image receptor, but the primary factor controlling radiographic contrast in analog imaging is the kilovoltage (kV). In any imaging, the kV controls the wavelength or the penetrating power of the beam. Increased kV will increase the energy of the x-ray beam. The result is greater penetrating power. Increased kV is necessary to penetrate more dense breast tissue; however, at high energies, Compton's effect will predominate, producing more scatter radiation. However, as the kV is reduced, for the same tissue thickness, the penetrating ability of the beam is also reduced, resulting in the need for higher mAs.

In analog imaging, the effect of higher kV is a reduction of radiographic contrast. Increased kV, however, needs lower mAs and therefore will lower the dose to the patient. Decreased kV will increase dose but will also increase contrast. The kV, therefore, can have a major effect on image contrast, exposure time, and tissue dose in analog imaging.

With analog imaging, the kV will ultimately control the exposure latitude and radiographic contrast. In digital imaging, image acquisition and display are separated, and the user can alter the contrast of the image after the exposure; as a result, controlling factors of contrast are not only the kV but also processing software and predetermined digital processing algorithms.

Automatic Exposure Control

AEC is used in all modern mammography units to control the actual exposure, therefore ensuring a uniform density of the final image.

- In analog imaging, the kV controls the subject contrast, exposure latitude, and radiographic contrast.
- In digital imaging, the controlling factors of contrast are not only the kV but also processing software and predetermined digital processing algorithms.

Analog breast techniques must be very exact because the system has a very narrow exposure latitude, and small variations in the density of the final image can alter visualization of subtle breast changes. If the film optical density is too low, there are inadequate contrast and an increased chance of missing small cancerous lesions in the breast. With AEC on dedicated mammography machines, the chance of repeats and additional radiation dose to the patient will be significantly reduced.

Repeats due to poor exposure factors are even less common in digital imaging, which has a linear response to x-ray over a wide range of exposure values. Also, in digital imaging, the image contrast, brightness, and spatial frequency of the image can be modified after the exposure.

Most modern analog AEC detectors employ a single unit capable of moving up to 10 positions behind the nipple depending on the size of the breast. AEC systems should be adaptable for different screen/film systems and for grid or nongrid work. The most common analog AEC device in use is the ionization chamber. An analog AEC system is designed to achieve an optical density of 1.40 to 2.00 when using the ACR Plexiglas standard mammography phantom under standardized exposure techniques. Within this range, there is maximum image receptor contrast, and glandular tissue will not be underexposed. The AEC device in mammography units is positioned below the image plate to minimize OID (Box A2).[1–3]

The most common cause of failure of the AEC in analog imaging is improper placement of the detector (Box A3). The AEC allows the unit to respond to different breast compositions and various breast sizes. If the detector is placed over fatty breast tissue, the glandular tissue will be underexposed. To produce an adequate exposure, the detector must be placed over the most dense or most glandular areas of the compressed breast. The optical density on the mammogram should never fall below 1.00 to 1.25 or exceed 2.50 to 3.00.

When using analog imaging, implants should be imaged using manual techniques unless sufficient breast tissue covers the first AEC detector.

QUALITY ASSURANCE

Quality assurance or continuous quality improvement (CQI) includes the total overall management of actions taken to consistently provide high image quality in the radiology department with the primary objective being to enhance patient care. Quality assurance is very important in mammography, both in meeting the standards of the US Mammography Quality Standards Act (MQSA) and in ensuring quality patient care. Every quality assurance program must include a QC component, which includes the collection and evaluation of data with a systematic and structured mechanism to control variables such as repeat radiographs, image quality, processing variations, patient selection parameters, technical effectiveness, efficiency, and in-service education.

> **Box A2.** Analog Automatic Exposure Control Device
>
> • The ionization chamber

> **Box A3.** Common Cause of Underexposure in Analog Imaging
>
> • The most common cause of underexposure is poor placement of the automatic exposure control detector—that is, the detector is not placed over the densest area of the breast.

Quality control involves actual testing to ensure that the standards of the quality assurance program are met. All QC programs have a few basic elements. There should be a *preventative maintenance program* (PM), which includes actions taken on a regular basis to prevent deterioration of image quality or any breakdown in the imaging chain. There should be *corrective maintenance* (CM) whenever there is an actual problem or to eliminate potential problems. There should be well-defined *standards*, used to check the accuracy of the measurement devices against a calibrated or known standard. The person performing the QC testing should be familiar with the *tolerance* or amount of measurement variation allowed under normal operating circumstances and should be aware of *trends*, which could be a pattern of increasing or decreasing reading that might lead to readings that are outside of the limits of tolerance. Finally, all QC should have a *feedback* mechanism or a process of implementing corrective action.[1]

Benefits of Quality Assurance

One of the benefits of having a good quality assurance/control program is the reduction of unnecessary radiation to the patient by reducing repeats. A quality assurance program will also improve overall efficiency of the delivery of service, which will result in improved patient satisfaction. Overall, the facility will benefit in the achievement of improved image quality as well as an increased consistency of image production, reliability, efficiency, and cost-effectiveness of equipment use.

Quality assurance generally has a positive effect on personnel morale and always results in overall efficiency and improved customer service.

QUALITY CONTROL TESTING

Quality assurance and QC tests are absolutely essential in producing quality images. Guidelines as determined by the MQSA and the American College of Radiology (ACR) provide the standard criteria. These criteria exceed the criteria for processing of radiographic studies done in a diagnostic radiology department. Under the MQSA standards, specific duties are required of the mammographer, the radiologist, and the medical physicist who monitors the mammography imaging equipment. These duties include the imaging, processing, and viewing of the mammograms.

In the past, there was one set of quality control tests for all mammography units in the United States. These tests were specific for the mammographer, radiologist, and the medical physicist.

Technologists

The mammographer plays a crucial role in QC and is usually solely responsible for the day-to-day testing details. The mammographer should be knowledgeable enough to seek help, whether from the medical physicist, radiologist, or service personnel at the first indication of potential problems with data from the QC tests. Specific tests of the mammographer are the following.

General Technologist Quality Control Tests

Daily Tests

Viewing conditions and viewboxes QC tests are daily tests to ensure that all viewing conditions and viewboxes are maintained at optimum level. Viewing conditions can be extremely critical in mammography because they have a direct effect on image brightness or contrast. The light in all mammography reading areas should therefore be kept low.

Analog mammogram viewing requires proper masking of each film (Fig. A7). There should be no light coming directly from the viewbox to the eye of the observer, and the same viewbox, with the same lighting conditions, using the same magnifier, with the same masking should be used for all mammograms so that the images are viewed under optimal conditions.[2]

Weekly Tests

Phantom image tests are weekly checks taken to ensure that image quality is maintained at optimum levels. The phantom image test should be carried out using the recommended mammography phantom. In general, the phantom is equivalent to a 4.2 cm thick compressed breast consisting of 50% glandular and 50% adipose tissue. Each phantom has fibers with diameters of 1.56, 1.12, 0.89, 0.75, 0.54, and 0.40 mm; specks with diameters of 0.54, 0.40, 0.32, 0.24, and 0.16 mm; and masses with decreasing diameters and thickness of 2.00, 1.00, 0.75, 0.50, and 0.25 mm. The phantom is often used with an acrylic disk (4 mm thick and 1 cm in diameter).[1]

In analog imaging, the phantom can be used to assess the film density, contrast, and uniformity and is a check of both the mammography unit and the processor. Changes in image quality can be due to the film, the cassette, the screen, the x-ray generator, the processing, or the

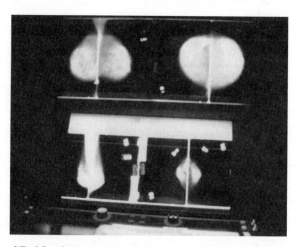

Figure A7. Viewbox masking will improve radiographic contrast.

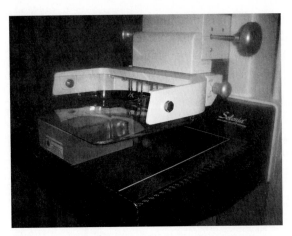

Figure A8. Imaging the phantom. The mammography phantom is equivalent to a 4.2-cm-thick compressed breast consisting of 50% glandular and 50% adipose tissue.

viewbox. If there is a change in phantom image quality, other tests will be needed to determine the cause of the change (Fig. A8).

Monthly Tests

Visual checklist is performed to ensure the mechanical integrity and safety of the mammographic equipment and accessory devices (Fig. A9). The system indicator lights, displays, mechanical locks, and detents are all checked to ensure that the system's mechanical rigidity and stability are optimum.

Quarterly Tests

Repeat/reject analysis is a systematic approach of collecting all repeated and/or rejected radiographs to analyze and categorize them to determine the cause of all repeat radiographs. Causes for repeat may be related to the competence of technical staff, equipment problems, specific difficulties associated with particular examinations, or a combination of these.

A reject is regarded as imaging performed for QC testing. An image that results in additional radiation dose to the patient is a repeat. Therefore, in the four-projection series, if the inframammary fold is missed on the mediolateral oblique (MLO) projection and the mammographer repeats the MLO but sends both the original and the repeated projection to the radiologist for interpretation, the additional MLO is considered a repeat and should be documented as such. The repeat/reject analysis program is an effective means of identifying problems and evaluating the effect of quality assurance programs on the quality of radiographic images.

Monthly Check List
Site _____   Room _____Tube _____

C-arm	SID indicator			
	Angulation indicator			
	Locks (all)			
	Field light			
	High tension cable/other cables			
	Smoothness of motion			
Detector	IP lock			
	Compression device			
	Compression scale			
	Amount of compression			
		Automatic		
		Manual		
	Grid			
AWS (Control Panel)	Hand switch placement			
	Window			
	Panel switch/light/meters			
	Technique chart			
Other	Gonad shield/ lead glove/ apron			
	Cleaning solution			

Pass=√	Month	
Fail = F	Date	
Does not apply = N/A	Initials	

Figure A9. Visual checklist chart.

A

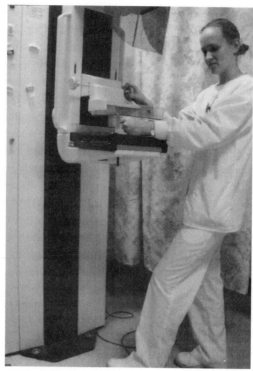

B

Figure A10. Adequate compression is provided by using both manual and automatic modes. **(A)** Applying automatic compression using a foot paddle. **(B)** Manually adjusting the compression force.

Semiannual Tests

Compression testing ensures that the mammographic system can provide adequate compression in both manual and automatic modes and that too much compression cannot be applied (Fig. A10). The compression should be adequate to separate breast tissue without causing injury to the patient. Under MQSA standards, both the minimum and maximum compression values are regulated in manual and automatic modes (Box A4).

Daily Tests

Darkroom. The processing of the mammography film starts in the darkroom. The darkroom functions to protect the film from white light and ionizing radiation during handling and processing. After a film has been exposed to light or ionizing radiation in the cassette, depending on the film emulsion, it can be as much as two to eight times more sensitive to white light than an unexposed film. This means that any accidental exposure to white light can destroy a diagnostic image. Films are also affected by excess heat, humidity, static electricity, pressure, and chemical fumes.[1]

Darkroom Environment. This is a carefully monitored environment and should be well ventilated to prevent the build-up of heat or humidity that can degrade the film. Ventilation is also necessary because the US Occupational Safety and Health Association (OSHA) and Environmental Protection Agency (EPA) consider fumes from the processing chemicals toxic, corrosive, and potentially

---

Box A4. Equipment Standards-Specific Requirements

All mammography equipment must provide:

1. An initial power-driven compression device (automatic compression)
   - The initial power drive must maintain a compression force of at least 25 lb (111 N) for the length of time it usually takes the technologist to either complete an average exposure or engage the fine adjustment control.
   - Automatic mode compression should not exceed 45 lb (200 N).

2. Fine adjustment compression controls (manual compression)
   - Manual compression should not exceed 45 lb (200 N).

3. Hands-free controls operable from both sides of the patient

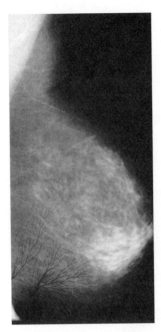

Figure A11. Static on the radiography due to low humidity in the darkroom.

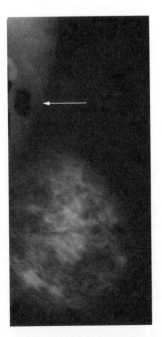

Figure A12. Fog on the radiograph.

carcinogenic. Ideal darkroom temperature should range from 65° to 75° F (18°–24° C). There should be no eating, drinking, or smoking in the darkroom. Even small food particles, water or liquid drops, or ashes from a cigarette can cause film artifacts, which could mimic pathology or degrade the diagnostic quality of the image. To prevent static electricity and static artifacts, all work surfaces should be well grounded. Wearing natural fiber such as cotton instead of synthetic such as nylon or polyester and keeping the humidity level in the range of 40% to 60% relative humidity can also prevent static (Fig. A11). Films should be stored in a vertical film storage bin (upright) versus horizontal storage, which can cause pressure artifacts. Most film storage bins are equipped with an interlock to prevent the opening of the darkroom door when the bin is open, potentially exposing the entire box of films. The bins are also usually mounted under the countertop to minimize the chance of being exposed to white light.

Darkroom Lighting. Darkrooms are usually equipped with two types of lighting overhead, which are the standard white light and the safelight. The safelight is a light source that emits wavelengths to which a particular type of film is not sensitive. A typical mammography film should be able to remain in safe lighting for at least 40 seconds without becoming fogged. Safelights should be mounted at least 3 to 4 feet from feed trays or loading counters[1-3](Fig. A12).

Darkroom Cleanliness. This is essential in order to minimize artifacts on radiographs due to bits of dust, dirt, or food between the screen and film. This is particularly important when using the single emulsion film of mammography imaging because artifacts will be more obvious and can mimic microcalcifications or even lead to misdiagnosis or repeat examinations. Generally, the basic construction of the darkroom can aid or hinder the ability to keep the area free of dust and dirt.

Processing. Processing is a very important component in mammography screening. After exposure to x-ray, the latent images are formed by the action of light on the crystal of the film emulsion. The exposed silver halide crystals become visible only after processing, a process that converts the crystals into silver. In the past, film processing was a messy manual task that could take up to 1 hour. With the introduction of the automatic processor in 1942, the processing of the x-ray film has become more streamlined.[1-3]

Processor QC is essential because a dirty processor will not function within established parameters and often will result in processor breakdown. Processor operating levels are established whenever a new processor is brought into service at a facility. Once established, the MQSA standard requires daily processor testing to ensure that the processor is working effectively. Processors, however, should also have regular scheduled cleaning on a daily, monthly, quarterly, and annual basis.

A properly run processor should be operating within established levels. Occasionally, however, a facility will have to establish new processor operating levels. Examples of when the facility may need to establish new operating levels include change in film brand or type, change in chemical brand or type, change in replenishment rates, and change of sensitometer or densitometer. Establishing new operating levels is a 5-day process. Facilities should not use the establishment of new operating levels to correct problems in the processing system, but should troubleshoot and solve the problem with appropriate corrective action.

While establishing new operating levels, the facility can continue to process mammograms but must perform and plot daily processor QC tests. New operating level values can be on the same graphs as the previous data or on a different graph and should be clearly identified as "establishing new operating levels" for inspections. When establishing new operating levels, the facility is exempt from having to stay within any processor action limits during this 5-day averaging period. The US Food and Drug Administration (FDA) recommends that during the 5-day averaging period, the facility should perform and evaluate a phantom image daily, as a means of monitoring image quality.

Because mammographic films are produced in batches with slight variations in film characteristic, whenever a new QC box of film is opened, a crossover must be performed. A crossover should be carried out only with seasoned chemistry that is operating within the ±0.10 of the control limits.

Automatic Processing. There are four basic steps in the processing of the mammographic film: developing, fixing, washing, and drying.

Developing. This is a critical stage of processing. During developing, the latent image is converted to a manifest image. That is, the silver ions of the exposed crystals are converted into metallic silver. Sometimes, a process called "wetting" is incorporated in the developing. In "wetting," a wetting agent, generally water, is used to penetrate through the gelatin of the emulsion, causing it to swell so that chemicals can penetrate easily and uniformly.

The composition of the developing chemistry and the developing temperature are both important in enhancing the contrast and film speed. Variation in the procession solution temperature can significantly affect radiographic contrast, optical density, and the visibility of recorded detail. Therefore, developer temperature should not vary by more than ±0.5° F (0.3° C) from the manufacturer's recommendations. It should be monitored at the beginning of the workday and then periodically throughout the day. Variations in developer time can have the same effect on image quality as variations in solution temperature. The developer time should be maintained within ±2% to 3% of the manufacturer's specifications. The time of development measures from the time the leading edge of the film enters the developer until it enters the fixer. In the past, extended processing, which extends developing time or uses higher developing temperature, was often used in mammography to enhance the image film contrast and film speed, therefore allowing a lower radiation dose to the patient. Most modern mammography films can now achieve high image contrast without the need for extended processing. Disadvantages of extended processing include the increased risk for processing artifacts and increase in film fog.[1]

Fixing. The fixing agent, clearing agent, or hypo removes the unexposed and undeveloped silver halide crystals from the film. This action stops the developing process and is controlled by the solution concentration and replenishment. The fixing process also hardens the gelatin portion of the emulsion.[1]

Replenishment. As the film is processed during the course of a workday, the processing solutions are constantly being used. Most film processors have replenishment rate flow meters that indicate the replenishment rate for each solution. Replenishment can affect the pH of both the developer and fixer solutions. To function and convert the latent image into a visible image, the developer solutions must maintain a pH range between 10 and 11.5. If the developer pH is too low, sometimes caused by underreplenishment or contamination, there is decreased radiographic contrast and optical density, whereas high pH will have the opposite effect. Underreplenishment or overreplenishment in the fixer will have the same effect on radiographic contrast and optical density. In most processing systems, the replenishment rate will depend on the processing time. Placing the film with the short side parallel to the rollers versus the long side parallel will, therefore, affect the replenishment rate because of the increase or decrease in processing time. To avoid inconsistent replenishment, some processors have *flooded* replenishment that turns on the replenishment pumps at regular intervals, regardless of the number of films being processed (Fig. A13).

A

B

C

Figure A13. **(A)** A mammographer removing the crossover rollers for cleaning. **(B)** The mammographer is lifting out the transport rollers while another mammographer is checking out the solution temperature. **(C)** Most modern processors have an automatic replenishing system. A buzzer will warn the mammographer whenever the tank needs refilling.

Washing. The film is washed with water to remove any remaining chemical in the emulsion. The washing can affect the archival quality of the film. Retention of fixing agents in the emulsion will cause oxidation by combining with silver to form silver sulfide, which gives the film a yellow-brown appearance.

Drying. This is the final step in film processing. The drying system consists of two or three heating units between 1500 and 2500 W, which are used to dry the film. The film must be dried to remove the water used to wash it and to further harden it, therefore making handling, viewing, and storing possible.

Transport System. A series of rollers transport the film through the various steps in the processing and in effect control the total development and total processing time. Films fed in the processor with the emulsion side up places the emulsion against the inner rollers. If the emulsion side is down, then the emulsion is in contact with the outer rollers.

Weekly Tests

Screens should be cleaned at least weekly, but also anytime that dust or other artifacts are noted by the mammographer or radiologist. Dirty or worn screens can also cause artifacts. Screens should be cleaned with a soft cloth and either a recommended screen cleaner or a mild soap-water solution. Screens must be thoroughly dry before they are returned to service.

Quarterly Tests

Analysis of fixer retention in film is a check of the amount of fixer (hypo) retention that is retained in any processed film. It will indicate the length of time that the film will retain its archival quality or image quality. Excess residual fixer can degrade the quality of the image and could indicate insufficient washing in the processor. The hypo estimator provides an estimate of the amount of residual hypo in the film in units of grams per square meter.

Semiannual Tests

Darkroom fog test will ensure that fogging of the film is not occurring as a result of cracks in the safelight or other light sources in the darkroom. Fog will reduce contrast or results in variations in optical density (Box A5).

Screen/film contact: A poor contact between the intensifying screen and the film reduces the contrast and recorded detail in the radiographic image. Dirty or worn screens can also cause artifacts. Cassettes and screens are tested periodically for poor contact, using a wire mesh test. It is important to be able to identify each screen/cassette combination. If a problem occurs with one of the cassettes, for example, if an artifact is detected in one of the cassettes, appropriate identification will allow the mammographer to locate the cassette and correct the problem. Each screen should be marked with a unique identification number near the left or right edge of the screen, using a marker approved by the screen manufacturer. The same identification number should be placed on the outside of each cassette. Cassettes must also be checked for worn latches, or hinges, and the frame inspected for warps and light leaks.

Box A5. Mammographer Quality Control Duties for Analog Imaging

1. Daily tests
   - Darkroom cleanliness
2. Weekly tests
   - Screen cleanliness
   - Viewboxes and viewing conditions
   - Phantom images
3. Monthly test
   - Visual checklist
4. Quarterly tests
   - Repeat analysis
   - Analysis of fixer retention in film
5. Semiannual tests
   - Darkroom fog
   - Screen/film contact
   - Compression

Mammography Equipment Evaluation

All mammography units must be specifically designed for mammography and meet the specific requirements of the MQSA. The components of the Mammography Equipment Evaluation (MEE) can vary by manufacturer or unit type (analog versus digital). Testing for all units must include the following:

Motion of the tube-image plate assembly—the system should not undergo unintended motion and should be capable of being fixed in any position.

Image detector size—the magnification system should not have grid use.

Light fields—the light beam shall have an average illumination of not less than 160 lux (15 foot-candles) at 100 cm or at the maximum SID, whichever is less.

Magnification assembly—the system should provide at least one magnification value within the range of 1.4 to 2.0.

Focal spot selection—if the system has more than one focal spot size, the system will display the selected focal spot before the exposure. If the focal spot selection is based on the exposure or on a test exposure, the system should display the target material and the focal spot size after the exposure.

Compression—the system should have manual and automatic compression controls on both sides of the patient. Both the initial power-drive compression (automatic) and the manual compression should have hands-free control. The system should have various-sized compression paddles to match the sizes of the full-field detector. Spot and compression paddles should also be matched with spot detectors. The compression paddles should be flat and parallel to the breast support, except they are specifically designed not to be flat or parallel to the breast support. The chest-wall edge of the compression device should be straight and parallel to the edge of the image plate.

Technical factors—the system should have manual and automatic modes. In the manual mode, the exposure factors should be displayed before the actual exposure. The automatic mode should display the actual exposure used during the exposure. The system should display the exposure factors in kilovolts (kV) and either tube current and exposure time (mA and ms) or as a product of the tube current and exposure time (mAs). The system should have a means of varying the optical density.

Film processing—in analog imaging, the chemical solutions used should meet the minimum requirement specified by the film manufacturer.

Lighting—analog systems should have special lighting for film illumination capable of producing light levels greater than that provided by the viewbox.

Film-masking devices—the illuminated viewbox should have a film-masking device that will limit the illuminated area to the region equal to or smaller than the exposed portion of the film.

Assessment of the Facility's Quality Control Program

At an MQSA inspection, the medical physicist's report is always checked. Most failures allow a 30-day period to provide corrections. The exceptions are a failure of glandular dose and phantom image quality, which must be immediately corrected. Because a facility cannot perform mammograms if either of these tests fails, it is always a good policy to have the medical physicist provide the facility with a preliminary report, allowing the facility time to correct any failures.

Mammographic Unit Assembly Evaluation

This is a check of the entire system to ensure that all locks, detents, angulation indications, and mechanical components for the x-ray tube, image plate, and holder assembly are operating properly. It is also a test of the breast thickness indicators and a check of the exposure technique chart. Although all modern mammography units have AECs, manual techniques will still be necessary on implants or very small breasts, and MQSA requires each unit to have a technique chart. If any test falls outside the action limits, the source of the problem should be identified, and corrective action should be taken within 30 days.

This test is not the same as the MEE test done by the medical physicist; however, this test must be performed for all units on installation and annually as a part of the MEE.

Collimation Assessment

This test ensures that the x-ray field aligns with the light field. The collimation must cover the entire image receptor and should not extend beyond any edge of the image receptor by more than 2% of the SID. The test also ensures that the chest-wall edge of the compression plate aligns with the chest-wall edge of the detector. If the test fails, corrective action needs to be taken within 30 days.

Evaluation of System Resolution.

This is an evaluation of the resolution of the entire mammography system, to include the effects of geometric blurring and including an assessment of the focal spot.

In analog imaging, the capability of the screen/film combination is also assessed. A slit camera or a pinhole camera can be used to determine the physical size of the focal spot, whereas a resolution test pattern or star test pattern is used to determine changes in focal spot over time and resolving capability. (Although still a useful tool, the slit camera and pinhole camera evaluation of focal spot are not acceptable under MQSA rules after October 28, 2002.)

Slit camera: This test cannot be used with high-frequency generators used in today's imaging world. The slit camera is actually a very thin sheet of highly attenuating metal such as tungsten, with a small slit of approximately 10 to 30 m wide in its center. The sheet is placed between the focal spot and the film, at least 100 mm from the focal spot. An exposure is made, which projects an image of the focal spot size, through the slit, and onto the film.

Pinhole camera: This test cannot be used with high-frequency generators used in today's imaging world. The pinhole is also a sheet of lead or tungsten with a tiny hole, approximately 0.03 to 0.08 mm in diameter. The diameter of the hole must be smaller than the focal spot to be measured; therefore, smaller focal spots (0.1–0.3 mm) must be measured by other means. The sheet is placed exactly halfway between the focal spot and film. The pinhole must be aligned with the central axis of the x-ray beam. An exposure is made, and the developed film should show the actual dimensions and shape of the focal spot.

Physicist QC test summary for Lorad:

- Mammographic Unit Assembly Evaluation
- Collimation Assessment
- Artifact Evaluation
- kVp Accuracy and Reproducibility
- Beam Quality Assessment Half-Value Layer (HVL) Measurement
- Evaluation of System Resolution
- AEC Function Performance
- Radiation Output Rate
- Phantom Image Quality Evaluation
- Signal-to-Noise Ratio (SNR) and Contrast-to-Noise Ratio (CNR) Measurement
- Diagnostic Review Workstation (RWS) QC
- DICOM (Digital Imaging and Communications in Imaging) Printer QC
- Detector Flat-Field Calibration
- Compression Thickness Indicator
- Compression

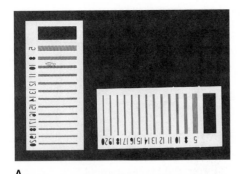

Figure A14. (A) Radiograph of an x-ray test pattern using resolution-testing tool. **(B)** Actual reaction testing tool.

Star pattern: The star test pattern resolution tool uses a sheet of highly attenuating metal with a pattern of slits diverging from a central point. The arrangement is in the form of a star. The separation between slits increases linearly with distance from the center. When an x-ray image is taken, the lines closest to the center will be seen separate from each other but farther from the center, the lines will blur together. The radius of the blur will be directly proportional to the size of the focal spot. The slits are typically in an orthogonal pattern, allowing the two orthogonal dimensions of the focal spot to be measured. For accurate measurement, the star pattern must be placed exactly parallel to the film, and in the central axis of the x-ray beam.

Resolution testing: The unit of resolution is line-pairs per millimeter (lp/mm) or cycles per mm (Fig. A14). A radiographic resolution tool consisting of a bar pattern with pairs of lines at a set distance from each other can be used to assess the resolution of a unit. The line-pair tool should be up to 10 lp/mm or greater. Two exposures are made: one with the pattern's bar parallel to the anode-cathode axis and the other with the pattern's bar perpendicular to the anode-cathode axis. The point at which the viewer can see the closest pair of lines separate from each other represents the lp/mm reading. MQSA states that each analog system should have a minimum resolution of 11 lp/mm when tested with the bars perpendicular to the anode-cathode axis. If the test pattern is aligned parallel to the anode-cathode axis, the minimum resolution should be 13 lp/mm. Any resolution test should be conducted using routine SID, kV, and AEC settings. The focal spot should be tested at 4.5 cm above the breast support using technical factors most commonly used for imaging.

If the test fails, corrective action needs to be taken within 30 days.

Automatic Exposure Control System Performance Assessment

This assessment of the unit's AEC system is necessary to maintain consistent image optical density according to the density control selector. The test also ensures that the unit can alter the optical density according to the breast tissue thickness and density. The AEC should be capable of maintaining an image optical density within ±0.15 over a thickness ranging from 2 to 6 cm and over the range of kV used in the clinical setting.

Uniformity of screen: This is an analog system test. The screen speed of all the cassettes in the facility should be tested, and the difference between the maximum and minimum optical densities should not exceed 0.30. This ensures that each cassette, imaged under the same AEC conditions, will produce approximately the same optical density.

Artifact Evaluation

In analog imaging, the artifact evaluation test can be combined with the uniformity of screen test to assess the degree and source of artifacts visualized on the mammogram phantom images. The system should be

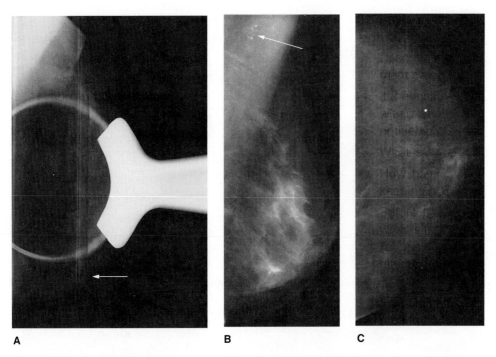

A   **B**   **C**

Figure A15. Processing, exposure, and handling artifacts. **(A)** *Processing artifacts* caused by roller marks. **(B)** *Exposure artifacts* caused by deodorant on the patient. **(C)** *Handling and storage artifacts* caused by fingerprints.

evaluated with a high-grade, defect-free sheet of homogenous material large enough to cover the mammography cassettes. All cassette sizes used in the facility should be tested. Artifacts may be placed into any one of the three categories: processing artifacts, exposure artifacts, and handling and storage artifacts (Fig. A15).

Analog processing artifacts are caused by, or occur during, the processing of diagnostic images. A processing artifact will have the same orientation with respect to motion through the processor. Examples include guide shoe marks (on the leading edge of the film), delay streaks, scratched emulsion, Pi lines (+ density lines), pick-off, static, hesitation lines, or roller marks.

Analog exposure artifacts are caused by the patient, the mammographer, or the equipment during a mammographic procedure. Examples include filter and mirror artifacts, hair, wigs, lotions, deodorant, or even skin lesions.

Handling and storage artifacts in analog imaging occur during darkroom handling or during storage before the film is used. Examples include fingerprints on the radiograph, kinks or creases, or pressure marks.

If an artifact is seen on the mammogram, the patient should not be radiated solely to eliminate an artifact that is not obscuring clinical information. Always practice ALARA (As Low As Reasonably Achievable) and avoid excessive radiation to the patient. The interpreting physician, however, has the final say on tolerable artifacts. If the test fails, corrective action needs to be taken within 30 days (Fig. A16A).

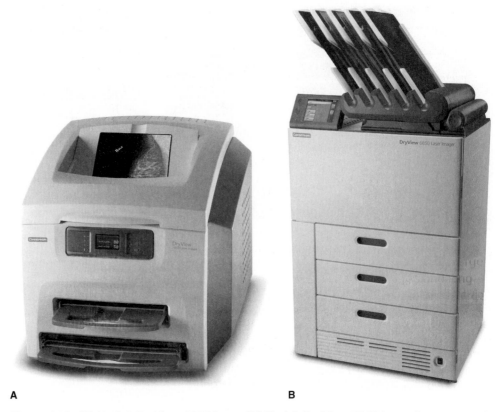

A   **B**

Figure A16. **(A)** Kodak DryView 6850 laser. **(B)** Kodak DryView 6850 laser. (Images used with permission from the Eastman Kodak Company.)

Image Quality Evaluation

This is an assessment of the mammography phantom used to determine and detect changes in image quality (Fig. A17).

In analog imaging the test actually measures all the components of imaging, the film, the cassette and the screen, the x-ray generator, added filtration, processing, and viewbox. Because of the broad range of this test, if a fault is detected in image quality, another test will be needed to localize the cause of the fault.

Analog imaging will require optimal viewing conditions, including low-level and diffuse lighting, and analog images should be masked to eliminate extraneous light. The optical density at the center of the image should be at least 1.20 when exposed under typical conditions, and the density should not change by more than ±0.20 from any previously established operating levels. The density difference (DD) between the background of the phantom and the added test object should not vary by more than ±0.05 from previously established operating levels. The phantom background optical density should be a least 1.40, and the density difference due to the 4.0 mm thick acrylic disk should be at least 0.40.

The phantom should visualize a minimum of four largest fibers, three largest speck groups, and three largest masses.

Subjective judgments are often difficult, and in phantom testing it is recommended that the same person should review the images each time, using the same criteria, including the same viewing conditions

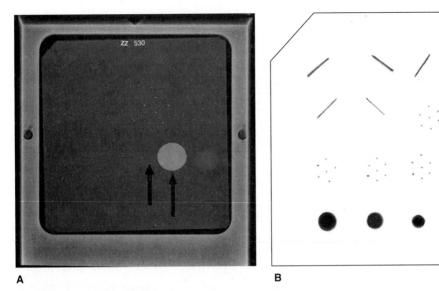

A   **B**

Figure A17. **(A)** An actual radiograph of a phantom with 1-cm diameter, 4-mm-thick disk. This disk is used for contrast measurement. *Arrows* indicate points where density measurements should be made. The image is scored based on what is seen on the film. **(B)** A schematic diagram of the phantom showing the relative position of the different objects embedded within the phantom. Each phantom will have *six fibers, five speck groups,* and *five masses.*

and magnification. A magnifying glass of 2 times or higher can be used to assess the analog image.

If the test results indicate a failure, the source of the problem must be identified, and corrective action must be taken before further examinations are performed or any images are processed (Fig. A18A and B).

Kilovoltage Accuracy and Reproducibility

The kV on the mammography unit should be accurate within ±5% of the indicated kV (equal to 1.5 kV at 30 kV) and should be reproducible with a variation of less than 0.02. Testing is performed at the lowest kV that can be measured on the testing device—the most commonly used clinical kV and the highest kV available. If the test fails, corrective action needs to be taken within 30 days.

Beam Quality Assessment (Half-Value Layer Measurement)

This test ensures that the HVL of the x-ray beam is sufficient to penetrate the breast with adequate contrast, yet with minimized dose to the patient. The more penetrating the beam, the higher will be its HVL (Fig. A19). Lower energy x-ray will only contribute to patient skin dose; therefore, the half-value layer will determine whether adequate filtration exists. It is the most important patient protection characteristic of any unit.

The HVL can be determined using the kV meter and exposure time meter or by using sheets of 0.1 mm of aluminum between the x-ray tube and the image plate. The aim is to start with no sheets, adding sheets until the reading is less than one half the original exposure reading. If the test fails, corrective action needs to be taken within 30 days.

Physicist QC test summary Siemens:

- Site Audit/Evaluation of Technologist QC Program
- Medical Inspection
- Acquisition Workstation Monitor Check
- Detector Uniformity
- Artifact Detection
- Collimation, Dead Space, and Compression Paddle Position
- AEC Thickness Tracking
- Spatial Resolution
- AEC Repeatability
- Image Quality
- Radiation Dose
- HVL and Radiation Output
- Tube Voltage Measurement and Reproducibility
- Printer Check
- RWS Tests

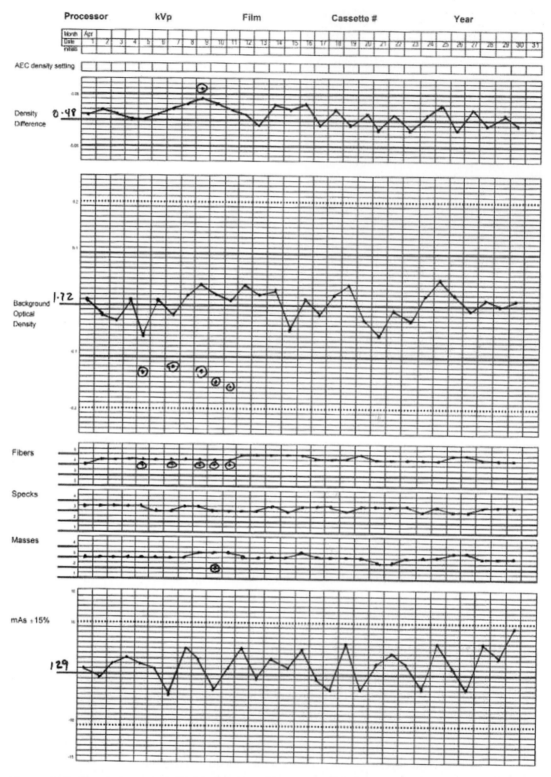

Figure A18. Phantom control chart is used to ensure optimal image quality and the film optical density. The background optical density and the density difference from the exposed image of the phantom are plotted. The numbers of specks, fibers, and masses and the mAs are also graphed.

MQSA REQUIREMENTS:

The HVL shall meet the specifications of FDA's Performance Standards for Ionizing Radiation Emitting Products (Part 1020.30) for the minimum HVL. These values, extrapolated to the mammographic range, are shown in the table below. Values not shown may be determined by linear interpolation or extrapolation.

X-Ray Tube Voltage (kilovolt peak) and Minimum HVL		
Designed Operating Range (kV)	Measured Operating Voltage (kV)	Minimum HVL (millimeters of aluminum)
Below 50	20	0.20
	25	0.25
	30	0.30

If the test results fall outside the action limits, the source of the problem shall be identified and corrective actions shall be taken within 30 days of the test date.

Figure A19. Half-value layer (HVL) requirements are mandated by MQSA.

Breast Entrance Exposure, Automatic Exposure Control Reproducibility, Average Glandular Dose, and Radiation Output Rate

A test of the typical entrance exposure for an average patient is assessed using a breast phantom representing 4.2-cm compressed breast thickness that is 50% adipose and 50% glandular.

The average glandular dose is calculated from the entrance exposure delivered on the FDA phantom and the HVL measured at the kV used for the breast entrance exposure. Generally, roentgen to mrad conversion charts are used to convert the breast entrance exposure to average glandular dose based on the indicated kV and measured HVL for that kV setting. On a single craniocaudal projection of an FDA-accepted phantom, which simulates a standard breast, the average glandular dose must not exceed 3.0 mGy (0.3 rad) per exposure with a grid or 1.0 mGy (0.1 rad) without a grid. If it falls outside of these limits, the source of the problem must be identified and corrective action taken before examinations are performed or any images processed.

Testing is also an assessment of the short-term AEC reproducibility and the air kerma rate. The AEC system should deliver a consistent optical density regardless of compressed breast thickness and composition. The coefficient of variation (standard deviation divided by the mean of exposure measurements) for both the air kerma and the mAs must not exceed 0.05. If the test fails, corrective action needs to be taken within 30 days.

The mammography system should be capable of producing a minimum output of 7 mGy air kerma per second (800 mR/second) when operating at 28 kV in the standard mammography (Mo/Mo) mode at any SID. The detector should be placed 4.5 cm above the breast support surface, and the compression plate should be between the source and detector. The system should also be capable of maintaining the required minimum output averaged over a 3.0-second period. If the test fails, corrective action needs to be taken within 30 days.

Physicist QC test summary for Carestream:

- Cassette Exposure Response Test
- Cassette/Phosphor Screen Artifact Test
- Erase Screen Test
- Scanner Uniformity Test
- Scanner Response Linearity Test
- Spatial Frequency Response Test
- Geometric Accuracy Test
- Image Plate Fog Test
- AEC System Performance/Constancy Test
- Mammography Unit Assembly Evaluation
- Mammographic Unit Collimation Assessment
- Beam Quality and Half-Value Layer
- kVp Accuracy and Reproducibility
- Breast Entrance Exposure, Dose and Radiation Output
- Phantom Image Quality Evaluation
- RWS Tests
- Laser Printer Test

> **Physicist QC test summary for General Electric:**
>
> - Flat Field
> - Phantom Image Quality
> - CNR Measurement
> - MTF Measurement
> - Automatic Optimization of Parameters (AOP) Mode and SNR
> - Collimation Assessment
> - Evaluation of Focal Spot Performance
> - Subsystem Modulation Transfer Function (MTF)
> - Breast Entrance Exposure, Average Glandular Dose, and Reproducibility
> - Artifact Evaluation and Flat-Field Uniformity
> - kVp Accuracy and Reproducibility
> - Beam Quality Assessment (HVL Measurement)
> - Radiation Output
> - Mammographic Unit Assembly Evaluation
> - RWS Tests

Viewbox Luminance and Room Illuminance

This test ensures optimum viewing conditions and includes a check of the luminance of the viewboxes used for breast interpretation or QC and the room illuminance levels. A photometer is used to measure both the luminance and illuminance. Luminance is the amount of light either scattered or emitted by a surface, measured in cd/m^2 (the unit candela per square meter is sometimes referred to as the *nit*). Illuminance is the amount of light falling on a surface, measured in lux (1 lux = 1 lumen/m^2). Other units used to measure luminance and illuminance are foot-lamberts and foot-candles. To convert foot-lamberts to cd/m^2, multiply the numerical value by $10.764/\pi$; from foot-candles to lux, multiply the number by 10.764.

The illumination levels should be 50 lux or less (equivalent to a moonlit room).

Viewboxes used for analog mammograms should be capable of producing a luminance of at least 3000 cd/m^2. The viewbox used by the QC mammographer to check the density and quality of the mammographic images should be similar to the reading viewbox in luminance and color of light, in an area where the ambient light is similar to that of the reading room.

The illumination levels should be 50 lux or less (equivalent to a moonlit room). If the test fails, corrective action needs to be taken within 30 days.[3]

Compression Paddle Alignment

All systems should have different-sized compression paddles that match the sizes of all full-field image plates provided for the system. This test ensures that the compression paddle fits flat and parallel to the breast support and is aligned to the image plate.

The chest-wall edge of the compression device should not extend beyond the chest-wall edge of the image plate by more than 1% of the SID. Also, a shadow of the vertical edge of the compression paddle should not be visible on the image.

Mammography Equipment Evaluation

This annual evaluation is performed whenever a new unit is installed, disassembled, or reassembled. The test is also triggered if there is a software upgrade. This test must be performed before any imaging. The actual components of the test include all the general MEE tests.

TESTING DETAILS—GENERAL

Phantom Images—Analog Imaging

Frequency. This test should be carried out initially with fresh chemistry in the processor, and weekly, or whenever equipment has been serviced.

Comment. Subjective judgment is always very difficult, and for this reason, the phantom images should be viewed under the same viewing conditions by the same person at the same time each day. The viewing should be in an area with minimal ambient light.

Equipment

1. A mammographic phantom (4–4.5 cm thick tissue-equivalent breast phantom), with an acrylic disk 4 mm thick permanently fixed on the phantom, so as not to obscure any phantom detail. The phantom should contain appropriate details ranging from visible to invisible.
2. Cassette and films. There should be a dedicated phantom-testing cassette to eliminate inconsistent results due to variation in the cassette or screen.
3. Magnifying lens (approximately 2 ft or higher).

Procedure

1. Load film into the cassette.
2. Place the cassette in the holder with the phantom on top of the cassette so that the phantom is positioned with the edge of the phantom aligned with the chest-wall side of the image receptor.
3. Bring the compression device into contact with the phantom, making sure that the AEC sensor is in the same location as for previous phantom images and under the center of the wax insert.
4. Make an exposure using technical factors most commonly used for a 4.5-cm compressed breast.
5. Note the mAs or exposure time if it is displayed after exposure.
6. Measure the optical densities in three locations. The background density measures the density in the center of the phantom image. The DD is the difference in the density in the area of the acrylic disk and the area directly adjacent to the disk.

Physicist QC test summary for Fuji:
- S-Value Confirmation
- System Resolution
- Computer Reader (CR) Reader Scanner Performance
- Mammography Unit Assembly Evaluation
- Collimation Assessment
- AEC System Performance Assessment
- System Artifact Evaluation
- Phantom Image Quality Evaluation
- Dynamic Range
- Primary Erasure (Additive and Multiplicative Lag Effects)
- Interplate Consistency
- kVp Accuracy and Reproducibility
- Average Glandular Dose
- Beam Quality Assessment and HVL Measurement
- Radiation Output
- Viewing and Viewing Conditions
- RWS Tests
- Printer Tests

Box A6. Analog Imaging Test for the Medical Physicist

	Frequency	Report
Mammographic unit assembly evaluation	Annually	Within 30 days
Collimation assessment	Annually	Within 30 days
Evaluation of system resolution	Annually	Within 30 days
Automatic exposure control (AEC) system performance	Annually	Within 30 days
Uniformity of screen speed	Annually	Within 30 days
Artifact evaluation	Annually	Within 30 days
Image quality evaluation	Annually	*Immediately*
kV accuracy and reproducibility	Annually	Within 30 days
Beam quality assessment (half-value layer)	Annually	Within 30 days
Breast exposure and AEC reproducibility	Annually	Within 30 days
Average glandular dose	Annually	*Immediately*
Radiation output rate and intensity	Annually	Within 30 days
Measurement of viewbox luminance and room illuminance	Annually	Within 30 days

7. Plot the background density and density difference on the control chart.
8. Determine the number of fibers, speck groups, and masses visible in the image and enter the number of each type of object on the control chart.
9. Count each fiber as one point if the full length of the fiber is visible. A fiber is counted as 0.5 if more than half but not the entire fiber is visible. The location and orientation must also be correct.
10. Using a magnifying lens, count each speck group (consisting of six specks each) as one point if four or more specks are visible. Count a speck group as 0.5 if only two or three specks in that group are visible.
11. Count a mass as one point if a density difference is seen in the correct location with a generally circular border, and as 0.5 if a density difference can be seen, but a generally circular shape to the mass is not visible.
12. Carefully examine the image for artifacts, nonuniform areas, or grid lines.

The optical density of the film should not be less than 1.20 with control limits of ±0.20. An optimal optical density of 1.40 is suggested, and the density difference (due to the 4.0 mm acrylic disk) should be approximately 0.40, with the control limits of ±0.05. It is essential that all units at one facility produce similar densities. The mAs noted on the generator should not change by more than ±15% for a given density control setting.

The present criteria for the number of objects to pass the ACR Mammography Accreditation are a minimum of four of the largest fibers, the three largest speck groups, and the three largest masses. If the test falls outside of the limit, the source of the problem must be identified and corrective action taken before any mammography examinations are performed or any radiographs are processed. Phantom images must be retained for a year

Visual Checklist—Analog

Frequency. This test should ideally be performed monthly, or after any service or maintenance.

Equipment. Visual checklist

Procedure

1. Review all the items listed on the visual checklist.
2. Date and initial the checklist when indicating the status of operation.

Although some of the items on the visual checklist are for the technologist convenience, many items are essential for patient safety and high-quality diagnostic images. Items not passing the visual checklist should be repaired or replaced immediately. Items missing from the room should be located or replaced immediately.

Repeat/Reject Analysis—Analog

Purpose. To determine the number and cause of repeated mammograms and to identify the "whys" to reduce patient exposure and reduce costs.

Frequency. This test should be performed initially, then at least quarterly. To be statistically meaningful, a volume of at least 250 patients needs to be measured.

Equipment

1. All rejected and repeated mammograms
2. A method of counting the total number of images taken
3. A sorting area for analysis
4. Data sheet

Procedure—Analog Imaging

1. Dispose of all existing rejected or repeated films or images in the department.
2. Inventory film supply as a starting point to determine the total number of films consumed during the test.
3. Collect all repeated images, with the reason for rejection written on the image, and continue to collect until there are at least 250 patients total.
4. Sort the rejected films into the categories listed on the repeat analysis sheet.
5. Tabulate the counts and determine the total number of repeated images, the total number of rejected films, and the total number of films exposed. Repeated images result in an exposure to the patient while rejected films could include clear films or films taken during QC testing. Rejected films are not included in the repeat analysis calculations.
6. Determine the percentage of repeats in each category by dividing the repeats in the category by the total number of repeated images from all categories and multiplying by 100.
7. The overall repeat percentage is determined by dividing the total number of repeated images by the total number of films exposed during the analysis period and multiplying by 100.

The overall repeat rate ideally should not exceed 2%, but a rate lower than 5% is acceptable when the quality assurance program is operational. The percentage of repeats from each category should be close. If one category is significantly higher than the others, it should be targeted for improvement.

If a facility does not image at least 250 patients in a quarter, the repeat analysis can still be performed to determine the reasons for repeat images (Fig. A20). If the test fails, corrective action must be taken within 30 days.

Viewbox and Viewing Conditions—Analog

Frequency. This test should be performed weekly.

Comment. Because intensity varies over time, the fluorescent bulbs should be changed every 18 to 24 months even though the typical fluorescent bulb has a rated life of 7500 to 9000 hours. Over time, heat from the fluorescent bulb can discolor inside the plastic front of the viewbox. Dirt and dust can form on both the inside and the outside of the plastic viewing surface, as well as the outside surface of

Mammography Repeat-Reject Analysis   Date From ___9/3/01___ To _12/28/01_

Reason for Reject	Projection Repeated Check one for each repeated film						Number of Films	% of Repeats
	Left CC	Right CC	Left MLO	Right MLO	Left Other	Right Other		
1. Positioning			✓✓	✓✓✓			5	19.2
2. Patient Motion		✓			✓		2	7.7
3. Light Films			✓	✓	✓		3	11.5
4. Dark Films	✓	✓					2	7.7
5. Black Films			✓				1	3.8
6. Static, Artifacts				✓	✓✓		3	11.5
7. Fog		✓	✓				2	7.7
8. Incorrect ID or Double Exposure							0	
9. Equipment Error			✓	✓			2	7.7
10. Technologist Error		✓✓	✓	✓			4	15.4
11. Good Film (No apparent reason)			✓✓				2	7.7

12. Clear Film	6
13. Wire Localization	10
14. Quality Control	7

		Number	Percentage
	Repeats (1–11)	26	1.17
	Rejects (All –1–14)	49	2.21

Total Films Used	2216

Comments	
Action taken	

Figure A20. A sample of the repeat/reject chart used by the mammographer to chart the reasons for rejects films. The final percentage of repeats and rejects will be calculated as a percentage of the total films used. The percentage of each category of repeated films will be calculated as a percentage of total repeat rate.

the fluorescent bulb. This can reduce light output by as much as 10% per year, which will decrease radiographic contrast. The intensity of all viewboxes within the department should be checked for consistency. With multiple viewboxes, replace all lamps at the same time and replace them with bulbs of the same manufacturer, production lot, and color temperature to maintain consistency. Corrective action must be taken to repair any problem before patient images are interpreted or comparisons reviewed.

Equipment

1. Window cleaner
2. Soft towels
3. Antistatic cleaner

Procedure

1. Clean viewbox surface using window cleaner and soft paper towel.
2. Check and remove all marks.
3. Visually inspect the viewboxes for uniformity of luminance.
4. Ensure that the masking equipment is functioning.
5. Ensure that extraneous lighting is not directly reflecting on the viewboxes.

In addition to regular tests, the monitor must be checked for dirt, scratches, and fingerprints and wiped clean using microfiber cloth or glass cleaning agents (Fig. A21).

Compression—Analog

Frequency. This test should be performed initially and semiannually thereafter.

Equipment

1. Analog bathroom scale. Digital scales designed specifically for this purpose can be used, but regular digital scales are not recommended because they may not respond well to additional pressure.
2. Several towels.

Procedure

1. Place a towel on the image plate to protect the device, and then place a bathroom scale on the towel with the dial in a position that is easy to read with the center of the scale directly under the compression device (Fig. A22).
2. Place one or more towels on top of the scale to protect the compression device, and activate the automatic compression device and allow it to operate until it stops automatically.
3. Read and record the compression and release the device.
4. Next, repeat steps 2 and 3, using manual compression until it stops.

Adequate compression on the initial power drive should range from 25 to 45 lb in automatic mode (111 to 200 N). If the test falls outside of

Figure A21. Cleaning the monitor.

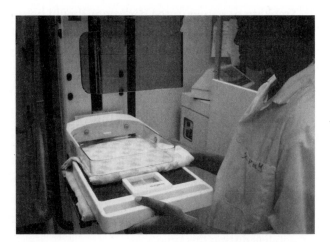

Figure A22. Compression testing using a conventional bathroom scale.

Figure A23. Coins are used to outline the light field and provide a visual check in a quick light-field test.

the limit, the source of the problem must be identified and corrective action taken before any mammography examinations are performed or any radiographs are processed.

Light Field and X-Ray Field Congruence—Analog

Purpose. To check the centering and perpendicularity of the beam, as well as the congruence or alignment of the collimated light field and the x-ray beam field. It is essential in limiting radiation to the patient that the x-ray field coincides with the light field at all times. A visual check of the light field using markers to outline the field should be made before starting.

Frequency. The ACR requires the physicist to perform this test at least annually, but the test is simple and can be performed by the mammographer if a problem is suspected. If the test fails, the facility has 30 days to correct the problem.

Procedure

1. Place a cassette or image plate on top of the image plate holder or detector of the mammographic unit.
2. Turn the light field on and collimate or use the appropriate settings to expose an area smaller than the size of the image plate (Fig. A23).
3. Place radiopaque objects such as paper clips or coins intermittently within the border of the light field, along its perimeter. Also, note the location of the light field with respect to the chest-wall edge of the cassette holder, ensuring that the light field does not go beyond the chest-wall edge or falls short of meeting the edge.
4. Make the exposure. If the border of the x-ray field does not align with the border of the light field and the difference is more than 2% of the SID, alignment adjustment may be necessary and the service engineer should be contacted.

Darkroom Cleanliness

Frequency. This test should be performed daily at the beginning of the workday, before processing any radiographs.

Comments. Smoking or eating in the darkroom should be prohibited. If screens are being cleaned properly and there is still a lot of dust artifacts appearing on radiographs, the darkroom could be the source, for example, poor ceiling or door design.

Equipment

1. Wet mop and pail.
2. Vacuum.
3. Lint-free towels.
4. Antistatic cleaning solution.

Procedure

1. Turn the processor power and water on so that developer temperature can stabilize during this time.
2. Damp mop the darkroom floor.

Figure A24. A sensitometer used in mammography quality control testing.

3. Remove all unnecessary items from countertops and work surfaces.
4. Use a lint-free damp towel to wipe off the processor feed tray and then the countertops and other surfaces in the darkroom.
5. Once a week, *before* cleaning the feed tray and the countertops, wipe or vacuum the overhead air vents and safelights.

Processor Quality Control

Frequency. Processor quality control should be carried out daily at the beginning of the day before processing any radiographs to confirm and verify that the processor-chemical system is working properly according to specifications.

Equipment

1. A 21-step sensitometer (Fig. A24)
2. A densitometer (Figs. A25 and A26)
3. Fresh box of film
4. Digital or metal thermometer (Do not use a thermometer containing mercury. With a break or leak, even minute amounts of mercury

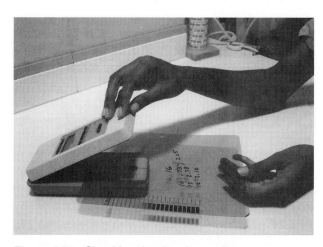

Figure A25. Checking the density readings on a sensitometer strip—using a densitometer.

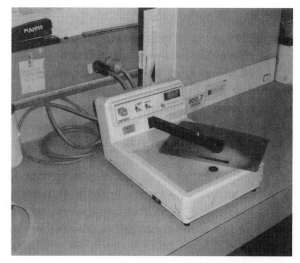

Figure A26. A densitometer used in mammography quality control testing.

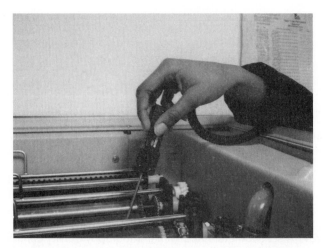

Figure A27. Digital thermometer used to check the developer temperature in quality control testing.

could contaminate the processor and cause inconsistent processing.) (Fig. A27).

5. Control charts

Comment. Processor QC control records should be saved for 1 year (Fig. A28). Sensitometric films should be saved for the last full month. There should be no delay between exposing and processing of the film because the latent image changes over time.

To establish beginning operating levels for the processor:

1. Select a fresh box of mammography film and reserve this box for quality assurance purposes only.
2. Drain the chemicals from the processor and thoroughly flush the racks and tanks with water.
3. Drain the replenishment tanks and refill with fresh replenisher.
4. Fill the fixer tank with fixer solution.
5. Fill the developer tank about one-half full with developer solution and add the specified amount of developer starter solution. Add sufficient developer solution to fill the developer tank or follow the manufacturer's specifications.
6. Set the developer temperature control at the temperature specified in the film manufacturer's literature. Set the fixer and water temperature if available. Dryer temperature should be set as low as possible while still providing good drying.
7. Set the replenishment rates for developer and fixer as specified by the film manufacturer.
8. When developer temperature has been stabilized, check the temperature with a digital or metal thermometer. Clean the thermometer stem after use.
9. Expose and process a sensitometric strip (known as a *control strip*). Repeat this step once each day for 5 days. (Before processing the strip, make sure that the temperature of the developer is correct. The strip is processed with the less-exposed end fed into the processor first and on the same side of the processor feed tray each time. Delay between exposure and processing should be similar each day to avoid any latent image changes that could occur with time.)

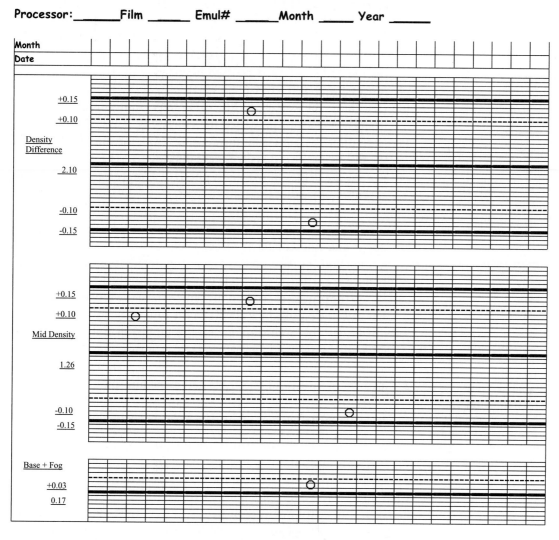

Figure A28. The sensitometer strip is exposed and processed each day, and the data are evaluated and plotted. The result is the processor control chart. Data out of the control limits are circled, and the test is repeated. The cause of the problem and corrective action are recorded in "remarks" section of the control chart.

10. Use a densitometer to read and record the optical densities of each step of the sensitometric strip, including an unexposed area of the film. Measurements should be taken in the center of each strip.

11. Decide which step has a density closest to but not less than 1.20. This is the mid-density (MD) step, sometimes recorded as the *speed index,* speed point, or speed step.

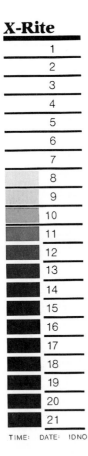

X-Rite

1
2
3
4
5
6
7
8
9
10
11
12
13
14
15
16
17
18
19
20
21

TIME: DATE: IDNO:

Figure A29. The sensitometer will produce a step-wedge pattern of 21 different optical densities. A densitometer is then used to record the base plus fog value and the high-, mid-, and low-density measurements. These results are plotted on the daily processor quality assurance chart.

12. Determine the average MD by using the densities for that step from the five strips processed for 5 days.
13. Decide which step has a density closest to 2.20, and which step has a density closest to, but not less than, 0.45. These are the high density (HD) and low density (LD). The difference in densities between these two is the DD (or *contrast index*). (Determining the DD in this manner is to be used only to observe consistency of film and processing and should not be used to compare different film types or processing at different facilities.)
14. Determine the average DD by using the readings from those steps from the five strips processed for 5 days (Fig. A29).
15. Determine the average of the densities from the unexposed area of the five strips. This density is base plus fog (B + F).
16. Record the values of MD, DD, and B + F on the centerline of the appropriate areas of the control chart.

Daily continuation of processor quality control:

1. Expose and immediately process a sensitometric strip, before processing clinical cases. Before processing, make sure developer temperature is correct. The strip is processed with the less-exposed end fed into the processor first, on the same side of the processor feed tray each day. The emulsion is always in the same orientation (processed with either the emulsion up or the emulsion down for single-emulsion films), and delay between exposure and processing is similar each day.
2. Read the densities of the indicated steps and plot MD, DD, and B + F.
3. Determine whether any of the data exceed control limits.

The MD and DD should remain within ±0.15 optical density units of the operating level, and B + F within ±0.03 of the operating level. If the MD or DD exceeds the ±0.15 limit, corrective action needs to be taken. Also, if B + F is ±0.03 of the normal, immediate corrective action must be taken. Changes in density or base plus fog can be due to chemistry, temperature, and replenishment. If the test falls outside of the limit, the outside of limit reading must always be plotted and circled. The source of the problem must be identified, corrective action taken, and a new reading plotted before any mammography examinations are performed or any radiographs are processed.

Establishing New Levels (FIG 30 – cross-over)
New operating levels would be established in the following circumstances:

- Change in film volume
- Change in brand or types of chemical used
- Change in film brand or type
- Change in replenishment rates
- Change in development time (normal to extended processor)
- New sensitometer or densitometer
- New or change in processor
- Film box empty and no crossover performed
- MD or DD step does not conform to established guidelines

Routine preventative maintenance is not considered a justification for changing QC operation levels. Establishing new operating levels follows the crossover procedure.

Crossover is used to determine the new operating level for the daily processor QC chart when a new box of quality control films is opened. The crossover is not used when clinical films are opened. The rationale is that variations in film batches (emulsion #'s) can affect the consistency of readings.

The crossover must be carried out only with a processor with seasoned chemistry that is operating within the ±0.10 of the control limits. The crossover must begin when there are at least five sheets of old QC films remaining.

1. On each of the 5 remaining days (corresponding to the five films in the QC film box), expose and process a QC strip using one film from the old QC box and one film from the new QC box.
2. Determine the MD, DD, and B + F from all 10 films processed (five from the old box and five from the new).
3. Record the value in the appropriate box of the crossover worksheet (Fig. A30).
4. Use the crossover worksheet to calculate the new operating levels. (The difference between the old and new values should be added [new–old], including the sign.) If the difference is positive, the new operating level is increased. If the difference is negative, the new operating level is decreased.
5. Record the new operating levels and control on a new control chart.

Comment. If the B + F of the new film exceeds the B + F of the old film by more than 0.02, the reason must be investigated.

CROSSOVER WORKSHEET

Site _____   Date _____
Film Type _____   Technologist _____

New Emulsion #					Old Emulsion #				
Film#	Low Density (LD) M Step #	Mid Density (MD) Step #	High Density (HD) Step #	B+F	Film#	Low Density (LD) M Step #	Mid Density (MD) Step #	High Density (HD) Step #	B+F
1	0.46	1.26	2.49	0.18	1	0.50	1.27	2.31	0.18
2	0.48	1.23	2.51	0.18	2	0.51	1.28	2.33	0.19
3	0.49	1.27	2.47	0.18	3	0.52	1.30	2.39	0.18
4	0.50	1.25	2.50	0.17	4	0.49	1.32	2.30	0.18
5	0.48	1.24	2.49	0.17	5	0.50	1.31	2.30	0.18
Average	0.48	1.25	2.49	0.18	Average	0.50	1.30	2.33	0.18
Average Density Difference: DD=HD-LD =2.01					Average Density Difference: DD=HD-LD =1.83				

MD difference between new and old film (New MD – Old MD)	-0.05
DD difference between new and old film (New DD – Old DD)	+0.18
B + F difference between new and old film (New – Old)	0

	MD	DD	B + F
Old operating levels	1.34	1.86	1.18
Difference between new and old film	-0.05	0.18	0
New operating levels	1.29	2.04	0.18

Figure A30. A sample crossover worksheet with calculations.

Screen Cleaning

Frequency. This test should be carried out at least weekly but is determined by the environment.

Equipment

1. Screen cleaner recommended by the manufacturer.
2. A lint-free gauze pad, camel's hair brush, or antistatic brush.
3. Canned air—it must be "clean" air. (Containing no oil, moisture, or any other contaminants.)

Procedure

1. Check screens for dirt, dust, lint, pencil marks, fingerprints, or other marks.
2. Clean screens according to the manufacturer's recommended materials and procedures.
3. Allow screens to air-dry, standing vertically before closing and using. Wait at least 15 minutes or follow the manufacturer's recommendations.

Analysis of Fixer Retention in Film

Purpose. To determine the quantity of the residual fixer in the processed film.

Frequency. This test should be carried out quarterly.

Equipment

1. Residual hypo test solution, available commercially
2. Hypo estimator (e.g., Kodak Hypo Estimator, publication N-405, or equivalent)

Procedure

1. Process one sheet of *unexposed* film through each processor being tested—label each film to identify the processor.
2. Place one drop of the residual hypo test solution on the emulsion side of each processed film, and allow the solution to stand on the film for 2 minutes.
3. Blot off the excess solution and place a sheet of white paper under the film.
4. Place the estimator that comes with the test kit over the film sample to help compensate for B + F in the film (when not in use, the estimator is kept in a sleeve for protection).
5. Compare the stain with the hypo estimator. This should be done immediately after the excess stain has been removed from the film. If the film is allowed to sit, external conditions could cause the stain to darken and this would alter the reading.

The hypo estimator provides an estimate of the amount of residual hypo in the film in units of grams per square meter. It should be 0.05 g/m^2 or less (5 mg/cm^2). If the stain indicates that the residual hypo has increased, the test should first be repeated. If the test fails, again the source of the problem must be identified and corrective action taken

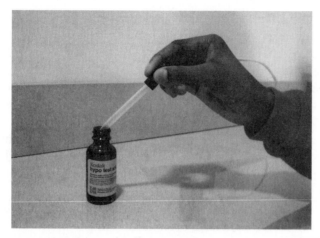

Figure A31. Hypo testing to determine the amount of fixer on the film.

within 30 days. Excess of hypo retained on the film can be due to the processor wash tanks and water flow rates in addition to fixer replenishment rates. All should be assessed and compared with recommended rates (Fig. A31).

Darkroom Fog and Safelight Testing

Darkroom fog failures can be isolated to safelight problems or room environment light leaks by repeating the test with the safelight off. To be effective, it is important to use film exposed to a clinical optical density falling between 1.40 and 2.00 measured at the chest-wall edge of the film. Films that are underexposed or overexposed will be less sensitive to the effects of darkroom fog.

Frequency. This test should be performed initially and semiannually thereafter. It should be performed in any new darkroom and anytime the safelight bulbs or filters are changed. It is essential that this test be carried out with the different types of films used in the darkroom.

Comments. Before beginning this test, checks should be made of the filter and safelight. Filter and safelight checks include a check to see if they are cracked or faded and a check of the wattage of the safelight and distance of the bulbs from work surface against the recommendation of the film manufacturer.

Equipment

1. Mammographic unit
2. Mammographic film (taken from a new box, not from the film bin)
3. Densitometer
4. A radiopaque card
5. A watch or timer

Procedure

1. Turn on the overhead lighting and inspect the countertops and processor feed tray for foreign objects, dampness, cleanliness, and sharp edges.

2. Locate the reserve fixer and developer tanks and check to ensure that they are properly ventilated and that the temperature of the area is within accepted limits.

3. Locate the film storage area and verify that the boxes of film are being stored vertically and under the proper temperature conditions. Also, check the age of the film and the visibility of the expiration date.

4. Use a ruler to verify the proper distance from the safelight to the work counters or feed trays. Also, check the wattage of light bulbs and inspect the filter for cracks or pinholes. Make sure that the filter type matches the film type.

5. Look for light leaks around doors, passboxes, processor, and ceiling and correct any light leaks before continuing with the test.

6. Turn off all safelights and overhead lights and wait approximately 5 minutes for the eyes to adjust.

7. Check for white light leaks, especially near the processor, darkroom doors, water pipes, ventilation ducts, suspended ceiling tiles, pass box, and processor and in the ceiling.

8. In total darkness, load the film from a new box of film into a cassette.

9. Return to the mammography room.

10. Place a phantom on the breast support surface and position it so that the edge of the phantom is aligned with the chest-wall side of the image receptor and centered.

11. Apply gentle compression, then expose the film using a technique that will produce a density of approximately 1.4 to 2.0 after processing.

12. In the darkroom, in total darkness, unload the cassette on the counter where cassettes are normally loaded and unloaded. Place the film with the emulsion side up. Cover one half of the film with the opaque card placed perpendicular to the chest wall. Keep this half covered for the remainder of the test.

13. Turn on the safelights and let the film and the opaque card lie on the countertop for 2 minutes.

14. At the end of 2 minutes, process the film.

15. Determine the density of each section of the film (fogged and unfogged) using a densitometer. Measure the density close to the edge separating the fogged and unfogged portions of the phantom image (if one is seen).

The optical density differences of adjacent portions of the exposed and developed film should not exceed 0.05. If any do, then a problem may exist. If the test falls outside of the limit, the source of the problem must be identified and corrective action taken before any mammography examinations are performed or any radiographs are processed (Fig. A32).

Screen/Film Contact and Identification

Frequency. This test should be carried out initially when the cassettes are new and semiannually thereafter or when a reduction in image sharpness is suspected.

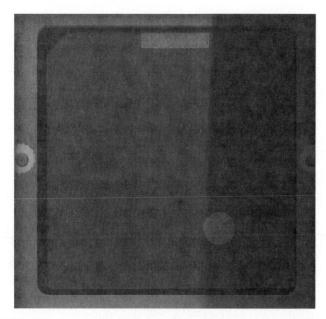

Figure A32. Phantom images showing acceptable levels of darkroom fog. A densitometer is used to measure close to the edge separating the fogged and unfogged portion of the phantom image. The density should not be measured over any test object in the phantom.

Equipment

1. Copper wire mesh screens with at least 40 wires/inch grid density. These are commercially available. The mesh can be placed between two thin sheets of acrylic to protect it.
2. Mammographic film
3. Densitometer
4. Acrylic sheets sufficient to provide 4 cm thickness

Procedure

1. Carefully clean the screens and cassettes to be tested using a cleaner recommended by the manufacturer and a lint-free cloth (Fig. A33).
2. Allow the screens to air-dry at least 30 minutes and load with film.

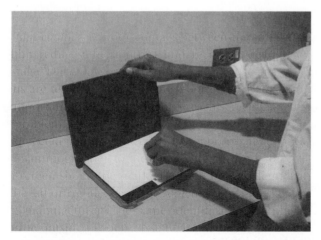

Figure A33. Proper screen-cleaning method.

A

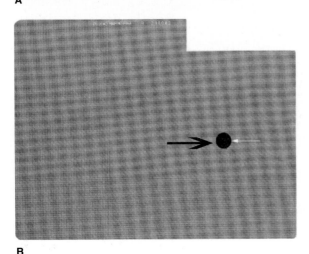

B

Figure A34. **(A)** Copper wire mesh used for screen/film contact testing. **(B)** A screen should be replaced if it has an area greater than 1 cm in diameter (*arrow*) of poor contact that cannot be eliminated. Multiple small areas are considered acceptable.

3. Wait 15 minutes so that any air trapped in the cassette has a chance to escape.
4. Place the cassette to be tested on top of the cassette holder. Do not use the grid.
5. Place the copper screen on top of the cassette.
6. Place the acrylic sheet, if needed, on top of the compression device and move the device up, as close to the x-ray tube as possible.
7. Select a manual technique (approximately 25–28 kV), which will provide a film density between 0.7 and 0.8 when measured with a densitometer over the film. (This should be measured near the chest-wall edge of the film.)
8. Expose and process the film, repeating step 7 for each cassette.
9. View the films on a viewbox at a distance of at least 3 feet to clearly visualize the wires. Look for areas of poor contact (darker areas in the mesh image).
10. Retest cassettes that do not pass the test.

Note: The acrylic is used to ensure a reasonable time for exposure but is not absolutely necessary. The optical density should be between 0.70 and 0.80 when measured near the chest-wall side of the film. If used, the acrylic is placed near the x-ray tube window but must project to cover the entire area of the cassette. The exposure should be at least 0.5 seconds, for good exposure reproducibility.

Small specks of dust can degrade the image for up to 1 cm or more away from the dust particle. Areas of poor contact will be dense and cloudy. Any cassette having a large area (<1 cm in diameter) of poor contact that cannot be eliminated should be replaced. Multiple small areas are considered acceptable.

If the test falls outside of the limit, the source of the problem must be identified and corrective action taken before any mammography examinations are performed or any films processed (Fig. A34).

Summary

Having a good quality assurance program is an integral part of maintaining quality mammography imaging and good patient care. Any quality assurance must have quality control (QC) components, and in mammography these QC tests are regulated under the MQSA.

QC tests for technologist can be divided into three parts: general test for all units and specific analog tests.

1. General tests
   • Visual checklist
   • Phantom images test
   • Repeat analysis
   • Viewboxes and viewing conditions
   • Compression force test

2. Analog QC tests
   - Darkroom cleanliness
   - Processor QCs
   - Screen cleanliness
   - Fixer retention test
   - Darkroom fog test
   - Screen/film contact

These medical physicist tests can include the following:

1. General QC tests
   - Mammographic unit assembly evaluation
   - Collimation assessment
   - Evaluation of system resolution
   - Automatic exposure control (AEC) system performance assessment
   - Uniformity of screen speed
   - Artifact evaluation
   - Image quality evaluation
   - kV accuracy and reproducibility
   - Beam quality assessment (half-value layer measurement)
   - Breast entrance exposure, AEC reproducibility, average glandular dose, and radiation output rate
   - Viewbox luminance and room illuminance
   - Assessing the mammography facility's QC program
   - Compression paddle alignment

2. Analog tests summary
   - Daily tasks: darkroom cleanliness, processor quality control
   - Weekly tasks: viewbox and viewing conditions, phantom images
   - Monthly task: visual checklist
   - Quarterly tasks: repeat analysis, analysis of fixer retention
   - Semiannual tasks: darkroom fog testing, screen/film contact, compression

REVIEW QUESTIONS

1. What type of maintenance includes actions taken on a regular basis to prevent deterioration of image quality or any breakdown in the imaging chain?

2. Give an example of how a good quality assurance program will benefit the patient.

3. Name one federal agency responsible for ensuring the safety and quality of imaging practices within the United States.

4. Identify some of the causes of repeats.

5. How often is the compression test performed?

6. What are the minimum and maximum allowed compressions?

7. Collimation must cover the entire image receptor but should not extend beyond any edge of the image receptor by more than what percentage of the source-to-image receptor distance (SID).

8. Who is responsible for medical auditing and patient communication?

9. Which assessment is used to ensure that the x-ray field aligns with the light field and the collimation covers the entire image receptor?

10. How is the effect of subjective judgment reduced when viewing the phantom image test?

11. For a statistically meaningful reject/repeat analysis, what is the recommended patient volume?

12. Ideally, the overall repeat rate should not exceed what percentage?

13. What is the MQSA's recommended next step to take if the compression test fails?

14. Which target material is best for imaging fatty breast tissue, especially in analog imaging?

15. At what exposures does the reciprocity law fail?

16. How will placing the AEC detector under the fatty portions of a breast with mixed fatty and glandular tissue affect the final radiograph in analog imaging?

17. How will grid use affect contrast and exposure in analog imaging?

18. Identify the single biggest advantage of using the single-emulsion system in analog mammography imaging.

19. Why is the toe of the characteristic curve never zero?

20. In relation to a film with a higher speed, where will the slower speed film be located on a characteristic curve?

21. Name two factors affecting resolution and sharpness in analog imaging.

CHAPTER QUIZ

1. Test on beam quality assessment (half-value layer measurement) is performed by the:
   (A) Mammographer
   (B) Medical physicist
   (C) Technologist
   (D) Radiologist

2. In quality control, the standards refer to:
   (A) Action taken on a regular basis
   (B) Action taken to eliminate potential problems
   (C) Reference calibrations or measurements to check accuracy
   (D) Measurement variations allowed

3. The person responsible for overseeing all the activities of the mammographer and the medical physicist is:
   (A) Quality control radiographer
   (B) State inspector
   (C) MQSA inspector
   (D) Radiologist

4. QC duties of the mammographer include testing or monitoring of all of the following except:
   (A) Phantom image
   (B) Screen/film contact
   (C) Collimation assessment
   (D) Compression

5. The ability of an x-ray tube to always produce the same intensity of radiation when the same set of technical factors is used to make an exposure is called:
   (A) Radiation output
   (B) Exposure time
   (C) Exposure reproducibility
   (D) kV accuracy

6. The MQSA standards were enacted:
   (A) Because mammography was over-regulated
   (B) To address the poor quality mammograms that were available
   (C) To enforce continuing education for radiologic technologist
   (D) To enforce continuing education for radiologist

7. The repeated image is:
   (A) Any image that is rejected
   (B) An image that is repeated and results in radiation dose to the patient
   (C) The magnified image requested by the radiologist
   (D) An image taken during quality control tests

8. The person responsible for performing evaluations of and generating reports on the mammographic unit is the:
   (A) Radiologic technologist
   (B) Mammographer
   (C) Medical physicist
   (D) Radiologist

9. A test used to determine whether the time of exposure indicated by the control panel is the same as that delivered by the x-ray tube is performed by the:
   (A) Technologist
   (B) Physicist
   (C) Quality assurance technologist
   (D) Radiologist

10. In analog imaging, AEC failure, resulting in an underexposed film, can be caused by
    (A) Processing deficiencies such as fluctuating developer temperature
    (B) Improper placement of the AEC detector
    (C) Decreased radiographic contrast
    (D) Processing a film when the processor temperature is too low

Answers

1. Preventative maintenance

2. Improve overall efficiency and delivery of service, patient satisfaction and reduction of repeats, radiation dose to the patients, and consistency in image production; and increase cost-effectiveness of equipment

3. Any of the following: OSHA, EPA, FDA, NCDRH, NRC, CDC, or MQSA

4. Causes for repeat may be related to the competence of technical staff, equipment problems, specific difficulties associated with particular examinations, or a combination of these.

5. Semiannually

6. 25–45 lb (111–200 N)

7. 2% of the source-to-image receptor distance (SID)

8. The interpreting physician—the radiologist

9. Collimation assessment

10. Phantom images should be viewed by the same person, using the same type of magnifier, at the same time of day, under the same viewing conditions.

11. 250

12. 2%

13. Identify the source of the problem and take corrective action before any additional mammography examinations are performed

14. Molybdenum

15. Very long and very short exposures, especially in analog imaging

16. The dense areas of the final radiograph will be underexposed if the AEC is placed under the fatty breast tissue in a breast with both fatty and dense tissues. The AEC detector selector should be positioned under the dense areas of the breast.

17. Grid use cleans up the image and will result in increased contrast. However, typically the use of a grid will need an increase in technical factors which will result in increased radiation dose to the patient.

18. Improve recorded detail

19. Blue dye added to the base, plus chemical fog will always result in a density reading.

20. A curve with slower speed will be positioned to the right of the curve with faster speed.

21. Motion, poor screen/film contact, increased focal spot size, increased OID, decreased SID, characteristic of the screen, relationship between OID and SID

Chapter Quiz Answers

1. Answer: B. Beam quality assessment in a general test performed by the medical physicist.

2. Answer: C. Standards are reference objects or devices with known values used to check the accuracy of other measurement devices.

3. Answer: D. It is the radiologist's responsibility to oversee the activities of all personnel responsible for the quality assurance process.

4. Answer: C. The collimation assessment tests are performed by the medical physicist.

5. Answer: C. The exposure reproducibility is a check of the AEC. The system should deliver a consistent optical density regardless of compressed thickness and composition. The kV accuracy ensures that the kV selected is within ±5% of the indicate kV. The radiation output is the selected kV and mA. The exposure time is the milliseconds of the actual exposure.

6. Answer: B. MQSA was enacted to ensure quality patient care and imaging for patients needing a mammogram. The MQSA involves regular checks of the training and continuing education of radiologist, technologist, and medical physicist.

7. Answer: B. Any projection or position that is repeated and result in an additional exposure to the patient is considered a repeated image. Images that are used in quality control and do not cause additional patient dose are considered rejects, not repeats.

8. Answer: C. It is the responsibility of the medical physicist to regularly check the mammography unit and component parts and generate reports as needed.

9. Answer: B. Tests on the x-ray tube are generally performed by the medical physicist.

10. Answer: B. The placement of the AEC device is critical in analog imaging. If the device is placed under fatty breast tissue that is a mixture of fatty and dense, the dense areas will be underexposed.

REFERENCES

1. Bushong SC. *Radiologic Science for Technologist.* 9th ed. St. Louis, MO: Mosby; 2008:193-203.
2. Papp J. *Quality Management in the Imaging Sciences.* St. Louis, MO: Mosby; 1998.
3. American College of Radiology (ACR). *Mammography Quality Control Manual.* Reston, VA: The American College of Radiology; 1999.

BIBLIOGRAPHY

Center for Devices and Radiological Health (CDRH). https://www.fda.gov/aboutfda/centersoffices/officeofmedicalproductsandtobacco/cdrh/ Accessed March 11, 2017.

Centers for Disease Control and Prevention. http://www.cdc.gov/. Accessed February 16, 2011.

Environmental Protection Agency. http://www.epa.gov. Accessed February 16, 2011.

Food and Drug Administration. https://www.fda.gov/radiation-emitting products/mammographyqualitystandardsactandprogram/. Accessed February 16, 2011.

Hendrick ER, Botsco M, Plott CM. Quality control in mammography: Breast imaging. *Radiol Clin North Am.* 1995;33(6):1041-1057.

Joint Review Committee on Education in Radiologic Technology home page. http://www.jrcert.org. Accessed February 2011.

Nuclear Regulatory Commission. What We Do. http://www.nrc.gov/. Accessed February 2011.

Peart O. *Lange Q & A Mammography Examination.* New York, NY: McGraw-Hill; 2008.

Answers to Review Questions

CHAPTER 1: THE HISTORY OF BREAST CANCER AND BREAST IMAGING

1. There were taboos, both religious and social, against human dissection. Scientists were forced to experiment with animals.
2. The Flemish physician, Andreas Vesalius (1514–1564)
3. There was no early detection tool; cancers were discovered by visual inspection at later stages, with surgery being the only option.
4. The German pathologist, Rudolf Virchow (1821–1902)
5. Radiation therapy became a treatment option.
6. In the 1960s, the Health Insurance Plan (HIP) discovered that routine screening of asymptomatic women would offer a 31% reduction in the mortality rate from breast cancer in women older than 50 years.
7. The imaging, processing, and interpretation
8. It was voluntary, there were no means of enforcement, there was no on-site evaluation of the facility, and it could result in a possible conflict of interest.
9. Women aged 20–39 years: clinical breast examination (CBE) every 3 years and monthly breast self-examinations (BSEs). Women older than 40 years: annual screening mammogram and CBE; monthly BSEs. Magnetic resonance imaging or ultrasound studies for high-risk women or those with dense breast tissue only.
10. On October 27, 1992, the US Congress passed the Mammography Quality Standards Act (MQSA)
11. Certification and accreditation of all mammography units are now required.
12. The US Preventive Services Task Force (USPSTF), a group of independent health experts, is now recommending that routine screening of average-risk women should begin at age 50 years, instead of age 40 years. The task force also suggests that screening should end at age 74 years, that women should be screened every 2 years instead of every year, and that BSEs have little value.

Chapter 1 Quiz

1-1. Answer: B. American Egyptologist Edwin Smith bought the ancient document in 1862. The document dates back to the 17th century BC (1600 BC).

1-2. Answer: A. There were social and religious taboos against human dissection. This forced Galen to dissect animals. Mammals include both animals and humans.

1-3. Answer: D. Andreas Vesalius was famous for his work on human anatomy and his book *The Seven Books on the Structure of the Human Body*, published in 1542.

1-4. Answer: A. Marie Curie (1867–1934)

1-5. Answer: D. Rudolf Virchow (1821–1902)

1-6. Answer: C. Xeroradiography

1-7. Answer: B. The Health Insurance Plan (HIP) of greater New York showed that the possible carcinogenic effects of mammography were outweighed by the benefits, namely a reduction in the mortality rate from breast cancer.

1-8. Answer: C. The first dedicated units in the early 1970s used a single-emulsion film and a single-emulsion calcium-tungstate screen. This system helped bring the radiation dose to acceptable levels.

1-9. Answer: D. The American College of Radiology (ACR) adopted a voluntary accreditation program in 1987. FDA is the US Food and Drug Administration; MQSA is the Mammography Quality Standards Act; NCI is the National Cancer Institute.

1-10. Answer: A. On October 27, 1992, the US Congress passed the Mammography Quality Standards Act (MQSA). ACS is the American Cancer Society, ACR is the American College of Radiology, and FDA is the US Food and Drug Administration.

CHAPTER 2: PATIENT EDUCATION AND ASSESSMENT

1. To avoid misdiagnosis, to avoid unnecessary surgeries or biopsies, to avoid the need for additional workup projections, and to help the radiologist make a final interpretation. Document any history of trauma, surgery to the breast, breast cancer abnormalities or symptoms, health problems, or risk factors for breast cancer.

2. The best way to combat misconceptions about mammography is communication. Technologists should communicate with their patients before, during, and after the mammogram. Technologists should always invite questions from the patient, then listen and encourage further comments. By communicating with the patient, technologists will be able to identify concerns and answer questions before the patient leaves.

3. Some of the benefits of communication include relaxing the patient, which reduces the chance of suboptimal imaging of the pectoral muscles; identifying sensitive breasts and the reason for the sensitivity; allowing for possible rescheduling of the mammogram if needed; and educating the patient by talking about any fears and misconceptions.

4. Examples of nonverbal communication include (1) paralanguage, which refers to the pitch, tone, stress, pauses, speed, rate, volume, accent, or quality of one's voice (paralanguage instills interest in conversation and avoids monotony); (2) body language, which can indicate interest in the patient's answers or an unwillingness to offer help; and (3) professional appearance and physical hygiene, which can help to project confidence and can inspire the patient's confidence.

5. Open-ended questions are nondirected and nonleading (e.g., "Where is the lump?"). A leading or directed question would be, "Is the lump in your left breast?"

6. All women have a lifetime risk of approximately 12%, meaning that women face a 12% probability of developing breast cancer. However, age is an important risk factor because as a woman ages, the incidence of cancer increases. It is estimated that although the probability of developing breast cancer is only 0.06% for a woman at age 20 years, that percentage jumps to 1.44% at age 40 years, 2.39% at age 50 years, and more than 3% for a woman at age 70 years.

7. Women older than 40 years: annual screening mammogram; MRI or ultrasound for high-risk or dense breast only.

8. No greater than 0.3 rad (300 mrad or 3 mGy) with a grid or 0.1 rad (100 mrad or 1 mGy) without a grid.

9. The sensitivity of mammography is dependent on breast glandularity plus the age and hormone status of the patient. Mammography understates the multifocality of a lesion. Inadequate position or compression can compromise the image.

10. Major risks are mainly outside a woman's control, carry a higher probability of breast cancer, and often cannot be changed; minor risks carry a lower probability for breast cancer and are often within a woman's control.

11. Gender, that is, being female.

12. Aging, genetic risks, family history of breast cancer, personal history of breast cancer.

13. Not having children, early menarche or late menopause, use of alcohol, use of hormone replacement therapy, not breastfeeding, lack of exercise.

14. Physical activity helps to reduce breast cancer risk. The American Cancer Society recommends that adults get at least 150 minutes of moderate intensity or 75 minutes of vigorous intensity activity each week (or a combination of these), preferably spread throughout the week.

15. Trauma or injury to the breast may not cause breast cancer. However, when breast tissue is injured, fat necrosis can occur, and as the body heals, a scar can develop, resulting in a lump. On discovering the lump, some women go for a mammogram only to find cancer. However, the cancer is unrelated to the lump due to the trauma.

16. The adolescent breast is very radiosensitive, and studies have suggested that past irradiation to the chest and breast tissue can lead to the development of bilateral breast cancer at an early age. Past irradiation can be the result of irradiation for cancer treatment, such as Hodgkin's disease or non-Hodgkin's lymphoma, or radiation during scoliosis imaging.

17. The Gail Model is a risk assessment tool used by the National Cancer Institute. This will estimate a woman's risk for developing invasive breast cancer during the next 5-year period and up to age of 90 years (lifetime risk), based on the woman's age and the risk factor information provided. For comparison, the tool will then calculate 5-year and lifetime risk estimates for a woman of the same age who is at average risk for developing breast cancer.

18. Studies show that women with dense breast tissue face a higher risk for missed breast cancer if the mammogram is the only screening tool used. The reason is that on the mammogram, the cancer is visualized as a white area within the background optical density of the breast. In a fatty breast, the cancer is easier to visualize. However, in dense glandular breast tissue, it is harder to locate the cancer. Breast glandularity is affected by age, menopausal status, the use of hormone therapy, pregnancy, and genetics.

Chapter 2 Quiz

2-1. Answer: B. Mammography measures the glandular dose to the deeper tissues of the breast because it is assumed that these cells are more radiosensitive.

2-2. Answer: D. All of these can be significant. A history of breast trauma can result in a hematoma—which could present as a lump. Painful lumps or masses can be benign but would need further evaluation. A sudden nipple retraction can indicate malignancy.

2-3. Answer: B. The mammogram will detect only approximately 90% of breast cancer under normal conditions. Mammography is also more accurate in postmenopausal women compared with menopausal women.

2-4. Answer: A. Objective data include signs and symptoms that can be seen or felt and such things as a lump. Subjective data are symptoms that are perceived by the affected individual only. If the patient feels a lump and the technologist does not feel the lump, the area should still be marked and the patient's clinical symptoms documented. The technologist should never disregard anything the patient says, even if it cannot be seen or felt.

2-5. Answer: B. Examples of nonverbal communication include (1) paralanguage, which refers to the pitch, tone, stress, pauses, speed, rate, volume, accent, or quality of one's voice; (2) body language or motions, such as nodding or shaking of the head, can indicate interest in the patient's answers or an unwillingness to offer help; (3) professional appearance and physical hygiene can help to project confidence and inspire the patient's confidence.

2-6. Answer: C. History taking should start with the use of open-ended questions (nondirected and nonleading; e.g., "Where is your pain?"). A leading question would be, "Is the pain in your left breast?" Whenever possible, the technologist can use facilitations, such as a nod, or say "yes," "okay," or "go on" to encourage elaboration. Silence can be used to give the patient time to remember, and it also facilitates accuracy. Probing questions must be used after the initial open-ended questions to get more details (e.g., "How long have you had the pain?"). Repetition is basically a rewording of the questions and can be used to clarify information. Repeating the patient's response will also help to verify that the patient has not changed her mind. In repeating, the technologist should avoid jargon and should use precise and clear words. Before ending, the technologist should summarize the patient's information to verify their accuracy.

2-7. Answer: B. Breast glandularity is affected by age, menopausal status, the use of hormone therapy, pregnancy, and genetics. Overweight or obesity, especially after menopause, increases breast cancer risk. Before menopause, most of the estrogen in your body comes from your ovaries with very little linked to fat tissue. After the ovaries stop estrogen production at menopause, most of a woman's estrogen comes from fat tissue. Fat tissue after menopause can therefore raise estrogen levels and increase the risk for breast cancer.

2-8. Answer: C. Minor risks are associated with hormonal changes or a woman's lifestyle choice. This would include the uses of hormone replacement therapy (HRT) or not having children. A family history of breast cancer is considered a major risk factor.

2-9. Answer: B. Major risk factors are considered those that are outside of a woman's control and contribute significantly to the possibility of developing the disease. These would include age and genetic risks. Use of birth control is considered a minor risk because of its association with hormones and personal choice.

2-10. Answer: C. Risk factors for breast cancer are considered anything that would increase a person's chance of getting the disease. Breast cancer risks are divided into major and minor risks.

CHAPTER 3: ANATOMY, PHYSIOLOGY, AND PATHOLOGY OF THE BREAST

1. The clavicle at the second or third rib
2. Fatty or adipose
3. Superior and medial
4. At the approach to the nipple
5. 8 o'clock
6. Cooper's ligaments
7. Veins
8. 2 cm
9. Acinus
10. Terminal ductal lobular units (TDLUs)
11. Involution
12. Multiparity
13. Klinefelter's syndrome
14. The alveoli and lactiferous ducts become dilated and distended with milk.

Chapter 3 Quiz

3-1. Answer: D. Hormones are the most important factor in breast growth and development; however, breast shape and size change with age and weight gain and loss.

3-2. Answer: D. The upper extent of the breast is at the clavicle at the second or third rib. The lowest extent is the abdominal wall at the level of the sixth or seventh rib. Medial extent is the midsternum, and lateral extent is the mid-axillary.

3-3. Answer: C. The breast is composed of approximately 15–20 *lobes*.

3-4. Answer: B. The retromammary space is filled with a layer of adipose or fatty tissue as opposed to the supporting and connective tissue (stroma), blood vessels, and various ductal structures that make up the glandular and fibrous tissues of the breast.

3-5. Answer: B. The breast is loosely attached to the fascia covering the pectoralis major muscle. Latissimus dorsi muscle is a broad triangular muscle located on the inferior part of the back. Pectoralis minor muscle is a thin, flat, triangular muscle lying below the pectoralis major. Serratus anterior is a large, flat, fan-shaped muscle on the lateral aspect of the chest.

3-6. Answer: A. Parity is the terminology used if a woman carries a pregnancy to a point of viability (20 weeks of gestation) regardless of the outcome.

3-7. Answer: A. Hormone therapy (HR) is mainly used to relieve the symptoms of menopause. It can also increase the proliferation of glandular tissue and can be used to control osteoporosis associated with lower estrogen.

3-8. Answer: B. Montgomery glands are specialized sebaceous type glands found on the areola. The TDLU is the terminal ductal lobular unit where milk is formed. The retromammary space (not gland) is the area immediately behind the breast, and oil cysts are benign structures found in the breast.

3-9. Answer: C. The milk enters the ductal system from the TDLU through the segmental or mammary ducts. As the ducts approach the nipple, it widens to form a dilated structure called the ampulla or lactiferous sinus before draining into the last portion of the ductal structure—the lactiferous duct.

3-10. Answer: C. The prepuberty breast is composed of very few lobules (glandular components). As a woman ages, the glandular components are replaced with fatty tissue in a process called involution. Only rarely do men develop glandular components within the breast.

CHAPTER 4: BENIGN AND MALIGNANT DISEASES OF THE BREAST

1. A milk-filled cyst with a high fat content. It is associated with lactation.
2. Breast where the skin is thicker than normal—described as thickened skin syndrome
3. Fatty tissue is seen as a dark area on the radiograph, whereas dense breast tissue, which represents fibroglandular areas, is seen as white. Lesions are also seen as white, which makes them difficult to discern in dense breast tissue.

4. A technique used to reduce the overall size of the breast for cosmetic or medical reasons

5. The halo is a complete or partial radiolucent ring seen predominately around circular or oval lesions. Typically, circular or oval lesions are benign, but the content of any lesion must be assessed.

6. Implant rupture or leaks

7. If the lesion is radiolucent, the breast parenchyma is visible through the lesion. An example is a cyst.

8. Radiolucent and radiopaque, low-optical density, halo sign, capsule, spherical or oval, smooth borders

9. Causes of benign unilateral skin thickening include infection that can result in axillary lymphatic obstruction, which blocks the lymphatic drainage from the breast. Benign conditions that can produce bilateral skin thickening include cardiac failure and chronic renal failure.

10. Dimpling or thickened skin syndrome

11. Could indicate a malignancy, especially if seen in the posterior aspect of the breast in a postmenopausal woman. The breast tissue structure should be a mirror image on both sides.

12. Mammographic characteristics: distinct central mass, sharp dense line radiating in all directions from the central mass, dimpling or skin thickening if the spicules reach the skin; can be associated with malignant type calcification. Ultrasound characteristics: spiculated outline, taller than wide, shadowing, ductal extension pattern, microlobulations

13. Smooth borders, high uniform tissue composition, evenly scattered, sharply outlined, spherical or oval, teacup calcifications on the mediolateral (ML), ring-like, eggshell-like, large bizarre size

14. Calcifications can be analyzed for density in terms of how radiolucent or radiopaque they are, distribution, change over time, number, morphology, and size.

15. Ductal carcinoma and lobular carcinoma

Chapter 4 Quiz

4-1. Answer: D. Reduction mammoplasty reduces the overall size of the breast. In general, all breast reduction will involve the relocation of the nipple.

4-2. Answer: A. *Peau d'orange* describes a thickening of the skin of the entire breast. Skin thickening can be caused by benign conditions or malignant disease.

4-3. Answer: D. Casting-type calcifications are produced when ductal carcinoma *in situ* fills the ducts. The calcifications are the result of cell necrosis at the center of the growth. Milk of calcium forms in the lobules as a result of calcified debris. Plasma cell mastitis is the result of increased cellular activity, and vascular calcifications are calcified arteries or veins.

4-4. Answer: D. All of the above are general features of a benign circular or oval lesion.

4-5. Answer: C. Architectural distortion represents areas of increased glandularity on one breast that are not mirrored on the other side. The cause can be benign or malignant. Stellate or spiculated lesions often represent malignancy.

4-6. Answer: C. The halo often will surround an oval or circular lesion. These lesions are often benign. However, a circular or oval lesion can hold a cancer within the lesion itself. In any evaluation, the content of the circular or oval lesion must be checked.

4-7. Answer: B. Oil cysts are benign eggshell-like calcifications.

4-8. Answer: A. Ductal carcinoma and lobular carcinoma are the two most common types of breast cancer. Metastasis refers to the spread of cancer beyond the breast.

4-9. Answer: C. Benign lesions are often circular or oval with few lobulations. Cancerous lesions are often spiculated with multiple lobules and high optical density.

4-10. Answer: D. Fibroadenomas are formed from stromal and epithelial elements. Although they are most often benign, a cancer can grow within a fibroadenoma.

CHAPTER 5: MAMMOGRAPHY EQUIPMENT

1. Compton's effect, photoelectric effect, coherent or classical scattering, pair production, photodisintegration

2. Characteristic or photoelectric radiation

3. Cathode

4. Line voltage from a wall outlet enters the x-ray circuit and is adjusted with a line voltage compensator. The line compensator measures the incoming voltage and adjusts that voltage to precisely 220 V. Incoming voltage can vary by as much as 5%, so the line voltage compensator ensures that the voltage going to the x-ray tube is always constant, with no variations or fluctuations.

5. Molybdenum and rhodium

6. Decreasing target angle will decrease effective focal spot.

7. Magnification

8. The cathode is positioned to the chest wall.

9. The tube window will filter the emerging beam. It is important that the tube not filter out the low-energy x-ray needed for mammography imaging.

10. Molybdenum

11. Very long and very short exposures, especially in analog imaging

12. 25–45 lb (111–200 N) are the minimum and maximum forces on initial power drive of the automatic compression.

13. The filament circuit is used to supply the lower voltage and higher current needed for the tube filament to heat the filament. The filament circuit therefore provides the electrons for interaction with the target. Adjusting the current being applied to the filament results in variations in the radiation intensity. When the tube current is varied, the number of electrons being supplied to the anode (target) varies. Increase mA will increase electron emission.

14. The filament circuit is used to create thermionic emission at the cathode. The filament is located at the cathode end of the x-ray tube, and it is a coil of wire approximately 2 cm in diameter and 1 to 2 cm long. Connections from the autotransformer provide voltage for the filament circuit. However, this is a high voltage. A filament step-down transformer ensures that the voltage to the filament is low and the current is correspondingly high. The step-down transformer converts high voltage/low current to low voltage/high

current. This high current causes the filament to glow and to emit heat and electrons. The cloud of electrons will remain in the vicinity of the filament before being accelerated to the anode when the high voltage is applied to the tube.

15. Extraneous light will compromise the perception of fine details.

16. The range is 50–80 cm with an average of 60 cm.

17. The anode rotates through the principle of electromagnetic induction. An induction motor consists of two parts separated from each other. The part outside the tube is called the *stator,* which is a series of electromagnets. The part inside is the rotor. Current in the stator winding induces a magnetic field that surrounds the rotor. The stator winding is energized sequentially so that the induced magnetic field rotates on the axis of the stator. The field interacts with the ferromagnetic of the rotor, causing it to rotate synchronously with the activated stator winding.

18. If a projectile electron strikes the target of the x-ray tube, it can interact with an inner shell electron, knocking the electron out of its orbit. The ejection of an inner shell or K-shell photoelectron results in a vacancy in the K shell, which is then filled by an outer shell electron. The movement of the electron from the outer shell to the inner shell results in the emission of an x-ray. The x-ray emitted, called *characteristic radiation,* has energy equal to the difference in the binding energies of the orbital electrons involved. A photoelectric interaction cannot occur if the energy of the incident electron is lower than the binding energy of the electron of the target. The higher the atomic number of the target, the higher the binding energy. This means that the characteristic radiation produced by each element will be different, depending on the atomic number of the element, and is not affected by the kVp.

19. A projectile electron striking the target of the x-ray tube can lose its kinetic energy as it interacts with the nuclear field of the target atom. Because the electron has a negative charge and the nucleus is positive, the electron is attracted to the nucleus. It is slowed and will change course as it loses energy. The loss of kinetic energy reappears as an x-ray photon. This type of x-ray production is called *bremsstrahlung x-ray.* Unlike characteristic radiation with specific energies, the energy of bremsstrahlung radiation can vary tremendously because the incident electron can lose some, all, or very little energy. The only certainty is that the final energy cannot exceed the initial energy.

20. Inherent filtration is represented by anything that the x-ray beam passes through before it enters the breast. The material used for the exit port or window in mammography tube is borosilicate glass or beryllium because a regular glass window would harden the emerging beam by eliminating the soft characteristic radiation. The oil in the x-ray tube, the mirror assemble of the collimation design, and the composition of the compression plate are all designed to attenuate very little of the x-ray beam.

21. Clear communication of instructions to patients.

22. Image brightness can be controlled by the processing software, which is a predetermined digital processing algorithm. In any digital image, the user can alter the brightness of the digital image after exposure.

23. Image contrast is controlled by the processing software, which is a predetermined digital processing algorithm. In any digital image, the user can alter the contrast after exposure.
24. Postprocessing options include:
    - Windowing to adjust image contrast and brightness on the monitor. The window width will control the contrast and the window level controls the brightness.
    - Smoothing the image by bringing the brightness value of adjacent pixels closer together
    - Magnifying or zooming all or parts of image
    - Edge enhancement to increase the brightness along the edges of structures to increase the visibility of the edges
    - Subtraction to remove background anatomy
    - Image reversal, whereby the dark and light pixel values of an image are reversed (e.g., from negative to positive)
    - Annotating the image with text or numbers

Chapter 5 Quiz

5-1. Answer: A. By having the same element as the mammography tube target, added filtration is designed to allow the K-characteristic x-rays of the specific target to be emitted while absorbing the higher and lower energy bremsstrahlung x-rays.

5-2. Answer: D. Aluminum is not used in analog imaging because this would absorb the low energies needed for breast imaging.

5-3. Answer: D. All these structures are in the part of the beam as it leaves the anode. There they will have an impact on the energy of the beam.

5-4. Answer: D. Mammography imaging uses smaller focal spot sizes than general x-ray imaging. Also, the material of the exit port window and the filtration used all are designed to minimally absorb the low-energy photons needed for breast imaging.

5-5. Answer: D. Magnification mammography is used to magnify one area of the breast to image calcification, lesions, or specimens. In general, magnification will not image the entire breast. The radiation dose to the patient, particularly to the patient's skin, is higher in magnification mammography because the breast is very close to the radiation source.

5-6. Answer (D). Mammography interpretation monitors should be a minimum of 4 megapixels, although 5 megapixels or above is recommended. Typical monitors are 21.3 inch, 5-megapixel, monochrome liquid crystal display (LCD). They are high contrast with a wide viewing angle and a luminance of up to 750 cd/m^2 (candelas per square meter). Many have a luminance equalizer and calibration function to enable smooth gray-scale display.

5-7. Answer: C. The exit port is made of borosilicate glass or beryllium because regular glass would harden the emerging beam by eliminating the low-energy photons needed for breast imaging.

5-8. Answer: A. The anode material produces characteristic x-ray in energies that will easily penetrate both fatty and dense breast tissue structures. After leaving the target, the beam is further filtered to remove low- or high-energy photons that are not needed in image production.

5-9. Answer: C. Image contrast is controlled by the processing software, which is a predetermined digital processing algorithm. In any digital image, the user can alter the contrast after exposure.

5-10. Answer: B. Windowing can adjust image contrast and brightness on the monitor. The window width controls the contrast, and the window level controls the brightness. Smoothing involves bringing the brightness value of adjacent pixels closer together. Edge enhancement increases the brightness along the edges of structures to increase the visibility of the edges. Subtraction is used to remove background anatomy.

CHAPTER 6: BREAST IMAGING MAMMOGRAPHY

1. Compression:
   a. Allows a more uniform attenuation of x-rays by flattening the base of the breast to the same degree as the more anterior regions, permitting optimal exposure of the entire breast in one image.
   b. Reduces dose to the breast by reduced tissue thickness.
   c. Brings lesions closer to the image plate for more accuracy when evaluating fine details.
   d. Decreases motion unsharpness by immobilizing the breast during the exposure.
   e. Increases contrast by reducing the amount of scattered radiation and by decreasing breast thickness.
   f. Separates superimposed areas of glandular tissue by spreading apart overlapping tissue, reducing confusion caused by superimposition shadows, and allowing visualization of the borders of circumscribed lesions.
2. Having fibrocystic breast or breast cyst
3. Medial and superior are more rigidly attached than the inferior and lateral margins
4. The final image is magnified, and the entire breast will not be imaged.
5. Tissue that is eliminated or poorly imaged on one projection will be imaged on another.
6. Anterior, central, medial, and posteromedial
7. The air gap replaces the grid. The grid will increase exposure time, tube loading, and motion artifact due to long exposure time.
8. The angle of the pectoral muscle
9. Extreme posterior, upper outer quadrant
10. Drooping breast, abdominal tissue in the image, missing or closed inframammary fold, skin folds in the axilla region
11. Cleavage (CV)
12. Lateromedial oblique (LMO)
13. Tangential (TAN)
14. Exaggerated craniocaudal lateral (XCCL)
15. The spatula is smaller than a finger, which would get in the way of the compression plate before adequate compression is achieved on the small breast.
16. Eight exposures

17. Radiation causes edema or coarsening of the fibrous or stromal elements of the breast, plus increased thickness and density of the ductal and glandular elements. These changes will diminish or resolve over time but will affect interpretation on the mammograms.

18. To localize a nonpalpable breast lesion when imaging spot compression or magnification

Chapter 6 Quiz

6-1. Answer: B. The craniocaudal (CC) best demonstrates the anterior, central, medial, and posteromedial portions of the breast but is poor at visualizing the lateral and posterior lateral breast tissues.

6-2. Answer: A. The XCCL demonstrates the most posterior aspect of the breast in the CC position. The mediolateral (ML) is a lateral projection and the LMO and mediolateral oblique (MLO) are both oblique projections.

6-3. Answer: C. The inframammary fold should be open when imaging the MLO projection.

6-4. Answer: D. Spot compression applies focused compression to one area using a small compression paddle. Cleavage images the medial breast, the tangential is used for skin calcifications or lesion, and the axillary tail images the tail of Spence.

6-5. Answer: B. The tangential is useful in projecting an area in question free of superimposition and is often used to locate skin calcification. The projection can be taken with the tube rotation in any direction.

6-6. Answer: A. Minimal compression is used when the implant is within the compression field. This avoids the possibility of implant rupture.

6-7. Answer: A. The superior-inferior oblique (SIO) directs the central ray from the upper outer aspect to the inferior medial aspect. A 45-degree tube angulation can be used to image the most posterior and inferior portions of the lower outer quadrant of the breast.

6-8. Answer: D. The main reason for breast compression is to maximize the visualization of breast tissue by imaging a uniform breast thickness.

6-9. Answer: A. Lateral lesions move down on the ML from their position on the MLO. Medial lesions move up on the latter from their position on the MLO. Central lesions will not change significantly.

6-10. Answer: D. The lateromedial (LM) directs the central ray from lateral to medial, placing the medial breast closest to the image plate. The medial breast will therefore be best seen.

CHAPTER 7: BREAST IMAGING—DIGITAL MAMMOGRAPHY, ULTRASOUND, AND MAGNETIC RESONANCE IMAGING

1. Dynamic range is the range of values over which a system can respond and is also known as the gray-scale range.

2. A 5×5 image matrix of cells will have less information stored per cell than a 10×10 matrix of cells.

3. Selenium (Se) is the common photoconductor used in direct flat-panel detector systems.

4. Computed mammography uses a flexible imaging receptor (IR), also called a photostimulable storage phosphor (PSP) because it is coated with storage phosphors.

5. Scintillation is the process by which certain materials scintillate or emit a flash of light in response to the absorption of ionization radiation.

6. Display methods include cathode ray tube (CRT), liquid crystal display (LCD), organic light-emitting diode display (OLED), and laser-printed film.

7. The wide latitude of the digital signals

8. Piezoelectric effect

9. They have greater resolution.

10. Artifactual echoes will be created, mimicking a malignant lesion.

11. To avoid obscuring the margins of the lesions

12. Rule out a malignancy or prove a benign lesion by showing that the cyst is fluid-filled versus solid

13. Smooth

14. Malignant

15. Benign

16. Malignant

17. One of each

18. Precession

19. To ensure superconductivity

20. Radiofrequency (RF) coil or RF probe

21. The switching on and off of the gradient coils

22. Longer T1 and shorter T2

23. Time to echo (TE)

24. Gradient coils

25. Gradient-echo pulse

26. Within 5 minutes

27. Aliasing artifact

28. Fibroadenomas, inflammation, recent scarring, surgical scars, fibrocystic changes, active glandular tissue

29. The magnetic properties of the MRI machine

30. Yes. Not all clips have magnetic properties. The clip should first be checked for magnetic properties.

31. Map tumor extent, screen dense breast, detect cancer reoccurrence, detect the spread of cancer, and assess the effects of chemotherapy or tamoxifen response, image implant to access leaks or rupture.

Chapter 7 Quiz

7-1. Answer: A. The data contained in the images will increase, but the image noise will also increase.

7-2. Answer: C. The range of values over which a system can respond. It is also known as the gray scale or the number of shades of gray that can be represented in each pixel.

7-3. Answer: A. The detector systems are direct or indirect. The direct system creates the electrical signal in one step, whereas the indirect system uses a two-step process similar to screen/film systems. The dynamic range refers to the range of values over which a system can respond, and the spatial frequency is the measure of the line pairs per centimeter.

7-4. Answer: D. During the reading process, light that is emitted from the photostimulable storage phosphor (PSP) is collected by a light guide and sent to a photomultiplier tube (PMT), photodetector (PD), or charge-coupled device (CCD). The PMT detects the blue light given off by the trapped electrons as they return to their normal neutral state. Signal from the PMT is sent to an analog-to-digital converter (ADC) and then to a computer for displays. The resultant digital information can now be electronically transmitted, manipulated, and more efficiently stored.

7-5. Answer: A. During an MRI examination, the patient is placed within a powerful magnet. Any metal objects on or around the patient can be a danger because they will be drawn to the magnet.

7-6. Answer B. Indirect digital systems use a two-step process. A scintillator, such as cesium iodide (CsI) doped with thallium (Tl), absorbs the x-rays and generates a light scintillation that is then detected by an array of the thin-film transistor (TFT) or thin-film diode (TFD). Amorphous silicon and amorphous selenium are photoconductors used to collect electrons in the TFT.

7-7. Answer: A. The MR image is a record in the difference in radio-frequency signals from different body tissues. A large value for T1 or T2 indicates a long, gradual relaxation time, whereas a small value indicates a rapid relaxation time.

7-8. Answer: B. Sound is a mechanical longitudinal wave that cannot travel through vacuum. It needs a medium for travel. Sound also travels very poorly through gasses.

7-9. Answer: D. These are the characteristics of a benign lesion. Malignant lesions tend to be irregular and spiculated margins. Intermediate lesions are harder to define, but they tend to have an echogenicity similar to fat.

7-10. Answer: A. The transducer consists of crystal made up of piezo-electric material capable of converting electrical energy into acoustic energy on transmission and acoustic energy into electrical energy on reception.

CHAPTER 8: BREAST IMAGING—ADJUNCTIVE MODALITIES

1. Digital tomosynthesis
2. Elastography
3. Magnetic resonance spectroscopy
4. Nuclear imaging or molecular imaging can include positron emission tomography/mammography, breast-specific gamma imaging (BSGI), scintigraphy or scintimammography, lymphoscintigraphy, or sentinel node mapping.
5. Radiopharmaceutical, radioactive tracers, radionuclide, radio-isotopes
6. Positron emission tomography or positron emission mammography
7. Positron emission tomography or positron emission mammography
8. ^{99m}Tc-sestamibi
9. The size of the gamma camera used and the position of the patient during the study. BSGI uses a small breast gamma camera and has the patient seated versus lying during the study.

10. Carpal tunnel syndrome
11. A blue dye—isosulfan blue vital dye
12. Angiogenesis
13. Optical imaging
14. Cone-beam CT imaging
15. Cannot detect microcalcification; reports a high false-positive rate; is difficult to image small breasts
16. Electrical impedance spectroscopy, transscan, or T scan

Chapter 8 Quiz

8-1. Answer: B. Scintigraphy or scintimammography injects the radiopharmaceutical into an arm vein; MRI uses the magnetic properties of elements; lymphoscintigraphy is used to track breast cancer as it leaves the breast.

8-2. Answer: D. Scintigraphy injects the contrast directly into a site of the lesion to track how cancer cells would leave the breast; MRI uses the magnetic properties of elements; PEM uses the principle that cancers need vast amounts of glucose to survive.

8-3. Answer: C. These will involve the injection of a radiopharmaceutical and are considered nuclear imaging. MRI uses the magnetic properties of elements.

8-4. Answer: B. MRS measures the functional breast cancer by-product, choline. Elastography uses ultrasound to scan the breast based on the principle that cancers are relatively hard compared with the surrounding tissue. BSGI used the radioactive compound ^{99m}Tc-sestamibi.

8-5. Answer: C. Scintigraphy or scintimammography injects the radiopharmaceutical into an arm vein; PET imaging uses a radioactive compound with the principle that most cancers need a vast amount of sugar to survive and grow.

8-6. Answer: C. Lymphoscintigraphy is used to track breast cancer as it leaves the breast. BSGI, or breast-specific gamma imaging, is the method of breast scintigraphy using specialized gamma cameras. Optical imaging uses infrared light to penetrate breast tissue.

8-7. Answer: A. Elastography uses ultrasound to scan the breast based on the principle that cancers are relatively hard compared with the surrounding tissue.

8-8. Answer: C. Fluorodeoxyglucose is used in PEM imaging, gadolinium is used in MRI imaging, and mammography uses radiation.

8-9. Answer: A. Elastography uses ultrasound to scan the breast based on the principle that cancers are relatively hard compared with the surrounding tissue. Lymphoscintigraphy is used to track breast cancer as it leaves the breast. BSGI, or breast-specific gamma imaging, is the method of breast scintigraphy that uses specialized gamma cameras.

8-10. Answer: C. Digital breast tomosynthesis uses radiation to create 3D images of the breast. MRI uses powerful magnetic and radio-frequency pulses to create image. PET uses a radioactive compound with the principle that most cancers need a vast amount of sugar to survive and grow.

CHAPTER 9: INTERVENTIONAL PROCEDURES

1. Cytological analysis refers to the study of cells; histological analysis takes larger core samples to assess the organ and breast tissues.
2. BI-RAD 4 and higher
3. Aspiration
4. Stereotactic
5. Less space to work, risk for vasovagal reaction, possibility of motion
6. More expensive, units have only one use, posterior breast hard to image
7. 22–25
8. The distance traveled by the needle: short throw, less than 2 cm; long throw, more than 2 cm
9. Open
10. Mammography
11. To ensure the lesion was removed, to ensure clean margins
12. To evaluate nipple discharge, filling defects, irregularities, duct expansions, or duct defects

Chapter 9 Quiz

9-1. Answer: C. A lesion can be definitively classified only with histological or cytological analysis.

9-2. Answer: A. Augmentation is the technique used to increase breast size. Mastopexy is an elective surgical procedure designed either to lift or to change the shape of a woman's breast. Nipple aspiration removes the content of the ducts for evaluation. Ductography is a technique used to evaluate the large ducts.

9-3. Answer: B. Cosmetic intervention, or mastopexy, is generally done at the convenience of the patient, which is unlike a medically needed breast reconstruction, which could be required because of breast cancer treatment or to relieve back or shoulder pain associated with large breasts.

9-4. Answer: D. Ductography is the technique used to check for ductal defects and nipple discharge. Axillary dissection removes the lymph nodes in the axilla. Aspiration biopsy removes the content of a cyst. Lumpectomy removes the lesion and surrounding margins.

9-5. Answer: D The open surgical biopsy has the highest complication rate and the lowest false-negative rate. Next would be any of the core biopsies regardless of the modalities. Fine needle aspiration biopsy (FNAB) is the least accurate but least expensive and has the lowest complication rates.

9-6. Answer B. Core biopsy will obtain tissue samples for histology analysis, the FNAB and ductal lavage are methods of obtaining cellular material, and ductography is a technique used to evaluate the large ducts.

9-7. Answer: B. The core and surgical biopsies both remove the tissue sample for histological analysis.

9-8. Answer: B. The mastectomy removes all breast tissues. The lumpectomy removes the lesion and surrounding tissue, the biopsy removes the lesion only, and the axillary dissection removes the axillary nodes.

9-9. Answer: D. If the lesion is only seen on MRI, it must be biopsied using MRI.

9-10. Answer: D. If the lesion is nonpalpable, it needs to be located before any biopsy or aspiration. The stereotactic technique is one method of locating nonpalpable lesions. MRI and ultrasound are detection methods, and if a lesion is found, it will have to be biopsied to determine its exact nature.

CHAPTER 10: TREATMENT OPTIONS

1. Surgery, radiation, drugs
2. A process of classifying the cancer and placing it into five stages depending on the extent of the primacy tumor, regional lymph node involvement, and/or distant metastasis.
3. Modified radical mastectomy removes the entire breast, including the nipple and areola and some underarm lymph nodes. Lumpectomy removes only the tumor and surrounding margins.
4. Radiation therapy and/or chemotherapy and/or hormonal therapy
5. External-beam radiation typically runs 5 to 7 weeks.
6. External- and internal-beam radiation
7. a. Advantages: reduce treatment time, fewer skin reactions
   b. Disadvantages: not enough clinical trial or controlled studies, dose falls rapidly from source, specific areas around the lesion may not get a high dose of radiation
8. The drug is distributed throughout the entire body and will affect all tissues and organs.
9. To stop or slow the spread of cancer to other parts of the body, to kill cancer cells, and to relieve symptoms of cancer.
10. Chemotherapy affects the bone marrow that makes blood cells.
11. For estrogen receptor–positive cancer, tamoxifen blocks estrogen to the cancer, therefore lowering the risk for recurrence.
12. Photodynamic therapy
13. They prevent the growth of new blood vessels necessary for cancer spread.
14. TRAM flap

Chapter 10 Quiz

10-1. Answer: C. Radiation therapy can be teletherapy (external), whereby the treatment is given once a day for 6 to 7 weeks, or brachytherapy (internal), whereby the treatment can last 5 to 9 days. External-beam radiation is sometime called proton-beam radiation because it is considered therapeutic radiation using photons (x-rays or gamma rays) or particulate radiation (electrons, protons, or neutrons).

10-2. Answer: D. The lumpectomy removes the lesion and surrounding margins. The mastectomy, in which the entire breast plus most of the chest wall muscle is removed, is no longer performed. The modified radical mastectomy removes most of the breast tissue plus the nipple and areola. Chemotherapy uses drugs to treat breast cancer.

10-3. Answer: A. Chemotherapy affects all the organs and tissues of the body. Complete blood cell count, including red cells, white cells, and platelets, will fall, and the effect of chemotherapy on the cells lining the gastrointestinal tract can cause nausea and vomiting.

10-4 Answer: B. While removing a lesion the surgeon must ensure that all margins are clean. A positive margin indicates that all the cancerous cells have not been removed and the surgeon will have to remove more tissues.

10-5. Answer: C. Neoadjuvant treatment is given before surgery to help shrink the size of the cancerous tumor. Adjuvant chemotherapy is given in addition to another cancer treatment.

10-6. Answer: D. The benefits of tamoxifen are only considered effective within a 5-year period.

10-7. Answer: A. Tamoxifen is an antiestrogen drug used to block estrogen from latching to the cancer receptor cells.

10-8. Answer: C. Brachytherapy is a method of internal-beam radiation therapy. With brachytherapy, the high-energy radiation beam is concentrated on the cancerous tissue, sparing the surrounding normal tissue. This allows reduction in treatment time, less irritation of normal tissue, and fewer skin reactions.

10-9. Answer: B. In reduction mammoplasty, the oversize of the breast is reduced. Augmentation will increase breast size. The core biopsy removes tissue sample for analysis, and ductography is a check of the larger ducts.

10-10. Answer: C. The TRAM flap is the transverse rectus abdominis myocutaneous muscle flap in which tissue and muscle from the abdomen are reconstructed to make a breast mound. In the latissimus dorsi flap, tissue and muscles from the back are used. The pedicle flap and free flap describe methods of any flap technique. With the pedicle flap, the removed tissue and/or muscle remain attached to their original blood supply and are tunneled under the skin to the breast area. In the free flap, the tissue and/or muscle are removed and reattached using microsurgery to reconnect the blood vessels.

CHAPTER 11: QUALITY ASSURANCE AND QUALITY CONTROL

1. Preventative maintenance
2. Good QA will improve: overall efficiency and delivery of service; patient satisfaction; reduction of repeats and, therefore, radiation dose to the patients; consistency in image production; and cost-effectiveness of equipment
3. There are a number of agencies responsible for safety and quality imaging in the United States. These include the Food and Drug Administration (FDA), the Occupational Safety and Health Administration (OSHA), the National Center for Devices and Radiological Health (CDRH), the Nuclear Regulatory Commission (NRC), and the Centers for Disease Control and Prevention (CDC).

4. Causes for repeat may be related to the competence of technical staff, equipment problems, specific difficulties associated with particular examinations, or a combination of these.
5. Semiannually
6. 25 and 45 lb (111 and 200 N)
7. Centers for Disease Control and Prevention (CDC) is recognized as the main federal agency ensuring the protection, health, and safety of people—both at home and at work. The CDC develops and applies disease prevention and control and promotes many educational activities designed to improve the health of the people of the United States.
8. Main patient care responsibilities of the technologist: patient positioning, compression, image production, image processing, and infection control
9. Starting in 2017, the FDA approved the use of the new *ACR Digital Mammography Quality Control Manual* for two-dimensional digital mammography systems. Facilities with tomosynthesis or contrast enhancement equipment must follow all manufacturers' quality control guideless for both the digital mammography application and the advanced technology application of their system. Any facility using the ACR digital mammography QC guideline must also use the ACR digital mammography phantom with all the manual's procedures.
10. 2% of the source-to-image receptor distance (SID)
11. The interpreting physician—the radiologist
12. Collimation assessment
13. Phantom images should be viewed by the same person, using the same type of magnifier, at the same time of day, under the same viewing conditions.
14. 250
15. 2%
16. Identify the source of the problem and take corrective action before any additional mammography examinations performed.

Chapter 11 Quiz

11-1. Answer: B. Beam quality assessment in a general test performed by the medical physicist.
11-2. Answer: C. Standards are reference objects or devices with known values used to check the accuracy of other measurement devices.
11-3. Answer: D. It is the radiologist's responsibility to oversee the activities of all personnel responsible for the quality assurance process.
11-4. Answer: C. The average glandular dose is an annual test done by the medical physicists. The phantom test is done weekly by the technologist. The visual checklist is done monthly by the technologist. The compression force test is done semiannually by the technologist.
11-5. Answer: C. The exposure reproducibility is a check of the AEC. The system should deliver a consistent optical density regardless

of compressed thickness and composition. The kV accuracy ensures that the kV selected is within ±5% of the indicated kV. The radiation output is the selected kV and mA. The exposure time is the milliseconds of the actual exposure.

11-6. Answer: B. MQSA was enacted to ensure quality patient care and imaging for patients needing a mammogram. The MQSA involves regular checks of the training and continuing education of radiologist, technologist, and medical physicists.

11-7. Answer: B. All images repeated that result in an additional exposure to the patient are considered repeated images. Images that are used in quality control and do not cause additional patient dose are considered rejects, not repeats.

11-8. Answer: C. It is the responsibility of the medical physicist to regularly check the mammography unit and component parts and to generate reports as needed.

11-9. Answer: B. Tests on the x-ray tube are generally performed by the medical physicist.

11-10. Answer: B. The signal-to-noise ratio (SNR) is a random disturbance that obscures or reduces image clarity. Decreased SNR will give the image a grainy or mottled appearance. The modulation transfer function (MTF) checks the consistency of the contrast-to-noise ratio (CNR) and ensures adequate contrast over a specific spatial frequency range. The detector evaluation ensures adequate and consistent quality images acquired by the detector and displayed on the acquisition workstation (AWS) or printer.

CHAPTER 12: THE US MAMMOGRAPHY QUALITY STANDARDS ACT (MQSA)

1. To establish minimal national quality standards for mammography

2. Accreditation is the initial step when the facility has to show that it meets the standards under the MQSA. Certification is the final step when the facility is granted a certificate allowing the facility to legally perform mammograms.

3. 6 months

4. States of Arkansas, Iowa, and Texas

5. 3 years. However, to maintain certification, the facility must be inspected annually by FDA (US Food and Drug Administration) or state inspectors; have specific mammography equipment that is periodically surveyed by a medical physicist; employ specially trained personnel to perform mammograms and conduct regular quality control tests; have a quality assurance program; have a system in place to track mammograms that require follow-up studies or biopsies; and have a tracking system in place to obtain biopsy results.

6. There are one or more deviations that could compromise the quality of the mammography service.

7. The facility must transfer the medical record to a facility where the patient will be receiving future care, to another physician, to a health care provider, or to the patient.

8. Currently, the FDA-approved accreditation bodies are not prepared to accredit 3D tomography because standards have not been

developed for this imaging modality. This situation is similar to when FDA approved FFDM for marketing, but the accreditation bodies had not yet developed standards and could not accredit FFDM when it first came on the market. FDA developed the certificate extension program to cover situations in which a mammography device has been approved for marketing but accreditation standards have not yet been established. The certificate extension program that is currently in effect so that facilities can legally use 3D tomography requires that an FDA-approved accreditation body accredit the 2D imaging component of the 3D tomography unit.

9. The following test must be performed each time the unit is moved: (1) ACR (DM) phantom image quality–after each move and before examining patients; (2) printer QC (for mobile film printers only)—after each move and before printing patient images; (3) radiologist workstation (RWS) monitor QC (for mobile RWS only)—after each move and before interpretation; (4) compression thickness indicator—after each move and before examining patients.

10. By posting the following notice, "If there are any questions or comments about your mammogram please contact . . ." listing a contact name, title, and telephone number or email.

11. Not less than 5 years or not less than 10 years if the patient has no additional mammograms at the site. State laws may impose higher restrictions.

12. Name of patient plus an additional patient identifier, date of the examination, name of interpreting physician, and overall assessment of findings

13. The finding has some but not all the characteristics of malignancy.

14. The physician should be notified of concern finding within 3 to 5 days.

15. A method of tracking positive mammography results to correlate pathology results with the interpreting physician's findings

16. HIPAA (Health Insurance Portability and Accountability Act) will not affect the keeping of medical audits.

17. 25

18. 960

19. 15 CEUs

20. Conduct regular reviews of the medical audit.

Chapter 12 Quiz

12-1. Answer: A. The pathology review involves tracking all concern findings and documenting the results from any biopsy. Nonconcern findings and elective surgical procedures, such as cosmetic intervention and mammoplasty, are not tracked.

12-2. Answer: D. A viable quality assurance program must include all of the above. The program must be evaluated periodically with a review of records, and there must be defined standards. Standards must have a tolerance value. A reading above or below such a value would be an indication that corrective measures are needed.

12-3. Answer: C. Manipulation of the final image is an option available with digital technology; however, digital technology is not a requirement for MQSA.

12-4. Answer: B. HIPAA regulation only allows release of information related to the patient breast imaging or biopsy. No other releases are allowed. In addition, the general consent signed by patients at many facilities allows release for insurance purposes and does not allow the sharing of patients' medical information between facilities.

12-5. Answer: D. Normal results must be sent to the patient within 30 days. Notice of suspicious finding must be sent within 3 to 5 days of the interpretation, to the patient's health care provider.

12-6. Answer: C. BI-RAD 0—needs additional evaluations or comparisons. BI-RAD 5—highly suggestive of malignancy. BI-RAD 6—patient had a prior positive biopsy.

12-7. Answer: B. The mammography records must be kept for not less than 5 years if the patient returns for mammograms at the facility. If the patient does not return, the facility must keep the records for not less than 10 years. Some state or local laws may require longer storage times.

12-8. Answer: C. The MQSA certificate must be visible to all patients coming to the facility. A general waiting area is an ideal place; however, if needed the facility can obtain additional copies for placement in individual mammography rooms.

12-9. Answer: B. 15 CEs in mammography are required during the 36 months immediately preceding the last inspection.

12-10. Answer: D. Initial qualification technologist after April 28, 1999 includes all of the above.

Index

Note: Page numbers with (*b*) are boxes, with (*f*) are figures, with (*t*) are tables

A

ABBI (advanced breast biopsy instrumentation), 337
Abdomen, protruding, 190
Ablation
  thermal, 387
  radiofrequency, 387
  laser, 387
Ablative hormone therapy, 383
Abortion, 52
Abscesses, 86, 248
Absorption efficiency, 215
Accessory breast, 24
Accreditation, 442, 445
  withdrawal, 449
  suspension, 451–452
  revocation, 451–452
Acinus, 64
Acoustic enhancement, 86, 217
Acoustic impedance, 218
Acquisition workstation (AWS) monitor, 146–147, 413–414, 415*f*, 429–430
Acrochordon, 91
Added filtration, 130–132
Adenosis/tumoral adenosis/adenosis tumor, 86
Adhesive capsulitis (frozen shoulder), 191
Adipose tissue, 73
Advanced breast biopsy instrumentation (ABBI), 337
Air kerma, 402, 516
ALARA (As Low As Reasonably Achievable), 43
Alcohol dehydrogenase (ADH), 53
Aliasing artifact, 267, 269
Aliasing or wrap-around artifacts, 267–268
Alternating current (AC), 118
American Academy of Family Physicians (AAFP), 34
American Cancer Society (ACS), 11
  breast examination recommendations, 34
  MRI recommendations, 272
  screening recommendations, 39
American College of Obstetricians and Gynecologists (ACOG), 34
American College of Radiology (ACR), 9, 12
  accreditation program of, 12–13
  bandwidth requirements, 151
  function of, 404
  guidelines, 29
  mammography standardization by, 163
  screening recommendations, 14, 17
  supplementary projections/positions recognized by, 175
American Registry of Diagnostic Medical Sonographers (ARDMS), 405
American Registry of Radiologic Technologists (ARRT), 175, 404–405
American Society of Radiologic Technologists (ASRT), 405

Amorphous calcifications, 105
Analog exposure artifacts, 537
Analog imaging, 517–525
  automatic processing QC test
    developing, 531
    drying, 533
    fixing, 531–532
    replenishment, 532
    transport system, 533
    washing, 533
  cassette, 520–521, 521*f*
  characteristic curve, 519–520, 520*f*
  crossover, 518*f*
  cross-section diagram, 519*f*
  daily QC test, 529–530, 531
  double *vs.* single emulsion, 517–518
  equipment standard-specific requirement, 529*b*
  mammography equipment evaluation, 534
  quality control duties, 512, 533
  repeat/reject analysis, 544–545
  screen/film systems, 521–524
  semiannual QC test, 529
  semiannual quality QC tests, 533
  shoulder, 522
  straight-line portion, 522
  testing details, 542–549
  underexposure in, 525*b*
  weekly QC tests, 533
Analog mammography, 222
Analog-to-digital converter (ADC), 215
Anaplastic large cell lymphoma (ALCL), 54
Anasarca, 86
Anastrozole, 382
Anechoic, 218
Anesthesia, 358
Anesthetic, 2
Angiogenesis, 284, 358
Angiosarcoma, 91
Anode, 122, 134–135
Anode angle, 118
Anode effect, 132–134
Anode heel effect, 136–137, 136*f*
Antiangiogenesis, 383–385
Antigens, 358–359
Antiperspirants, 55
Areola, 67
Areola reconstruction, 389
Areolar glands. *See* Montgomery's glands
Aristotle (384–322 BC), 5
Arterial circulation, 70
Artifacts, 429. *See also* Ferromagnetic material artifact; Motion artifacts; Ultrasound artifacts
  clinically significant, 409, 418
  evaluation of, 536
  IMRT artifact, 371–372
  reverberations as, 219
  score for, 411
Aspiration, 322
Assessment categories, 460–461
  MQSA, 460
Asymmetric breast tissue density, 86, 100–101
AT (axillary tail) projection, 182–183
ATM gene, 49
Atypical ductal hyperplasia (ADH), 86
Atypical epithelial hyperplasia, 92
Atypical hyperplasia, 86

Atypical lobular hyperplasia (ALH), 86
Augmentation mammoplasty, 93–94
  cancer risks, 53–54
  defined, 24
  history of, 347
  imaging of, 348
  materials used in, 93–94
  placement, 348*f*
  rupture of, 347–348
Autologous tissue reconstruction (ATR), 392–393
  TRAM, 392
  DIEP, 392
Automated Breast Ultrasound (ABUS) SomoVu, 296–297, 296*f*
Automated ultrasound, 296–297
Automatic exposure control (AEC), 140
  characteristics of, 426–427
  detectors, 522
  reproducibility, 541
  use of, 524–525
Automatic processing, 531–533
Automatic tissue excision and collection (ATEC), 337
Average glandular dose (AGC), 427, 541
Axilla, thick, imaging, 190–191
Axillary nodes
  dissection, 359
  drainage, 72–73

B

Background equivalent radiation time (BERT), 42
Background radiation, 24
Backup timer, 140
Balloon catheter, 374
Bandwidth, 219
Bannayan-Riley-Ruvalcaba syndrome (BRR), 215
Barrel chest, 192–193
Baseline mammogram, 2
Basement membrane, 69
Beam quality, 431
Beam restricting device, 142–143
Becquerel, Henry (1852–1908), 7
Benign and malignant lesions
  analysis of, 92–93
  characteristics of, 94–108
    asymmetric densities, 100–101
    calcifications, 101–108
    circular/oval, 95–97
    mammographic, 97*b*
    MRI, 99*b*
  ultrasound characteristics, 248–251, 248*b*
Bilateral breast cancer, 110
Biopsy, 54, 331–342
  controversies, 331–332
  core, 333–342
  fine needle aspiration (FNA), 332–333
  large core, 337
  mammography system for, 336–337
  mammotome, 336–337
  modality-based, 338–342
    equipment necessary, 340
    limitations, 342
    procedural considerations, 341
    ultrasound core biopsy, 340
  MRI-guided, 340
  radiofrequency, 335–336

Biopsy (*continued*)
  surgical, 338
  types of, 331*b*
  ultrasound, 338
Biotherapy, 386
BIRAD system (breast imaging reporting and data)
  assessment categories, 460*b*
  composition category, 50*f*
  defined, 322
Bisphenol A (BPA), 54
Bloch, Felix, 255
Body language, 32–33
Body shape conductivity and extension, 269
BRCA, 46
BRCA1, BRCA2 mutation, 48–49
Brachytherapy, 359
Breast. *See also* Dense breast tissue
  anatomy, 66*f*, 68–70
  arterial supply to, 71*f*
  branching distribution system, 69*f*
  characterization of, 66
  deep anatomy, 68–70
  examinations, 34–41
    average lump size, 39
    documentation sheet for, 35*f*
    guidelines, 39–41
    identifying abnormalities, 36
    noticing changes, 36–38
    patterns for, 38, 38*f*
    recommendations, 34–36, 39–40
    types of, 34
  hormones effect of, 75*b*
  location of, 66*f*
  surface anatomy, 66–67
Breast and Ovarian Analysis of Disease Incidence
    and Carrier Examination Algorithm
    (BOADICEA), 46
Breast cancer
  cell theory of, 6–7
  characterization of, 3–4
  common locations, 92*f*
  early incidence of, 4
  humoral theory of, 6
  preventive surgical options, 365
  research accomplishments, 15–16
  risk factors, 45–52, 48*b*
    age, 47–48, 49*b*
    assessment tool, 47*b*
    assessments, 45–47
    biopsy, 54
    chemical exposure, 54–55
    daily aspirin use, 53
    diet, 56
    dense breast tissue, 50
    injury, 56
    family history, 49–50
    genetic, 48–49
    implants, 56
    major, 47–51
    minor, 51–52
    parity, 78
    personal history, 50
  screenings, 13–17, 27*b*, 44
    frequency recommendations, 14, 17
    guidelines, 40*b*
    Malmö Mammography Study, 13
    risk factors, 17

survival by stage, 17, 17*t*
  treatment throughout the ages, 4–7
  treatments for, 361–388
  types of, 108–112
  ultrasound image, 250
Breast Cancer Assessment Tool, 45–46
Breast Cancer Detection and Demonstration
    Project (BCDDP), 11, 45
Breast changes, 91–94
Breast Imaging and Reporting and Data System
    (BI-RADS), 324
Breast imaging mammography, 163–166
  compression in, 164–166
  craniocaudal (CC) projections, 193, 198*f*
    detector location, 168
    function of, 166, 166–167*f*
    imaging evaluation, 173*b*
    patient positioning, 172–173
    positioning, 163–164
  four-projection series, 163–164
  imaging the nonconforming patient, 188–199
  MLO projections, 169–172, 193, 198*f*
  positioning guide, 168–169
  specific imaging problems, 199–205
  standard projections, 166–173, 166*f*
  supplementary projections, 173–188
Breast molecular imaging (BMI), 285, 305–306
  advantages of, 272–273*f*
  artifacts, 267–272
  augmented breast, 273, 274*f*
  examinations, 266
  irregular margins, 270*f*
  limits of, 273–276
  new technologies, 295
  patient preparation, 266
  quality assurance, 276
  risks/complications, 276
  sequence, 267
  standards, 267, 267*b*
Breast reconstruction
  delayed, 390
  immediate, 360, 390
  on stage, 360
Breast self-examinations (BSE), 14, 27
  conduction of, 37*f*
  feeling for changes, 37–38, 37*b*
  frequency of, 34*b*, 36
  harms of, 14*b*
  looking for changes, 36–37
  lump size found, 39*f*
  patterns, 38, 38*f*
Breast-specific gamma imaging (BSGI)
  analysis of, 308
  defined, 284
  radiation dose for, 310–311
  study, 307
  technology, 306–307
  uses of, 307
Breast tissue composition, 73–74
  factors affecting, 74–78
    hormone fluctuation, 75
Breast ultrasound imaging, 237–282, 237–299.
    See also Ultrasound
  accuracy of, 245*b*
  advantages of, 254
  artifacts, 252–253
  documentations, 245

image recording, 245
  imaging implants, 252
  interactions, 245
  lesion appearance on, 248–251
  limitations of, 253–254
  nonpalpable mass management, 246–247
  palpable mass evaluation, 247
  preparation for, 243
  problems/solutions, 254
  process of, 242–243
  risks of, 254
  scanning technique, 243–245, 244*f*
    contrast enhanced, 299
    full-field breast, 297–298
  uses of, 246
Breathing technique, 289
Bremsstrahlung radiation, 124–125, 125*f*
Bremsstrahlung X-rays, 119
Brightness, 119
Bureau of Radiation Health (BRH), 404
Bytes, 215

C

Caffeine, 162
Calcifications in the breast, 101–108
  amorphous, 105
  analysis of, 101
  benign, 102–103*f*
  cause of, 106–107
  characteristics of, 105*b*
  morphology of, 107*b*
  number of, 104
  types of, 104–105
Calipers, 241, 241*f*
Capacitive microfabricated ultrasonic
    transducers, 298–299
Capsular contracture, 322, 359
Carcinoma, 5
Carcinoma *in situ* stage O carcinoma, 92, 250
Carcinomatosis or carcinosis, 218
Carlson, Chester F., 9
Cartilaginous portions of the ribs, 254
Cassetteless digital systems, 227
Casting-type calcifications, 104–105
Catenary, 219
Cathode, 121–122
Cathode ray tube (CRT), 144–145, 145*f*
Cautery, 2
Cells
  cell cycle, 374
  circulating cancer cells, 382
  Search Epithelial Cell Kit, 384
Cellular Pathology, 6
Center for Medicare and Medicaid Services
    (CMS), 452–453
Centers for Disease Control and Prevention
    (CDC), 404
Central nervous system, 73
Certification, 445–446
Certificate placement, 458
Changes over time, 104
Characteristic or photoelectric radiation
  defined, 119
  elimination of, 130
  energy generated by, 125
  production of, 133
Characteristic x-rays, 130

Checklist, 428
CHEK2 gene, 49
Chemical shift artifacts, 269
Chemotherapy, 375–379
  cell cycle, 376, 376*b*
  defined, 2
  drugs used in, 376–377
  history of, 8
  regimens, 377
  side effects of, 377–379
  uses of, 376
Circular or oval lesions, 95–97
Circulating breast cancer cell, 384
Cleavage or valley view, 181
Clinical breast examinations (CBE),
    14, 39
  benefits, 14*b*
  defined, 24
  questioning techniques, 34–39
  sensitivity for, 36*b*
Clinical image, 446–448
Coarse calcifications, 104
Coherent or classical scattering, 119,
    127, 127*f*
Coils in MRI, 260–261
Collimation
  collimator, 205
    assessment, 432–433, 535
Collimator test, 432
Colloid cancer, 110, 111
Color Doppler ultrasound
  angle of, 248*f*
  defined, 218
  process of, 248
Columnar-structure of the Cl (T1), 282
Communication
  verbal and nonverbal, 32
  results to patient, 461–462
Comedo ductal carcinoma, 107
Complex, 218
Compression
  applying, 165–166, 204
  devices, 138
  force, 419–420, 502
  lossy/lossless, 152
  in mammography, 164–165
  reasons for, 165*b*
  spot, 188
  testing, 529, 547–548
  thickness, 503
Compression paddle/plate, 138–140, 139*b*
  alignment of, 542
  components of, 139*b*
  force, 417–418
  paddle deflection, 422
  thickness indicator, 405*t*, 409–410
Compton scattering, 119, 126, 126*f*
Computed mammography (CM). *See also*
    Mammography
  image quality, 154–155
    resolution, 154
    spatial resolution, 154–155
  test for systems, 429
Computed mammography (CM) erasure,
    411–412
Computed tomographic laser mammography
    (CTLM), 312–313

Computer-aided detection (CAD)
  advantages of, 154
  bandwidth needs, 151
  function of, 152–153
  limitations of, 152
  for MRI, 153–154
  as second reader, 294–295
Computer reader (CR), 215
Cone-beam computed tomography, 312
Connective tissue disorders, 86, 322
Consent. *See* Informed consent
Consumer complaints mechanism, 458
Continuing Education Unit (CEU), 443, 464
Continuous quality improvement (CQI), 66
Contraceptive and Reproductive Experiences
    (CARE) study, 46
Contralateral, 171
Contrast agent, 255, 256, 263
Contrast, 119, 204, 512
Contrast enhanced mammography, 284
Contrast-enhanced ultrasound, 284, 299
Contrast media mammography (CMM), 285, 295
Contrast resolution, 120, 155
Contrast-to-noise ratio (CNR), 230, 422, 426
Control charts, 516
Control crossover, 516
Conversion efficiency, 215
Cooper's ligaments, 64, 74, 248
Copper-conductivity bands, 261
Core biopsy, 322
Core biopsy gun, 329, 335
Corrective maintenance, 402, 516
Cosmetic intervention, 346
Cowden's syndrome, 215
COX-2 (cyclooxygenase-2), 385
Craniocaudal (CC) projections, 172–173, 198*f*
  detector location, 168
  function of, 166, 166–167*f*
  imaging evaluation, 173*b*
  large or mosaic breast, 190
  patient positioning, 163–164, 172–173
  patient with implants, 193
Cryotherapy, 388
Curie (Ci), 24–25
Curie, Jacques, 237
Curie, Marie (1867–1934), 7
Curie, Pierre (1859–1906), 7, 237
Cutaneous papillomas, 91
CV (cleavage or valley view) projection, 182
C-view, 289
Cyclooxygenase (COX), 53
Cystosarcoma phyllodes, 111, 215
Cysts
  aspiration, 325–326, 332*f*
  assessment of, 326
  defined, 64
  occurrence of, 86
Cytological analysis, 324
Cytology, 87

D

Damaged x-ray target, 133*f*
Darkroom
  cleanliness, 548–549
  environment, 529–530
  fog test, 533, 555–556
DDE (dichlorodiphenyldichloroethylene), 54

DDT, 54
De Humani Corporis Fabrica Libri Septem
    (The Seven Books on the Structure
    of the Human Body), 6
Dedicated mammography system, 120, 475
Deep inferior epigastric artery perforator (DIEP),
    390–391
Density compensator, 140
Dense breast tissue
  cancer risk, 44, 51
  characterization of, 50
  mammographic imaging of, 77
Deodorants, 55
Depo-Provera, 52
Detective quantum efficiency (DQE), 215, 230
Detector calibration, 506
Detectors, 162. *See also* Flat-panel detector
      system; Indirect flat-panel detector
      system
  information transfer from, 226
  nonremovable systems, 277
  positioning of, 187
  size of, 233
Diagnostic ultrasound, 244
DICOM (Digital Imaging and Communications
      in Medicine), 119, 150, 215
Diet, 56
DIEP flap, 392
Diffuse optical tomography (DOT), 311
Digital breast tomosynthesis (DBT) or Three
      dimensional mammography, 286–295
  accreditation/certification options, 483–484
  acquisition geometry, 287*f*
  defined, 284
  Fujifilm ASPIRE Cristalle, 291–292
  General Electric SenoCare, 290–291
  hologic 3D unit, 287–291
  image generation, 287*f*
  mammogram comparison, 294*f*
  Medicare reimbursement, 293
  Philips MicroDose unit, 292
  positioning in, 292–293
  potential uses, 293–294
  regulations for, 483–484
  reimbursement for, 484–485
  Siemens Mammomat Inspiration, 291
  specific training for, 293
Digital linear tapes (DLTs), 119, 148
Digital mammography (DM), 222–237. *See also*
      Mammography
  advantages of, 235–236
  approved units for, 224–225*b*
  cell matrix of, 226–227, 227*f*
  characteristic curve of, 236*f*
  defined, 25
  digital-to-analog conversion, 226, 226*f*
  direct flat-panel detector technology, 231–232
  disadvantages of, 236–237
  image acquisition in, 225–226
  image quality, 154–155
  indirect flat-panel detector technology, 232–233
  multislit scanning, 235*b*
  pixel size/pitch, 227–228, 228*f*
  print/storage options, 148–149, 149*b*
  process of, 128–129
Digital subtraction mammogram, 284–285, 295
Digital-to-analog converter (DAC), 215, 226

Digitally reconstructed radiographs (DRRs), 369
Diodes, 215
Dipole, 220
Direct current (DC), 119
Direct supervision, 443
Disaster recovery storage, 148
Display contrast, 215–216
DNA (deoxyribonucleic acid), 2, 369
Doppler effect or shift, 218, 247–248
Dosimetrist, 359
Dual-energy mammography, 285, 295
Duct ectasia, 87
Duct epithelial hyperplasia, 87
Ductal carcinoma *in situ* (DCIS), 46, 109, 366
Ductal ectasia, 90–91
Ductal hyperplasia, 87
Ductal lavage, 344–345
Ductal papillomas, 87
Ductal proliferation, 75
Ductography, 322, 344
Dussik, Karl Theodore, 237
Dynamic range, 216
Dysesthesia, 285

E

Echo time (TE) (time-to-echo or echo delay time), 220
Echogenic lesions, 218
Echogenic or hyperechoic, 218
Echogenicity, 241
Eczema, 25, 87
Edge shadowing, 254
Edwin Smith Papyrus, 3
Effective dose, 25
Effective focal spot, 119
Egan, Robert L., 9
Eggshell calcifications, 107
Eklund. See Implant imaging
Eklund technique, 193
Elastography, 298
Elderly or unstable patient, 191–192
Electromagnetic spectrum, 255
Electron interaction with matter, 124–128
Electronic health record (EHR), 150
Electronic medical record (EMR), 150
Electronic patient record (EPR), 150
Electrons, 121, 134
    conductors for, 122
    formation of, 127
    one-way flow of, 121
EMR (electronic medical record), 150–151
Endocrine disrupter, 54
Endocrine-producing glands, 76f
Enhancement, 218
Entrance exposure, 541
Entrance skin dose, 402, 516
Entrance skin exposure (ESE), 25, 41
Environmental Protection Agency (EPA), 402, 404
Epidermoid cyst, 87
Epithelial hyperplasia, 87, 92
EQUIP, 475–476
Equipment artifacts, 252
Erythema, 87
Estrogen, 87
    breast density and, 97
    declining levels, effects of, 76–78

function of, 75
    in men, 79–80
    only therapy, 79
    plus progestin therapy, 97
Exaggerated craniocaudal lateral projection. *See* XCCL
Exaggerated lateral craniocaudal medial projection. *See* XCCM
Exercise, 56
Exposure latitude
    in AEC, 524–525
    in analog system, 222
    defined, 162
    in digital systems, 229–230, 235–236
    function of, 216, 516
    kilovoltage range, 524
    in PSP system, 234
    screen/film system, 522
Exposure linearity, 402
Exposure time, 137–138
External-beam radiation therapy (EBRT), 369–370
Extralobular terminal duct (ETD), 68
Extremely large breast-mosaic breast imaging sectional imaging of, 190

F

Facilitation, 33
Facility review test, 419
False negative, 25
False positive, 25
Fast spin echo (FSE), 264
Fat necrosis, 87–88
Fat suppression or fat saturation, 271f
    in augmented breast, 273
    function of, 264
    imaging standards, 267
Fatty image categories, 447–448f
FB (from below) projection, 179–181
Feedback, 402, 516
Ferromagnetic material artifact, 271, 271f
Fibroadenolipoma, 88
Fibroadenomas, 88, 251
Fibrocystic breast, 56, 56f, 64
Fibrocystic change, 252
Fibrosarcoma, 91
Fibrous histiocytoma, 91
Fibrous nodules, 88
Field of view (FOV), 216, 261
Fill factor, 232
Film digitizers, 119
Film printer, 416–418, 430
Filtration, 129–130
    added, 130
    inherent, 130
Fine needle aspiration biopsy (FNAB), 324
    advantages/disadvantages, 342
    patient procedures, 332
    post-care/follow-up, 333
    uses of, 332
Fixing agents, 531–533
Flat-panel detector system, 222–223, 223f
Flex paddle, 191
Flip angle, 220
Flooded replenishment, 516
Fluorine-18 fluorodeoxyglucose, 319
Fluorodeoxyglucose (FDG), 216

analysis of, 304–305
    characterization of, 301–302
    use of, 303–304
Fluorodeoxyglucose-positron emission tomography, 301–302
Fluorothymidine (FLT), 300
FNAB (fine needle aspiration biopsy), 323
Focal architectural distortion (FAD), 100
Focal fibrosis, 88
Focal spot blur, 155, 155b
Focal spot size, 134–136, 143b
Focal zone, 218
Fog, 516
Follicle-stimulating hormone (FSH), 75
Food and Drug Administration (FDA), 404, 443
    cryoablative approval, 388
    DBT units approval, 287, 287–291
    nipple aspirate test approval, 344–345
    silicon implant approval, 347
    tumor cell vaccine approval, 386
Four-projection imaging, 164, 193
Free flap procedure, 359
Free induction decay or FID signal, 220, 239
Frequency, 238
Fresnel zone (near field), 218
Fry, William, 237
    Fujifilm ASPIRE Cristalle, 291–292
Full-field breast ultrasound, 297–298
Full-field digital mammography (FFDM), 223, 226
    approved units, 453–454b
    MQSA certification, 452–455
    printers/monitors, 455
    quality control, 455
    results and analysis, 432–433
    two-D images, 484
Fulvestrant (Faslodex), 381

G

Gadolinium, 220
    based MRI, 276, 278
    chelated-contrast with, 266, 311
    function of, 263–264
Gail Model, 2, 45
Gain, 219
Galactocele, 88, 251
Galactography, 344
Galen (130–217 AD), 5–6
GEMS General Electric Senographe method, 232–233
Gender, 56
Genetic risk, 48
Gene therapy, 384–385
General Electric SenoCare, 290–291
General patient preparation, 324–325
Genomics
    Center for Genomics (CGR), 9
Geometry
    system, 143
Gersten, Jerome, 237
Ghost image, 434
Ghosting, 402
Glandular dose, 402
    defined, 2
    function of, 25
    trends in, 9t

Glandular tissues, 74, 101
Gonadotropin-releasing hormones (GnRHs), 75
Gradient-echo pulse, 264
Gradient spoiling, 220, 264
Granular cell tumors, 88
Gray (Gy), 25
Greco, T.C., 4
Grids
    defined, 141, 141*f*
    frequency, 119
    function of, 142
    ratio, 119
Gun-needle method, 334
    method comparison, 341–342
    MRI-guided, 340–341
Gynecomastia. *See also* Male breast cancer
    defined, 64
    diagnosis, 80
    imaging techniques for, 189
    locations of, 251
    prognosis of, 80
    radiograph of, 79*f*
    signs/symptoms, 79
    treatment of, 80

H

Halation, 119
Half-life, 216, 285
Half-value layer (HVL)
    defined, 119
    in filtration, 129
    QC test, 431
Halo, 88, 95
Halsted, William, 7
Hamartoma or fibroadenolipoma, 88
Handling and storage artifacts, 537
Health Insurance Plan (HIP) of Greater
    New York, 11, 11*t*
Health Level Seven International (HLSI), 151
Hematomas or microhematomas, 88–89,
    251, 359
Hematopoietic stem cell transplantation (HSCT),
    385–386
HER2 (human epidermal growth factor
    receptor 2)
    cancerous, drugs for, 383
    defined, 2
    expression of, 359
    function of, 364
Herodotus, 44
Hertz (Hz), 89, 118
High-transmission cellular (HTC) grid, 142, 142*f*
HIPAA (Health Insurance Portability and
    Accountability Act), 443, 462–463
Hippocrates, 5
HIS/RIS, 150
Histological analysis, 324
Histology, 89
HL7 (health level seven international), 151
Hodgkin's disease, 25, 89
Hologic Selenia Dimensions, 287
Hologic tomosynthesis, 287–290
    breathing technique in, 289
    drawbacks of, 289–290
    two-dimensional images generated with, 289
Homogeneous, 219
Hormone markers, 323

Hormone replacement therapy (HRT)
    breast cancer risks of, 52
    defined, 25
    risks of, 77
    use of, 77
Hormone therapy, 381–383
Hormones, 64, 75
Hospital information system (HIS), 150
Humor, 2
HVL (half-value-layer), 431
Hyperechogenic lesions, 89
Hyperechoic, 219
Hyperplasia, 2
Hypesthesia, 285
Hypoechoic, 87, 219

I

Illuminance, 119, 402
Illumination
    x-ray beam limiting device, 422
Image acquisition, 154
Image brightness, 154
Image detector size, 140
Image-guided radiation therapy (IGRT), 359,
    372–373
Image-modulated radiation therapy
    (IMRT), 395
    advantages/disadvantages, 372
    defined, 359
    delivery of, 372–373
Image plate (IP), 119, 402, 501
Image quality, 157
    degradation of, 134, 227
    enhancement of, 223
    improvement of, 128
    MRI-based improvement, 267
Image receptor (IR), 119, 216, 518
Image resolution, 155
Imhotep, 4
Immunotherapy, 386
Implant imaging, 193–196, 194*f*, 195*f*
    delayed reconstruction, 390
    immediate reconstruction, 390
    surgery, 389
Implant or tissue expanders, 390*f*
Implant surgery, 360
Implantable venous access, 201–202
In situ, 2
Indirect flat-panel detector system,
    232–233
Indistinct calcifications, 105
Infection, 359
Infection control, 457
Inflammatory breast cancer (IBC), 110
Informed consent, 323, 325
Inframammary fold/crease, 64, 163, 170
    imaging, 193
    non-horizontal, 190
Inframammary lymph nodes, 254
Inherent filtration, 130
Injury to the breast, 56
Intensity-modulated radiation therapy (IMRT),
    359, 371–372
Intermediate lesions, 249
Internal beam radiation
    single catheter, 372
    muli catheter, 373

Internal echoes, 242, 247
International Breast Intervention Study
    (IBIS), 46
International Normalized Ratio (INR), 333
Interpreting physician, 443
Interventional techniques, 323–324
Intra, 89
Intraductal calcifications, 108
Intraductal papillomas (IDPs), 90
Intralobular terminal duct (ETD), 68
Intraoperative electron radiation therapy
    (IOERT), 375
Invasive papillary carcinomas, 111
Inversion recovery pulse, 264–265
Inversion time (TI), 220
Involution, 50, 77
Ionizing radiation, 124
Ipsilateral, 163, 216
Irradiated breast, 196–197
Ischemia, 2
Isoechoic, 219

J

Joint Commission (TJC), 405
Joint Commission on Accreditation of Healthcare
    Organizations (JCAHO), 405
Joint Review Committee on Education in
    Radiologic Technology (JRCERT), 405
Juvenile papillomatosis, 89

K

K-characteristic x-rays, 119
Keloid scars, 89
Keynes, Geoffrey, 7–8
Kilovoltage range, 137, 524
Klinefelter's syndrome, 64
kVp accuracy and reproducibility, 431–432
Kyphotic patient, 191

L

Labeling, 205
Lactation or milk secretion, 78
Lactiferous ducts, 64
Lactiferous sinus, 65
Lag, 402
Lapatinib, 383
Larmor frequency, 257, 257*f*
Large core breast biopsy, 337
Laser ablation, 387
Laser printed image, 148
Lateromedial (LM) projection, 175–177
Lateromedial oblique (LMO) projection, 177
Latissimus dorsi flap, 359
Latissimus dorsi muscle, 65
LCD (liquid crystal display), 144–145, 145*f*
Le Dran, Henri François (1685–1770), 6
Leptomeninges, 89
Lesions, 3. *See also* Oval lesions; Spiculated
    or stellate lesions
    ABUS identification, 297
    biopsy techniques, 331–342
        mammotome, 336–337
        MR-guided, 340–341
        MRI-core, 340
        open surgical, 338
        procedures, 339–340
    CAD analysis of, 154

Lesions (*continued*)
DBT/3-D mammography identification, 287–288, 294
elastographic identification, 298
gamma ray identification, 306
localization of, 326–331
magnification of, 143, 186
margins, 94*f*
MRI identification of, 272–273, 275
optical image identification, 329
PICO-SPECT identification of, 305
positron emission mammographic identification, 304
proton magnetic resonance identification, 299
scintimammographic identification, 319
shapes of, 95*f*
on tail of breast, 475
triangulation of, 164, 204*f*
ultrasound imaging of, 244–270, 270*f*
advantages, 254
benign appearance, 248–249
color Doppler appearance, 248
identification with, 245
intermedial appearance, 249
limitations of, 252–253
problems/solutions, 253–254
scanning techniques, 244–245
standards, 267
typical appearance, 249–252
Levels of finding, 449
Li-Fraumeni syndrome (LFS), 216
Lindstrom, Peter, 237
Line-focus principle, 134–136
Linear accelerator (LINAC), 368
Linear array ultrasound, 326
Linear artifacts, 253
Linear rod-like calcifications, 104
Lipoma, 89
Liposarcoma, 91
Liquid-crystal display (LCD), 144–145, 145*f*
Lithium fluoride (LiF), 27
LM (lateromedial) projection, 175–177
LMO (lateromedial oblique), 177
Lobular calcifications, 92*f*, 107–108
Lobular carcinoma *in situ* (LCIS), 46, 109
Lobular hyperplasia, 87
Lobule. *See* Terminal ductule lobular unit (TDLU)
Localization
preoperative, 329
mammogram, 330
Localization terminology
lower inner quadrant (LIQ), 67, 68*f*
lower outer quadrant (LOQ), 67, 68*f*
region, 68
upper inner quadrant (UIQ), 67, 68*f*
upper outer quadrant (UOQ), 67–68
Local area network (LAN), 151
Longitudinal and transverse magnetization, 220
Longitudinal relaxation time (T1), 221
Long-throw gun, 342
Low-exposure factors (kV and mAs), 155
Ludwig, George, 237
Luminance, 120, 402
Lumpectomy, 8, 365–367
Luteinizing hormone (LH), 75

Lymph nodes, 251
biopsy, 367
cancer of, 8
defined, 2
drainage from, 71*f*
groups, 72
Lymphatic drainage of, 71–73
axillary nodes, 72–73
lymphatic vessels, 72
pectoral nodes, 72
Lymphatic system, 71–73
Lymphedema, 25
Lymphoma, 110, 359
Lymphoscintigraphy (sentinel mode mapping), 285
analysis, 310
indications for, 309–310
positive uptake on, 309*f*
technology, 308–310

M

Magnetic resonance imaging (MRI), 25, 255–276
advantages, 272
artifacts, 267–268
augmentation mammoplasty, 348
brief history of, 255
CAD for, 153–154
computer parameters, 261–162
contrast agents, 263–264
disadvantages, 273–275
echo pulse sequence, 263*b*
electromagnetic spectrum, 255–256
equipment, 259–261
hormone therapy cession before, 275
limitations of, 262*b*
magnetic field, 256*f*
magnetism, 256
operating console, 261
parameters, 262
postprocessing technique, 265–266
principles of, 256–259
90-degree pulse, 258*f*
hydrogen nucleus, 257, 257*f*
RF signals, 258–259
T1/T2 relaxation or recovery, 258–259, 262–263
Z axis, 257*f*
pulse sequence in, 264–265
terminology, 219–222
Magnetic resonance spectroscopy, 299
Magnetism, 256
Magnification, 143, 186–187, 186*f*
factor, 120, 144*b*
mammography, 162
setup, 144*f*
Male breast cancer, 78–80
diagnosis, 80
general description, 79
prognosis, 80
risk factors, 79–80
signs/symptoms, 79
treatment, 80
Male breast, imaging and, 189
Malignant, 2
Malignant masses, 97*b*
Malmö Mammography Study, 13
Mammary line, 66

Mammogram, 8
accuracy, 52–53
baseline, 2
discomfort, 56–67
Mammographer, 532*f*, *527–529*, 533*b*
Mammographic units
assembly, 535
design of, 123
evaluation of, 535
Mammography. *See also* Digital mammography (DM)
advantages, 44
benefits of, 41–43
challenges, 474–475
cross-section, 142*f*
development of, 9
diagnostic options, 286
disadvantages, 44–45, 44*b*
effectiveness of, 286
equipment evaluation, 534, 542
patient satisfaction, 31*b*
quality *vs.* quantity, 11–12
radiation dose in, 41–42, 41–43*b*
risks of, 17, 41–43
standardization of, 12–13
Mammography Equipment Evaluation (MEE), 406–407, 422, 443
Mammography imaging, 128–129
Mammography interpreters, 93
Mammography magnification, 143–144
Mammography monitors, 144–148
Mammography Quality Control Manual, 407
Mammography Quality Standards Act (MQSA) in the United States, 443, 475
accreditation and certification, 445–446
assessment categories, 460–461, 460*b*
certificate placement, 458
certifying states, 445*b*
consumer complaints mechanism, 458
content of records and reports, 459
DBT imaging rules, 293
documentation, 463–474
by interpreting physicians, 463, 466–467, 468–470*t*
by medical physicists, 467, 470–471, 472–474*t*
by radiologic technologists, 465–466*t*
final inspection questions under, 485–492
final regulations, 445
HIPAA and, 462–643
HVL-requirements, 541*f*
identification and labeling, 168–169
interim regulations, 444–445
lawful services, 446
passage of, 13
procedure standardization role, 163
program assessment, 534–535
purpose of, 444
quality assurance program, 446–463
adverse reporting, 452
clinical images, 446–448
findings or noncompliance, 449, 449*b*, 450–451*t*
suspension or revocation, 451–452
withdrawal of accreditation, 449, 451
radiologists responsibilities identified in, 405–406
record keeping, 459

results rule, 29
supplementary projections/positions
    recognized by, 175
Mammography screens, 519, 530–531
Mammography tube, 129–140
  anode heel effect in, 136–137, 137*f*
  automatic exposure control in, 140
  back-up timer in, 140
  compression in, 138–140
  exposure time in, 137–138
  filtration system in, 120–130, 130*f*
  focal size in, 136
  general x-ray, 129
  kilovoltage range in, 137
  soft tissue in, 129
Mammography tube tilt, 135, 135*f*
Mammoplasty, 323, 346–348
  augmentation, 345
  reduction, 344
Mammotome biopsy system, 336–337
Manufacturer detector calibration,
    433–434, 506
Mastectomy
  complications of, 364–365
  defined, 2–3
  early history of, 7
  procedure, 364
  scar, 57
  types of, 365
    skin sparing, 363
    nipple sparing, 363
Mastitis, 89–90
Mastopexy, 323, 347
Matrix, 120, 216
Matrix size, 261
Medical audit, 462
Medical history, 31–34
  data in, 31–32
  documentation, 31*b*
  techniques for, 32–33
Medical physicist, 359–360, 406–407, 443
Mediolateral or 90 degrees lateral (ML)
    projection. *See* ML
Mediolateral oblique (MLO) projection. *See* MLO
Medullary carcinomas, 110
Melanoma, 90
Menarche, 76
Menopause, 75–77
Metaplasia, 3
Metastasis to the breast, 111, 251
Meyers, Russell, 237
MIC1, 3
Microcalcification, 249, 249*f*
Microsurgery, 360
Microwave imaging spectroscopy, 313–314
Milk of calcium, 106
Milk ridge or line, 25
Milliampere (mA), 137–138
Minimum HVL, 132
Minus density artifacts, 100*f*
ML (mediolateral or 90 degrees) projection, 175
MLO (Mediolateral Oblique) projections
  areas, 166*f*
  common problems, 171*b*
  CQ tests for, 528–529
  function of, 163–164, 166
  image evaluation, 171*b*

for large or mosaic breast, 190
  patient positioning, 169–172, 198*f*
  patient with implants, 193
  pictogram of, 170*f*
  poor positioning, 171*f*
  sector location, 168
Mobile unit, 456–457
Modulation transfer function (MTF), 216
Mole/scars, 201, 203*f*
Molecular imaging technologies, 299–310
Molecular treatment, 379–381
Moles (nevi, singular nevus), 90
Molybdenum filtration, 133
Molybdenum x-ray emission spectrum, 131*f*
Monitor, 146–147, 144, 455
  AWS, 144, 427, 405t
    test on, 44
  calibration QC, 148
  RWS, 427
    test on, 412–413
  QC, 453
Monoclonal antibody therapy, 383
Montgomery's glands, 65
Morgagni, Giovanni Battista, 7
Morgagni's tubercles, 65, 67
Morphology, 90
Mosaic imaging, 189–190
Motion, 155, 266, 427, 412–413
  avoiding, 266
  causes of, 252, 295
  in MR images, 271–272
  test on, 44
MQSA assessment categories, 460
MR-guided biopsy, 341
MRI. *See* Magnetic resonance imaging (MRI)
MRI-guided core biopsy and needle localization,
    340, 340*f*
Mucinous, 111
Mucinous cancers, 111
Multicatheter method, 375
Multicentric tumors, 360
Multifocal tumors, 360
Multiparity, 78
Multiparous woman, 77–78
Muscle, 65–67, 67*f*
Muscle-sparing TRAM flap technique, 360

N

National Cancer Institute (NCI), 11–12, 17,
    40–41
National Center for Devices and Radiological
    Health (CDRH), 405
National Comprehensive Cancer Network
    (NCCN), 35, 344
National Council for Radiation Protection and
    Measurement (NCRP), 42
Neal, Donald, 237
Near-infrared fluorescence (NIRF), 300
Near-infrared spectroscopy (NIRS),
    311–312
Necrosis, 360
Nephrogenic systemic fibrosis and nephrogenic
    fibrosing dermopathy (NSF/NFD),
    25–26, 276
Nephrogenic systemic fibrosis/nephrogenic
    fibrosing dermopathy (NSF/NFD), 220
Nervous system, 73

Neurons, 73
Neutron simulated emission computed
    tomography (NSECT), 314
Nipple, 67
Nipple and areola reconstruction
  guidelines for, 344
  with implants, 389
  procedures for, 364–365
  in profile, 198
Nipple aspiration, 344
Nipple not in profile, 199
Nipple-sparing mastectomy, 365
Nitrous oxide (N_2O)/laughing gas, 3, 6
Noise, 155, 204, 216, 230
  quantum noise, 155
Non-Hodgkin's lymphoma, 26, 110
Nuclear magnetic moment, 257*f*
Nuclear magnetic resonance (NMR), 255
Nuclear magnetic technology
  advantages/disadvantages, 301
  as molecular modality, 300
  radiation dose related to, 310–311
  T-scan, 313
Nuclear medicine, 300–301
Nuclear Regulatory Commission (NRC), 404
Nulliparity, 78
Number of excitation (NEW), 261

O

Obese upper arms, imaging, 190
Obesity, 57
Object-to-image distance (OID), 186
Occupational Safety and Health Administration
    (OSHA), 402, 404
Oil cyst, 88, 90
Operating console, 261
Operator-dependent artifacts, 252–253
Optical density (OD)/radiographic density, 3, 73
Optical disk, 266
Optical imaging, 285, 311–312
Orbital electrons, 121
Organic light-emitting diode (OLED), 144–145
Osteosarcoma, 91
Oval lesions, 98*f*
  characteristics of, 94
  defined, 98
  detection of, 96
  mammographic characteristics of, 97*b*
Ovarian hormones, 58

P

p16 gene, 385
Pacemakers, 201–202
PACS, 149, 216
Paget, Sir James, 7
Paget, Stephen, 7
Paget's disease, 3, 111
Pain management, 393–394
Pair production, 127
PALB2 gene, 48–49
Palliative treatment, 65
Palpable lumps, 34*f*, 94
Papanicolaou (Pap) stain, 324
Papillary carcinomas, 111
Papillomas, 90
Papillomatosis, 87
Parabens, 55

Paralanguage, 32
Parallax, 216, 285
Parenchyma, 90
Parity, 26, 78
Partial radiation, 373–375
Parturition, 26
Pathology review, 345
Patient care and communication
  compassion in, 27–29
  documentation, 33–34, 35*f*
  general preparation, 324–325
  medical history, 31–33
  MRI preparation, 266
  nonconforming patients, 188–199
    abdomen, 190–191
    adhesive capsulitis, 192
    axilla, 191
    barrel chest, 193
    elderly/unstable, 192
    imaging, 169–173
    inframammary fold/crease, 190, 193
    elderly/unstable, 192
    kyphotic, 191
    large breast, 190
    male breast, 189
    obese upper arms, 190
    pectus excavatum/pectus carinatum, 192
    small breast, 188–189
  results and findings, 461–462
  satisfaction of, 31*b*
  waiting room environment, 29–31
Peau d'orange, 90
Pectoral nodes, 72
Pectoralis major muscle, 65
Pectoralis minor muscle, 65
Pectus carinatum (pigeon chest), 192
Pectus excavatum (sternum depressed), 192
Penumbra, 120
Peri, 90
Periductal mastitis, 90–91
Peripheral nervous system, 73
Peritoneum, 90
PET/PEM imaging, 301–305
  advantages/disadvantages, 304
Phantom breast pain, 360
Phantom images
  ACD tests, 408–411, 410–411*f*
  QC form, 500
  radiography, 539*f*
  test for, 527–528, 542–544
Philips MicroDose Unit, 292
Photoconductor, 216
Photodiode/transistor flat panel, 233
Photodisintegration, 127
Photodynamic therapy, 386–387
Photoelectric interaction, 126–127
Photoelectric radiation, 124
Photomultiplier tube (PMT), 234
Photon-counting image, 225
Photon-counting image capture technology, 235, 235*f*
Photons, 370
  characteristics of, 369
  cohort scattering, 127
  Compton effect, 126
  film interactions, 517, 519
  low-level, 127

number striking detector, 157
  photoluminescent effect, 126–127
Photostimulable phosphor, 216
Photostimulable storage phosphor (PSP)
  technology, 233–235, 234*f*
Photostimulated luminescence (PSL), 217, 234
Phyllodes tumors, 90, 111, 251
Physical science, 443
Physician, 443
Physicist test, 422, 423*t*, 426–434
  medical, 404
PICO-SPECT, 305
Picture-archiving and communication systems
  (PACs), 152*b*
  advantages/disadvantages, 149–150
  description of, 149
Piezoelectric crystals, lead zirconate titanate
  (PZT), 237
Piezoelectric effect, 219
Pinhole camera, 535
Pixel bit depth, 228–229
Pixel density, 228, 228*f*
Pixel or picture element, 120, 217
Pixel pitch, 227
Plasma cell mastitis, 90–91
Plate-specific artifact, 429
Pleomorphic calcifications, 105
Pneumocystography, 323
Positioning guide, 168–169
Positron emission mammography (PEM)
  development of, 303
  factors affecting, 305
  radiation dose for, 310–311
  use of, 303–304
Positron emission tomography (PET) imaging, 302*f*
Posterior nipple line (PNL), 163
Postirradiated breast, 197
Postlumpectomy imaging, 196
Postmastectomy imaging, 196
Postmenopausal women
  benign tumors in, 97*b*
  cysts in, 86
  mammogram accuracy in, 53
  physical changes in, 76–77
Postprocessing technology, 265–266
Precession, 220
Prefetch, 149
Premenopausal, 76
Preoperative needle localization
  defined, 323
  mammography with, 330–331
  procedure, 329–331
Preprocessing technique
  codes for, 234
  options, 230
  process of, 266
Preventive maintenance (PM), 402–403
Preventive or prophylactic surgery, 360
Primary lymphoma, 110
Printer
  test, 416–418
  QC, 430, 455
Processing software, 154
Processor quality control, 549–553
  beginning operating levels, 550–552
  crossover worksheet, 553*f*
  equipment for, 549–550

establishing new levels, 552–553
  frequency, 549
Progesterone
  breast density and, 97
  levels, decline of, 76–77
  production of, 75
  replacement therapy, 52
Progestin, 77
Progressive Reconstruction Intelligently
  Minimizing Exposure (PRIME), 291
Projections, supplementary
  axillary tail (AT), 182–183, 182*f*
  cranialcaudal (CC), 172–173, 198*f*
  cleavage or valley view (CV), 181–182, 181*f*
  exaggerated lateral or medial craniocaudal
    projection (XCCL/XCCM), 178–179,
    179*f*, 180*f*
  from below (FB), 179–181, 180*f*
  lateromedial (LM), 175–177, 177*f*
  lateromedial oblique (LMO), 177, 178*f*
  magnification, 185–187, 186*f*
  mediolateral or 90 degrees lateral (ML), 175, 176*f*
  mediolateral oblique (MLO), 169–172, 170*f*, 171*f*
  rolled positions, 184–188, 184–185*f*
    rolled inferior (RI), 184
    rolled lateral (RL), 184
    rolled medial (RM), 184
    rolled superior (RS), 184
  spot compression, 187–188, 187*f*
  superior–inferior oblique (SIO), 183–184, 183*f*
  tangential (TAN), 178, 178*f*
Prone, 217
Prophylactic mastectomy, 57
Prophylactic surgery, 388–389
Prostate cancer, 65, 80
Proton density (PD), 220
Proton density weighted imaging, 220–221
Protooncogenes or tumor suppressor genes, 45
PSP
  uniformity, 429
  erasure, 411
Puberty
  breast development during, 66
  hormonal secretions during, 79
  onset of, 75
Pulsing time interval, 261
Purcell, Edward, 255

Q

Quality assurance, 3
  agencies responsible for, 404–405
  benefits of, 404
  components of, 525–526
  elements of, 526
  personnel responsibilities, 405–407
Quality control (QC), 3, 403, 497
Quality control (QC) testing, 407–434
  acquisition workstation monitor, 413–414
  adjustment repairs, 423*t*
  annual, 424–425*t*
  compression force, 419–420
  compression thickness indicator, 412
  computed mammography PSP erasure, 411
  digital for physicists, 407–408*t*
  facility's program assessment, 419, 534–535
  film printer, 416–418
  forms for, 500–514

function of, 407
general details, 408–411, 542–558
general technologist tests, 527–542
   image quality evaluation, 538–539
   monthly, 528
   quarterly, 528–529
   screens/film systems, 533
   semiannual, 529
   weekly, 527–528
image quality evaluation, 538–539
kilovoltage accuracy and reproducibility, 539
mammographer-specific, 527–529
manual for, 407
physicists tests, 426–434
   automatic exposure control, 426–427
   average glandular dose, 427
   beam quality, 431
   CM system, 429
   collimation assessment, 432–433
   compression paddle deflection, 422
   ghost image, 434
   kVp accuracy, 431–432
   manufacture detector calibration, 433–434
   monitors, 429–430
   signal-to-noise/contrast-to-noise ratio, 422
   site technologist program, 430–431
   spatial resolution, 426
   unit checklist, 428
   x-ray beam limiting device illumination, 422
radiologist workstation monitor, 414–416
radiologists' systems, 421–422
repeat analysis, 420–421
viewbox cleanliness, 418
visual checklist, 412–413
Quantum, 155
Quantum mottle, 524

R

Race, 57
Rad (radiation absorbed dose), 26
Radiation, 58
Radiation dose, 41
   in nuclear imaging, 310–311
   risks, 59
Radiation output rate, 541
Radiation therapy, 3, 7, 8, 368–375
   biological effects, 369–370
   characteristics of, 124f
   classifications of, 370–371
   external beam, 367–368
   internal beam, 371–372
   history, 368
   indications for, 370
   side effects of, 371
   technology, 368–369
Radical mastectomy, 7–8, 80
Radical scar, 91
Radioactive pellets (iridium-192), 375
Radiofrequency ablation, 387
Radiofrequency (RF) biopsy, 335–336, 336f
Radiofrequency (RF) waves, 255, 322
Radiographic noise, 524
Radiologist, 405–406
Radiologist
   interpreting physician, 406
   lead interpreting physician, 406
   training, 462–463

Radiologist image quality feedback, 421–422
Radiologists workstation (RWS) monitor, 414–416
Radiology information system (RIS), 150
Radiopharmaceuticals, 285
Radium (Ra), 3
RAID (redundant array of independent disks), 120
Raloxifene, 360, 381
RAR beta gene, 345
Ray emission spectrum for a molybdenum target with 30 mm Mo filter, 102f
Real time, 219
Real-time scan, 242
Reciprocity law, 120
Reconstruction
   delayed, 390
   immediate, 390
   nipple and areola, 389
   surgical, 389
Reconstructive surgery, 360
Record keeping, 459
Rectification, 120
Reduction, breast, 24
Reduction mammoplasty, 94, 346
Redundant array of independent disks (RAID), 148
Rejected images, 403
Relaxation time, 221
Rem (rem), 26
Repeat analysis
   daily counting form, 510
   procedures, 420
   summary form, 511
   tally sheet, 509
Repeat/reject analysis
   analog, 544–545
   mammography, 546f
   quarterly tests, 544–545
Repeated images, 403
Repetition time (TR) or time to repetition, 221
Replenishment, 532
Reproducibility, 431
Resolution. *See also* Spatial resolution
   aim of, 523
   defined, 120
   factors affecting, 523b
   focal size and, 154
   testing, 536
Retromammary space, 65
Returning echo, 239, 242, 242f
Reverberations, 219
Review workstations (RWS), 147–148, 427–428
Rhodium targets with rhodium filtration, 133
Risks factors for breast cancer, 58
Roentgen (R), 26
Roentgen, Wilhelm Conrad (1845–1923), 7
Rolled projections, 184–188
   rolled inferior (RI), 184
   rolled lateral (RL), 184
   rolled medial (RM), 184
   rolled superior (RS), 184
Room luminance, 542
Round calcifications, 104

S

SAC certification, 449, 452
Safelight, 517, 555–556

Saline-filled implant, 360
Sarcomas, 91, 111
SAVI applicator, 374f
Scan, 219
Scar markers, 33, 34f, 201
Scarring, 251
Scintigraphy, 285, 305–306
Scintillation, 217, 285
Scintillation phosphors, 217
Scintimammography, 285, 305–308, 308f
   defined, 285
   negative, 307f
   positive, 308f
   procedure, 305–306
Sclerosing adenosis, 86
Sclerosing duct hyperplasia, 91
Screen/film systems
   cameras in, 535–536
   cleaning test, 554
   fixer retention analysis, 554–555
   function of, 518–519
   image quality, 521–524
      cassette position, 521f
      latitude, 522
      radiographic noise, 524
      receptor contrast, 522
      resolution, 523, 523b
      sharpness, 523, 523b
      speed or sensitivity, 522
      visibility of detail, 523
   low quality cause, 522
   printer quality, 505
   testing protocol, 532–533
Screening mammograms, 27
Screening ultrasound, 244
Sebaceous glands, 65, 91
Sectional imaging, 189–190
SEER data, 46
Selective estrogen receptor modulators (SERM), 379, 381
Selenium (Se), 231
Self-referrals, 459
Semester hours, 444
Senographe 2000 D, 232f
Sensitivity, 3, 217, 286
Sentinel lymph node biopsy, 360
Seroma, 360
Serratus anterior, 65
Shadowing, 91, 219
Sharpness, 523, 523b
Short Ti inversion recovery (STIR), 265
Si unit, 25
Siemens Mammomat Inspiration, 287, 291
Sievert (Sv), 26
Signal-to-noise ratio (SNR), 420
   calculation of, 422, 426
   control of, 155
   defined, 230
Silhouette sign, 91, 95f
Silicone gel-filled implant, 360
Silicone or saline implants, 348
   ALCL from, 54
   connective tissue, 64, 66, 73–74, 79
   disorders from, 54, 86, 9–93, 108, 111, 322
   scarring from, 53
Single-entry multicatheter device, 374

Single-photon emission computed tomography (SPECT), 305
SIO (superior-interior oblique) project, 183–184
Skin fold, 199–200, 200*f*
Skin lesions
 breast lesions *vs.*, 96–97
 characterization of, 87
Skin markings, five-shape, 202*f*
Skin tags, 91
Skin thickening, 108
Skin-sparing mastectomy, 365
Slit camera, 535
Sloping surface, aim of using, 139
Small breast, imaging, 188–189
Smith, Edwin, 4
Solid, 219
Somatic nervous system, 73
Sound Navigation and Ranging (SONAR), 237
Sound-tissue artifacts, 253
Sound waves, 271
Source-to-object distance (SOD), 186
Spatial frequency, 217
Spatial resolution
 control of, 154
 defined, 120
 procedure for, 426
SPCA2 protein, 385
Specificity, 3, 217
Specimen
 analysis, 333–334, 364
 contaminated, 342
 large breast, 336
 margins, 366
Specimen radiography, 343–344
SPECT (single photon emission tomography) imaging, 305
Spiculated or stellate lesions, 91
 characterization of, 97
 recognition of, 99–100
 retraction of, 99*f*
 size of, 97, 99
 stage II, 99*f*
 ultrasound imaging of, 104*f*
Spin, 221
Spin-echo pulse, 265
Spin-lattice, 221
Spin-spin, 221
Spot projection, 187–188
Staging of breast cancer, 3, 360, 363–364, 363*t*
Standard scientific prefixes, 26
Standards, 403, 517
Star pattern test, 536
Steady state, 265
Stellate lesions, 91
Stem cell
 hematopoietic stem cell transplantation (HSCT), 385–386
Stereotactic breast localization and biopsy, 323, 326–329
 add-on units, 327, 327*f*
 blended technology, 328
 dedicated prone units, 328, 328*f*
 limitations, 329
 stereotactic procedure, 328–329
 types of, 331*b*
 unit comparisons, 329*b*

Sternum, 65
STK11 gene, 49
Storage phosphor screen (SPS), 119
Stretcher imaging, 197–198
Stroma, 91
Subject contrast, 163
Superior-inferior oblique (SIO) projections, 183–184
Supine, 197–198
Surgical biopsy, 338
Surgical reconstruction using implants, 389–392
 options, 390–392
 preparation for, 389
 procedures, 389
 risks and complications of, 392
Surveillance, Epidemiology, and End Results (SEER), 45
Suspensory ligaments. *See* Cooper's ligaments
Symptoms of breast cancer, 58
System gain, 241–242
System geometry (SID, OID, magnification), 143, 155
System QC for radiologists, 421–422
System-related artifact, 272

T

T1-weighted images, 221
T2-weighted images, 221
T2, 221
Tail of Spence, 65
Tamoxifen
 benefits of, 380–381
 defined, 379
 side effects of, 380
Tangential (TAN) projection, 178, 178*f*
Target or anode angle, 134–135
Targeted treatment, 379–383
Tc-sulfur colloid, 309
Technologist, 406
Technologist training, 464
Teleradiography, 151, 152*b*
Terminal ductule lobular unit (TDLU), 68
 characterization of, 69*f*
 epithelium of, 69
 size of, 69*f*
Tesla (T), 256
Thallium-activated cesium iodide phosphor, 217
Thermal ablation, 387
Thermionic emission, 120–121, 124
Thermoluminescence dosimeter (TLD), 27, 41
Thickened skin syndrome, 108
Thin-film diode (TFD), 231
Thin-film transistor (TFT), 120, 145
Three-dimensional breast ultrasound, 296
Three-dimensional FSPGR, 267, 268*f*
Three-dimensional imaging, 265, 483
Three-dimensional mammography, 284
Three-dimensional ultrasound imaging, 296–297
Through transmission, 219
Time to repeat (TR) interval, 264
Tobacco, 58
Tolerance, 517
Tomosynthesis. *See* Digital breast tomosynthesis (DBT) or Three dimensional mammography
TP53 gene, 49
TRAM flap, 360, 392

Transducer, 219, 323
Transducer focal zone, 242
Transmitting facility, 236–237
Transport systems, 533
Transscan or T scan, 313
Transverse rectus abdominis muscle (TRAM) flap, 391*f*
 common types of, 391–393
 defined, 360
 risks of, 393
Transverse relaxation time (T2), 221
Trastuzumab (Herceptin), 364, 382–383
Trauma, 56
Trends, 150
Triangulation, 202–204
Tubular carcinoma, 112
Tumor cell vaccines, 386
Tumors/lesions, 3
Tungsten targets with rhodium or silver filtration, 133–134
Two-dimensional imaging, 265
Two-stage reconstruction, 360

U

Ulcers, 27
Ultrasound, 27. *See also* Breast ultrasound imaging
 advantages, 254
 of breast cancer, 249
 history of, 237
 longitudinal wave, 238*f*
 principle of, 238–239
 risk, 254
 terminology, 217–219, 241–246
 transducer, 239–241
Ultrasound artifacts, 252–253
Uniformity density
 automatic exposure control, 140
 compression for, 136, 165*b*
Uniformity of screen, 536
Unit checklist, 428
Uranium (U), 3
US Preventive Services Task Force (USPSTF)
 breast screening recommendations, 34
 panel of experts, 14
 recommendations of, 13–14
 screening recommendations, 39–40

V

Vaccines, 386
Vacuum biopsy analysis, 336*b*
Vascular access device (VAD), 377
Vascular circulation, 70–71
Vasovagal reaction, 323
Vector, 221
Vein, 70, 107
Velocity
 blood flow, 248
 moving structure, 247
Velocity or speed in ultrasound, 219
Vendor neutral archives (VANs), 120, 151
Venous access systems or ports, 201–202
Venous drainage, 70–71
Vesalius, Andreas (1514–1564), 6
Viewbox
 cleanliness, 418–419, 513
 conditions, 545–547

luminance, 434, 541–542
  QC tests for, 527
Viewing conditions, 527
Virchow, Rudolf (1821–1902), 6
Visibility of detail, 523
Visual checklist, 407*t*, 412–413

W

Wavelength, 219
  formula, 238
  low, 312
  optical, 311
Wheelchair imaging, 198–199
Whole-body PET, 302*f*
  with fusion imaging, 304
  with PEM, 303

Wide area network (WAN), 151
Wild, John Julian, 237
Wrinkling breast, 199–200, 200*f*

X

X-rays
  beam limiting devise illumination, 422
  discovery of, 7–10
  energy, 133*b*
  interaction with matter, 125–128
  processing control chart, 551*f*
X-rays imaging system, 120–123
  anode in, 122
  cathode in, 121–122
  circuit, 122*f*
  heat dissipation in, 122

line compensator in, 120–121
  protective housing in, 123
  thermionic emission, 121
  time circuit in, 121
  tube in, 122–123
  window or exit port in, 123
XCCL (Exaggerated Craniocaudal lateral
    Projection), 178–179
XCCM (Exaggerated Lateral Craniocaudal Medial
    Projection), 179
Xeroradiography, 9, 457–458
XXY males. *See* Klinefelter's syndrome

Z

Zoom, 242